Activation, Deactivation, and Poisoning of Catalysts

Activation, Deactivation, and Poisoning of Catalysts

John B. Butt
CHEMICAL ENGINEERING DEPARTMENT
NORTHWESTERN UNIVERSITY
EVANSTON, ILLINOIS

Eugene E. Petersen
DEPARTMENT OF CHEMICAL ENGINEERING
UNIVERSITY OF CALIFORNIA
BERKELEY, CALIFORNIA

ACADEMIC PRESS, INC.
Harcourt Brace Jovanovich, Publishers

SAN DIEGO   NEW YORK   BERKELEY
BOSTON   LONDON   SYDNEY   TOKYO   TORONTO

This book is dedicated to Kay and Regina,
and to the memory of John B. Butt, Jr.

COPYRIGHT © 1988 BY ACADEMIC PRESS, INC.
ALL RIGHTS RESERVED.
NO PART OF THIS PUBLICATION MAY BE REPRODUCED OR
TRANSMITTED IN ANY FORM OR BY ANY MEANS, ELECTRONIC
OR MECHANICAL, INCLUDING PHOTOCOPY, RECORDING, OR
ANY INFORMATION STORAGE AND RETRIEVAL SYSTEM, WITHOUT
PERMISSION IN WRITING FROM THE PUBLISHER.

ACADEMIC PRESS, INC.
1250 Sixth Avenue
San Diego, California 92101

United Kingdom Edition published by
ACADEMIC PRESS INC. (LONDON) LTD.
24-28 Oval Road, London NW1 7DX

Library of Congress Cataloging-in-Publication Data

Butt, John B., Date
  Activation, deactivation, and poisoning of
catalysts / John B. Butt, Eugene E. Petersen.
  p. cm.
  Includes bibliographies and index.
  ISBN 0-12-147695-2 (alk. paper)
  1. Catalyst poisoning. 2. Catalysts.
I. Petersen, Eugene E., Date. II. Title.
TP156.C35B84   1988
660.2'995—dc19   87-15636
CIP

PRINTED IN THE UNITED STATES OF AMERICA
88 89 90 91   9 8 7 6 5 4 3 2 1

CONTENTS

PREFACE ix

I Deactivation of Catalytic Surfaces: Microscopic Processes

1 *Physical and Chemical Description of Deactivation*

I.	The Concept of the Active Center as Utilized in Catalysis	4
II.	Operational Definitions of Deactivation	10
III.	Regeneration, Rejuvenation, and Detoxification	18
IV.	Summary—Conventional Wisdom and Other Matters	19
V.	More Definitions	22
	References	25

2 *Mathematical Description of Deactivating Systems*

I.	Deactivation of Systems Having a Single Main Reaction	28
II.	Phenomenological Description of Catalyst Deactivation	48
III.	Comments on Obtaining Deactivation Data: Experimentation	55
IV.	Summary and Evaluation	60
	References	61

3 Deactivation by Fouling

I.	Coke Formation and Fouling Kinetics	64
II.	Chemistry of Coke Formation	83
III.	Coke Distribution in Catalyst Pores	97
IV.	Mechanism of Fouling	108
V.	Fouling of Reforming Catalysts	112
VI.	Summary and Evaluation	118
	References	119

4 Deactivation by Poisoning

I.	Some Beginning Ideas: A General Discussion	122
II.	Poisoning of Nonuniform Surfaces: True and Apparent	130
III.	Some Studies of Homogeneous Surfaces	139
IV.	Heterogeneous Surfaces: Site Strength Distributions	146
V.	Particle Size Dependence	149
VI.	Multifunctional Catalysts	160
VII.	Summary and Evaluation	166
	References	168

5 Deactivation by Sintering

I.	Sintering in Supported Metal Systems	172
II.	Particle Growth Models	205
III.	Crystallite Splitting and Redispersion	214
IV.	Metal–Support Interaction Effects on Sintering and Redispersion	225
V.	Summary and Evaluation	230
	References	232

II Deactivation of Catalyst Pellets: Macroscopic Processes

6 The Time Scale of Deactivation in Pelleted Catalysts

I.	General Formulation of the Problem	237
II.	Some Aspects of the General Solution	241
	References	244

7 Intraparticle Deactivation

I.	Limiting Types of Deactivation in Pellets	245
II.	Modeling the Deactivation of Pellets	252
III.	Nonuniform Distribution of Catalytic Materials in Pellets	270
IV.	Fouling of Hydrodesulfurization Catalysts	279
V.	Summary and Evaluation	288
	References	290

8 Direct Measurement of Reaction Nonuniformity in Catalyst Pellets

I.	The Single-Pellet Diffusion Reactor: Concentration Measurements	292
II.	The Single-Pellet Diffusion Reactor: Temperature Measurements	301
III.	Summary and Evaluation	318
	References	320

9 Regeneration of Coked Particles

I.	Reactions of Coke with Hydrogen and Oxygen	321
II.	The Intraparticle Problem	334
III.	Summary and Evaluation	342
	References	343

III Deactivation in Chemical Reactors: Global Processes

10 Deactivation in Fixed Beds

I.	Activity Distributions in Fixed Beds—A Snapshot	347
II.	Moving Reaction Zones—Theoretical Development for Isothermal Reactors	350
III.	Moving Reaction Zones—Theoretical Development for Nonisothermal Reactors	369
IV.	Some Experimental Results—Isothermal Reactors	378
V.	Scale-up to a Nonisothermal Industrial Reactor	390
VI.	Catalyst Poisoning and Nonisothermal Fixed-Bed Reactor Dynamics	395
VII.	Constant-Conversion Operation	413
VIII.	Summary and Evaluation	422
	References	424

11 Regeneration of Fixed Beds

I.	Regeneration as a Deactivation Problem in Reverse	427
II.	Some Quantitative Studies	429
III.	Summary—Deactivation in Reverse	447
	References	447

12 A Case History: Kinetic Lumping, Deactivation, and Reactor Models for Catalytic Cracking

I.	Kinetic Lumping and Deactivation	450
II.	Reactor Models—Fixed, Moving, and Fluidized Beds	452
III.	Reaction Kinetics and Model Evaluation	457
IV.	Some Comparisons of Reactor Performance	461
V.	Another Optimization Problem	463
VI.	Effects of Deactivation on Selectivity	468
VII.	Some Feedstock Correlations	473
VIII.	Summary Remarks	487
	References	488

INDEX 489

PREFACE

Catalysis as a subject continues to increase in importance each year as new processes are modified and refined to use improved catalysts. Among the most important criteria used to screen successful catalysts is mortality. Indeed, the length of time a given catalyst can maintain activity under reaction conditions ranks in importance with selectivity and initial activity. The economic consequences of losses of activity and selectivity have stimulated more interest in the circumstances that cause the phenomena, particularly in the last two decades. Numerous articles on catalyst activation, deactivation, and poisoning are to be found in the literature, but there is no reference in which this information is accumulated and presented in a common framework. As a consequence, the catalytic chemist, the researcher, the reactor designer, and the student of the subject are forced to go to the literature and dig the details from the original papers and two or three review articles. It is for the above group that this book is written, not only as a guide to the literature but also as a systematic framework in which to interpret research on activation, deactivation, and poisoning of catalysts. We fall somewhat short of this mark, however, and the reason is clear. The nature of the catalytic act is imperfectly understood even after many decades of study, and it follows that the details of the circumstances and mechanisms that cause catalysts to lose activity must also be imperfectly understood.

Despite these gaps in the theoretical understanding, further study of catalyst deactivation can be justified on the grounds that inquiry into the nature of these phenomena does indeed lead to improvements in the operation of processes depending on catalysts. Moreover, numerous studies have demonstrated that the study of deactivation is a powerful means of studying fundamental catalysis.

The book is divided into three parts. Part I is devoted to a systematic development of the manner in which catalysts are activated, deactivated, poisoned, and in some cases reactivated on a microscopic basis. An effort is made to present the results of numerous studies in a way that shows how these phenomena alter the number of catalytic events occurring. Our goal is to describe these events mathematically and physically in terms of intrinsic kinetic parameters. For reasons discussed above, this goal is achieved only in part, and some work is presented on a phenomenological basis.

In Part II we view the catalyst particle as an entity. Drawing on the results of Part I, we explore the problem of transport of heat and reactants and products in the interior of the particle coupled with chemical reaction therein. In this manner, we describe the macroscopic deactivation behavior of the catalyst particle in terms of fundamental kinetic deactivation phenomena and of parameters describing heat and mass transfer. The average activity of a particle is thus the main objective of this part.

In Part III we are primarily concerned with a collection of catalyst particles within the reactor. In this part we seek methods to describe the global activity of the reactor in terms of the macroscopic and microsocpic concepts of the two earlier parts. Modeling represents, perhaps, the most powerful way by which hypotheses concerning the nature of deactivation processes can be tested and added to the technology of catalytic processing. By far the most information is available on fixed-bed processes. In the last chapter, a pragmatic approach is presented that permits one to predict the design and performance of reactors containing a deactivating catalyst.

As the text unfolds, it becomes clear that the techniques for handling deactivating systems, not results per se, are the primary objectives of this book. For this we make no apology; indeed, to gain perspective on a subject from the literature is very time consuming. Moreover, to keep the length of the book within reasonable bounds, it has been necessary to use examples of representative literature in many instances.

These constraints lead to several practical considerations in our presentation. The first is that of notation. As indicated above, we use many examples from the literature. Consequently, we develop the topics using a nomenclature generally corresponding to that of the original. The discussion of prior literature leads to another concern. We attempt to integrate this material so that there is a unifying theme in each chapter, but this becomes a bit impersonal after a while. Therefore, we present at the end of most chapters something of a personal assessment titled "Summary and Evaluation." Although these assessments are brief, we hope they are thoughtful. One may choose to agree or disagree with our comments, but they represent our perspective on each chapter and may serve to suggest possible directions for future work.

We acknowledge fruitful interactions with our colleagues, A. T. Bell, R. L. Burwell, Jr., J. J. Carberry, and V. W. Weekman, Jr. We would also like to acknowledge former graduate students who have contributed to the subject of this book, in particular Jay Balder, Rustom Billimoria, Louis Hegedus, Larry Jossens, Peter Kehoe, Jon Lee, Mike Pacheco, Hans Weng, and Eduardo Wolf. We are also grateful to Nancy Monroe for her painstaking and artistic production of the figures.

PART I

Deactivation of Catalytic Surfaces: Microscopic Processes

The kinetics of catalytic reactions conducted on surfaces (heterogeneous) differ from rates of many homogeneous reactions in that they tend to decrease with chronological time under otherwise steady-state conditions. Such variations, thus, cannot be explained by changes in reactant concentrations or reaction temperature and are reflective of a decrease in the intrinsic activity of the catalytic surface. Changes in intrinsic activity are normally caused by chemical or physical processes occurring simultaneously with the main reaction, and the identification of these processes at the microscopic level will be central to later efforts to understand such phenomena in the larger scale of chemical reaction engineering applications.

Any process, physical or chemical, which decreases the activity of a catalytic surface we shall term *deactivation*. In Chapter 1 we describe various types of deactivation phenomena and how they may reasonably be classified. However, in a foreword it is appropriate to remark that one of the more difficult things in applied catalysis is to define a useful and representative measure of activity. Some comparison of various states of the same catalyst is normally required to identify deactivation, but many different measures have been employed in the literature. If we were to identify some of these in an order roughly increasing in the amount of fundamental information provided, a typical list might be:

(a) Temperature required for a given conversion.
(b) Conversion attained at set temperature and space velocity.
(c) Space velocity required for a set conversion at fixed temperature.
(d) Reaction rate under differential conversion conditions, either per unit weight, per unit surface area, or per surface active site (turnover frequency).
(e) Rate constants and adsorption parameters determined from extensive kinetic studies.

Most of these measures pertain to laboratory investigation and the resulting data; yet intercomparison among them is fraught with danger and different conclusions regarding activity may be arrived at using different norms for the "measurement" of activity.

Further, it must be realized that in many important cases the influence of deactivation on activity alone may not be of prime importance; the most important example of this is the case where there is more than one pathway for the main reaction and a particular product selectivity must be maintained. Here, a measure of the effect of deactivation might be more soundly based on product selectivity than on overall activity.

In either event—activity or selectivity determination—it is important to remember that the various measures are not equivalent, and whether we are attempting to make sense of literature data or devise a suitable measure for our own data, caution and (above all) consistency are always appropriate. Keep in mind that the measures described here refer only to the characterization of activity for a given state of the catalyst. Other measures have been devised to determine the rates of deactivation in various instances; this will be discussed in a later section.

CHAPTER 1

Physical and Chemical Description of Deactivation

> This alone gives an uncomfortable feeling...
> N. Pevsner

In order to understand better the nature of the deactivation, and regeneration, of catalytic surfaces it is appropriate first to review some of the concepts that have been employed over the years to provide some measure of general correlation for catalytic processes. Historically, research in catalysis has been primarily experimentally oriented, and attempts to make a general organization of information on catalysis would more properly be termed *correlation* rather than *theory*, although the concepts involved in such correlation rest on theoretical grounds.

Table 1-1 gives a summary of some of the concepts used during this century to correlate catalytic properties. The list is not exhaustive, and perhaps the chronology is not as well defined as it might be, but the variety of factors which people have thought to be important in catalysis provides a clue to the wide latitude possible in approaches to the problem. The table also indicates some periodicity in the relative favor with which the various correlations have been viewed. This is true to some extent, but in most cases there is considerable overlap between the different approaches, and it would be incorrect to attempt a sharp distinction between them. For example, the idea that lattice imperfections might be important in heterogeneous catalysis has been around almost since the beginning of x-ray crystallography, when it became clear that real crystals were not the geometrically neat things that artists' drawings tend to show. In metal crystals there can exist edge and screw dislocations and Schottky and Frenkel defects—any of which can introduce substantial changes in the geometry of the surface and simultaneously alter its microscopic electronic properties. Thus, geometric proper-

TABLE 1-1

Some Correlation in Catalysis over the Years since 1900

Date	Major concepts	Some associated names
~1900	Unstable surface intermediate compounds	Sabatier, Ipatieff
~1920	Lattice imperfections	Many
~1925	Surface active sites	Taylor
~1930	Geometric properties of the surface	Balandin, Beeck
~1950	Electronic properties	
	Metals	Boudart, Beeck
	Semiconductors	Hauffe, Volkenstein
~1960	Unstable surface intermediates and surface active sites	Sabatier, Taylor, Ipatieff

ties, electronic properties, and lattice imperfections are all involved in this type of correlation. Similarly, the defect structure of semiconductor oxides plays a large role in the electronic theory of the catalytic properties of those materials.

Since there is danger in attempting unnecessarily sharp distinctions between these correlations, we shall use as a unifying theme the combined concepts of surface intermediates and surface active sites; each of the approaches given in Table 1-1 plays a slightly different role in such a view of the catalytic event. The basis for such a point of view was established in an excellent summary by Boudart (4) and has been discussed in additional detail (6).

I. THE CONCEPT OF THE ACTIVE CENTER AS UTILIZED IN CATALYSIS

To an important extent the nature of the surface intermediate compound, and thus the course of the catalytic reaction, is established by the nature of the active site on the surface. A number of questions must be answered concerning the active site before we can attain any significant understanding of a given catalytic reaction or of the nature of possible deactivation mechanisms. Important among these are such questions as:

(a) What is the nature (chemical, geometric) of the active site and how is it related to surface structure?
(b) How many active sites are there per unit surface area?
(c) What is the area?

I. The Concept of the Active Center

(d) What is the chemistry of the reactions at the active site?
(e) What are the chemical properties of the active site?
(f) Are there specific chemical or physical properties of the active site that will permit investigation independent of a particular reaction?

If one had available answers to all these questions for, say, a sufficient number of related catalysts and reactions, then the establishment of reasonably reliable extrapolations to other, similar, catalysts and reactions might be possible on the basis of one or the other of the concepts given in Table 1-1. Similarly, such information would provide detailed insight into the possible ways in which the active site could be rendered inactive.

As an example, now classical, of the development of information concerning the chemical nature of active sites and their number on a given surface, we shall discuss the investigation of several oxides via chemical titration/chemisorption techniques. These methods are distinct from the physical adsorption methods employed for measurement of total surface area and internal volume-area distribution; such characterizations via physical methods are beyond the scope of the present discussion but have been reviewed in detail elsewhere (10, 11).

A. Chemical Titration of Acidic Oxide Surfaces

A number of oxides, used either as catalysts alone or as one function of a bifunctional formulation, have been shown to be acidic in nature. Silica-alumina is a prime example; if we consider the surface to be represented by the intrusion of aluminum atoms in a predominately silica structure, then the development of acidity can be viewed as occurring via the formation of two types of sites:

$$\begin{array}{c} Si-O-Al-O-Si \\ | \\ O \\ | \\ Si \end{array} \qquad \textbf{I-A}$$

and

$$\begin{array}{c} H \\ | \\ O \quad H^+ \\ | \\ Si-O-Al-O-Si \\ | \\ O \\ | \\ Si \end{array} \qquad \textbf{I-B}$$

in which the exposed Al in **I-A** acts as a Lewis acid (capable of accepting electrons) and the proton in **I-B** as a Brønsted acid. The development of such surface acidity is seen to be closely related to the extent of hydroxylation of the surface; in a partially dehydroxylated surface both types of acidity can be present. The acidic nature of the hydrated surface of **I-B** is apparent; however, the development of Lewis acidity may not be so clear. This acidity arises from the incomplete coordination of aluminum in the surface layer, in which one oxygen bond is lacking for formation of the stable eight valence electrons. Thus, the coordinatively unsaturated aluminum will accept a pair of electrons from suitable donor species, which, in turn, will be basic materials. Commonly encountered in catalytic applications are materials such as hydrogen sulfide, organic thiols and sulfides, and nitrogen-containing compounds such as pyridine, quinoline, and ammonia.

Whether one deals with Lewis or Brønsted acid sites, the possibility exists of strong and specific acid-base interactions with such compounds. In principle, this property of surface acidity should be characteristic of any oxide with coordinatively unsaturated metal atoms in the surface layer, although such characterization is normally applied to insulator oxides such as alumina, silica, and silica-alumina. Semiconductor oxides such as zinc or nickel oxide exhibit electronic charges on chemisorption of adsorbates that tend to becloud an interpretation on the basis of simple acid-base interaction. For the insulators, however, one can use data on the chemisorption of basic compounds as some measure of surface acidity. Indeed, if we can envision stoichiometry as illustrated in Fig. 1-1 for the example of

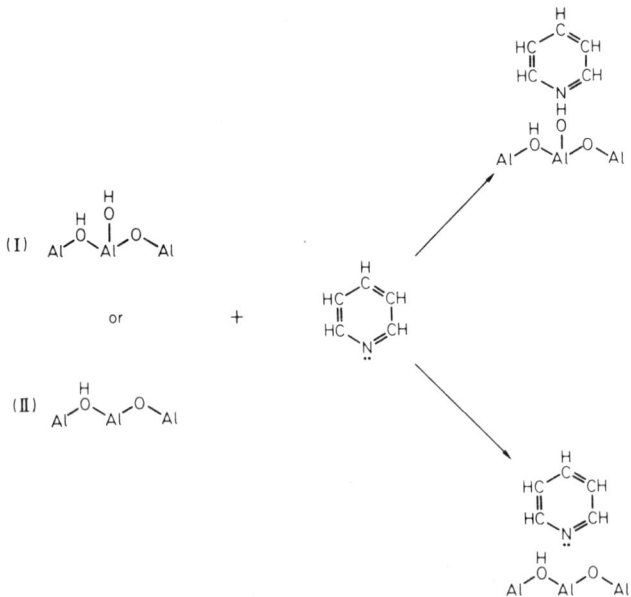

Fig. 1-1. Titration of acid sites on alumina with pyridine.

I. The Concept of the Active Center

pyridine on alumina, chemisorption data should provide information on the total number of sites, and, by varying the basicity of the adsorbent, a distribution of site acidity can be obtained[1]. Obviously, if the acid site is active for a catalytic reaction, that reaction will be of the class which we know to be promoted by acidic media. Reactions such as the cracking or isomerization of hydrocarbons are of this category, and it is well known that acidic oxides such as silica–alumina are extremely effective catalysts for them.

From this example we see that the concept of an active site has assumed some chemical reality; so also has the concept of surface intermediate if we recognize, for example, the complex as an entity differing in chemical identity from both the surface acidic site and the pyridine molecule before adsorption.

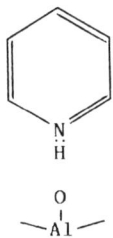

B. Active Sites on Metal Surfaces

The concept of active sites is also readily extended to another important class of catalytic surfaces, metals. In fact, on first glance the situation for metals seems simpler than that for oxides, since in this case we may associate active sites with surface metal atoms, either singly or in some combination depending on the crystal plane and adsorbate under consideration. The single best-understood example deals with the chemisorption of simple molecules such as carbon monoxide or hydrogen on group VIII metals. The chemisorption of hydrogen has been used widely as a means of characterizing these metals when supported on oxide surfaces, as they are most often employed in large-scale catalytic processes. Hydrogen chemisorption in the majority of cases is dissociative, with the hydrogen atom bonding to individual surface metal atoms. Similarly, carbon monoxide adsorption is generally nondissociative while bonding to surface metal atoms.

---

[1] The illustration indicates the existence of Brønsted acid sites on alumina. This has been disputed on the basis of infrared data, which give no indication of the existence of the pyridinium ion on alumina surfaces. The distinction between types of acidity is not important to the discussion here.

Even with simple molecules, however, identification of the surface metal site may not be simple. Figure 1-2 illustrates the existence of two types of carbon monoxide adsorbed on the surface of palladium as indicated by the infrared spectra of the surface. The linear adsorbate is associated with a single metal atom, while the so-called bridged species is associated with two atoms, although the molecule has not been dissociated.

A related problem pertaining to surface metal sites is that while we know that, say, hydrogen is dissociatively adsorbed on a platinum surface, this information may not be particularly useful in an interpretation of how a more complex molecule such as cyclopentane is adsorbed on the same surface. In addition, one is often concerned with the degree of crystalline perfection and its influence on the chemisorption properties of metallic surfaces. First, there is the possible effect of the defect structure such as vacancies, dislocations, and twins on the surface. Second, there is the effect of geometry (essentially size) on the nature of surface atoms and their number, that is, edge atoms versus corner atoms versus planar surface atoms. In either case we are concerned with coordinatively unsaturated species

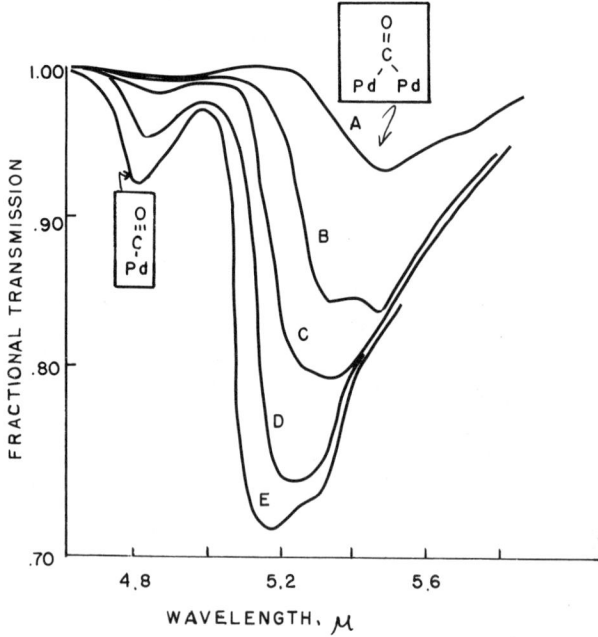

Fig. 1-2. Infrared spectra of linear and bridge-bonded CO on supported Pd, as a function of surface coverage: A, 0.20; B, 0.45; C, 0.65; D, 0.85; E, 1.0. [After R. P. Eischens and W. A. Pliskin, *Adv. Catal.*, **10**, 1 (1958). Reproduced by permission of Academic Press, Inc.]

I. The Concept of the Active Center

whose extent of coordination depends in detail on the immediate surroundings.

An example of the geometric effects associated with metal crystallite size is provided by the calculations of Van Hardeveld and Hartog (17) for typical arrangements of face-centered-cubic (fcc) metals. Figure 1-3 shows three types of surface atom configurations for a fcc octahedron together with the corresponding statistics for their frequency of occurrence as a function of the equivalent spherical diameter of the octahedron. It is apparent that the net degree of coordinative unsaturation is a strong function of size, particularly for smaller particles, and this has important bearing on some titration methods which have been proposed for supported metals. A particularly well documented example of this has to do with the hydrogen-oxygen titration method of Benson and Boudart (2). Oxygen is contacted with the reduced surface of the metal, say platinum, to form the surface oxide PtO. The titration proceeds according to

$$PtO + \tfrac{3}{2}H_2 \rightarrow Pt \cdot H + H_2O \qquad \text{II}$$

In principle, this method is more sensitive than a straightforward measure of hydrogen chemisorption, since three hydrogens are consumed for each surface Pt as compared to one in the latter case. However, the stoichiometry of **II** is apparently a function of particle size; more generally we may write

$$PtO_x + \left(x + \tfrac{y}{2}\right)H_2 \rightarrow Pt \cdot H_y + xH_2O \qquad \text{III}$$

in which x is related to the stoichiometry of the surface oxide and y to the stoichiometry of the titration reaction. As one decreases particle size, the values of x and y approach 1 and 2, respectively (13), which variation is

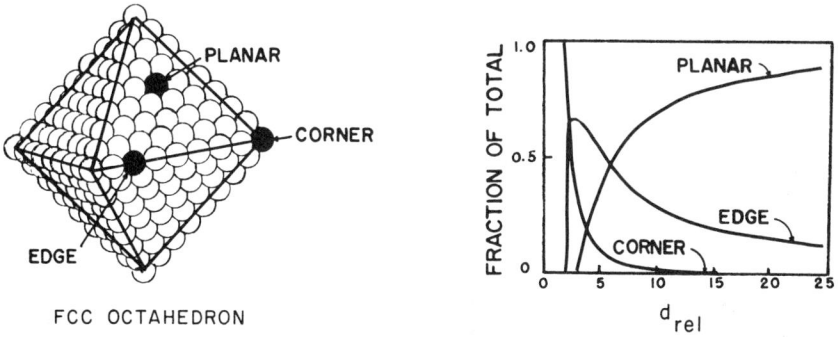

Fig. 1-3. Surface atom statistics as a function of metal crystallite size. [After Van Hardeveld and Hartog (17). Reproduced by permission of North Holland-Physics Publishing.]

most probably the result of variations in the composition of PtO_x with particle size (16).

In spite of difficulties in interpretation of measurements designed to characterize the numbers of atoms on the surface of a metal crystallite, we will conclude this short discussion by stating simply that, nonetheless, the concept of the active site as associated with single atoms or groupings of atoms on the surface is valid in principle.

II. OPERATIONAL DEFINITIONS OF DEACTIVATION

The reader may wonder why we expend the opening paragraphs in a book such as this on a discussion of active sites, a topic which must certainly be familiar to anyone who has more than a casual acquaintance with the field of catalysis. The point is, basically, to reemphasize how widely applicable the concept is and to reemphasize its reasonableness in both chemical and physical terms. Why? Well, almost all cases of deactivation can be viewed as the result of phenomena which remove active sites from the catalytic surface. Hence, an understanding of the chemistry of the active site not only provides understanding of what reactions it might be a useful catalyst for but also gives insight into how it might be deactivated—and regenerated. Further, we shall see that classical kinetic formulations based on active site mechanisms may easily be extended to include deactivation phenomena; this, of course, is the ultimate objective of most who are concerned with catalyst deactivation in practical application.

While deactivation is the overall result of the removal of active sites from the catalytic surface, this result can occur by a number of different mechanisms, both chemical and physical. We shall divide these into several more or less arbitrary classifications with the understanding that, just as in the case of correlations in catalysis, there can be considerable overlap. The three major categories of deactivation mechanisms are *poisoning, coking* or *fouling,* and *sintering.* These may occur singly or in combination, but their net effect always is to remove active sites from the surface. Other mechanisms can be important in certain instances; these include *pore blockage* via coke or metal deposition on the internal pore structure of a catalyst, *volatilization* of an active catalytic component, *destruction* of active surface via *detachment* from the support (by means other than volatization), and *incorporation* of the active component into the support structure in an inactive form. A brief description of what we mean by each of these terms is given below; extensive discussion and review of the various mechanisms of deactivation will be given in Chapters 3, 4, and 5.

II. Operational Definitions of Deactivation

A. Deactivation by Poisoning

Systematic investigation of catalyst poisoning had its origin rather recently if one considers the time span from the present day to that of Berzelius. However, some observations of poisoning, if we interpret them in modern terms, are indeed very old. Quoting here from a very interesting and informative survey of the history of catalysis [Burwell (5), p. 47], we offer:

> ...the following is clearly a catalytic experiment:
> Paracelsus sent his waiting man to deliver a paper containing a small amount of blood-red powder (unspecified) with the command that it be poured into molten lead and stirred well.... The Master of the Mint paid several thousand guilders for the resulting gold.

The following observations can be made regarding this:

(a) The experiment was probably not reproducible.

(b) There is no record of the fate of the Master of the Mint, but it is probable that payment in guilders was greater than gold received (in units of red powder plus lead).

(c) It is possible there was some poisoning or deactivation effect.

(d) There is also no record that the experiment reported was formally withdrawn.

On a more factual basis, we note the observations of Faraday (8), who, among other things, was interested in the reaction of hydrogen and oxygen in the presence of platinum of various forms. Such experiments were conducted in a glass apparatus and one would presume that in general the success of the experiment must have been measured by the force of the explosion destroying the apparatus. In one experiment, some ethylene was added to the hydrogen-oxygen mixture and nothing happened for a long time. Finally, the apparatus exploded in its customary fashion, presumably much to Faraday's satisfaction. One would say now that the ethylene was preferentially adsorbed on the platinum surface and only after it had been oxidized and removed could the violent coupling between H_2 and O_2 occur. In this case, then, ethylene was acting as a temporary poison for the hydrogen-oxygen reaction.

Perhaps the best-documented examples of catalyst deactivation have been associated with the removal of active sites via the strong chemisorption of impurities on the surface, thus blocking access of the reactants. In general, instances of catalyst poisoning share the common feature that they can be well characterized in a chemical sense. For instance, if we return to the example of titration of an acidic oxide surface with pyridine (Fig. 1-1), it is clear that the strongly chemisorbed molecule of pyridine would block access of any other molecule to the acidic site it occupies. Hence, if one were conducting an acid-catalyzed reaction, say isomerization, on this surface, pyridine would constitute a very effective poison for that reaction. Catalyst poisons can be classified in various ways, such as by their degree of affinity for the surface, as temporary or permanent, or as selective or nonselective. In particular, we will make a distinction between temporary and permanent poisons, a point which rests upon the degree of reversibility of chemisorption on the surface, and inhibitors of reaction rate which are either reactants or products of the main reaction being carried out. In formal kinetic formulations such as the Langmuir–Hinshelwood–Hougen–Watson (LHHW) methodology, the latter effects appear in the denominator of the rate equation as "adsorption" terms, and some authors refer to this as a "self-poisoning" or "self-inhibition" effect. We shall reserve this terminology for some cases of deactivation by coke formation, discussed below, and what we mean by *poisoning* excludes those effects arising from reactant or product inhibition.

The description of a poison as selective or nonselective is related to the nature of the surface and the degree of interaction of the poison with the surface. Here we are concerned with the collective properties of the ensemble of active sites that constitute the catalytic surface. Every student of kinetics knows that a basic assumption of LHHW formulations for catalytic reactions is that the surface appears uniform to the reactant molecules; in practical terms this means that chemisorption on the surface is not affected by the degree of coverage of the surface, among other things, and an observable result of this is the fact that the heat of chemisorption is independent of surface coverage. On a more microscopic scale, this implies a uniformity of surface sites and the absence of interactions between chemisorbed molecules. Much the same picture is implied when we characterize a poison as nonselective. In such a case the chemisorption of poison on the surface removes active sites in a uniform manner, such that the net activity of the surface is a direct function of the amount of poison chemisorbed. In essence, every active site looks like every other active site to the poison molecule. Conversely, in selective poisoning there will be some distribution of properties of the active sites (e.g., acid strength), which can be the result of any number of factors but which results in nonuniform deactivation of the surface. Often these will appear as exponential or hyperbolic relationships between the net activity of the surface and the amount of poison chemisorbed. A qualitative picture of selective and nonselective poisoning is given in Fig. 1-4. A word of caution: it is important in making such a distinction about the nature of the surface–poison interaction that the supporting experimental observations be reflective of the true intrinsic activity of the

II. Operational Definitions of Deactivation

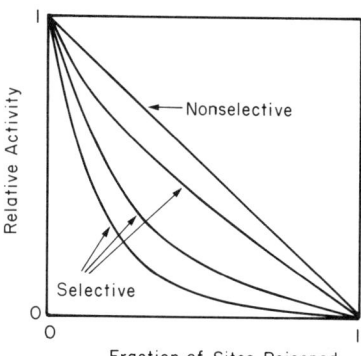

Fig. 1-4. Qualitative activity–poison content relationships for selective and nonselective poisoning.

surface. We shall see later that extraneous factors such as diffusion can mask this intrinsic relationship and lead to erroneous conclusions concerning the nature of poisoning.

In summary, we can describe the major distinguishing features of deactivation by poisoning as being those of a chemically well-defined event which normally can be thought of in terms of preferential competitive chemisorption of some component in the reaction mixture which removes the active site from access by the components involved in the main reaction. Since we are really considering this process in terms of chemisorption, various degrees of reversibility can be associated with poisoning under different circumstances or in different reaction systems. Conversely, the problem of *regeneration* of poisoned catalysts is also a function of the degree of reversibility involved in the poisoning process. Some examples of typical catalysts and poisons are given in Table 1-2.

In order not to sacrifice generality at this point, we should probably also include as poisons materials which may be present as impurities in the catalyst itself, independent of the reaction mixture. For example, the presence of small amounts of alkali metals such as Na in alumina can significantly affect its behavior in reactions demanding strong acid properties such as isomerization. Since one method of preparing "alumina" is via the alkali aluminates, the presence of such hidden impurities is not altogether uncommon (12) and there are plenty of unfortunate examples in the catalytic literature which attest to ignorance of such effects even for such a widely used material as this. Even more insidious may be the effect of trace metal impurities on the behavior of supported metal catalysts. The thorough study of Fuentes and Figueras (9) shows, for example, that even several hundred parts per million iron are quite sufficient to alter the catalytic properties of Pd supported on SiO_2 or Al_2O_3. Again, in practice such contaminants may be associated with the support material and, although there may not be

TABLE 1-2

Some Catalysts and Catalyst Poisons

Catalyst type	Reaction	Poisons
Supported or unsupported group III or IB metals	Hydrogenation at moderate temperatures	H_2S, H_3P, thiols, unsaturated rings with heteroatoms (i.e., thiophenes, pyridine, etc.)
SiO_2/Al_2O_3	Isomerization or cracking of hydrocarbons	Basic organic compounds
Supported Pt	Oxidation at lower temperatures	Sulfur compounds
NiO	Oxidation	CO
Supported Pt, Pd	Hydrogenolysis of cyclopropanes	CO, sulfur
Al_2O_3	Alcohol dehydration, isomerization	Na, K

much that can be done, it is important to recognize the existence of such effects. In the main part of what follows we shall be concerned with catalyst poisoning and not catalyst purity.

B. Deactivation by Coke Formation

While deactivation by poisoning has been the best documented and best understood from a chemical point of view, deactivation by coke formation would probably win the prize for sheer volume of words published. Several different terms are employed to describe generally the same phenomenon: *coking, fouling*, and *self-poisoning* are most commonly encountered. Since we have been at pains to define poisoning within a particular context above, we will try not to confuse the issue by avoiding the term "self-poisoning" in description of coke formation. What is coking? To start with, it is something which is generally a bit more difficult to describe than poisoning, because it is not so well defined chemically. We can begin by saying that in many reactions involving hydrocarbon molecules (or even carbon oxides) there are side reactions on the catalyst surface that lead to the formation of carbonaceous residues which tend to cover over the active surface. In many instances the "mechanism" of coke formation can be visualized as a kind of condensation polymerization on the surface resulting in macromolecules of empirical composition approaching CH. Now, having ventured this description, we must immediately soothe those readers who are outraged by it by saying that there are probably as many mechanisms for coke formation as there are reactions and catalysts in which the phenomenon is encountered.

II. Operational Definitions of Deactivation

The immense variety of environments in which coke formation occurs is reflected in the literature dealing with it. By contrast to poisoning, where the chemistry is often identified and kinetic effects can be handled in a rather specific manner, coke deposition effects are often interpreted *ex post facto*. Hence we will see in later discussions rather extensive use of time-on-stream correlations or related empirical representations of deactivation via coke formation. Further to this point, such correlations are most often those for amount of coke deposited on the catalyst and not for actual catalyst activity. For this reason one is often confronted with the necessity of developing the additional relationship between something termed "coke on catalyst" and something else termed catalyst activity.

Since coke deposition normally manifests itself as macroscopic deposits of carbonaceous material on the active surface, the quantity of such residues may be measurable on the same scale as the amount of catalyst—coked catalysts containing 15 or even 20 wt. % of the deposit (grams per gram of catalyst) are not unusual in certain services. By contrast, even micromoles per gram of certain poisons are sufficient to completely deactivate some catalysts. The large amounts of coke often encountered suggest two additional important factors associated with this mode of deactivation. First, the deactivation effect is accomplished primarily by the covering over of active sites by the residue; reactants are denied access to active sites by physical screening rather than by the competitive chemisorption at work in poisoning. Second, since the amounts of coke formed can be quite large, such deposits can eventually build to the point of blocking pores in the internal volume of the catalyst and further restricting access of reactive molecules to active surface.

There are some important applications in which coke formation of a very different nature from that described above occurs. Here the element carbon itself is more directly involved, rather than hydrogen-deficient carbonaceous residues. In Fischer–Tropsch synthesis with iron catalysts, for example, carbon is deposited on the surface which eventually is incorporated into the metal matrix as iron carbides of various types. Even more striking is the behavior of nickel in some high-temperature hydrocarbon reactions; here no carbide is formed, but the solubility and diffusivity of carbon in the metal are sufficient for the carbon to diffuse through the metal crystallite, accumulate between the metal and support, and eventually detach the metal from the support.

Although we have heavily qualified any general description of mechanisms for coke formation, there are certain instances in which the chemistry is understood in at least qualitative form. In many areas of hydrocarbon processing, and particularly in catalytic cracking as carried out in petroleum refining, there is much evidence to suggest that the aromatic content of the materials being processed is primarily responsible for coking. Typical of such materials are polycyclic materials such as naphthalene and anthracene and rings with various alkyl substituents. Since side chains are effectively removed first in cracking reactions, the bare ring structures seem to be the intermediates important in coke formation. Detailed results of some experiments based on this postulate (1) are shown in Fig. 3-27. Weight percent coke on catalyst is reported as a function of the individual feed materials as shown; coke formation is more pronounced in condensed rings (naphthalene → anthracene, etc.) than in linked rings (biphenyl → terphenyl, etc.), while with the heterocyclic materials the hydrocarbon analog is always most productive of coke. The very high activity of anthracene and similar condensed aromatics makes it inviting to use the interactions between benzene and naphthalene as a model for important reaction paths in coke formation. Such is shown in Fig. 3-28, where at each of the intermediate steps additional dehydrogenation or ring opening might lead to the formation of a "coke" molecule. Since it is the reactants or products of reaction that are responsible for this mechanism of deactivation, there is some basis in reason for the term "self-poisoning." Further details of this are given in Chapter 3.

C. Deactivation by Sintering

Normally, the term sintering is applied in describing the loss of active surface via actual structural alteration of the catalyst. Most commonly this is a thermally activated process and is physical rather than chemical in nature. Sintering can occur in both supported metal catalysts and unsupported materials such as zeolites or amorphous silica–alumina. In the former case, active surface is lost via the agglomeration of small metal crystallites into larger ones with smaller surface-to-volume ratios, while in the latter instance the process may involve actual collapse of the internal pore structure. Sintering of unsupported materials is often encountered in the regeneration procedure when, for example, coke deposits are burned off a cracking catalyst. Various mechanisms have been proposed to describe the sintering of supported metals, some chemically based and some physically based, but all consist of sequential steps of transport, collision, and agglomeration described in some terms of operating conditions. These will be discussed in more detail subsequently.

There has been relatively rapid advancement in our knowledge (if not understanding) of the sintering of supported metals over the past few years owing to the development of techniques such as hydrogen chemisorption or hydrogen–oxygen titration for characterization of metal particle size. With this has come increased appreciation of the fact that sintering

II. Operational Definitions of Deactivation

phenomena not only are important as far as catalyst deactivation is concerned but also play an important role in the various pretreatment methods commonly used for "activation" of supported metal catalysts. Further, there is an increased awareness of the opposite process to sintering, often called *redispersion*. In particular, the interaction between oxygen and commercially important noble metals such as platinum and palladium under certain conditions leads to the formation of species which are mobile on the surface and which reverse the process of agglomeration.

Because sintering is normally a physical rather than chemical process, the models and mathematical description employed differ in kind from those used to describe poisoning or coking. Also, the magnitudes of thermal activation (i.e., activation energy) are normally quite different; typical activation energies for sintering phenomena may be twice or three times those associated with the chemical processes in poisoning or coking. We should temper this point, however, since the numbers associated with such parameters are somewhat obscured by the inadequacies of the kinetic models which have been employed to describe sintering phenomena.

A final, extreme form of deactivation which can be viewed as a form of sintering is, with supported metals, the incorporation of the metal into the support. Many support materials such as alumina have spinel-type structures in which regular vacancies occur in the lattice. These vacancies are, of course, able to accommodate additional metal atoms, either of the parent metal (nonstoichiometric oxide) or others, depending on atomic or ionic dimension. These processes involve solid-state reactions which ordinarily occur only at very high temperatures and hence can be regarded as somewhat extreme cases. An example is nickel supported on alumina: exposure of such a catalyst with nickel loadings on the order of 1 to 5 wt. % to temperatures of about 1000°C is sufficient to incorporate completely the metallic nickel into the support as a nickel aluminate. Needless to say, nickel aluminate is not a very good catalyst for anything.

D. Other Mechanisms of Deactivation

Among other possible modes of catalyst deactivation, that of pore blockage or plugging is probably the most important. This has already been mentioned with respect to coke deposition, where, at very high levels of coke on catalyst, it may be possible for the coke itself to block off the pore structure. Other important examples of pore plugging are found in certain hydrotreating processes, where metals (principally Ni and V) in the feedstocks have a strong tendency to deposit on the catalyst near the external surface and block access to the interior. One mechanism involved in the deactivation of automobile exhaust catalysts is a similar blocking or coating by Pb deposits when fuel containing lead antiknock compounds is used. These pore-blocking mechanisms are partly chemical and partly physical in nature; in many cases the metal-containing molecules or the coke precursor molecules are large and subject to significant diffusional effects within the pore structure. Hence, it will be important in later analysis to understand the interrelation between reaction, deactivation, and transport rates in order to interpret such effects.

At the opposite end of the scale from the deposition of poisons or coke on the active catalytic surface is the loss of an active component via processes such as volatilization. The Deacon process for the production of chlorine by oxidation of by-product HCl is a classical example, since the active copper-based catalysts most effective for this oxidation form volatile $CuCl_2$ under the conditions of operation. In spite of a long and interesting history of work on this process and its catalyst, no one yet has really succeeded in stabilizing the Cu on a support to avoid the volatilization problem. Much the same story can be related concerning supported Ru, which at one time was considered a leading candidate for NO_x emissions control via catalytic reduction in various (but primarily automotive) applications. Unfortunately, if the supported Ru is exposed to oxidizing atmospheres at elevated temperatures, a mixture of the volatile oxides RuO_3–RuO_4 is formed and Ru disappears from the scene forthwith. Again, many efforts have been made to stabilize the Ru on various support materials, with small success. Volatilization also appears to be an important mechanism of deactivation for several types of supported homogeneous catalysts (i.e., homogeneous catalysts contained within the pore structure of a support). Experience with the Wacker oxidation catalyst, normally a $PdCl_2$–$CuCl_2$ mixture in acidic solution, has shown that when incorporated within a support structure the catalyst rapidly loses activity because of volatilization of chlorine.

III. REGENERATION, REJUVENATION, AND DETOXIFICATION

As soon as one has deactivated a given catalyst, by whatever means, the question arises of how it may be returned to its initial state of activity. Unfortunately, in many instances catalyst deactivation—or, more properly, catalyst activity—is not a state function in the thermodynamic sense and there is no pathway short of remanufacture to accomplish this. Consider the problem of a sulfur-contaminated supported nickel catalyst. The spent catalyst surface probably consists of a nickel sulfide-type layer (or layers), assuming the catalyst was used in a reducing atmosphere, and the problem of regeneration is very simply stated: remove the sulfur from the surface while at the same time returning the nickel to its original state on the support. There are difficulties in accomplishing both of these objectives simultaneously or, indeed, even separately. The Ni–S bond is a strong one and regeneration via either oxidation or reduction will require high temperatures. High temperatures easily lead to sintering; hence, even though the sulfur contaminant is removed, the metallic nickel resulting is not in its original state.

Although the "catch-22" cycle described above for the nickel–sulfur system is true for many potential *in situ* regenerations, such procedures are possible and routinely conducted for the regeneration of catalysts with coke deposits. In these cases the technique is essentially that of oxidation or burning of the residue off the surface. Again, careful temperature control must be exercised in order to avoid sintering; this is, in fact, possible in many cases and a large area of technology has been developed in application to the regeneration of coked catalysts—particularly in catalytic cracking. Some fascinating reaction engineering problems are connected with this area of catalyst regeneration, involving both chemical reactor analysis at a fairly high level of technical sophistication and operational strategy dictated more by economic considerations. We shall have a chance to investigate these in depth for both fixed-bed and fluid-bed operation later on.

Short of regeneration or rejuvenation procedures, often one can resort to some preventive medicine to retard or block deactivation. The use of surface area stabilizing agents is often encountered. One of the roles proposed for potassium in potassium-promoted iron ammonia synthesis catalysts, for example, is stabilization of the iron surface area under synthesis conditions. Volatilization problems may be overcome in some instances by inclusion of small concentrations of the material being volatilized into the process feed stream, as, for example, chlorine in the case of supported Wacker oxidation catalysts.

IV. SUMMARY—CONVENTIONAL WISDOM AND OTHER MATTERS

In his original discussion, Berzelius (3) wrote concerning catalysis: "It is, then, proved that several simple or compound bodies, soluble and insoluble, have the property of exercising on other bodies an action very different from chemical affinity. By means of this action they produce, in these bodies, decompositions of their elements and different recombinations of these same elements to which they remain indifferent." The problem is that he did not also note that the "simple or compound bodies" possess in common a property of living bodies, that of mortality. Perhaps the key to the entire topic of this book is the phrase "to which they remain indifferent." This is literally so only to a degree of approximation. The contrast between the real world of catalysis which we attempt to describe here, if only imperfectly, and the concepts expressed by Berzelius and many others since is very nicely expressed by the following philosophical dialogue, somewhat tongue in cheek, published almost 140 years later [excerpted from (7)];

...I. **By definition,** as the Freshman Theoretical Chemistry Philosopher teaches, catalysts endure forever—being regenerated *in situ* during each reaction cycle. For does he not declare in general that

$$A + X \leftrightarrows AX$$
$$B + X \leftrightarrows BX$$
$$AX + BX \to P + 2X$$

$$\overline{A + B \to P}$$

And so it would seem that insofar as catalytic research workers are concerned, they should bear witness to the everlasting life of their agent, the catalyst. Therefore, there does not exist a need for a school devoted to contemplation of the death of catalysts.

II. **Sed Contra**: I answer that, as the Philosopher of Experimental Chemistry teaches,

(a) few if any of those who labor in experimental catalysis have ever been witness to even modest catalyst life.

(b) those who seemingly bear witness to the promise of everlasting catalyst life generally find such evidence of immortality to be limited to o-p H_2 conversion at $-195°C$, at a celestial pressure of 10^{-6} Torr. And does not the Poet teach us that such conditions befit only the Prince of Darkness and those souls who failed to surmount the VII storey mountain of industrial Purgatory (Divina Commedia—Inferno, Canto XXXIV)?

III. Now insofar as catalytic reactions of real interest seize our heavenly aspirations, it would seem that

(a) realists should address themselves to reality, for does not the Philosopher teach that "all changes," not the least of which are catalytic sites,

(b) that those observers of changing catalytic dispositions must change in a manner commensurate with the dispositions of that saintly agent which guides them—the catalyst. For does not the modern sage declare "get with it, man" (Diable Commedia—Channel 7, Canto I-MMCXII)?

(c) the catalyst, subject as it admittedly is to the terrors of mortality, should be worthy of unique notice, study and compassion.

Ergo, given, then, the general, if not universal evidence of catalytic mortality (attrition, contrition, fouling, aging—in a word, poisoning), we deem it appropriate that all those communicants who labor in humble, indeed contrite, witness to mortal catalytic sites should reaffirm their obsequiousness to the Prince of Obfuscation.

Thus we assert that a subsection of the general field of heterogeneous catalysis be established with chapters East, West, North and South of Northwestern and termed The Lucrezia Borgia Catalysis Society.

For she, of happy memory, is surely a most knowing and sympathetic patron, ever devoted to sustaining our labors by interceding on our behalf with the gods, Coke,

IV. Summary—Conventional Wisdom and Other Matters

Sulphur, Heat and their Court, and so to assure us a livelihood and at least a few publications per annum....

Well, not only a few publications per annum. Each of your authors has had the opportunity to observe and participate in detail in "the VII storey mountain of industrial Purgatory" with particular regard to catalyst deactivation. The fact of the matter is that often the entire procedure of process development, pilot plant operation, scale-up in design, and commercial operation is dictated by considerations of activity and/or selectivity maintenance of the catalyst involved. We have before us an everyday reminder of this in the development and implementation of catalytic converters for automobile exhaust emissions control[2]. The required objective is maintenance of the emissions of carbon monoxide, hydrocarbons, and oxides of nitrogen below specified limits (normally written in terms of grams per mile, a wonderful mixture of units) for a specified total period of utilization (50,000 miles at this writing). The initial activity of the noble metal catalysts employed must be sufficiently high that enough residual activity will remain after this term of utilization still to meet the standards. This immediately involves one in a practice familiar to most of those readers who are engineers—something we might call "optimized overdesign."

The necessity for this procedure is almost universal in any conceivable application involving catalyst deactivation, much less the automotive emission problem. We have already mentioned that there are no state functions in the thermodynamic sense when it comes to deactivation phenomena. The current condition of a given catalyst is in general dependent on its entire previous history; to be precise, such measures as activity or selectivity levels as a function of time of utilization (or other related quantities) are functionals and may hinge on the most delicate details of pretreatment, operating conditions, operational upsets, and so on. It is not illogical in the applications that we are concerned with here to view the problem as always requiring shooting at a moving target (sometimes moving quite rapidly, too). If research in catalysis has primarily been experimentally oriented, investigation of catalyst deactivation must be even more so. The pure theory of diffusion-reaction-deactivation is of little comfort to process engineers who are concerned whether their hydrotreating unit will remain operational at required conversion for the year (or two) that their manager is counting on.

---

[2] A number of the topics to be treated later have had their origin in one or another of the aspects of the development of these devices. Fortunately, the existence of a review (14) which treats the topic of catalytic control of automotive emissions in more detail, and with more expertise, than we might provide relieves us of the task of discussing that noxious subject in specific terms here.

The thoughts above imply correctly that reliable experimental measurement is a primary, even controlling, factor in coping with deactivation. Just as we cannot generally predict the activity or selectivity properties of a new catalyst formulation for a given reaction, neither can we predict the deactivation behavior other than to say "this formulation will be susceptible to poisoning by bases" or "this may not be as thermally stable as alternative formulation X," and the like. What makes it so unfair is that even if we *could* do so quantitatively, we would still need to be able to look into the future in order to determine quantitatively what the history of a particular application might be.

Perhaps we sound a bit petulant in these remarks; not really so, for if we left matters at that we would have a jejune philosophy and could stop the effort right here. What we hope will come through to the reader in what follows is that examination of catalyst deactivation phenomena in all their various aspects—by various mechanisms and at various levels of magnitude—can provide us with the ability to cope with, if not eliminate, the problem. Thus, in the words of the Philosopher of Experimental Chemistry, "to assure us a livelihood and at least a few publications per annum."

V. MORE DEFINITIONS

We have defined a number of terms in this chapter to describe various phenomena in a very particular way. These terms will be used throughout the remainder of the book in the exact sense of their definition, and since the various definitions are somewhat scattered throughout the foregoing, it will be useful to summarize them here.

As the major mechanisms of deactivation, we will be concerned with poisoning, coking, or fouling, and sintering. *Poisoning* we visualize as the strong chemisorption of impurity materials on a catalyst surface which blocks access of reactants to active sites. Poisoning may be a reversible or irreversible process, dependent on temperature level and concentration, but in general we specifically exclude from this definition reactant or product inhibition effects which, in normal kinetic analysis, would be correlated by adsorption terms in the denominator of Langmuir–Hinshelwood–Hougen–Watson (LHHW)-type rate equations.

Coking or *fouling* is the process of formation of hydrogen-deficient carbonaceous residues on the surface of catalysts, most often in reactions involved in hydrocarbon processing. Here active sites are removed from the surface by covering with large molecules of something generally referred to as "coke." In general, the mechanisms of coke formation are less specific and chemically less well defined than those of poisoning. In addition, coke

V. More Definitions

can be formed in macroscopic amounts, sufficient to alter the internal pore structure of a porous catalyst or catalyst support and physically block access of reactants; by contrast, one envisions poisoning in terms of the monolayer surface coverage model, where damaging amounts may be at the micromole per gram level. The term self-poisoning is sometimes used to describe coke formation; although in general the meaning is clear enough to avoid confusion with impurity poisoning, we will try to avoid the term here.

Sintering is the loss of active surface area by the agglomeration of small metal crystallites into larger ones, in the case of supported metal catalysts, or the collapse of pore structure and loss of internal surface area in the case of supports or oxide catalysts. In contrast to poisoning and coking, sintering processes are ordinarily physical rather than chemical in nature and most often very highly thermally activated. Cases exist, however, in which sintering is *chemically assisted*, for example, via the formation of volatile oxides in some supported metal systems. Also, for supported metal catalysts we may think of the process of *redispersion* as something more or less the opposite of sintering.

Regeneration refers generally to any process or procedure which is employed to restore a deactivated catalyst to its original, or a substantial fraction of its original, activity. The term is most often used in reference to catalysts which have been deactivated by coking, where the carbonaceous deposits can be removed by oxidation (i.e., burning) under suitable conditions. Redispersion is, of course, a regeneration procedure, while the regeneration of poisoned catalysts is often termed *detoxification*.

We will also find it useful, in much of the simulation and analysis work concerned with either intraparticle problems or chemical reactors, to employ some very simple kinetic networks to represent the various mechanisms of catalyst deactivation. Little justification in detail can be provided for these networks in most instances, since they are the limiting cases of simplicity. However, they have the virtue of pointing the way very clearly to the general effects of deactivation in various types of catalytic reaction engineering problems; this can then be used in turn for the basis of more involved analysis.

As a prototype of a poisoning scheme, we will define a *type I*[3] network to consist of parallel main and deactivation reactions which can be represented as

$$A + S \rightarrow B + S \quad (k \text{ or } k_A)$$

$$L + S \rightarrow L \cdot S \quad (k_d \text{ or } k_L)$$

---

[3] These titles are carried over from the definitions of Wheeler (18) for analogous networks to define selectivity problems in nondeactivating systems.

where S is the active site, and L · S represents the removal of such sites via strong chemisorption of the poison L. The scheme can be varied, of course, to represent more complicated main reactions, reversibility in the poisoning step, and so on; however, the essential features here are the parallel and independent steps involving A and L with the surface.

In coking, it is normally a reactant, a product, or both that are involved in the deactivation step rather than an impurity. Hence we can envision two kinds of *type II* networks, one involving reactant as the coke precursor and one involving the product in that role:

$$A + S \to B + S \quad (k \text{ or } k_A)$$
$$A + S \to A \cdot S \quad (k_d \text{ or } k_{A,d})$$

and

$$A + S \to B + S \quad (k \text{ or } k_A)$$
$$B + S \to B \cdot S \quad (k_A \text{ or } k_{B,d})$$

Or, a general type II mechanism is just the sum:

$$A + S \to B + S$$
$$A + S \to A \cdot S$$
$$B + S \to B \cdot S$$

For sintering, which is normally not a chemical process, as we have pointed out, such reaction networks do not apply. In this case, other types of kinetic models, inferred from experimentally determined sintering rates, are generally employed.

Of course, it is perfectly possible for a catalyst to be deactivated by more than one of these mechanisms simultaneously. We have already alluded to the case of hydrotreating catalysis, where metals in the feeds tend to deposit on the catalyst external surface and block access to the interior. Typically, this process is going on at the same time that the catalyst is being deactivated by coke deposition. Parallel cases of poisoning and coking are not uncommon as well; cracking catalysts exposed to feedstocks containing basic organic compounds deactivate by both coke formation and acid–base reactions on the surface.

The development of the mathematical description of deactivation is the subject of Chapter 2; however, we do need to say a few words about kinetics here in the context of discussing definitions. The key word is *separability*, as coined by Szépe and Levenspiel (15). According to this concept, we may model the kinetics of deactivation processes according to simple power law rate models which are taken to be uncoupled from the equations describing the rate of the main reaction. Justification of this approach, at least in part, will be left to later development. In essence we separate our rate equations into two terms: reaction kinetic dependences, which are time-independent, and activity dependences, which are not. The rate of the main reaction at any time, $\mathcal{R}_t$, is given by

$$\mathcal{R}_t = f_1(a)f_2(C, T) \tag{1-1}$$

where $f_2(C, T)$ is the concentration- and temperature-dependent term, characteristic of the reaction in the absence of deactivation, and $f_1(a)$ is a separable factor expressing the current activity of the catalyst with reference to some standard condition. The kinetics of deactivation are expressed analogously in a second equation:

$$\mathcal{R}_d = f_3(a)f_4(C, T) \tag{1-2}$$

where $\mathcal{R}_d$ is the rate of change in activity, $f_4(C, T)$ the concentration- and temperature-dependent factor pertaining to the deactivation kinetics, and $f_3(a)$ a factor reflecting the fact that the rate of change of activity is often a function of the activity. Very often, in practice, the separable factor $f_3(a)$ is simply taken to be a normalized variable a, with $0 \le a \le 1$, where $a = 1$ would reflect the fresh catalytic surface and $a = 0$ the totally deactivated one. Similarly, $f_3(a)$ is often also represented by the separable factor a. We shall become much more fussy about the definition of activity and activity factors in the next chapter.

In a sense, the use of separable formulations for the kinetics of deactivation is analogous to the employment of ideal models such as the Langmuir-Hinshelwood approach to the kinetics of surface reactions. We will show that, in the limit of detail afforded by the theory of reactions on nonideal surfaces, separable formulations are not generally correct. Nonetheless, the approach has been used with considerable success over the years and is probably adequate to the precision that is characteristic of most kinetic data. As one might suspect, the amount of information available to test directly the assumptions involved in the separable formulation is scanty.

REFERENCES

1. W. G. Appleby, J. W. Gibson, and G. M. Good, *Ind. Eng. Chem. Process Des. Dev.* **1**, 102 (1962).
2. J. E. Benson and M. Boudart, *J. Catal.* **4**, 704 (1965).
3. J. J. Berzelius, *Ann. Chem. Phys.* **61**, 146 (1836).
4. M. Boudart, *Chem. Eng. Prog.* **57**, 33 (1961).
5. R. L. Burwell, Jr., *ACS Symp. Ser.* No. 222, 3 (1983).
6. J. B. Butt, *AIChE J.* **22**, 1 (1976).

7. J. J. Carberry, *CHEMTECH* Feb., p. 124 (1974).
8. M. Faraday, *Philos. Trans. R. Soc. London.* **124**, 55 (1834).
9. S. Fuentes and J. C. S. Figueras, *J.C.S. Faraday I* **74**, 174 (1978).
10. S. J. Gregg and K. S. W. Sing, "Adsorption, Surface Area and Porosity." Academic Press, New York, 1967.
11. W. B. Innes, *in* "Experimental Methods in Catalysis Research" (R. B. Anderson, ed.), Chap. 2. Academic Press, New York, 1968.
12. H. Pines and W. O. Haag, *J. Am. Chem. Soc.* **82**, 2471 (1960).
13. P. A. Sermon and G. C. Bond, *Catal. Rev.* **8**, 211 (1973).
14. M. Shelef, K. Otto, and N. C. Otto, *Adv. Catal.* **27**, 311 (1978).
15. S. Szépe and O. Levenspiel, *Chem. React. Eng., Proc. Eur. Symp., 4th, 1968,* p. 265 (1971).
16. T. Uchijima, J. M. Herrmann, Y. Inoue, R. L. Burwell, Jr., J. B. Butt, and J. B. Cohen, *J. Catal.* **50**, 464 (1977).
17. R. Van Hardeveld and F. Hartog, *Surf. Sci.* **15**, 189 (1969); *Adv. Catal.* **22**, 76 (1972).
18. A. Wheeler, *in* "Catalysis" (P. H. Emmett, ed.), Vol. 2, p. 105. Reinhold, New York, 1955.

CHAPTER 2

Mathematical Description of Deactivating Systems

> Some aspire to gain immortality through their work. I want to be immortal by not dying.
>
> *Woody Allen*

The approach to be taken here is largely an extension of that used by catalytic chemists endeavoring to understand a reaction taking place on a heterogeneous catalyst. It is necessary to identify the reactions occurring, then to learn the mechanism (i.e., the detailed path) that the molecules follow during reaction, and finally to measure the rates at which the various reactions occur. If, in addition, the deactivation characteristics of the catalyst are to be studied, then an effort of a similar nature is needed to identify the reactions causing deactivation, to learn the detailed manner in which these reactions interact with the main reaction, and to measure the rates of the deactivating reactions. Successful achievement of the above is the basis for the formulation of a mathematical expression to represent and faithfully reproduce the experimental measurements.

The above information is available to varying degrees, depending on the reaction considered. Some reactions have been thoroughly studied, others have only preliminary rate data available. In some processes it is even difficult to identify the reactions occurring, in particular when dealing with complex materials such as petroleum or coal. Thus, in common with more traditional catalytic problems, it is often necessary to describe processes involving deactivation even though information is incomplete.

Accordingly, in this chapter we shall develop methods for describing deactivating systems from a knowledge of the mechanism. We consider single and multiple main reactions, each on homogeneous and heterogeneous surfaces. In a later section, we consider empirical or semiempirical forms of deactivation functions and, where possible, relate them to the more fundamental forms developed in the earlier sections.

I. DEACTIVATION OF SYSTEMS HAVING A SINGLE MAIN REACTION

Our first task is to define a quantity that we shall call the activity, a, and to establish how this factor appears in the rate expression for the main reaction. We must then determine how to formulate an additional rate expression that will enable us to describe how the activity factor changes under the conditions of the reaction.

The specific rate of a chemical reaction on a heterogeneous catalyst is expressed as a unique function of the intensive variables of the system, the concentrations of the chemical species present and the temperature. The rate is made specific by referring to a unit area, a unit mass, or a single site. The rate of the main reaction is, in general, for a system of constant activity

$$\text{rate of reaction} = f(c_1, c_2, c_3, \ldots, c_i, T)$$
$$= f(c_i, T) \tag{2-1}$$

When we consider deactivating systems, Eq. (2-1) represents the initial reaction rate, that is,

$$\begin{bmatrix} \text{reaction rate of} \\ \text{catalyst initially} \end{bmatrix} = \mathcal{R}_0 = f'_1(c_i, T) \tag{2-2}$$

If, now, after some deactivation of the catalyst has occurred, the rate of the main reaction is measured again at the same values of the concentrations and temperature to obtain

$$\begin{bmatrix} \text{reaction rate of} \\ \text{catalyst after} \\ \text{deactivation} \end{bmatrix} = \mathcal{R}_t = f'_2(c_i, T) \tag{2-3}$$

it is convenient then to define the activity as

$$a \equiv \mathcal{R}_t / \mathcal{R}_0 \tag{2-4}$$

Thus the activity is an operational parameter of considerable utility in characterizing the changes in the reaction rate of a catalyst as it deactivates, and it is obtained easily and directly from the experimental results.

Equation (2-4) is equally applicable to the activation of a catalyst if $\mathcal{R}_0$ is thought of as the "fully" activated state and $\mathcal{R}_t$ is the unactivated state.

I. Deactivation of Systems Having a Single Main Reaction

This, however, calls attention to a problem that is difficult to resolve. We know that $\mathcal{R}_0$ depends on the activation conditions and the choice of R_0 depends on the intended use of the activity. Moreover, $\mathcal{R}_0$ is not measured in practice, but rather rates are measured at some time interval from the start-up of the reactor. The length of time necessary to get the measurement depends largely on the design of the reactor and the analytical equipment utilized. The quantity $\mathcal{R}_0$ is generally obtained by extrapolation to zero on a time plot having a time scale selected to be suitable for the purpose. Although this point is better developed in a later section, it is important to call attention to the difficulty in obtaining a meaningful normalizing factor $\mathcal{R}_0$ for Eq. (2-4).

In the usual kinetic experiment, the activity is thought of as a function of time even for flow reactors. However, from a mechanistic point of view, activity depends on time implicitly. The activity is explicitly related to the population of active sites at a specified time and, in addition, to the condition of concentrations and temperature of the reactor in a manner similar to the dependence of the rate of the main reaction on these variables. In fact, the most fruitful approach views deactivation reactions in competition with the main reaction. Thus, the discussion must now consider reactions in terms of mechanism.

We have seen that a fruitful concept in heterogeneous catalysis, as well as in kinetics in general, is to consider that the reaction takes place at *active centers* and that the rate of reaction depends on the number of active centers in the system.[1] As shown in Chapter 1, in heterogeneous catalysis such active centers are located on the surface of the catalyst. Thus a procedure to activate a catalyst is one that produces large numbers of active centers at which reactions may occur. This simple model of a catalyst surface is complicated by the fact that not all of the centers produced on the catalyst surface will be able to interact with reactants to produce products; therefore only a fraction of the centers are active centers. Moreover, not all active centers are identical. The situation is further complicated by the fact that centers with different properties can interact with the reaction mixture either in parallel or in series—for example, a dual-function catalyst. These ideas form the framework in which catalytic rates are expressed in terms of an extension of the Langmuir isotherm to Langmuir-Hinshelwood-Hougen-Watson (LHHW) kinetics.

The concept of an active center not only is useful in developing mathematical expressions to describe rates of reaction but also offers an attractive

---

[1] In a system at constant reaction conditions (i.e., constant concentrations and temperature), the rate is proportional to the number of sites in the macroscopic sense that if we double the mass of catalyst, we double the rate. However, the intrinsic rate may depend in a nonlinear way on the number of sites in the system.

and logical way to account for reactivation, poisoning, and deactivation processes. To explain these phenomena qualitatively it is necessary only to recognize that the number of active centers can be decreased by physically removing them from the system by sintering or rendering them inactive by having nonreacting molecules (poisons or coke) adsorbed on them, thereby preventing their further use by reacting molecules. Thus, from a fundamental viewpoint, a useful parameter for the interpretation of deactivation phenomena would be related to N_t, the number of active centers per unit mass of catalyst at any stage of deactivation. This number is logically compared to N_0, the number of active centers of a fully activated catalyst, to give

$$\alpha \equiv N_t/N_0 \qquad (2\text{-}5)$$

We shall call α the fraction of the active centers not deactivated or the active fraction.

The task of the catalytic chemist interested in deactivation phenomena is to relate a and α. This is where we must become more fussy in our treatment of the concept of activity than we were in Chapter 1. At first, one might be tempted to argue that they are proportional, and in many cases this is true. But, as we shall show in the next section, a can be proportional to α^2 or even α^3. Still later we shall show that no simple relationship may exist and indeed, because of heterogeneity of the surface, a single parameter such as α is insufficient to characterize the system.

These ideas can be stated in another way, by incorporating α into Eqs. (2-2) and (2-3):

$$\mathscr{R}_0 = f_1(c_i, T, 1) \qquad (2\text{-}6)$$

$$\mathscr{R}_t = f_2(c_i, T, \alpha) \qquad (2\text{-}7)$$

where we have indicated in Eq. (2-7) that the rate of reaction of the deactivated catalyst is dependent on α. Now if we can rewrite Eq. (2-7) in the form

$$\mathscr{R}_t = f_3(\alpha)f_4(c_i, T) \qquad (2\text{-}8)$$

then it is referred to as separable in the sense of Chapter 1.

In the final section of this chapter we shall explore theoretical and empirical relationships between a and α which have been utilized in describing deactivating systems.

A. Deactivation of Homogeneous Catalytic Surfaces

The LHHW methods for describing heterogeneous catalytic reactions of constant activity can be extended readily to describe systems of changing

I. Deactivation of Systems Having a Single Main Reaction

activity. The appeal of these methods for describing deactivating systems is due not only to the simplicity of the concepts but also to the relative ease and flexibility with which they can be applied to a given catalytic situation. Some of the same objections which are commonly raised in the interpretation of LHHW models of catalytic systems of constant activity are also present when these models are applied to deactivating systems. In particular, the resulting mathematical expressions have a sufficiently large number of parameters to permit easy adjustment to the experimental rate data. The primary difficulty with the LHHW framework as a means of getting fundamental information is that very little information is known about the nature of the active centers and their numbers. In fact, the study of deactivating systems using poisons has been one of the most powerful methods of studying active centers and their populations, as seen already in Chapter 1.

The LHHW models which we shall choose as examples will be relatively simple and should be familiar to catalytic chemists. Simple models are chosen for two reasons: first, a general mathematical framework for representing these ideas would be unnecessarily complex and cumbersome to follow; second, the purpose of these examples is to reveal general principles of the implications of deactivation processes rather than the more familiar mechanistic considerations and this is more easily accomplished by using simple models.

Let us first consider the case of parallel fouling reactions represented by the stoichiometric equations

$$A \rightleftarrows B$$
$$A \rightleftarrows W$$
(2-9)

where A is the reactant, B the product, and W a surface residue. The term parallel fouling is used here because the reactant has two paths: it can go to products via the main reaction or to a surface species that blocks an active center from further use in the main reaction, namely a poison or coke. The above scheme can be represented mechanistically by the following sequence of elementary steps (28):

$$A + S \underset{k_{-1}}{\overset{k_1}{\rightleftarrows}} [A \cdot S] \underset{k_{-2}}{\overset{k_2}{\rightleftarrows}} [B \cdot S] \underset{k_{-3}}{\overset{k_3}{\rightleftarrows}} B + S \quad\quad (2\text{-}10)$$
$$\downarrow k_4 \uparrow k_{-4}$$
$$W$$

where S is an empty site, $[A \cdot S]$ an adsorbed active site, and $[B \cdot S]$ an adsorbed product. As in the LHHW methods generally applied in catalysis, it is customary to postulate a rate-limiting step, which in this case we choose to be the conversion of adsorbed reactant to adsorbed product. This can be stated mathematically in terms of the relative magnitudes of the rate constants:

$$k_{-2} \ll k_2 \ll k_1, k_{-1}, k_3, k_{-3} \qquad (2\text{-}11)$$

In most catalytic situations the rate of poisoning is small compared to the rate of reaction[2], which leads to a mathematical statement of the form

$$k_{-4} < k_4 \ll k_2 \qquad (2\text{-}12)$$

If the initial concentration of active centers is X_0 (N_0 sites/cm^2) then

$$X_1 + X_2 + X_3 + X_4 = X_0 \qquad (2\text{-}13)$$

where X_1 is the concentration of empty centers corresponding to S, and X_2, X_3, and X_4 are, respectively, the concentrations of adsorbed reactants [A · S], adsorbed products [B · S], and adsorbed residue, W.

The sequence of Eqs. (2-10) and (2-11) leads to the familiar LHHW result, which is

$$\text{Reaction rate} = \mathcal{R} = k_2 X_2 \qquad (2\text{-}14)$$

From Eq. (2-11), it follows further that

$$X_2 = K_1 C_A X_1 \qquad (2\text{-}15)$$

$$X_3 = K_3 C_B X_1 \qquad (2\text{-}16)$$

and, using Eq. (2-13), these relations lead to

$$X_1 = \frac{X_0 - X_4}{1 + K_1 C_A + K_3 C_B} \qquad (2\text{-}17)$$

The form of Eq. (2-17) requires some comment. It has, of course, the familiar denominator terms of the LHHW treatment. It differs, however, by the appearance of X_4 in the numerator. The difference $X_0 - X_4$ corresponds to the unpoisoned surface; accordingly, Eq. (2-17) shows that the concentration of empty sites is proportional to the number of unpoisoned sites, and it follows that the reaction rate according to Eq. (2-14) is also proportional to $X_0 - X_4$.

This result obtains from our initial assumptions regarding the magnitudes of the rate constants, in particular, Eq. (2-12): $k_2 \gg k_4$. In physical terms we are saying that on the time scale necessary to establish the steady state on the catalyst surface for the main reaction, essentially no deactivation

---

[2] For example, a reaction first-order in reactant with a turnover number of 1 molecule reacted per site per second that loses half of its activity by a deactivation reaction first-order in reactant in about $\frac{1}{2}$ hr has a ratio of $k_2/k_4 \approx 10^3$.

I. Deactivation of Systems Having a Single Main Reaction

takes place. This is tantamount to saying that X_4 is essentially constant during this relaxation period. Had we included the terms for the fouling in the LHHW treatment, added terms would have appeared in the denominator which would then have been eliminated as being small compared to the other terms.

However, X_4 is not a constant during the course of the reaction. In fact, it varies with time, and we are obliged to write some form of an auxiliary rate equation to account for its time dependence. The auxiliary rate equation can be considered to be a site population balance equation. According to Eq. (2-10) this equation is

$$dX_4/dt = k_4 X_2 - k_{-4} X_4 \qquad (2\text{-}18)$$

where we can drop the second term on the right because of Eq. (2-12). From Eqs. (2-15), (2-17), and (2-18),

$$\frac{dX_4}{dt} = \frac{k_4 K_1 C_A (X_0 - X_4)}{1 + K_1 C_A + K_3 C_B} \qquad (2\text{-}19)$$

The original rate of the main reaction is proportional to X_0 and the reaction rate at any time is proportional to $X_0 - X_4$; therefore by the definition of Eq. (2-3)

$$a \equiv (X_0 - X_4)/X_0 \qquad (2\text{-}20)$$

Similarly, from Eq. (2-4)

$$\alpha = (X_0 - X_4)/X_0 \qquad (2\text{-}21)$$

whereupon

$$d\alpha/dt = -(1/X_0) \, dX_4/dt \qquad (2\text{-}22)$$

and on substitution into Eq. (2-19)

$$-\frac{d\alpha}{dt} = \left(\frac{k_4 K_1 C_A}{1 + K_1 C_A + K_3 C_B}\right) \alpha = K' \alpha \qquad (2\text{-}23)$$

Under conditions of constant temperature and concentration this leads to an exponential decay in activity with time, and

$$a = \alpha \qquad (2\text{-}24)$$

$$\alpha = e^{-K't} \qquad (2\text{-}25)$$

This result, of course, depends on the initial assumptions regarding the mechanism of the main reaction and the mechanisms of the deactivation reactions. This treatment, however, illustrates several points which are quite general. Just as the LHHW approach is useful for describing catalytic systems of constant activity, so it can also be extended quite readily to describe deactivating systems. Extensive application of this relationship is illustrated in Chapter 12.

In the LHHW framework, the rate of reaction can be decreased by poisons in two ways: (*i*) as the well-known rate inhibition term which appears in the denominator as the product of equilibrium constant and concentration terms and (*ii*) as a term, $X_0 - X_4$ in this case, in the numerator which contains the number of active sites. This second form follows naturally from an assumption regarding the relative relaxation times for the main reaction and the deactivation reaction; the former is nearly always much shorter than the latter. The form of this term permits us to determine the dimensionless activity of Eq. (2-20), which is useful in setting up the governing rate equations.

In quantitative terms, the main reaction takes place with a relaxation time of $1/k_2$, whereas the relaxation time characteristic of the deactivation reaction is $1/k_4$.

It may be of some interest to modify the deactivating function of Eqs. (2-9) and (2-10) to be of the form

$$2A \rightleftarrows W$$

whereupon

$$2[A \cdot S] \underset{k_{-4}}{\overset{k_4}{\rightleftarrows}} W \qquad (2\text{-}26)$$

which might be thought of as a polymerization reaction. Completing the analysis above to incorporate this new assumption, we obtain

$$-\frac{d\alpha}{dt} = k_4 X_0^2 \left(\frac{K_1 C_A \alpha}{1 + K_1 C_A + K_3 C_B} \right)^2 = K'' \alpha^2 \qquad (2\text{-}27)$$

Under constant conditions of temperature and concentrations, this leads to a hyperbolic form of activity decay. In the previous example

$$a = \alpha \qquad (2\text{-}28)$$

but

$$\alpha = 1/(1 + K''t) \qquad (2\text{-}29)$$

Returning to the original example of Eq. (2-10), a different modification is of interest. Suppose the main reactions of Eq. (2-9) are, respectively, of the form

$$2A \rightleftarrows B$$
$$A \rightleftarrows W \qquad (2\text{-}30)$$

I. Deactivation of Systems Having a Single Main Reaction

In this case:

$$\text{Reaction rate} = \mathscr{R} = k_2 X_2^2$$

and a similar analysis using a Langmuir-Hinshelwood treatment gives

$$\mathscr{R} = k_2 X_0^2 \left(\frac{K_1 C_A}{1 + K_1 C_A + K_3 C_B}\right)^2 \alpha^2 \qquad (2\text{-}31)$$

whereupon

$$a = \alpha^2 \qquad (2\text{-}32)$$

$$\alpha = e^{-K't} \qquad (2\text{-}33)$$

These examples bring out one method for obtaining a relationship between a and α. As we have seen, modeling the system using the LHHW method provided the needed relationship. In addition, the models considered above bring out another very important characteristic of deactivating systems in common with modeling of nondeactivating systems; that is, the activity a may not be linear in fraction of active sites remaining, α. The kinetic interpretation of this is that more than one active center may be involved in the fouling reaction; in the second example, Eq. (2-30), two sites were involved, hence the activity for the main reaction varies directly with α^2.

Homogeneous surfaces that display separability can therefore be expressed in the form

$$\text{Reaction rate} = \mathscr{R}_t = f_4(\alpha) f_5(c_i, T) \qquad (2\text{-}34)$$

All of the above examples satisfy the separability requirement. However, not all homogeneous surfaces satisfy the separability criterion, as shown in the following example based on an idea suggested by Spall and Reiff (21). Consider

$$A + 2S \underset{k_{-1}}{\overset{k_1}{\rightleftarrows}} [A \cdot S] \underset{k_{-2}}{\overset{k_2}{\rightleftarrows}} [B \cdot S] \underset{k_{-3}}{\overset{k_3}{\rightleftarrows}} B + 2S \qquad (2\text{-}35)$$

$$k_4 \downarrow \uparrow k_{-4}$$
$$[W]$$

As before, assume that

$$k_{-2} \ll k_2 \ll k_1, k_{-1}, k_3, k_{-3} \quad \text{and} \quad k_{-4} < k_4 \ll k_2$$

The rate expression is given by

$$\text{Reaction rate} = k_2(X_2) \qquad (2\text{-}36)$$

and the active center balance is again

$$X_1 + X_2 + X_3 + X_4 = X_0$$

where now X_1, X_2, X_3, and X_4 are, respectively, the concentrations of empty centers, adsorbed reactant, adsorbed product, and adsorbed foulant.

Now from the equilibrium assumptions

$$\left. \begin{array}{l} X_2 = K_1 C_A X_1^2 \\ X_3 = K_3 C_B X_1^2 \end{array} \right\} \qquad (2\text{-}37)$$

An active center balance, therefore, becomes

$$X_1 + 2K_1 C_A X_1^2 + 2K_3 C_B X_1^2 + 2X_4 = X_0$$

from which

$$X_1 = \frac{-1 \pm \sqrt{1 - 8(K_1 C_A + K_3 C_B)(2X_4 - X_0)}}{4(K_1 C_A + K_3 C_B)} \qquad (2\text{-}38)$$

We select the positive root to make X_1 positive. As before, we identify the activity

$$a = (X_0 - 2X_4)/X_0 \qquad (2\text{-}39)$$

Substituting back into Eq. (2-36) and letting $K_1 C_A + K_3 C_B = K$, we get

$$\text{Reaction rate} = K_2 K_1 C_A(-1 + \sqrt{1 + 8KaX_0/4K}) \qquad (2\text{-}40)$$

Equation (2-11) cannot be put in the form of Eq. (2-34) and is therefore nonseparable.

Comparing this example with the previous examples, we discover that multiple site involvement in the "fast steps" leads to nonlinearities in the site balance equation and loss of separability in the final rate expression as deactivation proceeds. Note that in the earlier examples, multiple site involvement in the rate-limiting steps and in the fouling step did *not* lead to loss of the separability property.

B. Deactivation of Heterogeneous Catalytic Surfaces

The key assumption in Section I,A is that all active centers on the catalyst are identical and equivalent and thus constitute an ideal surface. There are, however, many ways to show experimentally that not all sites on the surface of a catalyst are equivalent. The evidence most often cited is the variation of the heat of formation of adsorbed complexes with surface coverage. Other evidence more directly related to the subject of this book is that of a catalyst which can be poisoned for one reaction but not for another (4).

I. Deactivation of Systems Having a Single Main Reaction

Changes in selectivity with poisoning of either a mono- or bifunctional catalyst further support the contention that not all active centers on a catalyst are identical. Yet it does not follow rigorously that evidence of the latter type invalidates the assumption of equivalent sites for catalytic reaction. Indeed, in certain cases, only sites in a selected narrow range of heats of formation for the particular adsorbent can act to adsorb, react, and desorb products at a significant overall rate. That is, if the heat of adsorption for the reaction is too small, the activation barrier for adsorption will be high and the rate of adsorption will be rate-limiting. Similarly, if the heat of formation is too high, the activation barrier for the decomposition of the intermediate will be large and the corresponding rate will become limiting. It is argued, then, that the sites having moderate activation barriers for both adsorption and decomposition can be restricted to a narrow energetic range and treated as equivalent. Experimentally, this has been demonstrated by "volcano" correlations of reaction rate versus heat of chemisorption of reactant across a series of catalysts. See, for example, the results of Rootsaert and Sachtler (20) for formic acid decomposition or, more recently, those of Vannice (22) for CO hydrogenation.

The other extreme is, of course, the case where most of the sites participate in the reaction but do so at different rates. These are cases where the kinetics of catalyst decay cannot be treated by the simple methods of the previous section and must be treated taking into account the heterogeneity or nonideality of the surface (2, 14). As stated earlier, this leads to a situation in which the activity is not a unique function of the fractional coverage or, more generally, the amount of poison on the surface. Additional parameters that characterize the surface homogeneity are needed to determine the distribution of the poison over the surface. Obviously, if different elements of the catalyst surface exhibit unequal specific reaction rates, then the deactivation resulting from a given amount of poison depends on how the poison is distributed among the surface elements. Our purpose now is to explore these ideas quantitatively.

Let us identify ensembles $n_1, n_2, \ldots, n_j$ of active centers, each having identical thermodynamic and kinetic properties. We can visualize each ensemble n_j as the number of sites per square centimeter having a turnover number for the main reaction of N_j and assume that N_j is distributed continuously on the surface. For example, we can visualize a square centimeter of catalyst surface having many different crystallographic surfaces, each with a turnover number of N_j and a population of n_j sites. The overall specific reaction rate per square centimeter is a continuous summation of the ensembles given by

$$\mathscr{R}_N = \int N_j \, dn_j \qquad (2\text{-}41)$$

where $\mathcal{R}_N$ is the overall reaction rate per square centimeter of nonideal surface.

Since each ensemble is by definition composed of identical sites, deactivation of the ensemble can be described by methods developed in the previous section. That is, on the jth ensemble the turnover number is, from Eq. (2-34),

$$N_j = \frac{1}{n_0} \{f_4[\alpha(n_j)] f_5[C_i, T, k(n_j)]\} \qquad (2\text{-}42)$$

where α_j is the corresponding fraction coverage of poison and n_0 the total number of sites per square centimeter of catalyst surface. Attention should be called to the dependence of f_5 on n_j through the rate constant k_j of the jth ensemble:

$$\mathcal{R}_N = \frac{1}{n_0} \int f_4[\alpha(n_j)] f_5[C_i, T, k(n_j)] \, dn_j \qquad (2\text{-}43)$$

But, from Eq. (2-43), the initial overall rate of reaction (i.e., before deactivation) is

$$\mathcal{R}_{N_0} = \frac{1}{n_0} \int f_4(1) f_5[C_1, T, k(n_j)] \, dn_j \qquad (2\text{-}44)$$

Going back to Eq. (2-3), we can calculate the activity from Eqs. (2-43) and (2-44):

$$a = \frac{\mathcal{R}_N}{\mathcal{R}_{N_0}} = \frac{\int f_4[\alpha(n_j)] f_5[C_i, T, k(n_j)] \, dn_j}{\int f_4(1) f_5[C_i, T, k(n_j)] \, dn_j} \qquad (2\text{-}45)$$

Clearly, from Eq. (2-45), the relationship between a and α_j is indeterminate unless we know functional forms of $\alpha(n_j)$ and $k(n_j)$. Moreover, the general condition for separability would be that f_5 does not depend on n_j, whereupon

$$a = \langle f_4(\alpha) \rangle \equiv \frac{\int f_4[\alpha(n_j)] \, dn_j}{\int dn_j} \qquad (2\text{-}46)$$

Note that if f_5 does not depend on n_j, the implication is that k_j does not depend on n_j and all sites have identical kinetic properties for the main reaction, a characteristic of a homogeneous surface. If the kinetic properties of sites are identical for the main reaction, it seems reasonable that the kinetic properties for deactivation would also be identical, so Eq. (2-45) reduces unequivocally to Eq. (2-26) and the condition for separability is trivially satisfied. But these conditions are not general and we are forced to conclude that the property of separability is a characteristic only of certain systems on homogeneous surfaces. There may be special choices of the functional forms of $k(n_j)$ and $\alpha(n_j)$ for a nonideal surface which satisfy separability, but the property is certainly not general.

I. Deactivation of Systems Having a Single Main Reaction

The important question then is the magnitude of the error made in describing a nonideal surface using methods for ideal surfaces. This question has been answered in part by the work of Butt et al. (6). In this work, they essentially chose the forms of α, k, and n as functions of the heat of chemisorption q. The activation of the surface reaction was taken as proportional to the heat of chemisorption:

$$E = \beta q \tag{2-47}$$

whereupon

$$k = k_0 \, e^{-\beta q/RT}$$

Site distribution functions were those of Tempkin and Freundlich in addition to the Gaussian distribution. Activity functions were taken as either linear or exponentially dependent on q. The reaction model was similar to that given in Eq. (2-10) without the reaction to W. The results were obtained by largely numerical means, although certain limiting cases yielded to analytic forms.

The detailed results will not be presented here, but they show that the ratio of the rates obtained using Eq. (2-43) and a rate obtained from the product of the averages of f_4 and f_5 differ by as much as an order of magnitude in some cases, but factors of 30–40% are the usual deviation. This is not surprising because with such distributions one cannot expect the average of the product to be equal to or even close to the product of the averages.

The general conclusion is that if one is studying the deactivation of a nonideal surface, the separability assumption may not lead to a satisfactory description. More modeling of the nonideal surface in time using the full Eq. (2-43) and comparison with the results on an ideal surface would be useful in determining the extent of error involved.

Some experimental evidence concerning the magnitude of such errors has been provided for the nickel-catalyzed hydrogenation of benzene under conditions of thiophene poisoning (14). Rates were determined on a fresh catalyst and on catalysts which had been deactivated to 16 and 6% of the initial activity. At each activity level a detailed investigation of the kinetics was carried out, and experimental data were fit to the model

$$(-r_\theta) = \frac{kK \exp[(q-E)/RT] P_A P_B}{1 + K \exp(q/RT) P_B} \quad \frac{\text{gmol}}{\text{g-sec}} \tag{2-48}$$

Nonlinear least-squares analysis of the data in the range $60 \leq T \leq 200°C$ and $0.04 < x_B \leq 0.24$ gave the parameter values for Eq. (2-48) listed in Table 2-1. The small range of E indicates that the rate-determining step is not affected by poisoning. However, K does not change proportionately with

TABLE 2-1

Kinetic Parameters for Benzene Hydrogenation on Ni/Kieselguhr

Activity		E	K	q
(%)	(gmol/g cat-sec-torr)	(kcal/mol)	(torr)$^{-1}$	(kcal/mol)
100	0.98	13.0 ± 0.9	3.5 × 10^{-7}	11 ± 0.6
16	0.49	13.5 ± 1.5	1.3 × 10^{-6}	8 ± 0.3
6	0.16	14.0 ± 1.9	2.9 × 10^{-5}	7 ± 0.4

activity, as postulated in the separable formulation, and q decreases by almost 40% over the activity range investigated. Figure 2-1 shows a comparison of the fit of data by Eq. (2-48) (with parameters of Table 2-1) and prediction assuming separability, using fresh catalyst parameters at $(0.16k)$ and $(0.06k)$, respectively. Clearly there are significant deviations from the model of kinetic separability.

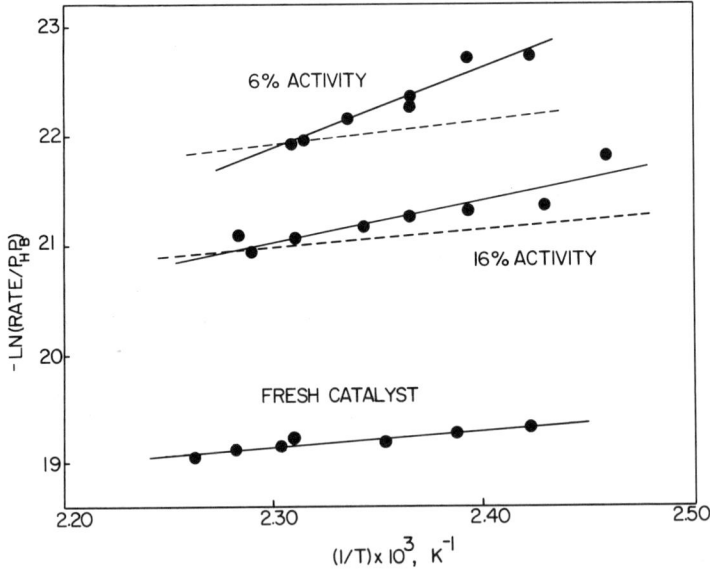

Fig. 2-1. Experimental data and model prediction for thiophene poisoning of Ni benzene hydrogenation catalyst. (———), Model prediction with experimental parameters; (- - - -), computed on assumption of kinetic separability of poisoning.

C. Deactivation of Systems Having Multiple Reaction Networks

A multiple reaction network, as used herein, refers to a reacting system that requires two or more stoichiometric equations for its description. We found earlier that an activity parameter was sufficient to describe a system involving a single stoichiometric equation, that is, a system having a single main reaction. The deactivation of the latter system was treated by relating the activity function to the fractional coverage, and the fractional coverage was determined by a site population balance equation. To treat multiple reaction networks, the methods are similar but more complex. For example, a network of two stoichiometric equations representing reaction on a bifunctional catalyst having two homogeneous surfaces will require an activity parameter and a selectivity parameter. In addition, a site population balance equation and a relationship between activity and site density are required for each function of the catalyst. It is evident that this complexity leads to a very large number of possible situations, so large in fact that, if many cases were solved, the results would be difficult and cumbersome to present even if they existed. What follows, then, illustrates the technique by which such systems are treated, utilizing a few examples. Additional examples of applications are given in subsequent chapters.

Let us consider reactions on two homogeneous ensembles, each having different sites. The simplest consecutive reaction network is

$$A \xrightarrow{k_1} B \xrightarrow{k_2} C \tag{2-49}$$

For convenience, assume that the reaction $A \to B$ occurs on a group of identical sites called X sites and correspondingly the reaction $B \to C$ occurs on identical Y sites. We adopt an LHHW model for each of these reactions similar to that in Section I,A represented by the elementary steps

$$A + S_1 \underset{k_{-1}}{\overset{k_1}{\rightleftharpoons}} [A \cdot S_1] \xrightarrow{k_2} [B \cdot S_1] \underset{k_{-3}}{\overset{k_3}{\rightleftharpoons}} B + S_1 \tag{2-50}$$

$$\Big\Updownarrow{}^{k_4}_{k_{-4}}$$
$$W_1$$

and

$$B + S_2 \underset{k'_{-1}}{\overset{k'_1}{\rightleftharpoons}} [B \cdot S_2] \xrightarrow{k'_2} [C \cdot S_2] \underset{k'_{-3}}{\overset{k'_3}{\rightleftharpoons}} C + S_2 \tag{2-51}$$

$$\Big\Updownarrow{}^{k'_4}_{k'_{-4}}$$
$$W_2$$

Making assumptions similar to those used to derive Eqs. (2-11)–(2-13) for sites X and Y, the concentrations of X_1 and Y_1 are

$$X_1 = \frac{X_0 - X_4}{1 + K_1 C_A + K_3 C_B} \qquad (2\text{-}52)$$

$$Y_1 = \frac{Y_0 - Y_4}{1 + K_1' C_B + K_3' C_C} \qquad (2\text{-}53)$$

In terms of our model, the rate of reaction of A to B is

$$\mathscr{R}_A = -k_2 X_2 \qquad (2\text{-}54)$$

but by a development similar to that used to derive Eq. (2-15)

$$X_2 = X_1 K_1 C_A \qquad (2\text{-}55)$$

whereupon the rate is

$$\mathscr{R}_A = -\frac{k_2 K_1 C_A (X_0 - X_4)}{1 + K_1 C_A + K_3 C_B} \qquad (2\text{-}56)$$

Similarly, the rate of the reaction of B to C is

$$\mathscr{R}_C = \frac{k_2' K_1' C_B (Y_0 - Y_4)}{1 + K_1' C_B + K_3' C_C} \qquad (2\text{-}57)$$

At this point, it is convenient to define

$$\alpha_1 \equiv (X_0 - X_4)/X_0 \qquad (2\text{-}58)$$

$$\alpha_2 \equiv (Y_0 - Y_4)/Y_0 \qquad (2\text{-}59)$$

To describe this system we need to define an activity and a selectivity. Similar to the earlier definition of activity, we define the activity for this system in terms of the disappearance of A, leaving the selectivity to account for the distribution of reacted A into products B and C. Thus, the activity becomes

$$a \equiv \mathscr{R}_A / \mathscr{R}_{A_0} \qquad (2\text{-}60)$$

where $\mathscr{R}_A$ is the rate of production of A at any time and $\mathscr{R}_{A_0}$ is the initial rate of production of A. Note that, as defined, both $\mathscr{R}_A$ and $\mathscr{R}_{A_0}$ are negative quantities. The selectivity follows as

$$s \equiv \mathscr{R}_B / \mathscr{R}_C \qquad (2\text{-}61)$$

where $\mathscr{R}_B$ and $\mathscr{R}_C$ are, respectively, the rates of production of products B and C at any time.

An analysis of this simple system yields

$$\mathscr{R}_A + \mathscr{R}_B + \mathscr{R}_C = 0 \qquad (2\text{-}62)$$

I. Deactivation of Systems Having a Single Main Reaction

which, on rearrangement and substitution of Eq. (2-61) gives

$$s = (-\mathcal{R}_A)/\mathcal{R}_C - 1 \qquad (2\text{-}63)$$

Note at this point that the use of the LHHW method had effected a simplification in that we have from our model a relationship between a and α_1 and a relationship between $\mathcal{R}_C$ and α_2 which lead to

$$a = \alpha_1 \qquad (2\text{-}64)$$

$$s = \frac{X_0 K_5 \alpha_1}{Y_0 K'_5 \alpha_2} - 1 \qquad (2\text{-}65)$$

where

$$K_5 \equiv \frac{k_2 K_1 C_A}{1 + K_1 C_A + K_3 C_B}$$

$$K'_5 \equiv \frac{k'_2 K'_1 C_B}{1 + K'_1 C_B + K'_3 C_C} \qquad (2\text{-}66)$$

To complete this example, we must now formulate a site population balance equation for X and Y sites. Paralleling the development of Eqs. (2-18) and (2-19),

$$dX_4/dt = k_4 X_2 \qquad (2\text{-}67)$$

to give, on substitution of Eqs. (2-52), (2-55), (2-58), (2-66), and (2-67),

$$d\alpha_1/dt = -k_4 K_5 \alpha_1 \qquad (2\text{-}68)$$

Similarly,

$$dY_4/dt = k'_4 Y_2 \qquad (2\text{-}69)$$

yields

$$d\alpha_2/dt = -k'_4 K'_5 \alpha_2 \qquad (2\text{-}70)$$

For constant temperature and concentration of species A, B, and C, Eqs. (2-68) and (2-70) integrate directly to give

$$\begin{aligned}\alpha_1 &= \exp(-k_4 K_5 t) \\ \alpha_2 &= \exp(-k'_4 K'_5 t)\end{aligned} \qquad (2\text{-}71)$$

Equations (2-64), (2-65), and (2-71) complete the analysis.

This particular result is of no special importance, except to note that the selectivity of the system depends strongly on the values of the constants in Eq. (2-71). However, the example has brought out some features characteristic of multiple reaction networks and some features characteristic of bifunctional catalysts. Our system had two independent stoichiometric equations;

therefore, we needed a selectivity parameter in addition to the activity parameter. In general, for a network of n stoichiometric equations, we would require $n-1$ selectivity parameters and an activity parameter. In assuming a bifunctional catalyst composed of two ensembles of homogeneous sites we were obliged to choose two parameters representing the fractional coverage of each of the ensembles. The values of these two parameters were obtained by solving a population balance equation for each parameter.

In keeping with our stated purpose of illustrating methods of handling deactivation of systems having multiple reaction networks, let us now explore how certain modified situations can be managed.

First, consider multiple reactions on a homogeneous surface. This case is treated in a manner similar to a single reaction on a homogeneous surface presented in Section I,A, the main difference being that the kinetic problem is more complex. An example of this type for a system of constant activity was reported by Wangmier and Jungers (24) for the liquid-phase hydrogenation of tetralin and p-xylene on a Raney nickel catalyst. The modification necessary to extend their treatment to a deactivating system is to include an appropriate deactivation step for each reaction of a form such as Eq. (2-50) or a poison from the feed stream competing for active centers. A mechanistic model will then permit the rate to be expressed as a function of α. A population balance equation for α completes the analysis. In this case the deactivation should not change the selectivity.

A more complex situation involves two parallel reactions on a heterogeneous surface. An important example of this type was reported by Bond (4) for the hydrogenation of acetylene to ethylene without further hydrogenation to ethane. Although the mechanistic details of the action of the poisons on the catalyst are uncertain, it seems fruitful to explore the changes in the heterogeneity of the surface with poisoning for the hydrogenation of acetylene and ethylene. This would involve an extension of the methods of Section I,B. If the sites used by the acetylene and the ethylene are different and energetically separated so that the overlap is negligible, this system can be treated by the method developed in the first part of this section.

Just how complicated reaction networks can become in the description of deactivating systems is illustrated in the kinetic model reported by Kmak (10) for the reactions involved in catalytic reforming. For a given carbon number, this scheme, shown in Fig. 2-2, provides for five reactant components—normal paraffins and isoparaffins, alkylcyclohexanes and alkylcyclopentanes, and aromatics. All reactions, except for hydrocracking and dealkylation, are reversible and hence potentially restricted by thermodynamic equilibrium. From the general pattern shown in the figure a complete

I. Deactivation of Systems Having a Single Main Reaction 45

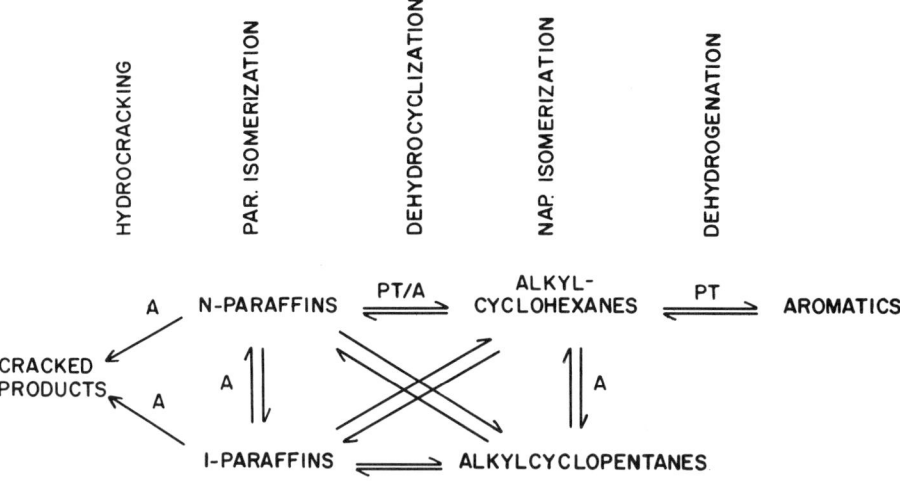

Fig. 2-2. Main reforming reaction. PAR., paraffin; NAP., naphthalene. [After Kmak (10). Reproduced by permission of the American Institute of Chemical Engineers.]

kinetic network was developed for individual carbon numbers C_1 to C_7 and C_8^+ lump as shown in Fig. 2-3. Included are 22 hydrocarbon components (see Table 2-2 for notation) and 35 individual reactions. The reactions are categorized into the four main types detailed in Fig. 2-2, with each type having its own characteristic rate expression. A typical rate equation, illustrated for isomerization of n-heptane to i-heptane, is

$$(\mathscr{R}_{C_7}) = \frac{kL[p_{nP_7} - p_{iP_7}/K]p_{H_2}^a}{1 + \sum_i K'_i p_i} \qquad (2\text{-}72)$$

where p_{nP_7}, p_{iP_7}, p_{H_2}, and p_i represent the partial pressure levels of n-heptane, i-heptanes, hydrogen, and hydrocarbon component i, and k, K, and K' are isomerization reaction rate, thermodynamic equilibrium, and adsorption equilibrium constants, respectively. The temperature dependence of the reaction rate, thermodynamic equilibrium, and absorption equilibrium constants is defined in terms of their respective energy of activation, heat of reaction, and heat of adsorption. L defines the number of active catalyst isomerization sites available and is therefore the activity

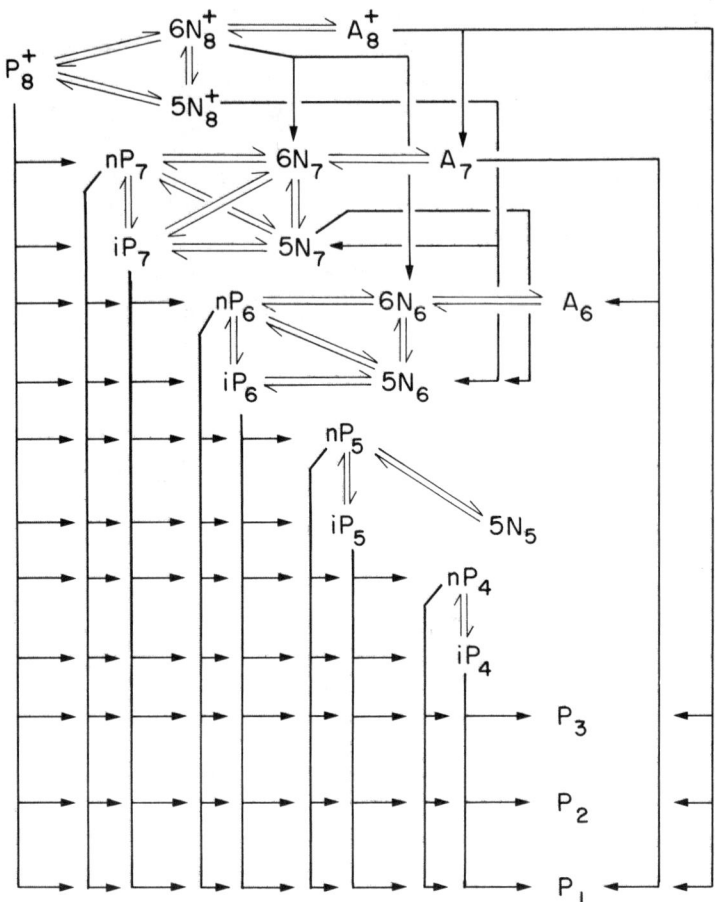

Fig. 2-3. Reforming kinetic model reaction paths. [After Kmak (10). Reproduced by permission of the American Institute of Chemical Engineers.]

factor in this model. The denominator accounts for the equilibrium adsorption of hydrocarbon components according to the Langmuir–Hinshelwood model. Hydrogen is treated independently with a Freundlich exponent.

With similar rate expressions developed for each of the n-heptane reactions, an n-heptane component balance according to the scheme of Fig. 2-3 yields

$$d(nP_7)/dt = -\mathcal{R}(nP_7 \rightleftarrows iP_7) - \mathcal{R}(nP_7 \rightleftarrows 6N_7) - \mathcal{R}(nP_7 \rightleftarrows 5N_7)$$
$$- \mathcal{R}(nP_7 \to Cr) + \mathcal{R}(P_8^+ \to nP_7) \qquad (2\text{-}73)$$

where Cr are cracked products.

I. Deactivation of Systems Having a Single Main Reaction

TABLE 2-2

Kinetic Reaction Model Components

Symbol	Component	Symbol	Component
P_8^+	C_8^+ paraffins (average carbon number)	$6N_6$	Cyclohexane
$6N_8^+$	C_8^+ alkylcyclohexanes (average carbon number)	$5N_6$	Methylcyclopentane
$5N_8^+$	C_8^+ alkylcyclopentanes (average carbon number)	A_6	Benzene
A_8^+	C_8^+ aromatics (average carbon number)	nP_5	n-Pentane
nP_7	n-Heptane	iP_5	i-Pentanes
iP_7	i-Heptanes	$5N_5$	Cyclopentane
$6N_7$	Methylcyclohexane	nP_4	n-Butane
$5N_7$	Dimethylcyclopentanes	iP_4	i-Butane
A_7	Toluene	P_3	Propane
nP_6	n-Hexane	P_2	Ethane
iP_6	i-Hexanes	P_1	Methane
		H	Hydrogen

Catalyst deactivation in reforming reactions is primarily via coke deposition, and the rate of deactivation is a strong function of hydrogen partial pressure. Thus an appropriate rate expression for coke formation defines a net coking rate based on the difference in the rate of formation from precursor species and the rate of removal by hydrocracking from the catalyst surface. In general form, then, coking is represented by

$$dC/dt = \mathcal{R}_c - \mathcal{R}_{c\text{-Hc}} f(p_{H_2}) \quad (2\text{-}74)$$

In the language of the original paper, "the first term depends upon operating conditions, catalyst state and the surface concentration of coke precursors. The precursor concentration is related to the reactant composition and particularly to the polymer content." Essentially, thus ends the exposition of Kmak. Now the reader will understand that this still leaves us a long way from direct application of the methodology developed earlier. What would remain to be specified? Essentially we need to do three things:

(1) Identify the nature of coke precursors.

(2) Determine the relationship between coke content and the fraction of active sites α [which is related to L in Eq. (2-72)].

(3) Determine the dependence of the rehydrogenation rate on p_{H_2} and also determine whether the reverse rate $\mathcal{R}_{c\text{-Hc}}$ has the same impact as the forward rate on the active fraction α.

II. PHENOMENOLOGICAL DESCRIPTION OF CATALYST DEACTIVATION

Among the preachers of catalyst mortality there are those who advocate the phenomenological description of deactivation in preference to detailed mechanistic descriptions. Their views parallel the views of those who favor the phenomenological description of systems of constant catalytic activity, as opposed to mechanistic descriptions. These differing viewpoints for systems of constant activity have been voiced in the literature (5, 26).

Under the title of phenomenological description, we shall explore some of the mathematical forms of the deactivation function used in the literature. As will become evident, the blend of theory and empiricism varies considerably among the different investigators. However, there are several threads of commonality among these illustrations, the most important of which is perhaps that all methods seek the simplest description that fulfills and satisfies the intended purpose. The descriptions all seek a relationship between a and α and utilize a population balance equation to relate α and time or its equivalent. For example, sometimes a is empirically related to time directly. Lastly, all examples assume separability.

A. Deactivation Rate Equations

The rate of the main reaction is, as before, given by

$$\mathcal{R}_t = a\mathcal{R}_0 \tag{2-75}$$

where, according to Eq. (2-7),

$$\mathcal{R}_t = f_2(\alpha)f_3(c_i, T) \tag{2-76}$$

and

$$\mathcal{R}_0 = f_2(1)f_3(c_i, T) \tag{2-77}$$

for separable forms.

In terms of the concentration of poison or foulant on the catalyst surface, C_p, and the ultimate capacity of the catalyst for poison or foulant, C_{p0}, we can write an expression for the fraction of poison or foulant, f, on the catalyst surface:

$$f \equiv (C_{p0} - C_p)/C_{p0} \tag{2-78}$$

In many cases f corresponds to the parameter α discussed earlier in this chapter. The correspondence is most likely when C_{p0} corresponds to monolayer coverage. In all cases f replaces α in Eq. (2-76) for the purpose of describing deactivation.

II. Phenomenological Description of Catalyst Deactivation

An equation can also be written for the accumulation of poison or foulant on the catalyst and takes the form

$$dC_p/dt = \mathcal{R}_p(C_p, C_i, T) \quad (2\text{-}79)$$

where the function $\mathcal{R}_p$ depends on the concentration of poison or foulant or its precursor and the temperature.

It is common to write a power law for the dependence of $\mathcal{R}_p$ on foulant concentration of the form

$$\mathcal{R}_p = k(C_{p0} - C_p)^n \phi(c_i, T) \quad (2\text{-}80)$$

which means that the rate of foulant accumulation is proportional to some power of the unused capacity of the catalyst to adsorb foulant. Combining with Eqs (2-78) and (2-79) and rearranging, we get

$$df/dt = -kC_{p0}^{n-1} f^n \phi(c_i, T) \quad (2\text{-}81)$$

Integrating for constant values of $\phi(c_i, T)$ and C_{p0} yields

$$\begin{aligned} f &= \left[\frac{1}{1+(n-1)k'C_{p0}^{n-1}t}\right]^{1/(n-1)} & n \neq 1 \\ f &= \exp(-k'C_{p0}^{n-1}t) & n = 1 \end{aligned} \quad (2\text{-}82)$$

where $k' = k\phi(c_i, T)$.

This simple treatment yields the linear exponential and hyperbolic forms for the accumulation of foulant or poisons with time. Note that the integration of Eq. (2-80) was possible because $\phi(c_i, T)$ was constant. In principle, the rate could also depend on concentrations of reactants and/or products. When we consider a catalyst pellet and later a bed of catalyst pellets, $\phi(c_i, T)$ will depend on the local conditions because, in general, c_i and T vary with position as conversion is taking place and, at a point, c_i and T vary in time because of deactivation of the catalyst. However, if we assume that at the point where deactivation occurs the concentrations do not change with time, then we obtain the solutions given in Eq. (2-82).

Having the relationship between f and t does not give the deactivation rate, because we need a relationship between a and f. This relationship often takes the form

$$a = f^P \quad (2\text{-}83)$$

which when combined with Eq. (2-82) yields

$$\begin{aligned} a &= \left[\frac{1}{1+(n-1)k'C_{p0}^{n-1}t}\right]^{P/(n-1)} & n \neq 1 \\ a &= \exp(-pk'C_{p0}^{n-1}t) & n = 1 \end{aligned} \quad (2\text{-}84)$$

Equation (2-84) readily gives many of the frequently used forms of the deactivation function, viz.: If $P/(n-1) = -1$ (the most likely combination being $P = 1$, $n = 0$) we get the linear form

$$a = 1 - kt \qquad (2\text{-}85)$$

If $P/(n-1) = 1$ (the most likely combination being $P = 1$, $n = 2$), we get the hyperbolic form

$$a = 1/(1 + kt) \qquad (2\text{-}86)$$

If $P/(n-1) = 2$ ($P = 2$, $n = 2$), we get another hyperbolic form

$$a = [1/(1 + kt)]^2 \qquad (2\text{-}87)$$

If $n = 1$, P any value, the exponential form appears:

$$a = e^{-Pkt} \qquad (2\text{-}88)$$

and so on for many others. In the above equations k is a constant characteristic of that form.

Investigators have also assumed forms for the relationship between a and f other than that represented by Eq. (2-83). Some of these will be considered in the next section.

B. Selected Examples Illustrating Various Deactivation Functional Forms

There are no generally accepted methods of treating deactivation functions in the literature. Indeed, various authors treat the phenomenon of deactivation using widely differing functional forms. Even the variables chosen to represent deactivation are not directly equivalent. In this context, the majority of the literature considers activity to be either a unique function of the amount of poison or coke on the catalyst or a unique function of time of exposure to reaction conditions. The relationship between these two approaches has already been discussed in Section I. Lastly, it is reemphasized that the overwhelming preponderance of authors assume separability, either explicitly or implicitly.

The simplest representation of poisoning is through the use of a linear function relating activity and amount of poison on the catalyst. Maxted (12) found that many of his experiments on metal-catalyzed hydrogenation reactions could be correlated by a function of the form

$$a = 1 - k_p(1 - f) \qquad (2\text{-}89)$$

where f is defined by Eq. (2-78) and k_p is a proportionality constant. In some cases this correlation was useful only during the first 80% of deactivation. Clay and Petersen (7) found that Eq. (2-89) was obeyed over the entire

II. Phenomenological Description of Catalyst Deactivation

range of activity for arsine poisoning of cyclopropane hydrogenolysis on Pt films. Although the simplest function, it does not correlate deactivation data as frequently as would be supposed. From the LHHW treatment in this chapter we saw that

$$a = \alpha^n \tag{2-90}$$

where $n = 1, 2$ and perhaps 3. This equation is probably valid for many systems, but the problem is that α does not correspond directly to f as a generality. In fact, it may be a rather infrequent happenstance because for it to be true one would have to postulate that the poison or foulant must restrict its adsorption to those sites that are active centers for the main section. There is reason to believe that this circumstance might be rare. An argument illustrating this contention, based on surface heterogeneity but homogeneous active centers, goes as follows: for a site to be an active center, the energy of adsorption must be strong enough to change the molecular configuration of the adsorbed reactant that in turn transforms into an adsorbed product molecule. But a further requirement is that the active center also release the adsorbed product in order that it may return to its original form for reuse. Of course, a poison also must be strongly adsorbed in order to remain on an active center; however, there is no requirement that it be released again—indeed, if the substance adsorbed were released readily it would not be a poison. Quite clearly, the poison does adsorb on active centers, because it removes some of them from active participation in the system, but not necessarily exclusively—it can also adsorb on other sites. A good example of this is provided in the results of Lyubarskii *et al.* (11) for the thiophene poisoning of nickel benzene hydrogenation catalysts. Qualitatively, their results, presented as relative activity for hydrogenation versus amount of thiophene adsorbed, appear as shown in Fig. 2-40. We note that adsorption of thiophene up to the level indicated by the dashed line in Fig. 2-4 results in the vast majority of the total change

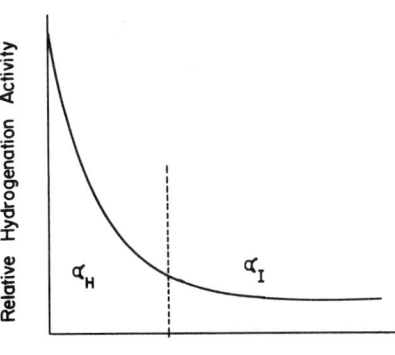

Fig. 2-4. Dual site adsorption of thiophene on Ni in benzene hydrogenation.

in activity noted. To the right of this line, a considerable additional amount, even the majority of the total, can be adsorbed with very little additional tax on catalyst activity. One might well separate the surface into two active fractions, α_H, which is active for both thiophene chemisorption and hydrogenation, and α_I, which is active for chemisorption but contributes little to hydrogenation activity.

Another important point regarding the relationship between a and α is also illustrated by the thiophene-Ni system. At lower temperatures (ca. 100°C) thiophene is bonded to Ni in a flat, planar configuration in which the π bonds in the ring interact with the metallic d orbitals in addition to a sulfur-metal bond. Hence an ensemble of perhaps six nickel atoms is involved in this chemisorption. As the temperature is increased, the thiophene molecule tends to be chemisorbed more through an upright or anchor configuration involving only S-Ni bonds, and the number of nickel atoms involved in the chemisorption of a single molecule of poison decreases significantly. Hence, the a-α relationship can change with operating conditions for the same reaction-catalyst-poison system.

These points bring out the major difficulty in applying the LHHW method to describe deactivation phenomena—that there is as yet no reliable general method for measuring the population of active centers on a catalyst surface or for characterizing changes in their nature or response to the reaction-deactivation process as experimental conditions are changed.

Heterogeneity of active centers can also be a reason for invalidating Eq. (2-90). Note that this leads to a relationship between a and f that could be far more nonlinear than Eq. (2-90). For these and perhaps other reasons, many functional forms have been used to describe deactivation, and representative examples are presented below.

Pease and Stewart (16) correlated the poisoning of a copper catalyst for ethylene hydrogenation using carbon monoxide by the relationship

$$a = e^{-k(1-f)} \tag{2-91}$$

This function was able to represent their data adequately; however, the choice was not based on a theoretical model.

Organic nitrogen poisoning of SiO_2-Al_2O_3 cracking catalysts gave almost exponential decay in poison concentration according to the data of Mills et al. (13). Later Froment and Bischoff (9) chose the same functional form to correlate fouling data for their theoretical study of catalytic crackers. They also chose the hyperbolic form

$$a = \frac{1}{1 + K(1-f)} \tag{2-92}$$

Both choices were entirely empirical.

II. Phenomenological Description of Catalyst Deactivation

Petersen (17) and Weekman *et al.* (25) used an empirical relationship between activity and time of the form

$$a = e^{-kt} \qquad (2\text{-}93)$$

The two exponential forms of Eqs. (2-91) and (2-93) are very different, as is clear from the earlier discussions in this chapter, yet both forms are apparently useful in describing fouling in reactors. Many others have assumed these forms, and some of these will be considered in detail in subsequent chapters. Indeed, application of Eq. (2-93) to deactivation of cracking catalysts is the topic of Chapter 12.

The attractiveness of the exponential form as an empirical representation of deactivation data is perhaps twofold: the function has the general shape exhibited by deactivating catalysts and can readily be adjusted to fit experimental data, and the function has many pleasant mathematical properties that endear it to model-builders. It is also worthy of note that Eq. (2-93) is readily obtained from the LHHW model as shown by Eq. (2-55), as well as from the phenomenological treatment leading to Eq. (2-23).

The hyperbolic form also readily adjusts to represent deactivation data. This form arises directly from the phenomenological treatment of Section II,A which is similar to that presented by Wojciechowski (27). Pozzi and Rase (19) arrive at a hyperbolic form using an extended LHHW treatment similar in many ways to the treatment leading to Eq. (2-29), and Corella and Asua (8) have presented a mechanistically based theory reminiscent of Eqs. (2-81) and (2-82).

The hyperbolic form, in common with the exponential form, is a popular function for the representation of deactivation data, as we shall see in later chapters. As we saw earlier with the exponential form, the hyperbolic form is expressed in two ways:

$$a = 1/[1 + K(1-f)] \qquad (2\text{-}94)$$

$$a = 1/(1 + kt) \qquad (2\text{-}95)$$

The right-hand sides of both of these equations can be raised to powers similar to the form of Eq. (2-84). Equations (2-94) and (2-95) are very different functional forms; yet both have been used successfully to describe deactivating systems and they will be considered in detail in later chapters.

Reciprocal power forms have had their greatest application in fouling of catalytic cracking catalysts. The most celebrated example of the use of this form is the work of Voorhies (23), who described the kinetics of coke formation in terms of the reciprocal first power of the amount of coke. The success of this approach is developed in detail in Chapter 3, and the relationship of this form to the exponential function is discussed in Chapter 12.

A final example of a functional form used to represent deactivation, mostly for oxide catalysts, is the Elovich equation. This function is used somewhat more rarely and requires more detailed exploration. An example that serves to show the utility of this approach was described by Parravano (15). In a study of carbon monoxide oxidation over a nickel oxide catalyst, the deactivation of the catalyst was rapid at first, followed by a period of relatively constant activity. Parravano interpreted the phenomenon in the following way. The activation procedure produced a number of active centers initially equal to N_0. On exposure to the reaction mixture, the number of active centers rapidly decreased as oxygen was removed from the surface faster than it could be replaced until a new steady value N_f was reached, whereupon the activity of the catalyst remained constant. Parravano envisioned the mechanism to be:

$$S \cdot O + CO \rightarrow S + CO_2 \qquad \text{I}$$

$$S + \tfrac{1}{2}O_2 \rightarrow S \cdot O \qquad \text{II}$$

Initially, the number of $S \cdot O$ sites was greater than could be maintained by the second reaction above.

Mathematically, the deactivation was represented by the following. The rate of change of adsorption sites with time is given by

$$dN/dq = -\beta N \tag{2-96}$$

an assumption equivalent to the Elovich isotherm, where N is the number of adsorption sites at coverage q and q is the amount of adsorbate. On integration, this leads to

$$N = N_0 \, e^{-\beta q} \tag{2-97}$$

where N_0 is the number of adsorption sites at $q = 0$.

Applied to the CO oxidation example, the rate of oxidation r_{co} is

$$r_{co} \propto N \propto dq/dt$$

or

$$dq/dt = kN_0 \, e^{-\beta q} \tag{2-98}$$

This can be integrated to give

$$kq = \log(t + 1/k\beta) + \log k\beta \tag{2-99}$$

Parravano found that his initial data followed the form of Eq. (2-99) and then began to deviate from the Elovich form to a conventional first-order plot, indicating stabilization of the catalyst activity.

In this section we have considered examples of the functional relationship among the variables activity, amount of poison or foulant on the surface of the catalyst, and time. In summary, we have found four useful functional forms relating the activity and the amount of poison or foulant:

Linear	$a = 1 - k_p(1-f)$	
Exponential	$a = \exp[(-k(1-f))]$	(2-100)
Hyperbolic	$a = \left[\dfrac{1}{1-K(1-f)}\right]^n$	
Reciprocal	$a = 1/f$	

Two relationships were found between activity and time:

Exponential	$a = \exp(-kt)$	(2-101)
Hyperbolic	$a = [1/(1+kt)]^n$	

Finally, a single relationship was found between the activity and the amount of reactant adsorbed:

Elovich	$a = \exp(-kq)$	(2-102)

III. COMMENTS ON OBTAINING DEACTIVATION DATA: EXPERIMENTATION

The form of the deactivation function developed in the previous two sections of this chapter demonstrates that the rate of reaction depends in a complex way on the concentrations of the various species in the reaction mixture and the temperature. That is, these variables affect both the rate of the chemical reaction and the rate of deactivation, and as a consequence it is, in general, difficult to distinguish deactivation in a system where the concentrations in the reaction mixture and the temperature are changing in the reactor. Another problem concerns the computation of the activity. As defined by Eq. (2-4), it is necessary to know a value of the initial reaction rate in order to compute the activity. In a system where deactivation occurs very rapidly, this poses very serious problems experimentally. These problems have been discussed by Petersen and Pacheco (18). The purpose of this section is to discuss briefly these two points and their relationship to the type of equipment used to study deactivation.

Integral reactors, illustrated in Fig. 2-5, are frequently used to obtain data on catalyst deactivation on the pilot plant scale. This type of reactor

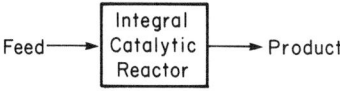

Fig. 2-5. Deactivation in an integral reactor.

serves a very important role industrially in improving the operation of existing reactors. In utilizing the information obtained from this type of reactor it is usually necessary to assume similarity between the pilot-scale reactor and the plant reactor. For reasons stated above, it is very difficult to extract fundamental information from data obtained in such reactors. The spatial variations of the concentration of the reaction mixture and the temperature force modeling of the system and curve fitting to extract deactivation data.

In nondeactivating systems, the spatial variation and its effects on the reaction kinetics can be obviated by using a recycle reactor as shown in Fig. 2-6. The system is designed to permit only a differential amount of reaction to take place per pass. In this manner, integral conversion takes place; however, each catalyst particle in the system is exposed to the same concentration and temperature history. Nevertheless, as the catalyst deactivates, the conversion and possibly the temperature change and again it is necessary to separate the effects of these changes on the activity and on the rate of reaction.

The ideal reactor system would be one in which the concentrations in the reaction mixture and the temperature remain constant throughout the deactivation period. A reactor system that accomplishes this is diagrammed in Fig. 2-7. The principle here is to decrease the flow rate just enough to compensate for the loss in activity during the deactivation period. In this way, the conversion is maintained constant. This is accomplished experimentally as illustrated in Fig. 2-7, by using an infrared spectrometer signal to position a motor valve to adjust the flow rate to the required value. In principle, then, the concentration of the reactant and the temperature are maintained constant throughout the deactivation period, and the activity and the kinetics of deactivation are obtained directly if the kinetic rate law remains unchanged and the selectivities for other reactions also remain unchanged.

This technique cannot, however, ensure that concentrations within catalyst pellets are constant or that profiles of concentrations remain invariant with time. The technique merely keeps the bulk concentrations constant for a system involving a single stoichiometric equation.

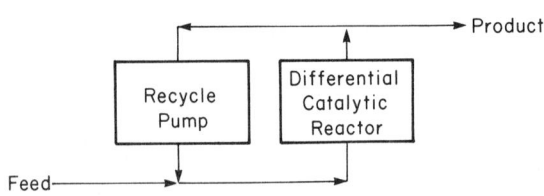

Fig. 2-6. Deactivation in a recycle reactor.

III. Comments on Obtaining Deactivation Data: Experimentation

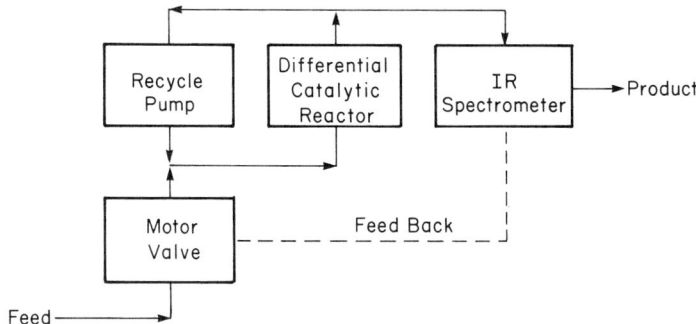

Fig. 2-7. Deactivation in a recycle reactor with flow control.

The initial rate of reaction is needed to calculate the activity as defined by Eq. (2-4). As implied by the name, the initial rate is the rate measured at time zero. For systems that deactivate relatively slowly, that is, systems where the deactivation time constant is small compared to the time constant for making concentration measurements, there is relatively little difficulty in extrapolating the rate to zero time. An example of this type of system is shown in Fig. 2-8. An example of a system in which considerable deactivation takes place before the first conversion measurements can be made is shown in Fig. 2-9. These data were taken by Blanding (3) and they illustrate how rapidly a cracking catalyst deactivates. The catalyst loses more than one order of magnitude in activity during the first second on stream. The following treatment is an attempt to model such a rapidly deactivating system in order to get a reasonable extrapolation to zero time.

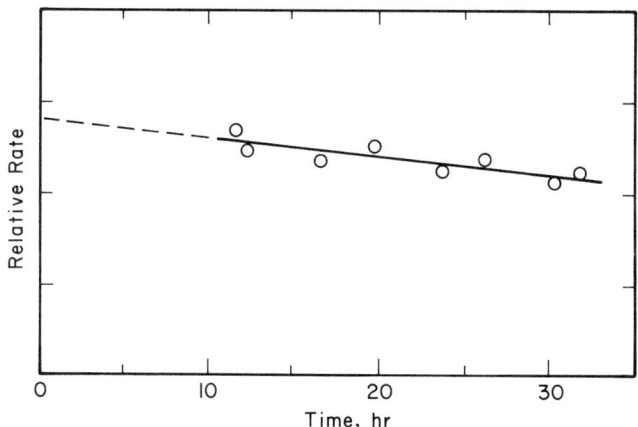

Fig. 2-8. Determination of initial activity in a slowly deactivating system.

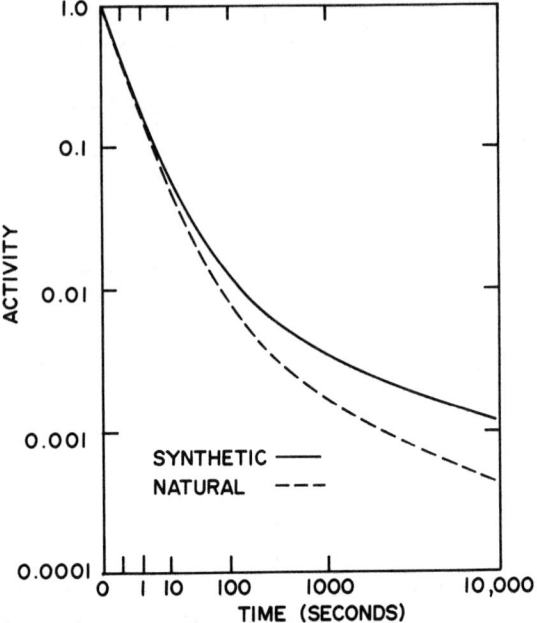

Fig. 2-9. Determination of initial activity in a rapidly deactivating system. [From Blanding (3). Reproduced by permission of the American Chemical Society.]

Consider the parallel deactivation scheme of Eq. (2-103):

$$A \xrightarrow{k_m} B$$
$$A \xrightarrow{k_f} F \quad (2\text{-}103)$$

where A is the reactant, B the desired product, and F the foulant or undesired product. The transient conservation equations in a recycle reactor with a high recycle ratio are

$$V_R \frac{dC}{dt} = Q(C_0 - C) - a_m W k_m C^n \quad (2\text{-}104)$$

$$\frac{df}{dt} = a_f k_f C^m \quad (2\text{-}105)$$

(when $t = 0$, $C = 0$ and $f = 1$), where a_m and a_f are the catalyst activities for the main and fouling reactions, C is the concentration of A, f is the fraction of surface unfouled, a_m and k_f are rate constants for the main and fouling reactions, Q is the volumetric flow rate of reaction, t is mixture time, V_R is reactor volume, and W is weight of catalyst.

III. Comments on Obtaining Deactivation Data: Experimentation

These equations can be transformed to dimensionless form to get

$$dX/d\tau = -X + \theta_m f(1-X)^n \tag{2-106}$$

and

$$df/d\tau = -\phi_d f^6 (1-X)^m \tag{2-107}$$

where

$$X \equiv \frac{C_0 - C}{C_0}, \quad \phi_m \equiv \frac{W k_m C_0^{n-1}}{Q}, \quad a_m = f, \quad a_f = f^6 \tag{2-108}$$

Equations (2-106) and (2-107) have been solved for $m = 0$, 1, and 2 and for $\theta_d = 0.1$, 1, and 100. These solutions are shown in Figs. 2-10 and 2-11. Figure 2-10 reveals that the activity after five time constants is insensitive to m, the order of the reactions with respect to the reactant concentration. However, the activity is quite sensitive to the magnitude of θ_d. Although not shown here, the activity is also very sensitive to the dependence of a_f on f. Figure 2-11 shows that the magnitude of the maximum in the product concentration is insensitive to the reactant concentration dependence of the fouling reaction. These curves have been developed by adjusting θ_m to give a product concentration of 0.3 at five time constants.

We observe that initial activity can be extrapolated using Eqs. (2-106) and (2-107) if we know the dependence of a_m on f and if we know the history. A general approach to finding the initial selectivity is beyond the scope of this section; however, the subject is treated again in somewhat

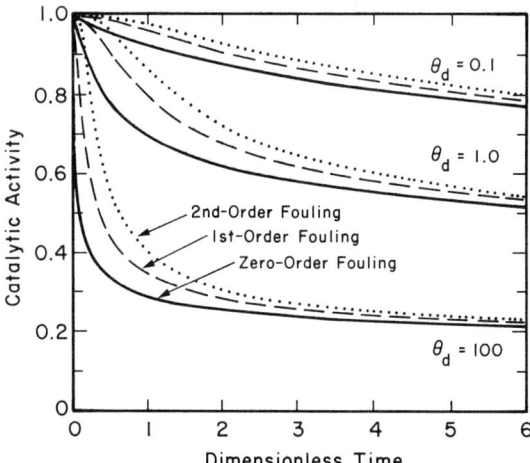

Fig. 2-10. Dependence of activity on θ_d.

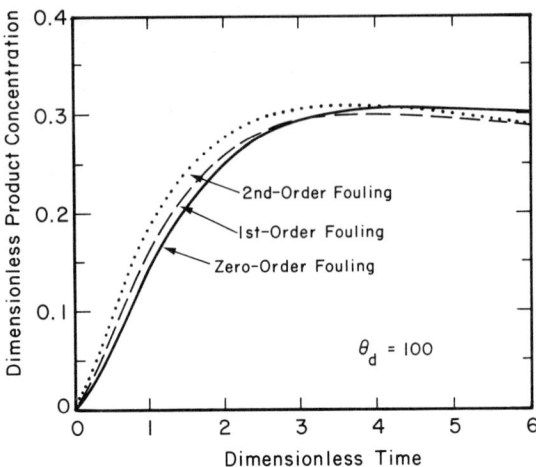

Fig. 2-11. Insensitivity of the product maximum to m.

more detail in Chapter 3. This question of extrapolation to zero time (initial activity) has also been considered from a somewhat different viewpoint from that above, and in some detail, by Absil *et al.* (1).

IV. SUMMARY AND EVALUATION

The purpose of this chapter was to explore the theoretical framework in which deactivation processes can be interpreted. The framework chosen was an extension of the LHHW method. In terms of a known mechanistic model on homogeneous sites, description of the deactivation becomes a straightforward exercise, just as the corresponding description of a known mechanism on a similar catalyst surface would be for a system of constant activity. However, for the same reasons that many mechanistic models described by LHHW methods lead to rate expressions difficult to differentiate by experiments, unique mechanisms for deactivating systems also cannot easily be resolved. A further complication owing to heterogeneous active centers was explored to learn that the property of separability, characteristic of homogeneous active centers, is not a general property of deactivating systems. Because no entirely satisfactory theoretical framework exists as yet, the latter part of this chapter was devoted to an exposé of the types of deactivation functions deriving from phenomenological approaches as applied in the solution of practical problems. This led to a number of commonly used mathematical representations all based on the property of separability.

In the next three chapters, we shall examine in detail the most common causes of deactivation, namely poisoning, fouling, and sintering. Our purpose will be to examine each of these phenomena in terms of fundamental processes as well as the methods that have been used to describe them.

REFERENCES

1. R. L. P. Absil, J. B. Butt, and J. S. Dranoff. *J. Catal* **85**, 415 (1984).
2. K. R. Bakshi and G. R. Gavalas, *AIChE J.* **21**, 494 (1975).
3. F. H. Blanding *Ind. Eng. Chem.* **45**, 1168 (1953).
4. G. C. Bond, "Catalysis by Metals," p. 296. Academic Press, New York, 1962.
5. M. Boudart, *AIChE J.* **2**, 62 (1956).
6. J. B. Butt, C. K. Wachter, and R. M. Billimoria, *Chem. Eng. Sci.* **33**, 1321 (1978).
7. R. D. Clay and E. E. Petersen, *J. Catal.* **16**, 32 (1970).
8. J. Corella and I. M. Asua, *Ind. Eng. Chem. Process Des. Dev.* **21**, 55 (1982).
9. G. F. Froment and K. B. Bischoff, *Chem. Eng. Sci.* **16**, 189 (1961); **17**, 105 (1962).
10. W. S. Kmak, *AIChE Natl. Meet. Houston, 1971*, paper 14e.
11. G. D. Lyubarskii, L. B. Andeeva, and N. U. Kul'kova, *Kinet. Katal.* **3**, 123 (1962).
12. E. B. Maxted, *Adv. Catal.* **3**, 129 (1951).
13. G. A. Mills, E. R. Boedeker, and A. G. Oblad, *J. Am. Chem. Soc.* **72**, 1554 (1950).
14. I. Önal and J. B. Butt, *Proc. Int. Congr. Catal. 6th, Tokyo, 1980* Part B, p. 1490 (1981).
15. G. Parravano, *J. Am. Chem. Soc.* **75**, 1448 (1953).
16. R. N. Pease and L. Stewart, *J. Am. Chem. Soc.* **47**, 1235 (1925).
17. E. E. Petersen, *AIChE J.* **6**, 488 (1960).
18. E. E. Petersen and M. A. Pacheco, *ACS Symp Ser.* No. 237, 363 (1984).
19. A. L. Pozzi and H. F. Rase, *Ind. Eng. Chem.* **7**, 1075 (1958).
20. W. J. M. Rootsaert and W. M. H. Sachtler, *Z. Phys. Chem.* **26**, 16 (1960).
21. K. B. Spall and E. K. Reiff, private communication
22. M. A. Vannice, *J. Catal.* **37**, 469 (1975).
23. A. Voorhies, Jr., *Ind. Eng. Chem.* **37**, 318 (1945).
24. J. P. Wangmier and J. C. Jungers, *Bull. Soc. Chim. Fr.* **10**, 1280 (1957).
25. V. W. Weekman, Jr., *Ind. Eng. Chem. Process Des. Dev.* **8**, 385 (1969); V. W. Weekman, Jr. and D. M. Nace, *AIChE J.* **16**, 397 (1970).
26. S. W. Weller, *AIChE J.* **2**, 59 (1956).
27. B. W. Wojciechowski, *Catal. Rev.* **9**, 79 (1974).
28. E. E. Wolf and E. E. Petersen, *J. Catal.* **47**, 28 (1977).

CHAPTER 3

Deactivation by Fouling

> Meddle with dirt and some of it will stick to you.
>
> *Anon.*

In a large number of catalytic processes carbonaceous deposits are formed from the reaction mixture on the surface of the catalysts with the passage of time. These strongly adsorbed carbonaceous deposits form large polynuclear aromatic structures, apparently through polymerization and condensation. This process is generally referred to in industry as *coking*. In general, a deactivation process occurs concomitantly with coking, and the overall process will be referred to in this book as fouling, although in the literature the process is also called self-deactivation or self-poisoning. However, although deactivation and coking are often observed to occur simultaneously, it does not follow that there is a simple and direct causal relationship between the two observations. Indeed, particularly for supported catalysts, there is a substantial body of information indicating that the relationship between coke content and activity is very complex.

There are two very good reasons why this should be so. First, coke is not a well-defined substance; normally it has an empirical formula approximating CH, but the chemical nature depends very much on how it is formed. For example, coke is known to develop in filamentous or whisker-like structures or encapsulating-type structures on metals and in pyrolytic-type structures on acidic surfaces. The coke also varies depending on the conditions of temperature and pressure. Moreover, these structures change as they age on the catalyst surface (a good example of this is given for hydrodesulfurization catalysts in Chapter 7). Thus, it is apparent that there are great variations in the morphology of the coke depending on the catalyst and its history. Second, industrial hydrocarbon feeds are complex mixtures of compounds and the amount and type of coke formed depend on the chemical nature of the feed and products formed. The activity of the catalyst depends on the number of active centers that are available to carry out the main reactions. A simple relationship between the activity and the amount of coke on the catalyst then depends on whether the coke is formed on the same active centers and, if so, whether the coke formed actually blocks the centers from further use. In some cases, such as catalytic cracking, it is thought that the coke formed is not the primary deactivation agent but that other compounds in the feed, notably organic bases, contribute in a major way to the deactivation process (25). In another example, reforming catalysts have shown substantial activity even when the coke content is equivalent to many monolayers of surface coverage.

In a review article on the quantitative deactivation of a catalyst by coke formation, Froment (13) bases his analysis on two principle assumptions: (*i*) the rate of deactivation of a catalyst depends on the local composition and conditions of the reaction mixture and (*ii*) the catalyst activity is a direct function of the amount of coke on the catalyst. The first assumption represents a very fundamental approach and is entirely consistent with the suggestions of Chapter 2. However, the second assumption poses some conceptual difficulties as a generality in view of the discussion above. Accordingly, in this chapter we shall discuss the topic of fouling in two parts: the kinetics of coke formation and deactivation during fouling. In general, we shall not be able to relate directly activity and amount of coke on the catalyst surface; therefore it must be viewed as a special circumstance when coke formation and deactivation are directly and causally related.

I. COKE FORMATION AND FOULING KINETICS

In a classical early paper, Voorhies (42) described coke formation as a function of time on stream during catalytic cracking in both fixed and fluidized beds. The work is important because to this day it remains a useful approach to coking kinetics. The essence of his proposal relating coke to time on stream is given by Eq. (3-1):

$$C_c = At^n \tag{3-1}$$

where C_c is the weight percent carbon on the catalyst, t the time on stream, A a constant depending on feedstock, reaction conditions, and reactor type, and n an exponent with a value close to 0.5.

The applicability of Eq. (3-1) to coke formation data is shown in Figs. 3-1 and 3-2. Values of n range from 0.38 to 0.53 on these plots. Perhaps the most remarkable observation that Voorhies made from these data was that the amount of coke formed on the catalyst after a given time interval

I. Coke Formation and Fouling Kinetics   65

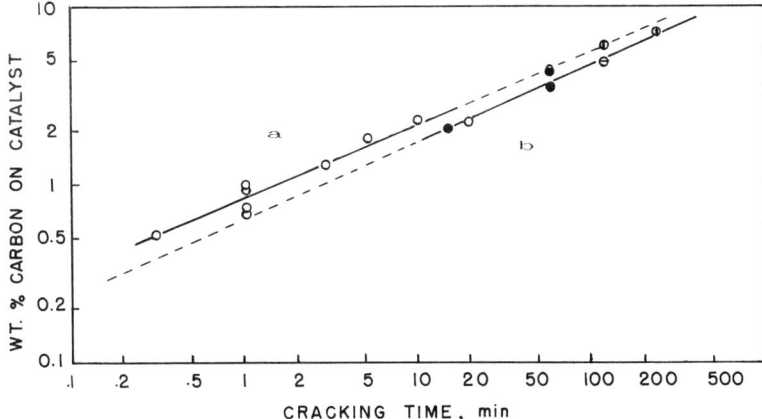

Fig. 3-1. Carbon formation versus cracking time in fixed-bed catalytic cracking. (○) West Texas gas oil, natural catalyst, 1.2 v/v-hr. East Texas gas oil, synthetic catalyst: (⊕) 0.3 v/v-hr; (⊖) 0.6 v/v-hr; (●) 1.2 v/v-hr. Curve a: $C_c = 0.85 t^{0.41}$. Curve b: $C_c = 0.65 t^{0.44}$. [From Voorhies (42). Reproduced by permission of the American Chemical Society.]

appeared to be independent on the hydrocarbon feed rate. He interpreted this to mean that, although the reaction mixture was changing as it progressed through the reactor, the products formed were gasoline and gases which tended to carbonize more than the feed. For the feeds used in his study, there appeared to be a fortuitous balance between the two classes of products to give a coke formation tendency that was the same as that of the original feed. This happenstance makes it relatively easy to predict coke

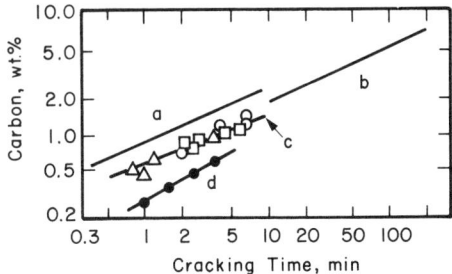

Fig. 3-2. Carbon formation versus time in fluid catalytic cracking compared to fixed-bed cracking. East Texas gas oil, natural catalyst: (△) reactor A; (□) reactor B; (○) reactor C. (●) Tinsley gas oil, synthetic catalyst—reactor D. Curve a: Fixed bed, West Texas gas oil, natural catalyst. Curve b: Fixed bed, East Texas gas oil, synthetic catalyst. Curve c: Fluid bed, natural catalyst. Curve d: Fluid bed, synthetic catalyst. [From Voorhies (42). Reproduced by permission of the American Chemical Society.]

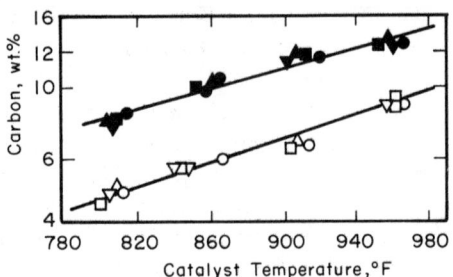

Fig. 3-3. Carbon formation versus cracking temperature for 2-hr cracking periods. Natural catalyst: (○) 0.6, (△) 0.8, (□) 1.0, (▽) 1.2 v/v-hr. Synthetic catalyst: (●) 0.6, (▲) 0.8, (■) 1.0, (▼) 1.2 v/v-hr. [From Voohries (42). Reproduced by permission of the American Chemical Society.]

formation but exceedingly difficult to interpret the coking process in a fundamental manner.

The temperature coefficient of the coking process as found by Voorhies is given in Fig. 3-3. The rate doubles for an increase in cracking temperature of about 200°F, corresponding to an apparent activation energy of about 7 kcal/mol. This is a very low value for a chemical reaction, particularly a carbon-forming reaction.

For a given catalyst, feedstock, and temperature, there is a good correlation between feedstock conversion and carbon yield based on feed, as shown in Fig. 3-4. Following Voorhies, the correlations of Figs. 3-1 and 3-4 can be used to interrelate feedstock conversion as a function of time, as shown in Fig. 3-5.

Voorhies recognized that Eq. (3-1) could be obtained from a differential equation of the form

$$dC_c/dt = A/C_c \tag{3-2}$$

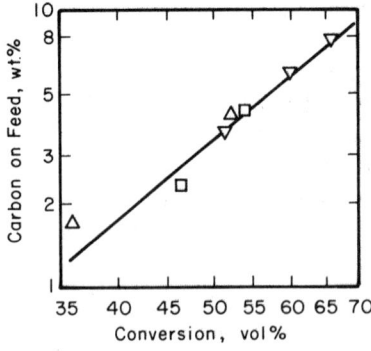

Fig. 3-4. Carbon yield versus feed conversion for fixed-bed catalytic cracking of an East Texas gas oil with synthetic catalyst. (□) 0.3, (△) 0.6, (▽) 1.2 v/v-hr. Correlation: $C_c = (3.55 \times 10^{-5}) V^{2.93}$. [From Voorhies (42). Reproduced by permission of the American Chemical Society.]

I. Coke Formation and Fouling Kinetics

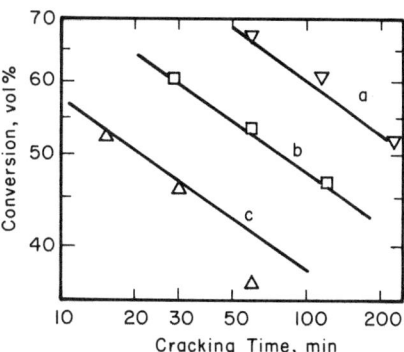

Fig. 3-5. Feed conversion versus cracking time and feed rate for fixed-bed catalytic cracking of an East Texas gas oil with synthetic catalyst. Curve a, 0.3; curve b, 0.6; and curve c, 1.2 v/v-hr. Correlation: $V = 96(SV)^{-0.34}\theta^{-0.19}$, where V is volume percent conversion, SV is space velocity, v/v-hr, and θ is cracking time, min. [From Voorhies (42). Reproduced by permission of the American Chemical Society.]

This result was obtained using a heuristic argument involving the coke as a diffusion barrier to further coke formation. The low value of the activation energy of the coke-forming process suggested the diffusion.

A final result from the work of Voohries is that the relationship between conversion and coke yield suggests implicitly a direct correspondence between activity and the amount of coke on the surface. As stated before, we have difficulties with this, and more modern work is supportive of such difficulties. However, the direct relationship has been accepted by many workers in the field and certainly requires more detailed consideration here. Further, the correlation of coke as a function of time on stream is widely used in the literature to represent not only coking but other deactivation processes as well.

Another signal early work is that of Blanding (5), who studied the changes in activity of cracking catalysts as they fouled in fixed-bed reactors. By measuring conversions as a function of time and using a model of the cracking reaction, Blanding was able to obtain activity as a function of time from a few seconds to several hours. These data were extrapolated to zero time to get an initial rate constant using a special abscissa coordinate. The activity decreased very rapidly, with the catalyst decaying by 99% after about the first minute on stream. The decay in activity for synthetic and natural catalysts is shown in Fig. 3-6. One is immediately impressed by the rapidity with which cracking catalysts lose their activity under cracking conditions. Blanding found curves of similar shape at both higher and lower temperatures. As expected, the cracking rate constant increases markedly at higher temperatures, the activation energy being about 11 kcal/mol after 10 sec on stream. However, the activation energy after 2 hr is about 19 kcal/mol. This leads to the conclusion that the catalyst deactivates more slowly at higher temperatures, as shown in Fig. 3-7.

The data of Fig. 3-6, when plotted on the log–log coordinates of Fig. 3-8, give a rate constant that decreases faster than the 0.5 power of time.

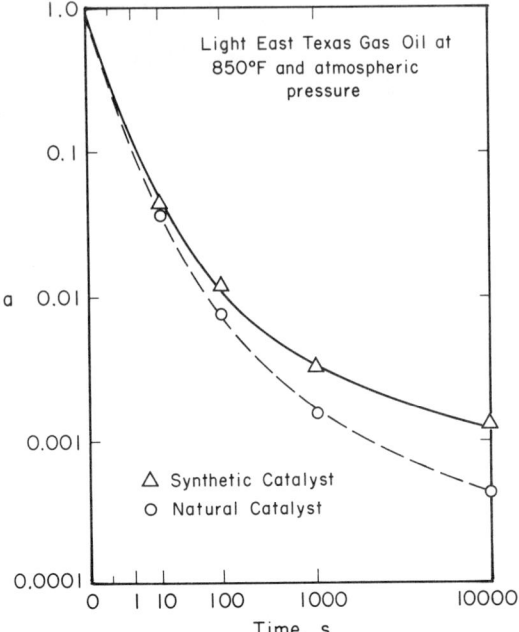

Fig. 3-6. Activity decay in catalytic cracking for natural and synthetic catalysts. [From Blanding (5). Reproduced by permission of the American Chemical Society.]

If one combines this result with that of Voorhies discussed earlier, it is tempting to conclude that the decrease in rate constant is approximately proportional to the coke on the catalyst.

This conclusion does not stand close examination, however, for two reasons. First, the activation energy for coke formation is about 10 kcal/mol, whereas the activation energy for cracking is about 19 kcal/mol. Second, and more convincing, the fraction that the catalyst deactivates *decreases* with increasing temperature whereas the amount of coke *increases*, leading to an inverse correlation between activity and coke on the catalyst. However, another interpretation of this is given in a later section.

Some insight into this situation is given in the previously cited paper of Mills *et al.* They found an excellent correlation between the amount of organic base in feedstock and the rate of deactivation of the cracking catalyst. Specifically, the activity declines exponentially with the amount of base on the catalyst. They also found a correlation between quinoline chemisorption and activity, but no correlation between activity and the amount of coke on the surface. Lastly, their experiments showed that the catalyst surface was nonuniform and that the active area of the catalyst was

I. Coke Formation and Fouling Kinetics

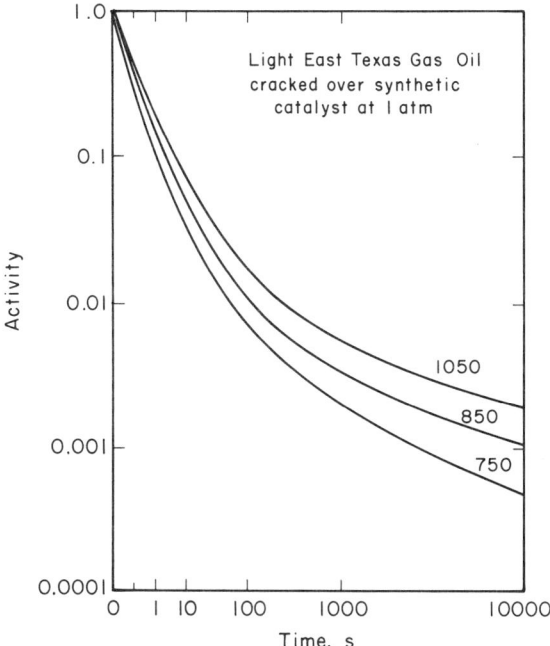

Fig. 3-7. Effect of temperature on rate of decay of synthetic cracking catalyst. [From Blanding (5). Reproduced by permission of the American Chemical Society.]

only a small part of the total surface area. This work of Mills *et al.* is discussed in more detail in the next chapter, but it suffices here to illustrate that organic bases can be the predominant cause of deactivation of cracking catalysts. In some cases these catalysts can be reactivated by heating in an inert atmosphere to remove desorbable foulants, leaving the coke level unchanged (32). This work will also be discussed in more detail later.

The discussion above of the classical papers of Voorhies and Blanding is not intended in any way to underrate the importance of those pioneering studies. Indeed, in what follows, we shall see that these ideas have been extended, modified, and generalized to a very useful framework for handling coking problems.

In further study of the apparent independence of coke formation on hydrocarbon feed rate reported by Voorhies, Eberly *et al.* (11) used *n*-hexadecane and light East Texas gas oil (LETGO) as feedstocks and cracked them at 500°C over a 13% alumina–87% silica catalyst of 382 m^2/g surface area. Their coking kinetics followed the Voorhies equation very well, but the constants in the equation depended on the reaction conditions and feed compositions. Representative values are shown in Table 3-1. Note par-

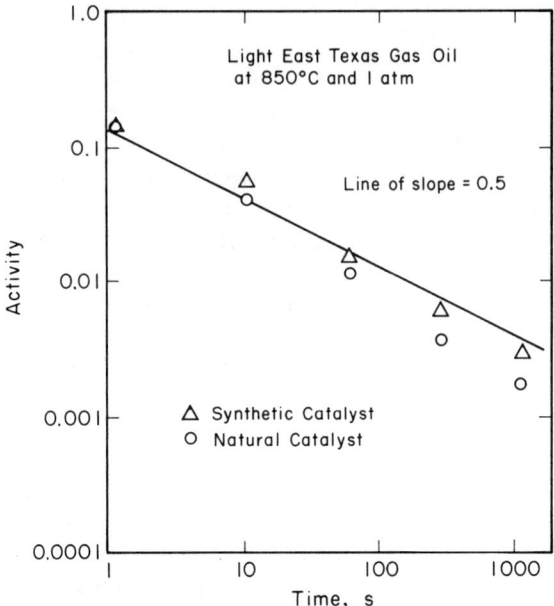

Fig. 3-8. Time-on-stream correlation of the data of Blanding. [From Blanding (5). Reproduced by permission of the American Chemical Society.]

TABLE 3-1

Constants in $C_c = At^n$ for a Fixed Bed of 13% Al_2O_3–87% SiO_2 at 500°C[a]

Feed	Effective space velocity (v/v-hr)	A (wt. %)	n
n-Hexadecane	0.2	0.049	0.97
	0.5	0.11	0.86
	1	0.16	0.78
	2	0.22	0.70
	5	0.29	0.60
	10	0.31	0.52
	20	0.29	0.44
Light East Texas gas oil	0.5	0.25	0.82
	2	0.52	0.70
	5	1.05	0.42
	10	0.90	0.41

[a] Data of Eberly *et al.* (11). Reproduced by permission of the American Chemical Society.

I. Coke Formation and Fouling Kinetics

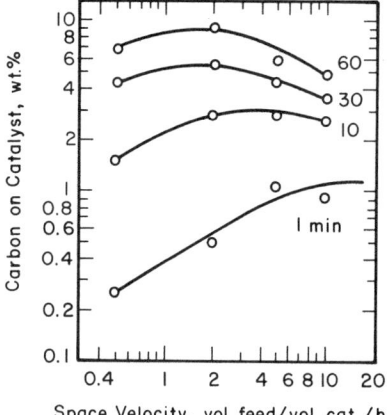

Fig. 3-9. Carbon formation on 13% Al$_2$O$_3$–87% SiO$_2$ from cracking of East Texas light gas oil at 550°C. [From Eberley et al. (11). Reproduced by permission of the American Chemical Society.]

ticularly that values of n range from 0.41 to 0.97. Values of n close to 0.5 were interpreted as diffusion limitation by some workers; however, the variation shown in this table is evidence against this mechanism. Furthermore, Eberly et al. investigated particle size effects on carbon deposition and found no dependence in the size range 75–2400 μm. Of course, this particle size test would not invalidate the original Voorhies diffusion model.

Eberly et al. investigated the effect of space velocity on coking kinetics, and the results are shown in Figs. 3-9 and 3-10 for LETGO and n-hexadecane, respectively. Note first that the results are similar for the two

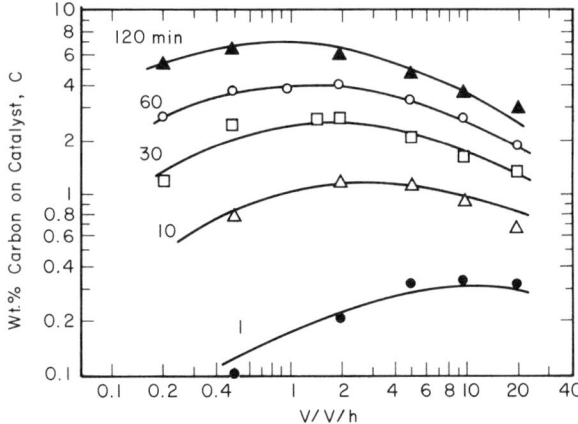

Fig. 3-10. Carbon formation on 13% Al$_2$O$_3$–87% SiO$_2$ from cracking of n-hexadecane at 500°C. [From Eberly et al. (11). Reproduced by permission of the American Chemical Society.]

feeds. Also note that the weight percent carbon for a given run length is relatively insensitive to the space velocity at more than about 20 min, although certainly not constant. Voorhies' work was carried out at space velocities between 0.3 and 2.4, and his observation that the coke laydown was independent of space velocity was a first-order approximation.

Figure 3-11 shows the weight percent carbon on the catalyst as a function of bed length for n-hexadecane at 500°C and a 60-min cycle time. These curves obtain from the observation that the effective space velocity [volume of feed per volume of catalyst per hour (v/v-hr)] decreases from a very high value at the entrance to a very much lower value at the outlet. One must be careful not to interpret the profiles in Fig. 3-11 as conforming to any given run. This figure does, however, show the carbon profiles that are calculated for a bed of a reactor having a given space velocity. The profiles are not uniform, but over the values of the space velocity with which Voorhies worked, the variation in coke laydown is remarkably constant in each third of the bed. It is clear, then, that the careful work of Eberly *et al.* is a strong confirmation of Voorhies' work but indicates ranges of space velocity where quite nonuniform coke laydown is observed.

Rudershausen and Watson (37) studied the variables affecting the activity of a commercial molybdena–alumina (10% molybdena as the trioxide and a surface area of 90 m^2/g) for the dehydrogenation of cyclohexane. They measured deactivation and coke formation as functions of time in an integral reactor and found the results shown in Fig. 3-12. The lines can be represented approximately by

$$\langle \bar{a} \rangle = 0.57 t^{-0.40} \qquad (3\text{-}3)$$

$$\langle \bar{a} \rangle = 0.68 C_c^{-0.65} \qquad (3\text{-}4)$$

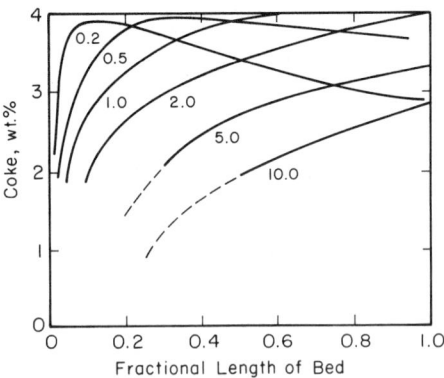

Fig. 3-11. Weight percent carbon on catalyst as a function of bed length. n-Hexadecane at 500°C and 60 min cycle time. Parameter: v/v-hr. [From Eberly *et al.* (11). Reproduced by permission of the American Chemical Society.]

I. Coke Formation and Fouling Kinetics

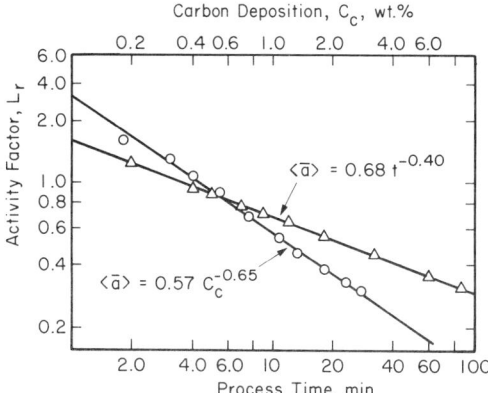

Fig. 3-12. Decline in activity with process time and with carbon deposition ("standard" conditions, except time on stream). Curve a: $\langle \bar{a} \rangle = 0.68 t^{-0.40}$. Curve b: $\langle \bar{a} \rangle = 0.57 C_c^{-0.65}$. (From Rudershausen and Watson (37). Reproduced by permission of Pergamon Press, Ltd.]

leading to a Voorhies-type correlation of the form

$$C_c = 0.76 t^{0.61} \tag{3-5}$$

Rudershausen and Watson found a strong temperature dependence of coke formation kinetics and accordingly sought a (semi) mechanistic model for this. They attempted to justify Eq. (3-5) on a theoretical basis utilizing an analysis suggested to them by Villiers-Fisher. The analysis assumed a polymerization process on the active centers to form the coke and gives rise to a term which is envisioned to dominate the denominator of a Langmuir–Hinshelwood expression for the rate of coke formation, thereby giving Eq. (3-2). However, if the coke dominates the active centers, the catalyst must have only a small fraction of its original activity or else this mechanism cannot explain the coking kinetics of the partially deactivated catalyst.

The effect of hydrogen partial pressure on the coking rate was investigated in this study also, and the results are shown in Fig. 3-13. The linear relationship was interpreted by the authors to result from the hydrogenation of olefinic coke precursors. They also observed a decrease in conversion at higher hydrogen pressures.

The activity, $\langle \bar{a}_{\text{approx}} \rangle$ as measured in this discussion[1], is difficult to interpret in any fundamental way because it represents an average of the

---

[1] The meaning of approximate activity is defined and discussed in detail subsequently. This average activity was defined by Hougen and Watson (18). The activity used by Rudershausen and Watson is not strictly $\langle \bar{a} \rangle$ as used in this book but it is close enough for the purpose of this discussion.

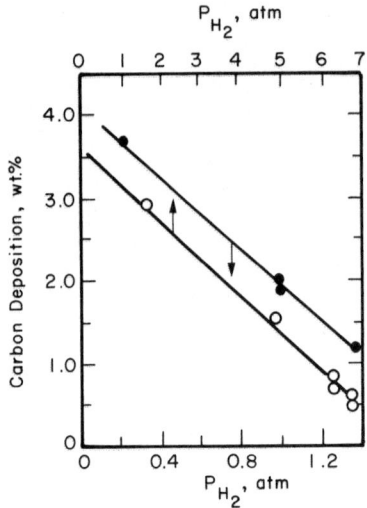

Fig. 3-13. Effect of partial pressure of hydrogen on carbon deposition ("standard" conditions except total pressure and H_2/HC molar feed ratio). (●) 1.7 atm abs., (○) 7.8 atm abs. [From Rudershausen and Watson (37). Reproduced by permission of Pergamon Press, Ltd.]

catalyst activity along the reactor bed as well as within an individual pellet. This activity was presented previously as a function of percentage of coke on the catalyst, in which the functional form depended on the partial pressure of hydrogen.

Prater and Lago studied the fouling of a 90 wt. % silica–10 wt. % alumina catalyst having an area of 350 m^2/g during the cracking of cumene and LETGO. In this study, the primary thesis with respect to coke formation was that carbonaceous deposits are composed of two limiting types: strongly chemisorbed inhibitors that deactivate the catalyst and are not easily desorbed, and harmless carbonaceous materials that do not affect catalyst activity. The inhibitor they selected for model studies was cumene hydroperoxide. An experiment to demonstrate the concept of an inhibitor and its reversibility is shown in Table 3-2. Pure cumene was passed over a cracking

TABLE 3-2

Reversibility of Inhibition of the Catalyst by Cumene Hydroperoxide[a]

Reactant	Time (sec)	Rate × 10^9 (mol/m^2-sec)	Coke (wt. %)
Pure cumene	0–4,000	269	<0.1
Cumene + 0.08 mol % cumene hydroperoxide	4,000–8,000	73	1.1
Pure cumene	8,000–12,000	252	1.2

[a] From Prater and Lago (32). Reproduced by permission of Academic Press, Inc.

I. Coke Formation and Fouling Kinetics

catalyst in a differential reactor for 4000 sec. Then cumene with 0.08 mol. % cumene hydroperoxide was passed for another 4000 sec. The result was a greatly deactivated catalyst and a significant increase in coke formation. In the final 4000 sec pure cumene was again passed through the reactor to recover approximately the original activity, but at about the same coke level. These results demonstrate dramatically, as did the work of Mills *et al.*, that coke level as such is not uniquely related to activity.

In a similar experiment using cumene with 10% LETGO, they demonstrated that some strongly adsorbed component in the LETGO acted as an inhibitor. The results in Fig. 3-14 show the initial deactivation during the first 4000 sec. The catalyst was then flushed with nitrogen (oxygen-free) for 1 hr at 420°C and its activity measured in pure cumene. There was little change in activity. A final flush for 1 hr in nitrogen at 510°C showed partial recovery of activity.

Finally, in a similar set of experiments using an integral reactor containing 1000 cm^3 of catalyst and passing 1.2 cm^3/min of LETGO to give a space velocity v/v-hr of 0.72 and a temperature of 420°C, they found that an initially deactivated catalyst could recover over 90% of its initial activity by a 1-hr flush in nitrogen at 650°C without changing the coke level.

These results support their original contention that two types of carbonaceous deposits develop during cracking: a deposit on active centers that can be desorbed and a coke that is relatively innocuous with respect to its effect on catalyst activity. This conclusion is harmonious with those of Mills *et al.* and leads to the view that there are at least two types of sites on a cracking catalyst surface; active centers on which the main cracking reaction occurs and nonactive centers. Active centers occupy only a small part of the surface; the remainder is composed of nonactive centers. The active centers may also be the sites on which coke forms, after which it desorbs to the inactive part of the surface, as suggested by Plank and Nace

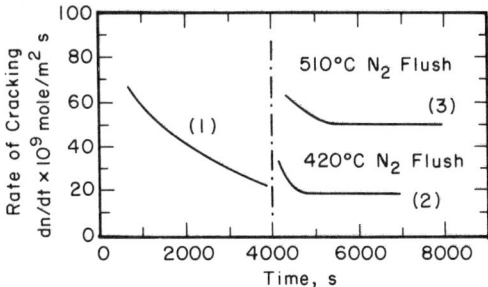

Fig. 3-14. Catalytic reactivation by desorption of inhibitors deposited on the catalyst by LETGO. [From Prater and Lago (32). Reproduced by permission of Academic Press, Inc.]

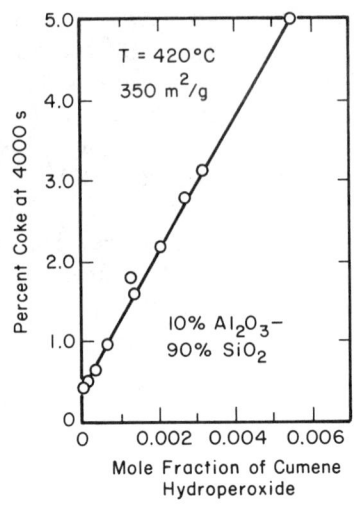

Fig. 3-15. Weight percent coke on catalyst at 4000 sec as a function of mole fraction cumene hydroperoxide in cumene. [From Prater and Lago (32). Reproduced by permission of Academic Press, Inc.]

(31) and Prater and Lago. Alternatively, coke may be made directly on sites inactive for the main reaction as suggested by the results of Heinemann (16), who found that benzene in excess hydrogen greatly reduced coke formation during the dehydrogenation of methycyclopentane. He suggested that benzene poisoned the coke-forming sites and that deactivation, according to this view, results when carbonaceous structures form on the active centers that cannot desorb under reaction conditions.

This model leads us to the conclusion that there is no cause-and-effect relationship between the activity and the coke level of a cracking catalyst and certain other catalysts. It suggests further that the apparent relationship observed by several workers arises because both the coke level and the adsorption of inhibitors are well-defined functions of time for a given set

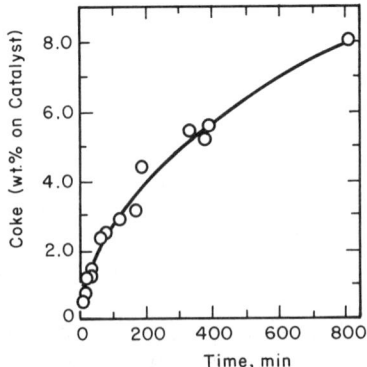

Fig. 3-16. Weight percent coke on catalyst as a function of time for cumene containing 0.26 mol % cumene hydroperoxide. [From Prater and Lago (32). Reproduced by permission of Academic Press, Inc.]

I. Coke Formation and Fouling Kinetics

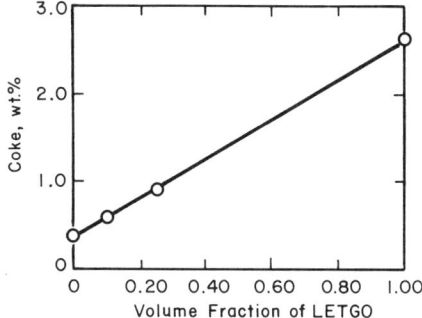

Fig. 3-17. Percengate of coke by weight as a function of volume fraction of light East Texas gas oil present in cumene. [From Prater and Lago (32). Reproduced by permission of Academic Press, Inc.]

of conditions. Therefore, under very well defined experimental conditions, the activity appears to be correlated with coke level, but when critical experiments are designed to test the relationship, such as those described above, no correlation is found. Apparently, then, for many reactions coke is not the culprit but an innocent bystander.

One further point about coking kinetics as observed by Prater and Lago is related to coke formation as a function of concentrations of inhibitors and of time. The former is shown in Figs. 3-15 and 3-17 and the latter in Figs. 3-16 and 3-18. For a given cracking period and conditions, the coke laydown is a linear function of the inhibitors cumene hydroperoxide and LETGO, respectively. Also, for a given cracking period and conditions, the coke laydown is proportional to the square root of time, in excellent agreement with the Voorhies equation.

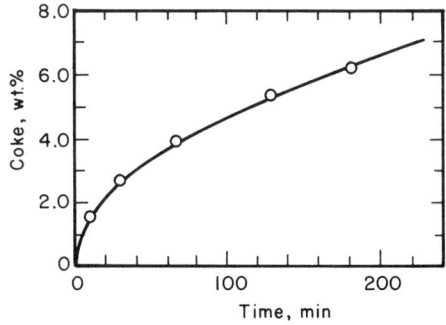

Fig. 3-18. Effect of amount of coke deposited on catalyst as a function of time for cracking of light East Texas gas oil. The solid line represents the relationship. $C_c = 0.47(t)^{1/2}$. [From Prater and Lago (32). Reproduced by permission of Academic Press, Inc.]

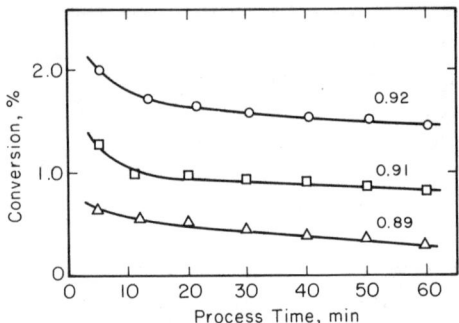

Fig. 3-19. Relation of conversion to process time. [From Ozawa and Bischoff (27). Reproduced by permission of the American Chemical Society.]

Ozawa and Bischoff (27) measured the cracking of ethylene over a commercial silica–alumina catalyst at 350–500°C. They observed characteristics common to many deactivation curves—an initial rapid decay in conversion followed by a slow linear decay with process time. The characteristic time between the rapid and slow decay periods was about 10 min, as shown in Fig. 3-19. Simultaneous measurements of coke laydown were obtained directly and continuously by monitoring the weight changes of the total catalyst with time. Attributing the increase in weight to coke, they obtained the curves of Figs. 3-20 and 3-21. These Voorhies-type curves also have a significant change in slope at about 10 min process time, indicating a change in the coking rate at about the same time as the change in regimes of rapid and slow deactivation. This suggests a relationship between coke

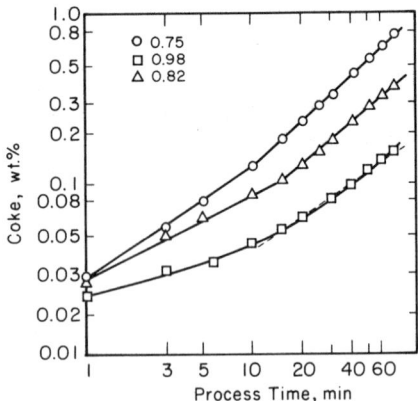

Fig. 3-20. Relation of weight percent coke on catalyst to process time (catalyst B). [From Ozawa and Bischoff (27). Reproduced by permission of the American Chemical Society.]

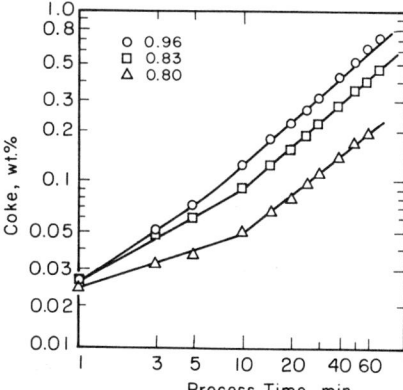

Fig. 3-21. Relation of weight percent coke on catalyst to process time (catalyst C). [From Ozawa and Bischoff (27). Reproduced by permission of the American Chemical Society.]

laydown and activity of the catalyst. The Voorhies constants, shown in Table 3-3, increase with increasing temperature, which is opposite to the direction expected if a diffusion mechanism is important.

A change in the slope of Voorhies plots was also observed by Butt et al. (8). They studied the cracking of cumene on an H–mordenite extrudate in the temperature range 230–350°C. The catalyst was either 2 × 6 mm extrudate cylinders or size 12–16 mesh and the space velocity was 0.20–0.65 g/g-hr. The cumene was purified to remove traces of cumene hydroperoxide. The activity of the catalyst falls off rapidly with time during the first hour at the

TABLE 3-3

Exponent in the Voorhies Correlation

Run number	Reaction temperature (°C)	C_2H_4 flow rate (cm^3/min)	n
077	499	100	0.902
096	499	75	0.855
075	498	50	0.913
094	500	30	0.881
097	450	75	0.742
082	449	50	0.820
083	450	75	0.841
098	400	75	0.625
080	399	50	0.736
081	399	30	0.605
099	350	50	0.554

[a] Data of Ozawa and Bischoff (27) for experiments with reactor A. Reproduced by permission of the American Chemical Society.

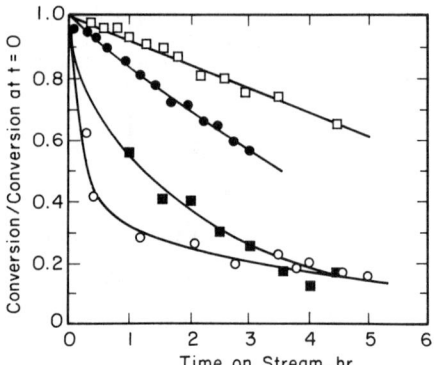

Fig. 3-22. Change of H-mordenite activity for cumene cracking at various temperatures, $SV = 0.33$ g/hr-g. (○) $T_p = 260°C$, $T_R = 230°C$; (■) $T_p = 260°C$, $T_R = 260°C$; (●) $T_p = 350°C$, $T_R = 350°C$ (2×6 mm); (□) $T_p = 350°C$, $T_R = 350°C$ (12–16 mesh). T_p, Pretreatment temperature in flowing He, 12 hr; T_R, reaction temperature. [From Butt et al. (8). Reproduced by permission of Academic Press, Inc.]

lower temperatures, as shown in Fig. 3-22, whereas at higher temperatures the activity appears to drop off nearly linearly with time. The corresponding Voorhies plots exhibit a distinct change in slope after about the first hour on stream, as shown in Fig. 3-23. Note that, although the slopes of the low-temperature curves during the first hour are smaller, the absolute level

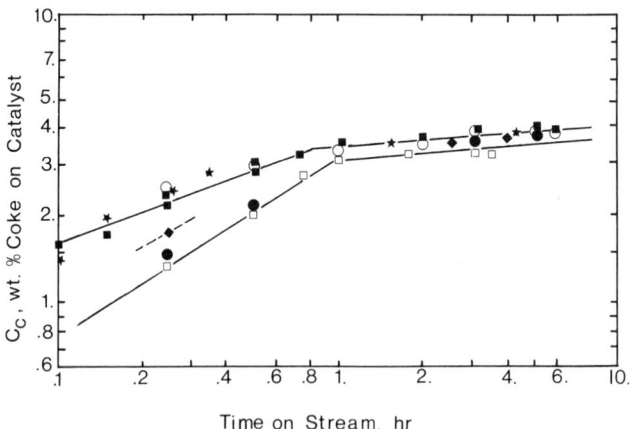

Fig. 3-23. Voorhies correlation of coke deposition at various temperatures, $SV = 0.33$ g/hr-g. (★) $T_p = 350°C$, $T_R = 260°C$; (◆) $T_p = 350°C$, $T_R = 300°C$. Other symbols as for Fig. 3-22. [From Butt et al. (8). Reproduced by permission of Academic Press, Inc.]

of coke is higher than for the high-temperature curves. There are also two regimes of coking as a function of space velocity, as shown in Fig. 3-24. Shorter times on stream indicate a rather strong dependence on this variable, with coke content increasing with increasing space velocity. However, again after about 1 hr on stream, this dependence disappears. Reasons for these changes in the dependence of coke deposition on both temperature and space velocity can be found by examining in more detail possible changes in intraparticle transport of reactants and coke precursors, as described below.

The effective diffusivity, using sulfur hexafluoride, was also measured by the chromatographic technique of Eberly (10), and the results are shown in Fig. 3-25. The diffusivities fall off nonlinearly with coke level. Furthermore, micrographs reveal a heavily coked outer shell, virtually no coke at the center, and intermediate amounts in between. In sum, these results of Butt *et al.* demonstrate the influence of pore diffusion on the cracking reaction, that the deposition of coke alters the magnitude of the effective diffusivity, and that the coke is nonuniformly distributed throughout the catalyst. This complex situation suggests that the relationship between coke level and activity of a catalyst for the main reaction must also be complex. There is an apparent correlation between effective diffusivity and coke content (Fig. 3-26), but this must be viewed with caution since the coke is nonuniformly distributed.

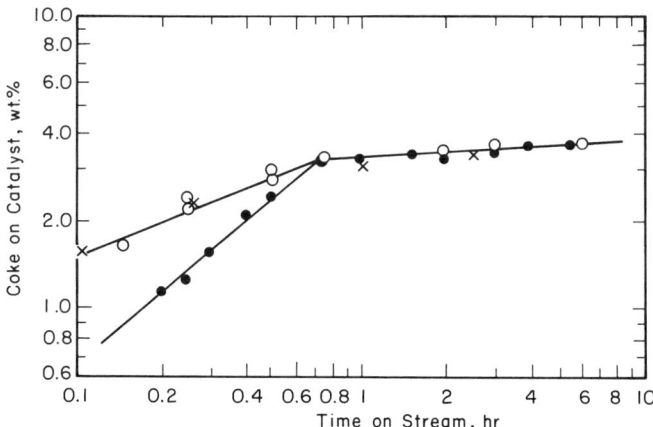

Fig. 3-24. Effect of space velocity on coke deposition at 260°C. Runs at $T_p = 260°C$, $T_R = 260°C$. (●) 0.20, (○) 0.33, and (×) 0.65 g/hr-g. [From Butt *et al.* (8). Reproduced by permission of Academic Press, Inc.]

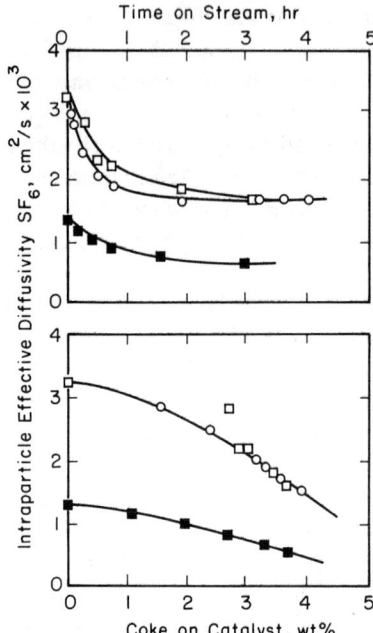

Fig. 3-25. Changes in effective diffusivity (SF_6, 258°C) of H-mordenite on coking in the temperature range 230–350°C. (□) $T_R =$ 230°C, (○) $T_R = 260$°C, and (■) $T_R = 350$°C. [From Butt et al. (8). Reproduced by permission of Academic Press, Inc.]

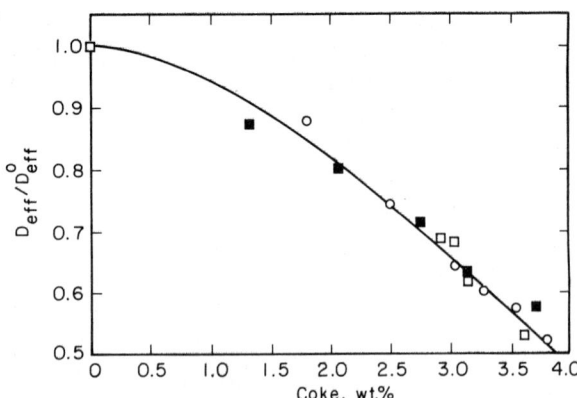

Fig. 3-26. Diffusivity relative to that of fresh catalyst. Symbols as for Fig. 3-25. [From Butt et al. (8). Reproduced by permission of Academic Press, Inc.]

II. CHEMISTRY OF COKE FORMATION

Coke is a common product of almost all hydrocarbon catalytic processes and typically consists of polyaromatic condensed-ring structures. These carbonaceous deposits tend to be pseudographitic or aromatic in nature and the hydrogen-to-carbon ratio of the deposits depends on the specific conditions of their formation. Moreover, the deposits are refractory in a chemical sense, as evidenced by the lack of carbon tracer exchange between the deposits and the reacting hydrocarbons.

The chemical nature of carbonaceous deposits has long been of scientific importance. Voge et al. (41) found a direct relationship between the amount of coke deposited on a silica-alumina catalyst and the aromatic and polynuclear aromatic content of the feed.

Extensions of this work by Appleby et al. (2) to a series of cracking experiments involving paraffinic, naphthenic, aromatic, and heterocyclic hydrocarbons on a synthetic silica-alumina-zirconia (86.2/4.3/9.4 wt. %; UOP, type B) or a synthetic silica-alumina (90/10 wt. %, American Cyanamid) catalyst have been reported in detail. The reaction conditions were usually 500°C, 1 atm pressure, and flow rates of 10-13 mol/liter-hr, and the time on stream was generally 15 min except for the light hydrocarbons.

In a series of runs using light hydrocarbons it was found that n-butane formed little or no coke, whereas n-butene and 1,3-butadiene formed significant amounts of coke. It is clear that, although these light hydrocarbons are of low molecular weight, by polymerization and dehydrocyclization reactions they form higher-boiling aromatics and coke.

At the expense of some redundancy with the remarks of Chapter 1, we represent the coke formation tendencies of polycyclic aromatics and some heterocyclic compounds in Fig. 3-27. Numbers in the figure represent coke deposition under the conditions of the experiments described above. Evidently, coke laydown is greater in the condensed-ring systems benzene (not much), naphthalene, anthracene, and phenanthrene than in the linked systems benzene, biphenyl, and terphenyl. The striking difference in coke deposition between anthracene and phenanthrene is noteworthy. The same is true for pyrene and chrysens, except that the difference is not so pronounced. The heterocyclics produced less coke than the hydrocarbon analogs. In order of decreasing coke laydown, they are fluorene, carbazole, dibenzofuran, and dibenzothiophene. Cracking a mixture of 1-pentene with methylnaphthalene produced more coke than cracking either of the two pure components, evidencing an interaction between them to enhance coke production, as shown in Table 3-4.

An interesting experiment on the cracking of benzene and naphthalene mixtures yielded the results shown in Table 3-5. The coke deposition was

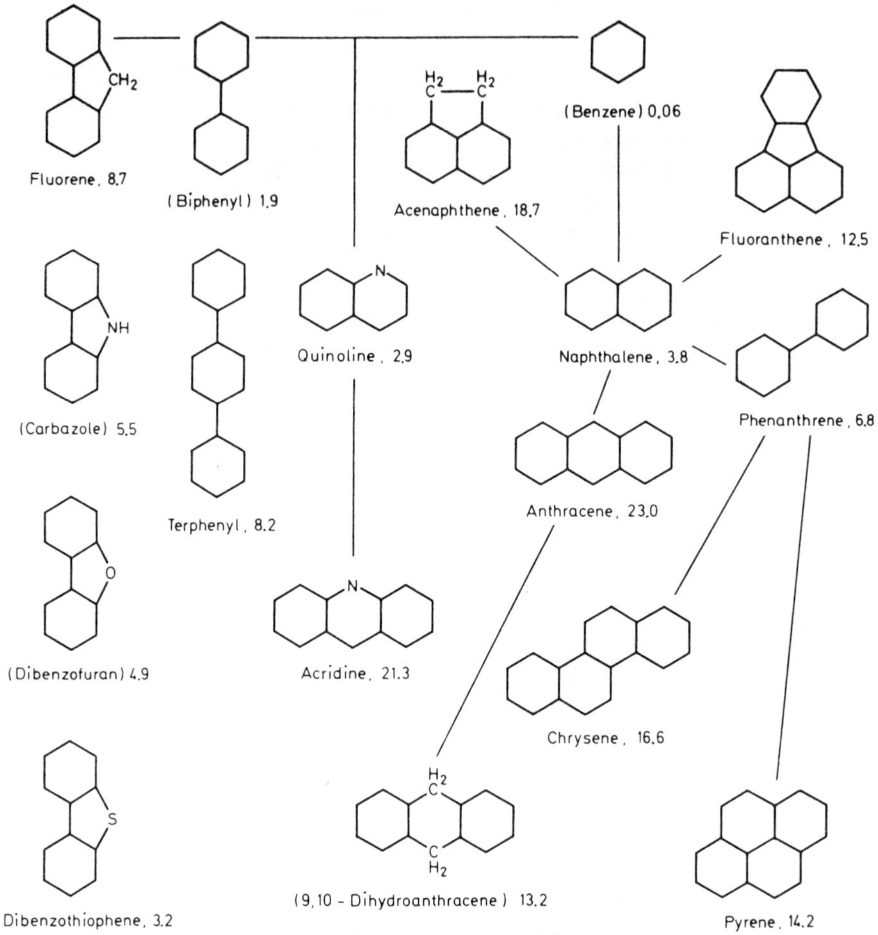

Fig. 3-27. Ranking of coke formation from polycyclic and heterocyclic aromatics: weight percent coke on silica–alumina after 15 min at 773 K. [From Appleby et al. (2). Reproduced by permission of the American Chemical Society.]

Compounds and coke yields shown: Fluorene, 8.7; Biphenyl, 1.9; Acenaphthene, 18.7; Benzene, 0.06; Fluoranthene, 12.5; Carbazole, 5.5; Quinoline, 2.9; Naphthalene, 3.8; Terphenyl, 8.2; Phenanthrene, 6.8; Dibenzofuran, 4.9; Acridine, 21.3; Anthracene, 23.0; Chrysene, 16.6; Dibenzothiophene, 3.2; (9,10-Dihydroanthracene), 13.2; Pyrene, 14.2.

about that expected from the pure components, with small deviation. It appears that possible products from the direct coupling of the reactants, not expected, are present in small amounts in the product stream. Benzene, naphthalene, and higher products are cracked to gaseous and lower hydrocarbon products, also with a minor amount of direct-coupled products. A possible growth process, involving the carbonium ions of the primary reactants, is shown in Fig. 3-28. This envisions coke formation as more of a polymerization process than a degradation process. Overall, such results and interpretation suggest that coke precursors are the simpler feed

II. Chemistry of Coke Formation

TABLE 3-4

Coke Formation from Hydrocarbon Mixtures[a,b]

Feed	1-Pentene	Methyl-naphthalenes	1-Pentene plus methyl-naphthalenes
Temperature, °C	520	500	500
Flow rate, mol/liter-hr	26.8	20.2	26.4
Olefin in feed, mol %	100	0	20 (11 wt. %)
Conversion, wt. %	23.9	10.9	—
Lower boiling, wt. %	18.2	7.9	—
Higher boiling, wt. %	10.4	4.8	6.0[c]
Coke, wt. % of			
Feed	5.7	3.0	4.2
Converted products	16.6	19.1	24.8
Catalyst	4.1	3.4	5.5

[a] From Appleby et al. (2). Reproduced by permission of the American Chemical Society.
[b] Process period, 15 min; pressure, atmospheric; catalyst, fresh silica-alumina.
[c] Fraction boiling above methylnaphthalenes.

TABLE 3-5

Reaction of a Benzene–Naphthalene Mixture[a–c]

		Distillate[d]			Residue above 492°C	Coke
Fraction	Gas	Below 161°C	161–261°C	261–492°C		
Charge, wt. %	0.01	69.52	23.79	0.18	2.26	2.63
Analysis						
Carbon	—	—	93.7	93.4	95.2	85.2
Hydrogen	—	—	6.3	6.6	4.8	2.9
Ash[e]	—	—	—	—	—	11.9
Loss on heating[f]	—	—	—	—	—	1.5

[a] From Appleby et al. (2). Reproduced by permission of the American Chemical Society.
[b] Composition, 78 mol % benzene (stock B) + 22 mol % naphthalene (stock B); temperature, 500°C; process period, 1 hr; pressure, atmospheric; flow rate, 6.0 mol/liter-hr; catalyst, fresh silica-alumina.
[c] Some polycyclic aromatics identified in residual oil from benzene–naphthalene mixture (relative molar quantities given in parentheses): fluoranthene (6), perylene (4), benzofluoranthene (1), 1,2-benzanthracene (3), chrysene (3), and pyrene (1).
[d] An effort was made to separate roughly the benzene, combined naphthalene, and methylnaphthalene fractions, and higher-boiling distillate. The temperatures listed are of the vapor in a Claisen flask. The quantities of material involved were too small to avoid some remixing in the take-off manifold.
[e] By combustion of coke remaining after NaOH dissolution of catalyst.
[f] In nitrogen at 200°C.

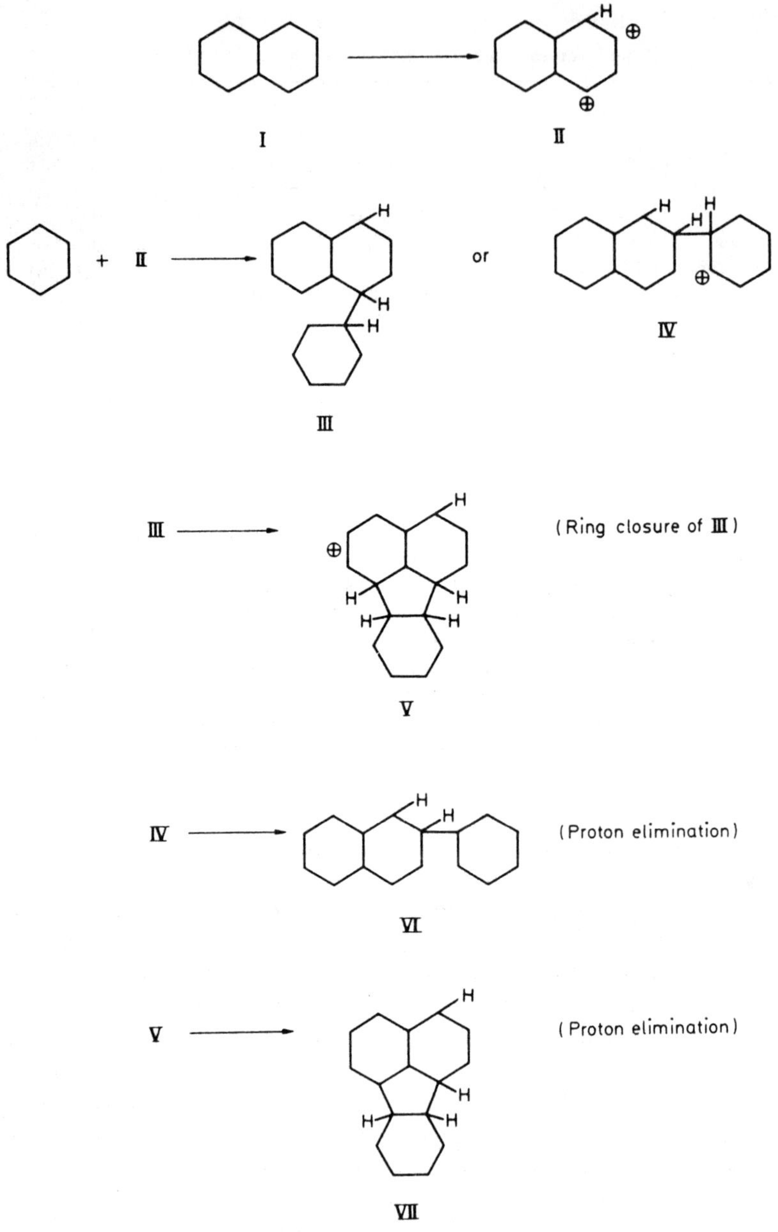

Fig. 3-28. Carbonium ion mechanism for formation of higher aromatics from benzene and naphthalene. [From Appleby *et al.* (2). Reproduced by permission of the American Chemical Society.]

molecules and that fragments, via polymerization or dehydrogenation reactions, form larger polynuclear aromatic intermediates characteristic of structures found by x-ray examination of coke.

In what has been said before, and in much of what is to come, it will be clear that there is, in spite of present evidence, no consensus on the mechanism of coke formation. It is a preliminary statement to make at this point, but the overall view is probably true.

Another series of experiments using alkylbenzene and alkylnapthalenes is summarized in Table 3-6. The coke laydown tendencies appear to be related to both structure and molecular weight. The unsubstituted nucleus determined the minimum coke-forming tendency. Coke increased with the number of alkyl carbon atoms in side chains rather than the number of alkyl substitutions.

Appleby et al. present the rough correlation between the basicity as defined by Mackor et al. (23) and the coke on the catalyst shown in Fig. 3-29. A relationship would be expected in view of the acid nature of the cracking catalyst and is presumably based on the ease of formation of a carbonium ion from the various aromatic species.

The work of Eberly et al. (11) also gives some insight into the structure of coke as determined from infrared spectroscopy. Figures 3-30 and 3-31

TABLE 3-6

Effect of Alkyl Substitution on Coke Formation[a,b]

Aromatic	Number of carbon atoms in alkyl group	Feed	Coke (wt. %) Converted product	Catalyst
Benzene	0	0.6	63	0.04
Toluene	1	2.9	20	0.22
Ethylbenzene	2	8.1	14	0.74
p-Xylene	2	8.8	26	0.81
n-Propylbenzene	3	9.8	11	1.1
Isopropylbenzene	3	9.6	10	1.0
Pentamethylbenzene	5	17.6	20	2.1
Triethylbenzenes	6	11.3	12	1.7
Hexaethylbenzenes	12	7.4	8	1.6
Naphthalene	0	23.4	89	2.6
2-Methylnaphthalene	1	27.1	51	3.5
2,3-Dimethylnaphthalene	2	41.2	57	5.5

[a] From Appleby et al. (2). Reproduced by permission of the American Chemical Society.
[b] Temperature, 500°C; process period, 15 min; pressure, atmospheric; feed rate, ca. 2.4 mol/liter-hr; catalyst, fresh silica–alumina.

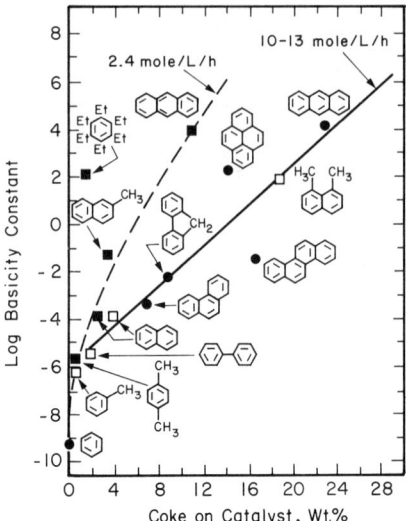

Fig. 3-29. Correlation of coke formation in catalytic cracking as a function of hydrocarbon basicity. [From Appleby *et al.* (2). Reproduced by permission of the American Chemical Society.]

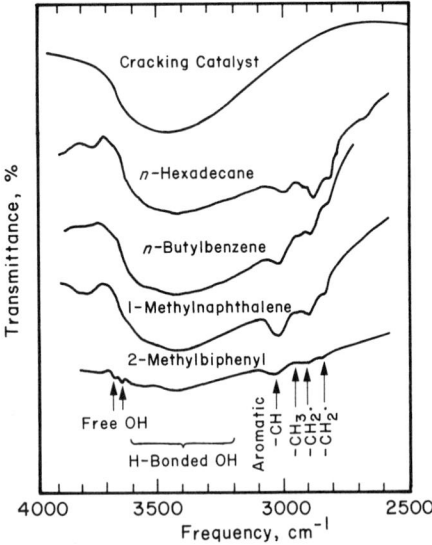

Fig. 3-30. Infrared spectra of coke deposits on 13% Al_2O_3-87% SiO_2 obtained by cracking various compounds at 445°C and $SV = 0.4$ v/v-hr. [From Eberly *et al.* (11). Reproduced by permission of the American Chemical Society.]

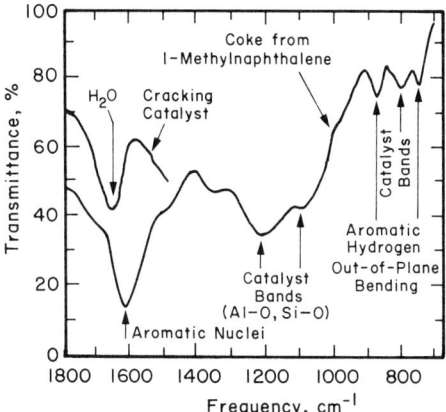

Fig. 3-31. Infrared spectrum of 1-methylnaphthalene coke deposit. [From Eberly et al. (11). Reproduced by permission of the American Chemical Society.]

show spectra resulting from various pure feeds. The most interesting features of these spectra are the fundamental C–H stretches in the range 3100 to 2800 cm^{-1}. Aromatic –CH absorption is at 3050 cm^{-1} and methylene groups at 2930 and 2860 cm^{-1}, and in some cases hydrogen atoms in methylene groups were evident at 2970 cm^{-1}. In Fig. 3-31 strong absorption in the region 1580–1590 cm^{-1} is shown, characteristic of aromatic skeletal frequencies. The ratio of the absorbance of infrared bands at 3050 cm^{-1} to that at 2930 cm^{-1} is a measure of the aromaticity of the coke. These values are shown in Table 3-7. The authors conclude that coke deposits are highly condensed aromatics of low hydrogen content and that infrared spectroscopy is useful in characterizing coke structures even though hydrogen contents are low. The data of Eberly et al. as plotted by Gates et al. (14), shown in Fig. 3-32, demonstrate the general increase in coke formation with conversion for pure feeds, but clearly the coke level depends on feed structure.

The use of tagged reactants has given some additional insight into the chemistry of coke formation. Blue and co-workers (6, 17) used tagged 1-butene with decalin and tetralin to assess the importance of hydrogen transfer reactions in coke formation during cracking on coprecipitated silica–alumina in the range 60–90% silica. To ensure that they were observing hydrogen transfer and not hydrogen exchange, they also studied (17) the activity for the latter reaction and found the catalysts to be relatively inactive under the conditions used in the cracking experiments. They found that the isobutene-decalin system reacted faster than the butene-decalin system, as shown in Fig. 3-33. Moreover, the isobutene-tetralin system coked faster

TABLE 3-7

Preparation of Carbon Deposits on 13% Al_2O_3–87% SiO_2 by Catalytic Cracking of Pure Hydrocarbons at 445°C and 0.4 v/v-hr[a]

Feed	Decalin	Tetralin	n-Butyl-benzene	1-Methyl-naphthalene	2-Methyl-biphenyl	n-Hexa-decane	1-Hexa-decene
Cycle time, min	32	28	32	30	34	60	60
Conversion, wt. %	98	91	78	86	67	71	100
Product distribution, wt. % on feed							
Gas	9	11	22	3	5	51	65
Liquid other than feed	64	74	51	67	54	13	23
Coke	25	6	5	16	8	7	12
Carbon on catalyst (C), wt. %	7.80	1.72	1.20	4.50	2.77	2.39	4.64
Coke aromaticity, A_{3050}/A_{2930}	1.03	0.92	1.35	1.23	1.55	0.51	0.47

[a] From Eberly et al. (11).

Fig. 3-32. Extent of coke formation in the cracking of pure hydrocarbons. (■) SiO_2-Al_2O_3, 445°C, LHSV = 0.4: (1) decalin, on-stream time 32 min; (2) tetralin, on-stream time 28 min; (3) n-butylamine, on-stream time 32 min; (4) 1-methylnaphthalene, on-stream time 30 min; (5) 2-methylbiphenyl, on-stream time 34 min; (6) n-hexadecene, on-stream time 60 min; (7) n-hexadecene, on-stream time 60 min. (△) SiO_2-Al_2O_3, 500°C, on-stream time 60 min. (○) SiO_2-Al_2O_3, 550°C, on-stream time 60 min. (1) n-Hexane, (2) n-octane, (3) n-dodecane, (4) decalin (500°C), (5) n-hexadecene (400°C), (6) cumene (isopropylbenzene), (7) n-hexadecane. [From Gates et al. (14). Reproduced by permission of McGraw-Hill Book Co.]

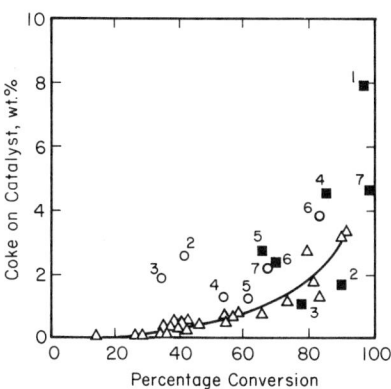

than the isobutene–decalin system, as shown in Fig. 3-34. It is also interesting that the coke level was nearly independent of space velocity but depended on temperature and time in approximately the same manner as found by Voorhies.

Of mechanistic importance, they determined the amount of coke formation in the reaction of tagged butene with tetralin and with decalin. The amount of radioactivity appearing in the coke is then a direct measure of the extent to which 1-butene contributed to the coke formation. The results are shown in Fig. 3-35, in which the fraction of coke derived from 1-butene, f, is plotted versus $1/X$, where X is the average paraffin in contact with the catalyst. The linear relationship shown implies a form according to the

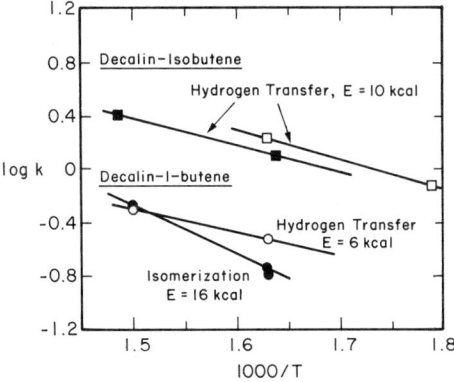

Fig. 3-33. Effect of temperature on rates of reaction. [From Blue and Engle (6). Reproduced by permission of the American Chemical Society.]

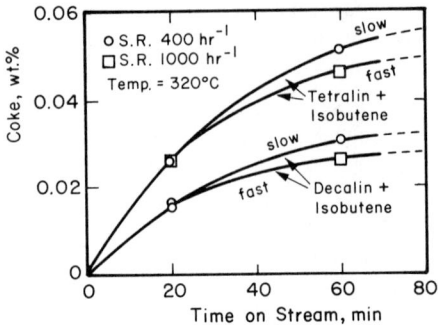

Fig. 3-34. Accumulation of the coke deposit on catalyst. [From Blue and Engle (6). Reproduced by permission of the American Chemical Society.]

following:

$$f = (A/X) - A$$

or

$$f = A(1-X)/(X) \tag{3-6}$$

which in turn suggests that f is proportional to the fraction of 1-butene in the mixture and inversely proportional to the conversion of 1-butene to butane. The coke deposit was much less radioactive than the 1-butane, revealing that some other reactant or product contributed primarily to the deposit.

All of the experiments show that the tetralin forms more coke than decalin; however, both compounds form more coke when hydrogen transfer is more extensive. This is shown in Figs. 3-35 and 3-36. These results suggest

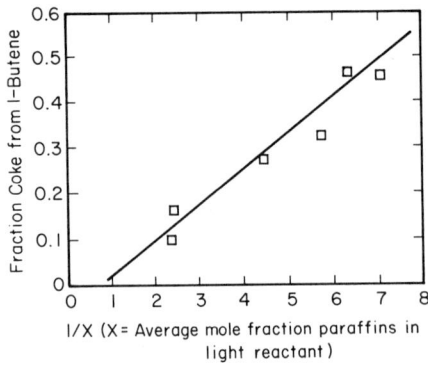

Fig. 3-35. Relation of coke percentage to hydrogen transfer. [From Blue and Engle (6). Reproduced by permission of the American Chemical Society.]

II. Chemistry of Coke Formation

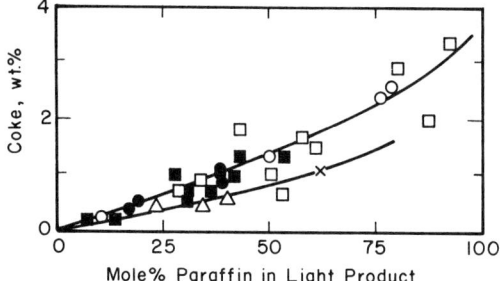

Fig. 3-36. Relation of coke production to hydrogen transfer. (□) Decalin-isobutene, space velocity (S.R.) 350; (■) decalin-isobutene, S.R. 1000; (○) tetralin-isobutene, S.R. 400, all at 340°C. (●) Tetralin-isobutene, S.R. 1000; (×) decalin-isobutene, S.R. 1000; (△) decalin-isobutene, S.R. 2000, all at 400°C. [From Blue and Engle (6). Reproduced by permission of the American Chemical Society.]

that coke forms most rapidly when a hydrogen acceptor such as the olefin is present to remove hydrogen from the adsorbed material. The mechanism proposed by Blue and Engle can be summarized as follows:

$$n\text{-butene} \rightleftharpoons \text{isobutenes} \qquad \text{I}$$

$$\text{isobutenes} + \text{decalin} \rightarrow \text{butane} + \text{tetralin} \qquad \text{II}$$

The reaction

$$n\text{-butene} + \text{decalin} \rightarrow \text{butane} + \text{tetralin} \qquad \text{III}$$

is much slower than **II** above, and the reaction

$$\text{butane} \rightleftharpoons \text{isobutane} \qquad \text{IV}$$

is also slow, so the route by **III** and **IV** is viewed as kinetically unimportant. Moreover, the route via **I** and **II** produces a concentration of isobutane that exceeds the equilibrium of **IV**; therefore the reaction path cannot include that reaction in the forward direction.

Further tracer studies of coke formation during cracking have been reported by Walsh and Rollman (44), who used radioactively labeled mixtures of hexane and aromatics over mordenite and Y-zeolites at 100–150 psi hydrogen pressures. This work is significant with respect to the hydrogen transfer mechanism because it is known that the hydrogen transfer activity of zeolites is greater than that of the silica-alumina catalysts used in the previous investigations, yet the zeolites in general make *less* coke. These experiments used a paraffin and an aromatic as a feed mixture in order to appraise simultaneously the participation of the cracking, olefin polymerization, and alkylation of aromatics as main reactions in the deactivation process. The high hydrogen pressure retarded the coke deposition, thus permitting an opportunity for selectivity in the deposition. The experiments assessed the influence of feed composition, reaction conditions, and silica-alumina ratio of the zeolites on coke formation. The primary experimental findings were that the contributions to coke deposits on mordenite were about in proportion to the amount of aromatic and paraffin in the feed. The percentage contribution of each remained essentially independent of the hydrogen partial pressure. However, the amount of coke increased substantially as the hydrogen partial pressure decreased. An increase in temperature sharply reduced the contribution of aromatics to coke formation and the coke yield. Only about 5% of light gas products came from the aromatic.

By contrast, in a Y-zeolite, aromatics were the major contributor to coke formation, the fraction being essentially independent of the silica-alumina ratio, hydrogen partial pressure, and time. Much higher percentages of light gas came from the aromatics. As with mordenites, a sharp decrease in the fraction of coke from aromatics and in the coke yield was observed when the cracking temperature was increased. Lastly, the origin of coke and light gas was insensitive to the structure of the aromatic constituent.

All of these results are plotted in Fig. 3-37, where the abscissa is a measure of the aluminum atom area density for each of the mordenites and Y-zeolites used. Coke yields and light gas yields generally increase with increasing Al density. The contribution of aromatics to coke formation also generally increases with Al density and appears to level off asymptotically. From

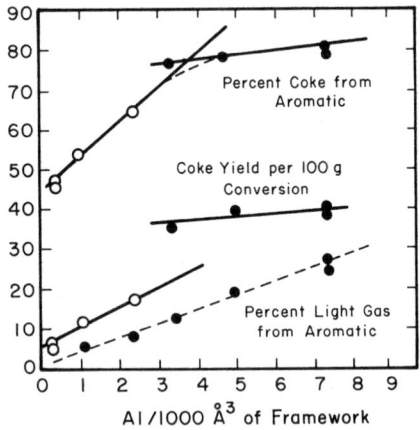

Fig. 3-37. Aromatic participation in coke formation and yield per 100 g of conversion for (●) Y-zeolite and (○) mordenite catalysts of varying aluminum density. [From Walsh and Rollman (44). Reproduced by permission of Academic Press, Inc.]

II. Chemistry of Coke Formation

these experiments Walsh and Rollman suggest that the initiating reaction in coke formation from paraffin–aromatic mixtures over large-pore zeolites is the alkylation of the aromatic. However, they also assign a major role to hydrogen transfer reactions in the continuing transformation of these deposits into coke.

As stated before, carbonium ion chemistry has been valuable in explaining a number of aspects of coke formation, particularly in catalytic cracking. The details of coke-forming reactions obviously vary with the constituents of the reaction mixture, the operating conditions, and the catalyst used, but one can conjecture how specific compounds combine, rearrange, and dehydrogenate into coke-type structures via carbonium ion-type reactions. A short discussion of this has already been given for the benzene–naphthalene combination, as shown in Fig. 3-28. Further consideration is appropriate at this point. It is seen that these reactants can combine to form fluoranthene, a prominent polycyclic component of the residual oil product from this mixture, as shown in Table 3-5. Figure 3-28 suggests that naphthalene is strongly adsorbed on the catalyst, whereupon it is protonated preferentially in the alpha position to form a carbonium ion having concentrated charge density as shown. The addition of benzene to structure **II** results in **III** (or **IV**), which undergoes ring closure to structure **V**. To produce coke, another molecule could add at the indicated point of high charge density, and so on into a very large condensed-ring system. (Otherwise, a condensation can occur, as shown in the figure.) Upon addition, the molecule will eventually terminate by a disproportionation reaction represented by structure **VII**. Hydrogen transfer reactions illustrated by the further reactions of structure **V** complete the formation of fluoroanthrene, and hence a product of residual oil instead of coke.

A similar type of chemistry has been discussed by Gates *et al.*; simply stated, once aromatics are present they can react to higher hydrocarbons and coke, formed both by condensation reactions and by combination of aromatics with other aromatics or coke. They present an interesting analysis of a sequence beginning with benzene (even though this is, according to Appleby *et al.*, not a prominent coke former), written in a formal sequence as a typical chain addition reaction. The reaction steps are as follows:

Step 1: initiation:

Step 2: propagation:

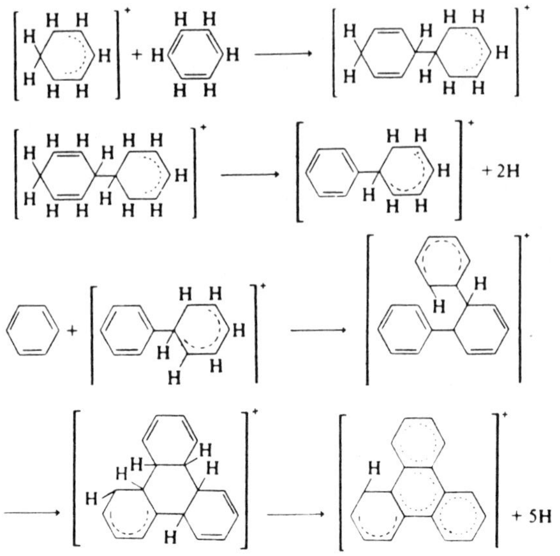

[chemical reaction schemes showing propagation steps with aromatic carbonium ion intermediates, producing +2H, then combining with benzene, and finally yielding a triphenyl-type carbonium ion + 5H]

and so forth.

Step 3: termination.

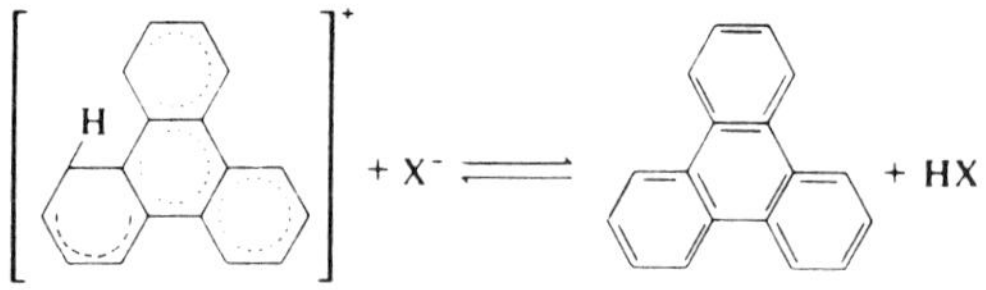

[chemical equilibrium showing the carbonium ion + X⁻ ⇌ triphenylene + HX]

Because of the high stability of polynuclear carbonium ions, the intermediates can have a relatively long lifetime on the surface. One thus has a situation similar to addition polymerization, in which all propagation rate constants are the same. Presumably one could analyze the kinetics of this sequence using the pseudo-steady-state hypothesis for the chain-growth molecules, but to our knowledge this has not been done.

In summary, in this section we have examined a number of papers that are related by their common interest in the chemical aspects of coke formation during hydrocarbon processing. The general chemical patterns that emerge from this work indicate that coke is made by the same kinds of reactions that account for the main products of cracking. Thus, carbonium ions can undergo addition, dehydrogenation, and hydrogen transfer to grow into large, immobile, and tightly adsorbed molecules that eventually are perceived as coke. Carbonium ions can also crack to form small fragments and higher fragments that can further participate in the coke-forming process as hydrogen transfer agents. In general, then, large polynuclear aromatics in feed mixtures tend to be coke-formers. Olefinic compounds promote coke formation, possibly either by direct interaction or by acting as hydrogen acceptors.

III. COKE DISTRIBUTION IN CATALYST PORES

As coke accumulates within catalyst pores, it occupies some of the space and as a consequence the effective diameter of the pore decreases. Under certain conditions the decreased size of the pores increases the resistance to the transport of reactants into and products out of the pore structure and is, in effect, a barrier. Some aspects of this have been alluded to before in the work of Butt *et al.* (8). It is apparent that for a given coke level the effectiveness of the coke deposit as a barrier depends very strongly on its distribution within the pore structure. For example, if a given amount of coke is concentrated near the pore mouths it will be more effective as a barrier than the same amount uniformly distributed. In this section, we shall be concerned with the distribution of coke in pellets and any changes in the effective diffusivity caused by coke deposition.

In cracking of a gas-oil, Ramser and Hill (33) reported that the coke formed was not uniformly distributed in the catalyst. This result was based on the observation that the decrease in pore volume was twice the volume of the coke deposited. Under similar conditions, however, Haldeman and Botty (15) found that the pore volume decrease corresponded very well to the volume of coke formed. Further light was thrown on this by a series of very nice experiments on silica–alumina cracking catalysts by Levinter *et al.* (22), who studied coke formation during reactions with a number of individual hydrocarbons and with a kerosene gas-oil fraction. They allowed the experiment to proceed until the coke formation leveled off, a value referred to as the limiting coke formation, C_{lim}. The theoretical limiting value of this quantity for their catalyst would have been 68% if all of the pores were filled. In none of their experiments did C_{lim} reach the theoretical value. Representative results for a number of individual hydrocarbons are shown in Fig. 3-38, where for styrene C_{lim} was about 12%. Apparently, the coke forms in plugs to isolate interior regions that are presumably still active. This contention was verified by splitting the particles, whereupon a considerable additional coke deposit was observed on them when subsequently exposed to the same reaction conditions.

In similar experiments at low temperatures (400°C) with a number of compounds, as shown in Fig. 3-38, they also found low values of C_{lim}, but for the case of isobutene, *n*-hexadecane, and kerosene gas-oil no further

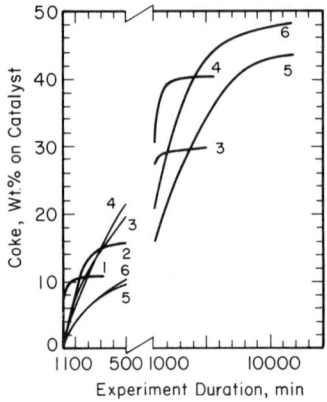

Fig. 3-38. Relation of coke deposition on silica–alumina catalyst to experimental time in cracking of (1) styrene, (2) butadiene, (3) α-methylnaphthalene, (4) isobutene, (5) hexadecane, and (6) kerosine gas-oil fraction. [From Levinter et al. (22). Reproduced by permission of the American Institute of Chemical Engineers.]

particle splitting effects. In these cases, apparently, there is no diffusional hindrance and the catalyst surface was deactivated at low values of C_{lim}. Increasing the temperature to 500°C resulted in partial blocking of the pores.

Figure 3-39 shows that the penetration and hence the utilization of the catalyst increase as the rate of coke formation decreases. Lastly, evidence of internal diffusion influence is shown in Fig. 3-40, where styrene was cracked over catalysts of a wide range of particle sizes. The figure clearly shows that the interior surface of the catalyst becomes more accessible as the size of the catalyst particle decreases.

Richardson (34) measured coke concentration profiles in coked catalysts using a clever microcombustion technique wherein the particles were regenerated under diffusion-controlled conditions. To obtain diffusion control, it is necessary to regenerate under conditions where the rate of reaction between oxygen and coke is very high. The rate at which the overall process

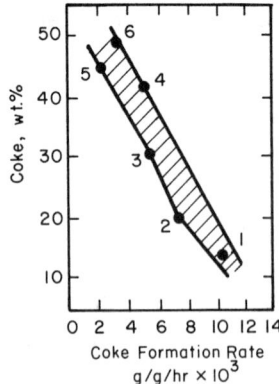

Fig. 3-39. Relation of maximum coke deposition to the coke formation rate after the first hour of the experiment. Symbols as for Fig. 3-38. [From Levinter et al. (22). Reproduced by permission of the American Institute of Chemical Engineers.]

III. Coke Distribution in Catalyst Pores

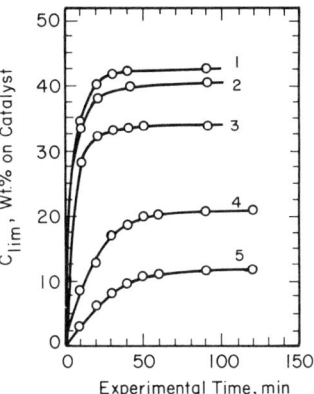

Fig. 3-40. Kinetics of coke deposition in the cracking of styrene on silica–alumina catalysts of various grain sizes (in millimeters): (1) 0.05, (2) 0.3, (3) 0.8, (4) 2, (5) 3. [From Levinter *et al.* (22). Reproduced by permission of the American Institute of Chemical Engineers.]

takes place, therefore, depends on how rapidly oxygen can be supplied to the coke. Since transport of oxygen is by diffusion in the pores of the catalyst, the maximum rate corresponds to an oxygen concentration of essentially zero where it comes in contact with the coke. Under these conditions the catalyst regenerates according to the shell model as described by Weisz and Goodwin (45). In a physical sense, the catalyst regenerates to maintain a sharp moving boundary between a completely regenerated outer shell and a completely coked inner core. The boundary moves progressively inward until the inner core is regenerated. We give some aspects of this in the following; the topic is reviewed in more detail in Chapter 9.

The catalyst used by Richardson was a 3/16-in. spherical cobalt–molybdenum–alumina (Harshaw Chemical Co.) and was deactivated and coked using metal-free coal-derived liquids. The spheres were regenerated for different lengths of time and then were cut open and examined under a microscope to determine the location of the sharp front between regenerated and nonregenerated zones as described above. The amount of coke burned was determined by measuring the CO_2 and H_2O during the partial regeneration. These results are shown in Figs. 3-41 and 3-42. In the former figure we see the regeneration history along with a pictorial representation of the microscope results at various stages in the regeneration process. In Fig. 3-42, the cumulative carbon content is displayed as a function of the radial position within the catalyst particle, from which the carbon concentration profile is calculated. Clearly, the carbon concentration profile is very nonuniform, with approximately 18% carbon at the exterior of the particle and only about 0.9% at the center.

The regeneration time is short for this sample because the bulk of the coke was deposited near the surface, where it is readily accessible to the oxygen used to remove it. It is interesting to compare this regeneration time

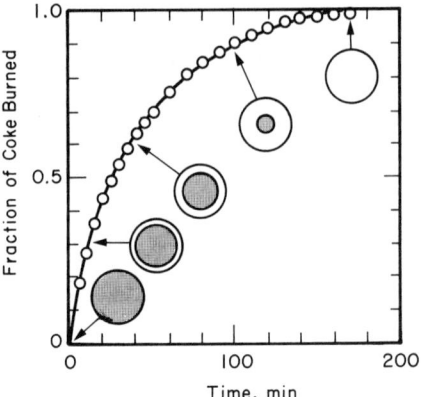

Fig. 3-41. Shell-progressive regeneration of fouled pellet. [From Richardson (34). Reproduced by permission of the American Chemical Society.]

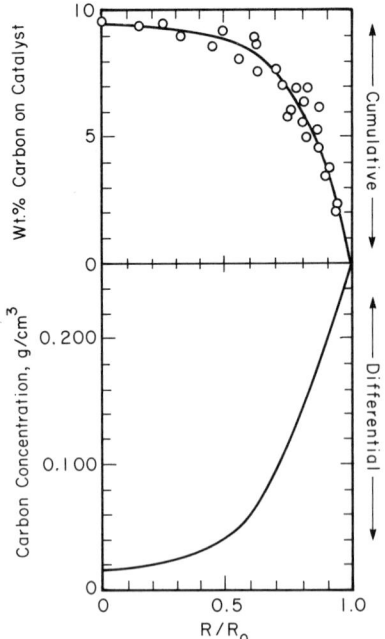

Fig. 3-42. Measurement of carbon profile. (Upper) Experimental data; (lower) calculated profile, parallel fouling mechanism. [From Richardson (34). Reproduced by permission of the American Chemical Society.]

III. Coke Distribution in Catalyst Pores

with that of a hypothetical catalyst having the same properties but a uniform coke distribution at the same average coke level.

The shell model can be developed very simply by referring to Fig. 3-43, which shows a coked particle that has been regenerated to a certain radius smaller than the particle radius $\bar{r}$. A material balance for oxygen at r is

$$-4\pi \bar{r}^2 D_{\text{eff}} \left(\frac{dC}{dr} \right)_{r=\bar{r}} = a 4\pi \bar{r}^2 \left(\frac{d\bar{r}}{dt} \right) \qquad (3\text{-}7)$$

where D_{eff} is the effective diffusivity of oxygen, square centimeters per second; C is the oxygen concentration, moles per cubic centimeter; and a is the moles of oxygen required to remove the coke from a unit volume of catalyst.

In words, Eq. (3-7) equates the rate of diffusion of oxygen through the boundary to the rate of consumption of the coke. However, the region between $\bar{r}$ and r_0 is described by the quasi-steady-state diffusion equation:

$$\frac{1}{r^2} \frac{d}{dr} \left(D_{\text{eff}} r^2 \frac{dC}{dr} \right) = 0 \qquad (3\text{-}8)$$

having the boundary conditions that at $r = r_0$, $C = C_0$; and at $r = \bar{r}$, $C = 0$. In dimensionless coordinates, this equation will be solved in Chapter 7, Eqs. (7-53a) to (7-53f) to give

$$(d\psi/d\xi)_{\xi=\bar{\xi}} = 1/(1-\bar{\xi})(\bar{\xi}) \qquad (3\text{-}9)$$

where $\psi \equiv C/C_0$, $\xi \equiv r/r_0$, and $\bar{\xi} = \bar{r}/r_0$. Now, making Eq. (3-7) dimensionless and substituting Eq. (3-9), we obtain

$$d\tau = -(1-\bar{\xi})(\bar{\xi}) \, d\bar{\xi} \qquad (3\text{-}10)$$

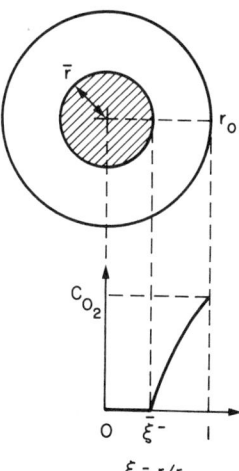

Fig. 3-43. Coked particle being regenerated according to a shell model.

where

$$\tau \equiv (D_{\text{eff}}C_0/r_0^2 a)t \qquad (3\text{-}11)$$

Integrating Eq. (3.10) gives

$$\tau = \frac{(1-\bar{\xi}^2)}{2} - \frac{(1-\bar{\xi}^3)}{3} \qquad (3\text{-}12)$$

For a uniformly coked particle, the fraction of coke burned, F, is

$$F = 1 - \bar{\xi}^3 \qquad (3\text{-}13)$$

and F is plotted versus τ in Fig. 3-44. This is the regeneration history of a uniformly coked particle. To compare this to the nonuniform case shown in Fig. 3-41, we must calculate the proportionality constant between τ and t of Eq. (3-11):

$$\frac{D_{\text{eff}}C_0}{r_0^2 a} = \frac{[(0.029)(40/32)^{1/2}(873/273)^{3/2}][(0.03)(273)/(22400)(873)]}{(0.24)^2(0.013)}$$

$$\approx 10^{-4} \text{ sec}^{-1} \qquad (3\text{-}14)$$

where D_{eff} has been corrected for molecular weight and temperature, and a is evaluated at its average value. Using this value of the proportionality constant, the nonuniform curve of Fig. 3-41 is plotted on the coordinates of Fig. 3-44. The expected result is, of course, that it takes much longer to regenerate the catalyst having uniformly distributed coke than the catalyst with the coke concentrated near the outer part of the pellet. Figure 3-45

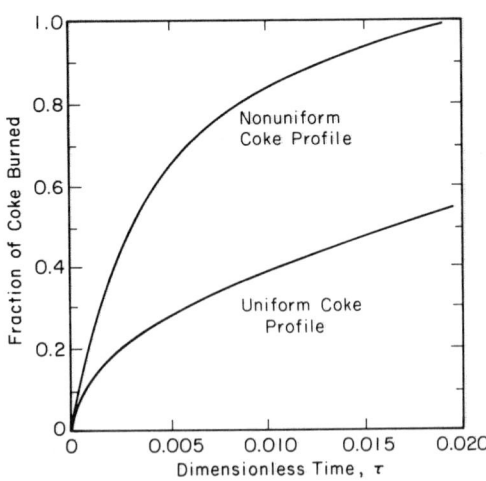

Fig. 3-44. Comparison of regeneration times for uniform and nonuniform coke distributions.

III. Coke Distribution in Catalyst Pores

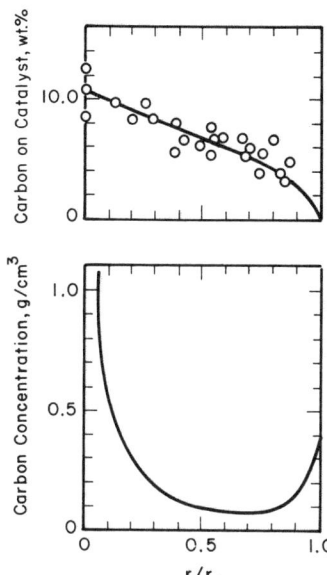

Fig. 3-45. Measurement of carbon profiles, series fouling mechanism. [From Richardson (34). Reproduced by permission of the American Chemical Society.]

shows the results of Richardson's technique on a $\frac{1}{8}$-in. extrudate cobalt–molybdenum catalyst used for petroleum resid desulfurization. This appears to be a series fouling mechanism (see Chapter 7) with metal deposits near the pore mouths.

We have seen that the distribution of coke in the pores of catalysts can be nonuniform and that, under certain circumstances, an active area can be made inaccessible owing to pore blockage. It is of further interest to know how coke deposits can affect the transport rates through a reduction in effective diffusivity. Of course, it is obvious that as we approach the limit of pores filled with coke, transport (rate-limiting) effects must be observed; however, in the range of much lower coke deposition experienced with practical catalysts, are transport effects observed? The work of Butt *et al.* discussed earlier in this chapter gives evidence that this is indeed so. In addition, the observations of Levinter *et al.* on pore plugging provide more circumstantial evidence to this effect. Ozawa and Bischoff found that during the cracking of ethylene on a commercial silica-alumina catalyst, the diffusivity and surface area changed very little up to about 1% coke level as shown in Table 3-8. However, Suga *et al.* (40) found quite different behavior during the dehydrogenation of *n*-butane over a chromia-alumina catalyst and at significantly larger coke levels, as shown in Fig. 3-46. The diffusivity is seen to decrease to approximately half its initial value at the higher coke level. The implication of this result is shown in Figs. 3-47 and 3-48. Three cases

TABLE 3-8

Effect of Coke on Diffusivity[a,b]

Sample	Wt. % coke on CAT	Before regeneration diffusivity (cm^2/sec)	After regeneration at 450°C diffusivity (cm^2/sec)
CAT-C	0.395	0.0104	0.0094
CAT-C	0.521	0.0104	0.0104
CAT-C	0.954	0.0090	0.0086
CAT-B (fresh)	0.0	0.0217	—
CAT-B	0.201	0.0175	0.0192
CAT-B	0.584	0.0174	0.0196
CAT-B	0.274	0.0188	0.0188

Surface area of fresh and coked catalyst

Wt. % Coke on catalyst	Surface area (m^2/g)
0.0	211
0.20	208
0.39	202
0.58	200
0.95	213

[a] Hydrogen at room temperature.
[b] CAT-B, Particle diameters between 4.76 and 3.36 mm; CAT-C, particle diameters between 3.36 and 2.38 mm. From Ozawa and Bischoff (27).

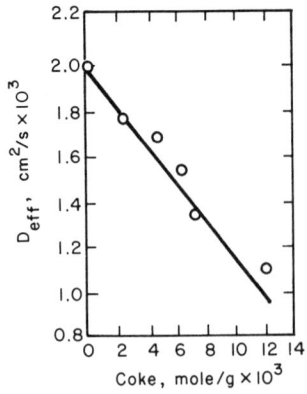

Fig. 3-46. Effect of coke level on effective diffusivity, dehydrogenation of n-butane on chromia–alumina. [From Suga et al. (40). Reproduced by permission of the American Institute of Chemical Engineers.]

III. Coke Distribution in Catalyst Pores

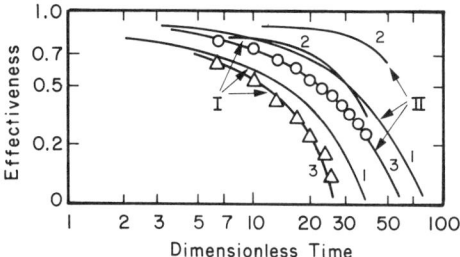

Fig. 3-47. Relation between effectiveness and time for *n*-butane dehydrogenation. (1) Constant diffusivity, variable activity; (2) constant activity, variable diffusivity; (3) variable activity and diffusivity. (I) 4.85 × 4.45 mm; (II) 2.85 × 3.20 mm. [From Suga *et al.* (40). Reproduced by permission of the American Institute of Chemical Engineers.]

are depicted: (1) coking affects only the catalytic activity, (2) coking affects only the diffusivity, and (3) coking affects both activity and diffusivity. The experimental data are consistent with case 3 for *n*-butane and *n*-butene, respectively, in the two plots.

A similar result was observed in the experiments of Richardson (35) using the $\frac{1}{8}$-in. cobalt–molybdenum extrudate and a reactant consisting of a coal-derived liquid under hydrotreating conditions. The effect of coke level on diffusivity is shown in Fig. 3-49, and further indication of pore blocking is given by the temperature dependence of diffusivity shown in Fig. 3-50. In the latter, the temperature coefficient for the fresh catalyst is about $T^{3/2}$, decreasing to $T^{1/2}$ for the fouled material. This reflects a transition from bulk to Knudsen diffusion in the pore structure as coke accumulates resulting from a decrease in pore dimension. The pore structure measurements listed in Table 3-9 indicate that the surface area decreases significantly with coking. These apparent changes in diffusivity on coke deposition are important to keep in mind when reviewing the developments

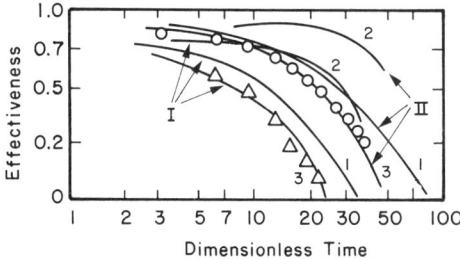

Fig. 3-48. Relation between effectiveness and time for *n*-butene dehydrogenation. Legend as for Fig. 3-47. [From Suga *et al.* (40). Reproduced by permission of the American Institute of Chemical Engineers.]

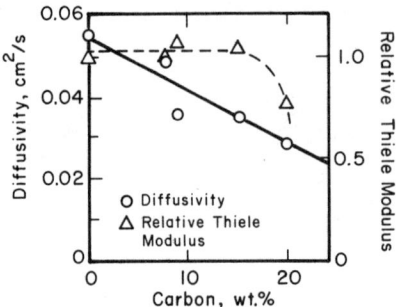

Fig. 3-49. Effect of carbon content on diffusion parameters. [From Richardson (35). Reproduced by permission of the American Chemical Society.]

of intraparticle deactivation presented in Chapter 7, since most of those analyses do not envision changes in transport properties.

Diffusional influence, though, is apparently not always necessary to obtain pore mouth plugging. It has been shown by Androutsopoulos and Mann (1) that, for complex pore structures, pore plugging may occur even in the absence of diffusional limitation. Experimental evidence for this, however, is scarce.

Chou and Hegedus (9) have devised a transient diffusivity measurement technique that is valuable in indicating the extent of pore plugging. Steady-state techniques are generally insufficiently sensitive and measure some complex average that is not very meaningful in the modeling of pores.

To this point we have learned that coke can deposit nonuniformly within catalysts and, when deposited in amounts of the order of only a few percent, can decrease significantly the magnitude of the effective diffusivity. If the reaction is in the transport-influenced regime initially, then the effectiveness factor can be decreased by approximately the square root of the ratio of $D_{\text{eff}}/D_{\text{eff}}^0$. (Some estimates of changes in effectiveness are provided by Butt

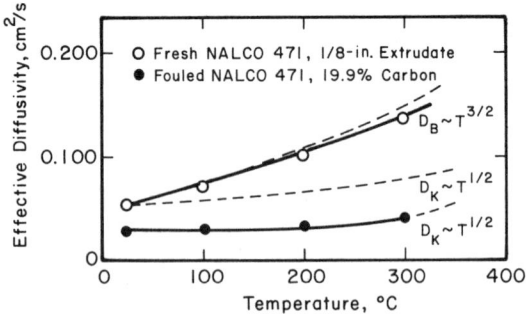

Fig. 3-50. Temperature dependence of diffusivity. [From Richardson (35). Reproduced by permission of the American Chemical Society.]

TABLE 3-9

Effect Diffusivities at 25°C for Fouled Catalysts, Argon–Helium Mixtures[a,b]

	ρ_p (g/cm^3)	d (g/cm^3)	S (m^2/g)	V_g (cm^3/g)	θ_t	D_{eff} (cm^2/sec)
Fresh	1.284	3.356	270	0.47	0.617	0.0546
Pyridine-extracted, heated at 600°C						
7.96% C	1.395	3.069	219	0.22	0.545	0.0485
9.01%	1.410	3.078	180	0.23	0.542	0.0357
15.3% C	1.498	2.976	160	0.21	0.297	0.0350
19.9% C	1.564	2.619	49.2	0.08	0.403	0.0281

Radius (Å)	Pore vol. (cm^3/g)
>1000	0.027
300–1000	0.041
100–300	0.051
35–100	0.265
15–35	0.085

[a] Catalyst: $\frac{1}{8}$-in. extrudate, Nalco 471 cobalt molybdate–Al$_2$O$_3$.
[b] From Richardson (35). Reproduced by permission of the American Chemical Society.

et al.) The exact manner in which changes in the effective diffusivity affect the overall rate of reaction, however, is very complex and depends in part on whether the reaction is diffusion-influenced initially and how coke deposits affect the local kinetics and the local diffusivity. El-Kady and Mann (12) discuss some of the theoretical implications of how the pore structure of a catalyst is modified by fouling, in particular by the influence of pore mouth plugging. In essence, they consider as a limiting case a reaction system in which a compountd A, a foulant precursor, is decomposed sufficiently rapidly that there is a concentration gradient within the pore. The main reaction, however, is slow so that its rate is proportional to the total available unfouled area of the catalyst. Since fouling is proportional to the concentration of the coke precursor, then according to this model coke builds up more rapidly at the pore mouths and eventually seals off the active area. Such a system thus behaves similarly to a pore diffusion-influenced main reaction with pore mouth fouling. From a mathematical viewpoint, this model certainly gives the behavior discussed above; from a physical point of view, though, one must be concerned about what happens to all the coke precursor that is developed.

IV. MECHANISM OF FOULING

In earlier sections of this chapter much of the key literature on deactivating systems was presented with the objective of showing how some catalysts deactivate on stream and what correlation (or lack thereof) could be made with the carbonaceous deposition that occurs concomitantly. In this section we shall try to unify much of this work in a common framework to correlate and perhaps predict catalyst deactivation rates. Let us begin by recalling Eq. (2-84), repeated below for reference:

$$a = [1/(1+Kt)]^{P/(n-1)} \qquad (3\text{-}15)$$

where $n = 1$ *and* $K = (n-1)k'C_{P0}^{n-1}$. When $Kt \gg 1$, then

$$a = Kt^{-P/(n-1)} \qquad (3\text{-}16)$$

From this we learn that the slope of a plot of log a versus log t is equal to $-P/(n-1)$ for the conditions under which the original equation was derived. A plot of the data of Jossens and Petersen (20) on these coordinates for a modest amount of deactivation is shown in Fig. 3-51. The slope is about -0.2, for which Pacheco and Petersen (28) speculated that P is 1, leading to a value of $n = 6$. From the model on which Eq. (3-15) is based, n corresponds to the number of bonds that the foulant employs in attaching itself to the surface of the catalyst.

IV. Mechanism of Fouling

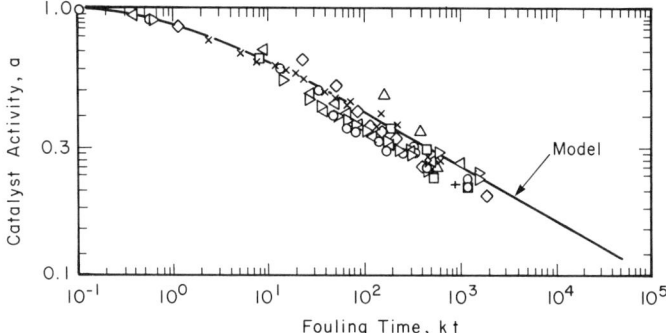

Fig. 3-51. Six-site fouling model. Fouling data for the dehydrogenation of methylcyclohexane over a Pt/Al$_2$O$_3$ reforming catalyst at shorter fouling times. [From Pacheco and Petersen (28). Reproduced by permission of Academic Press, Inc.]

A more complete set of data from the same source is shown in Fig. 3-52. In terms of this development the change in slope shown in this figure suggests that the number of catalyst sites involved in the deactivation reaction decreases with progressive deactivation. After a short time on stream, the number of sites appears to be six, whereas after considerable

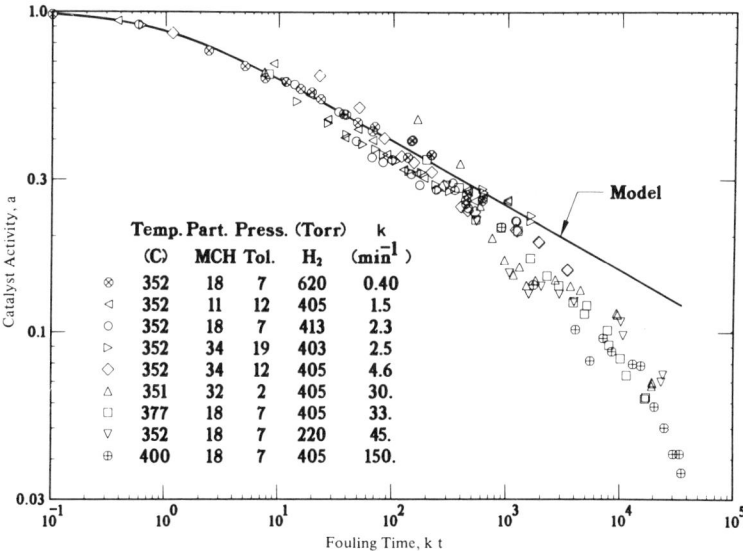

Fig. 3-52. Six-site fouling model. Fouling data for the dehydrogenation of methylcyclohexane over a Pt/Al$_2$O$_3$ reforming catalyst. [From Pacheco and Petersen (28). Reproduced by permission of Academic Press, Inc.]

deactivation the number of sites appears to be between two and three. Intrigued by this finding, Pacheco and Petersen put all of the data they could find in the literature for hydrocarbon reactions on a similar plot, shown in Fig. 3-53. This plot appears to correlate all of these data remarkably well.

In order to understand this correlation, in further work Pacheco and Petersen (29) modeled the system as a collection of parallel deactivating reactions as shown in Fig. 3-54, assuming that the activation energy for each of the parallel reactions depended on the number of bonds to the surface. For each bond to the surface, the activation energy was correspondingly lowered by the energy of that bond. This, of course, forces the six-bonded configuration to be formed first. However, since the rate of that reaction depends on the coverage to the sixth power, the six-bonded reaction slows down and the reactions with higher activation energies begin to dominate, until eventually reactions involving three and two sites predominate. At this time the overall rate of deactivation of the catalyst is much lower than initially. The mathematical formulation of these ideas is outlined briefly below.

The rate at which each of the deactivation processes takes place is represented by a power law model. The exponent representing the dependence of the rate on the active site density is assumed equal to the number

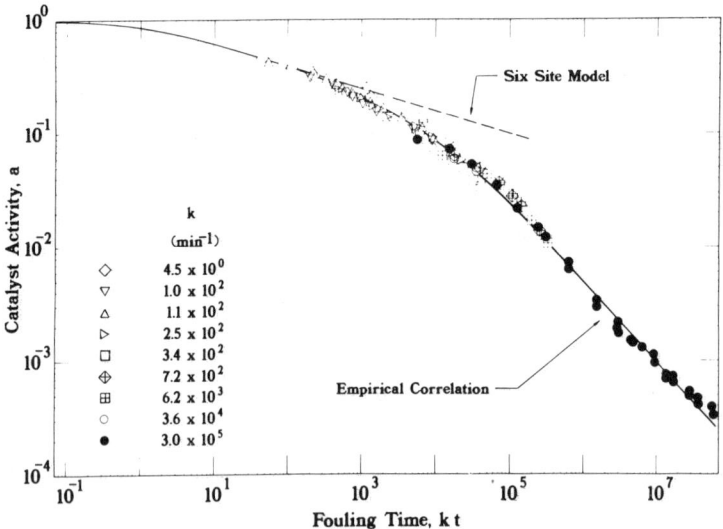

Fig. 3-53. Dimensionless fouling correlation. Variable reaction order fouling correlation for a wide range of specific reaction rates. [From Pacheco and Petersen (28). Reproduced by permission of Academic Press, Inc.]

IV. Mechanism of Fouling

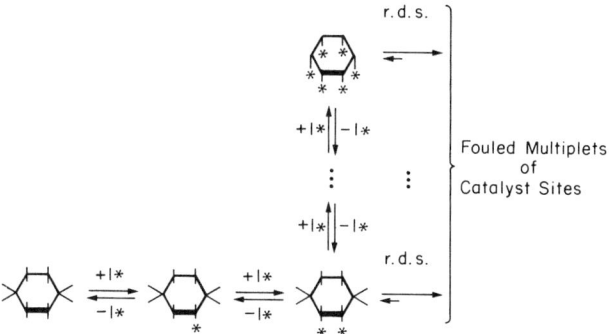

Fig. 3-54. Network of fouling reactions for multiplet fouling model (r.d.s., rate determining step). [From Pacheco and Petersen (29). Reproduced by permission of Academic Press, Inc.]

of active sites involved in the rate-determining step of each process. Summing all these rates together, the deactivation rate expression is written as

$$\frac{dS_T}{dt} = -\sum_m k_m S_v \left(\frac{S_v}{S_{T0}}\right)^{m-1} \tag{3-17}$$

where S_T is the number of active sites, S_v the number of vacant sites, and S_{T0} the number of active sites initially. This expression is based on the assumption that the rate of each process is proportional to the number of vacant sites times the fraction of the adjacent sites raised to the $(m-1)$ power. The apparent rate constants k_m, which incorporate the effects of temperature and composition of the reaction mixture, are indeed constant provided the experiments on deactivation are carried out under constant conversion and temperature conditions (see Section III in Chapter 2). Defining the nondimensional variables

$$\nu \equiv \frac{S_v}{S_T} \qquad f \equiv \frac{S_T}{S_{T0}} \qquad \tau \equiv 5k_6 \nu^6 t$$

and substituting into Eq. (3-17) yields

$$\frac{df}{d\tau} = -\frac{1}{5} \sum_m \left(\frac{k_m}{k_6}\right) \nu^{m-6} f^m \tag{3-18}$$

This quantity is a constant for systems having a linear site balance and, although arbitrary, is defined to correspond to the definition used to develop Fig. 3-53. Assuming an Arrhenius form for each of the deactivation rate constants leads to

$$\frac{df}{d\tau} = -\frac{1}{5} \sum_{m=2}^{6} \left(\frac{A_m}{A_6}\right) \nu^{m-6} \exp\left[\frac{-(E_m - E_6)}{RT}\right] f^m \tag{3-19}$$

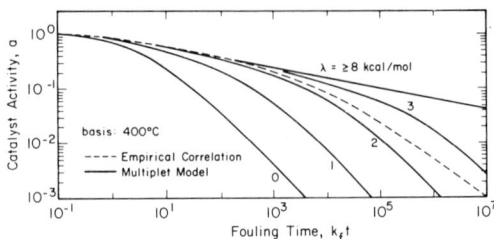

Fig. 3-55. Comparison of model equation with correlation. [From Pacheco and Petersen (29). Reproduced by permission of Academic Press, Inc.]

where
$$(A_m/A_6)\nu^{m-6} = 1 \tag{3-20}$$

$$(E_m - E_6) = \lambda(6 - m) \tag{3-21}$$

and λ is the decrease in activation energy per additional site involved in the rate-determining step of fouling. The simplifications allow Eq. (3-19) to be written as

$$\frac{df}{d\tau} = -\frac{1}{5} \sum_{m=2}^{6} \omega_m f^m \tag{3-22}$$

where
$$\omega_m = \exp[(m-6)\lambda/RT]$$

Setting Eq. (3-20) to unity is a matter of convenience to reduce the number of adjustable parameters. This is probably not correct in detail, but it implies that the only adjustable parameter is λ in Eq. (3-21).

Integration of Eq. (3-19) for various values of λ gives the results shown in Fig. 3-55. Comparing the curves obtained by integration of the model equation with the curve representing the correlation, one obtains a value for λ of 2.5 kcal/mol, indicating the energy for each bond with the surface. This value is not unreasonable, but it is lower than would have been expected *a priori*. The ratio of the rate constants for each of the reactions in Eq. (3-17) is probably less than one because the entropy term in k_6 is less than in k_m. Accordingly, the actual bond energy would be greater than the apparent one found by curve fitting.

V. FOULING OF REFORMING CATALYSTS

The reforming of naptha is carried out commercially on a bifunctional catalyst consisting of a metallic function and an acidic function. For many years the platinum/alumina catalyst was the basis of the most successful

V. Fouling of Reforming Catalysts

process. This catalyst has now been largely replaced by the bimetallic platinum–rhenium/alumina catalyst, not primarily on the basis of higher activity or enhanced selectivity, but because of significantly better activity maintenance. It is well established that the presence of rhenium materially affects the rate at which the catalyst ages in use, perhaps not so much by decreasing the amount of coke formed but by altering the nature of the coke deposits. Reforming reactions include dehydrogenation and isomerization taking place on the metal and acid sites, respectively. Obviously, with the multitude of possible reactions and the introduction of at least three possible reaction sites, the system is very complex. The reader is referred again to the reaction network for reforming proposed by Kmak, already discussed in Chapter 2, as an example. The deactivation is also very complex, since not only coke but also sulfur is involved in the overall deactivation process typical of commercial operation, and one would logically expect some interaction between the two. In this section we shall allow ourselves some speculation as to the mechanism(s) by which the platinum–rhenium/alumina catalyst fouls.

In an early paper, Myers *et al.* (26) discussed the aging of platinum catalysts. Carbonaceous deposits form on reforming catalysts by the processes discussed in Section II. At low hydrogen partial pressures these reactions occur rapidly; however, the rates go down rapidly at elevated hydrogen partial pressures. In fact, the fouling reactions can be reversed in hydrogen—the rate depending on how aged the catalysts are. Myers *et al.* speculated on the mechanism and kinetics of fouling following experiments on the fouling behavior of a number of specific compounds. In summary, they found that small aromatics and cyclohydrocarbons in the presence of hydrogen tend to age catalysts only a small amount, whereas cyclopentane ages the catalyst rapidly, apparently by producing an intermediate not easily formed from *n*-pentane. Large paraffins also aged the catalyst rapidly. From these experiments they suggested the following aging mechanism. Unsaturated reaction intermediates such as monocyclic diolefins are formed first, mainly at the platinum sites. These initial foulant precursors are then reversibly adsorbed on the platinum and can therefore migrate to the acid sites, where they are also reversibly adsorbed. Just how the migration occurs is not clear, but gas-phase transfer is suggested. These precursors can then polymerize to form polycyclic compounds having several double bonds per molecule. This latter process is viewed as the rate-limiting step, being slower than the platinum-catalyzed reactions and also slower than the transport process transferring the precursors to the acid sites. The mechanism is consistent with the observation that reactivation can occur in either nitrogen or hydrogen atmospheres and that the rate of reactivation is faster in hydrogen. The model is summarized in Fig. 3-56.

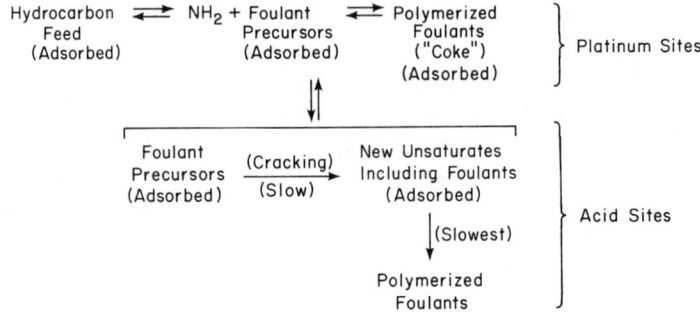

Fig. 3-56. Reversible reaction paths producing unsaturated foulants causing catalyst aging.

The introduction of the platinum–rhenium catalyst in 1968 (21) stimulated much research into trying to identify the precise role of rhenium in stabilizing this catalyst. In a somewhat oversimplified sense, the results of the bulk of this research can be summarized in terms of two hypotheses. The first envisions the role of rhenium as a simple scavenger. Thus, according to the suggestion of Bertolacini and Pellet (3) the rhenium acts to remove some precursor from the reaction system. The strongest evidence for this mechanism is that mechanical mixtures of platinum/alumina catalysts have about the same deactivation behavior as a platinum–rhenium/alumina catalyst. This observation appears to preclude synergistic effects between the platinum and the rhenium. A German patent issued to Roberts and Westfield (36) reveals that the activity as measured by an increase in octane number and stability of a physical mixture of Pt/alumina and Re/alumina catalysts was superior to the activity of either the Pt/alumina or the bimetallic Pt-Re/alumina when used to process heavy naphthas under normal reforming conditions. Also, Mieville (24) found that temperature-programmed reduction (TPR) experiments on physical mixtures of Pt/alumina and Re/alumina catalysts gave results similar to those for their bimetallic counterpart only if the catalysts were very finely ground.

A second hypothesis was suggested by Biloen *et al.* (4, 38) whose experiments indicated that rhenium atoms, when sulfided, divide the metal surface on platinum catalysts into ensembles containing smaller numbers of platinum atoms than on Pt alone. The resulting catalyst has a high selectivity for mild dehydrogenation. Obviously, this hypothesis requires that platinum and rhenium be either alloyed, in the form of bimetallic clusters, or in some other way interacting synergistically. The fact that other metals such as gold, iridium, and tin behave in a similar fashion supports this interpretation. Additional evidence is supplied by results of Shum *et al.* (39) for *n*-hexane and *n*-heptane conversion. The basic question is whether the platinum and

V. Fouling of Reforming Catalysts

rhenium exist separately or in close proximity on the surface of the catalyst. Central to this is the further question that, even if it is possible to produce platinum–rhenium in close proximity, is it possible that subsequent treatment can cause segregation of the two? Bolivar et al. (7) used temperature-programmed reduction to study the mobility of ReO on the surface of the catalyst. They suggested that water of hydration influences the migration rate. These ideas have been further discussed by Wagstaff and Prins (43) and by Isaacs and Petersen (19) based on additional temperature-programmed reduction studies.

Two mechanisms to explain the possible effects of degree of hydration are that it affects the rate of hydrogen spillover and that it influences the actual mobility of the ReO species on the surface. One can hardly be enthusiastic about either of these; certainly Isaacs and Petersen argue that spillover rates are too low to account for the phenomenon. In sum, the TPR results are strong circumstantial evidence for platinum and rhenium being in close proximity on the surface of the bifunctional catalyst.

Additional evidence for proximate interactions is supplied by the work of Isaacs and Petersen using the chemisorption of hydrogen and oxygen as well as the method of metal titrations. They showed that the behavior of a physical mixture of platinum/alumina and rhenium/alumina is not the same as that of the bimetallic platinum–rhenium/alumina catalyst and that there is direct evidence for the two metals being in close proximity. This is supported by the studies of Shum et al., who showed that for n-hexane conversion at 500°C and 1 atm (H/HC = 11/1) coimpregnated Pt–Re/alumina, unlike a physical mixture, displayed a very high selectivity for cracking, which is characteristic of alloy formation. Typical activity versus time plots are shown in Fig. 3-57 for the bimetallic catalyst, both sulfided and unsulfided, and are to be compared with similar data for Pt/alumina

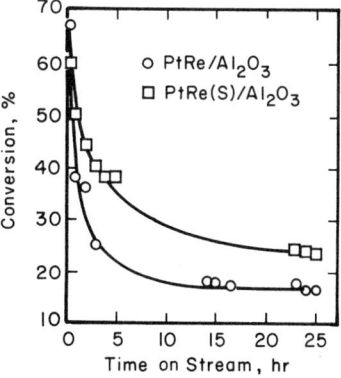

Fig. 3-57. Total conversion of n-hexane versus time on stream for unsulfided and pre-sulfided Pt–Re/alumina at 500°C. [From Shum et al. (39). Reproduced by permission of Academic Press, Inc.]

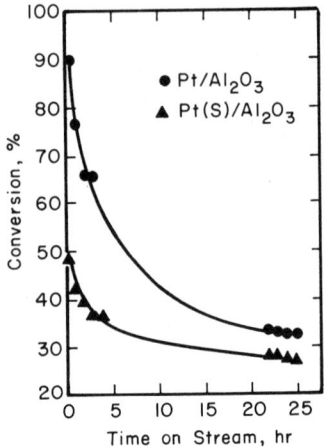

Fig. 3-58. Total conversion of n-hexane versus time on stream for unsulfided and presulfided Pt/alumina at 500°C. [From Shum et al. (39). Reproduced by permission of Academic Press, Inc.]

shown in Fig. 3-58. It is seen that, in fact, the activity of the bimetallic is lower; however, this is stabilized by 25 hr time on stream, while the platinum is still decaying at a noticeable rate. The effects of poisoning by sulfur and deactivation by fouling are in competition such that after some time on stream the sulfided Pt-Re displays significantly higher activity than the unsulfided material, while the opposite is the case for Pt alone. In both cases the selectivity for hydrogenolysis is significantly decreased on sulfidation. Similar experiments with these catalysts at 15 atm yielded generally corresponding trends although with much lower rates of deactivation, as might be expected. It was concluded that the *combined* action of Re and S is responsible for the differences in performance between sulfided Pt/alumina and Pt-Re/alumina. On the basis of thermodynamic data (38) it is expected that sulfur adsorbed on the Pt-Re surface should be preferentially bound to the rhenium, thus causing changes in the selectivity patterns characteristic of a reduction in the size of Pt ensembles. The superior activity maintenance of the sulfided Pt-Re/alumina is consistent with a model which assumes that sulfur fixed on Re, as shown in Fig. 3-59, impedes the reorganization of carbonaceous fragments into pseudographitic entities that cause irreversible deactivation of the platinum function.

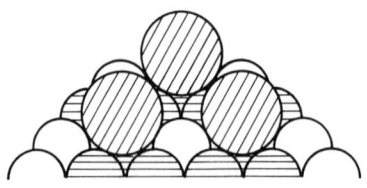

Fig. 3-59. Schematic model of sulfided Pt-Re alloy surface with S(ads) on Re. (○) Platinum; (◐) rhenium; (◍) sulfur. [From Shum et al. (39). Reproduced by permission of Academic Press, Inc.]

V. Fouling of Reforming Catalysts

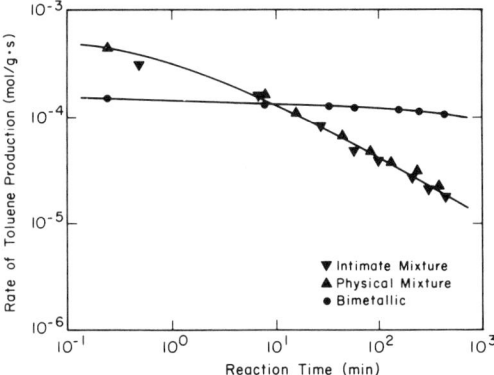

Fig. 3-60. Deactivation profiles for Pt and Pt–Re catalysts, methylcyclohexane dehydrogenation. [From Pacheco and Petersen (30). Reproduced by permission of Academic Press, Inc.]

Another set of reaction experiments indicating synergism between platinum and rhenium has been reported by Pacheco and Petersen (30), who compared differences in characteristic deactivation curves for Pt/alumina and Pt–Re/alumina catalysts for the dehydrogenation of methylcyclohexane. Figure 3-60 shows the behavior of sulfided catalysts on a real-time abscissa, while Fig. 3-61 shows the characteristic curves on

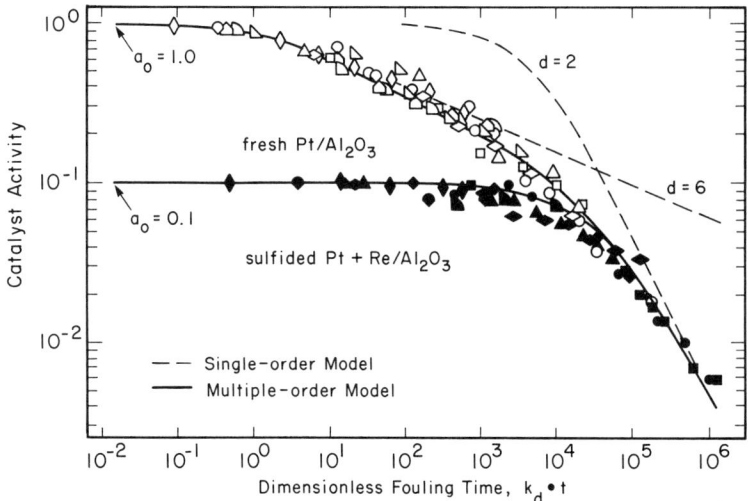

Fig. 3-61. Dimensionless fouling curves for Pt and Pt–Re catalysts, methylcyclohexane dehydrogenation. [From Pacheco and Petersen (30). Reproduced by permission of Academic Press, Inc.]

dimensionless coordinates. In terms of the interpretation presented in Section IV, it appears that the platinum catalyst follows the general correlation, whereas sulfiding appears to have lowered the initial activity of the Pt–Re catalyst more than the Pt catalyst and it deactivates essentially with a slope corresponding to a two- to three-site mechanism. This is consistent with the conjectures of Biloen *et al.* and Shum *et al.* on the formation of smaller ensembles, greatly suppressing the fouling via a six-site mechanism. The characteristic deactivation curves for a physical mixture of the Pt/alumina and Re/alumina catalysts, the Pt/alumina, and the bimetallic Pt–Re alumina were not the same, as seen in Fig. 3-60.

The apparent contradictions among the above experiments in establishing the role of rhenium and other additives to catalysts to stabilize them must await further experiments and interpretations for resolution. The list seems inexhaustible. For example, some hold the view that the stabilization by rhenium is at least partly due to a fraction of incompletely reduced metal, which would surely have a different interaction with sulfur than as pictured for Re^0. Whatever the complete picture may be, from a technological viewpoint the widespread use of catalysts with more than one metallic constituent is a testimonial to their enhanced stability.

VI. SUMMARY AND EVALUATION

The deposition of coke, which we have referred to generally as fouling, is a near-universal phenomenon in most catalytic reactions involving carbon-containing compounds. As perhaps can be inferred from the presentation in this chapter, and particularly by comparison with the material to follow in the next chapter, the chemistry of coke formation and the chemistry of the foulant deposited are not as well defined as that in most cases of impurity poisoning. It has been said that there are as many different types of coke as there are reactions in which it is formed—perhaps an overgeneralization, but not without a grain of truth.

One sees at the beginning of this chapter the heavy reliance of early work on various time on stream correlations. While these in general have been quite successful and are still used in many instances (aside from the examples here one should also refer to Chapter 12), they in large part ignore the chemistry of coke formation. One then sees in subsequent work increasing interest in coke chemistry, dealing with the indentification of significant foulant precursors, as in the investigation of Appleby *et al.* Such identification then provides the basis for more generalized ideas concerning the possible mechanisms of coke formation, in the sense of both appropriate carbonium ion chemistry, again from Appleby *et al.*, and overall kinetic networks, as envisioned in the multiplet fouling model of Pacheco and Petersen.

It is important to realize, however, that in a particular process or reaction environment the chemistry of coke formation, or at least the nature of the foulant, can be evolutionary with time. In the examples of both fouling of reforming catalysts, presented in this chapter, and deactivation of hydrotreating catalysts, which awaits the reader in Chapter 7, there is an evolution of foulant structure with time on stream toward more refractory graphitic structures. Knowledge of this allows one to devise strategies to cope with catalyst fouling in some significant way. For example, it is proposed here that one of the important features of M–Pt reforming catalysts is that addition of the second metallic function in effect reduces the size of Pt ensembles on the surface that are required for the formation of graphitic carbon, which is the major long-term foulant. Thus, while the total amount of foulant may not be substantially affected, its chemical nature is. Again, this highlights a weakness of the time on stream correlations, since even though a relation between amount of foulant and time is obtained, it is necessary further to obtain a correlation between amount deposited and activity that may itself vary with time. Clearly, a more fundamental approach is required.

In addition to these questions concerning correlations versus chemistry, we have seen that coke deposition can have significant effect on the transport processes occurring within porous catalyst particles. While the detailed consideration is reserved to Chapter 7, one cannot avoid the fact that many apparent effects of foulant deposition are as much related to physical as to chemical phenomena. Pore blocking, as demonstrated by Levinter *et al.*, can be a major factor. Indeed, one can argue in many cases that it is not useful to worry in detail about the chemistry of foulant deposition when the majority of active surface is denied to both reactant and coke precursor and the surface reaction mechanism has no opportunity to become a significant factor. One sometimes hears the statement that "coke begets coke," which in this case may not be far from the truth. So-called shell processes are important in both fouling and regeneration, with one process being nearly the mirror image of the other.

Although we have tried to be internally consistent in our presentation, those interested in the interaction between chemical and physical phenomena in coke deposition/regeneration will find that some iteration among chapters 3, 7, and 9 is inevitable, but we hope not indigestible.

REFERENCES

1. G. T. Androutsopoulos and R. Mann, *Chem. Eng. Sci.* **33**, 673 (1978).
2. W. G. Appleby, J. W. Gibson, and G. M. Good, *Ind. Eng. Chem. Process Des. Dev.* **1**, 102 (1962).
3. R. J. Bertolacini and R. J. Pellet, *Stud. Surf. Sci. Catal.* **6**, 73 (1980).

4. P. Biloen, J. N. Helle, H. Verbeek, F. M. Dautzenberg, and W. M. H. Sachtler, *J. Catal.* **63**, 112 (1980).
5. F. H. Blanding, *Ind. Eng. Chem.* **45**, 1186 (1953).
6. R. W. Blue and C. J. Engle, *Ind. Eng. Chem.* **43**, 494 (1951).
7. C. Bolivar, H. Charcosset, R. Frety, M. Primet, L. Tournayan, C. Betizeau, G. Leclerq, and R. Maurel, *J. Catal.* **39**, 249 (1975).
8. J. B. Butt, S. Delgrado-Diaz, and W. E. Muno, *J. Catal.* **37**, 158 (1975).
9. T. S. Chou and L. L. Hegedus, *AIChE J.* **24**, 255 (1978).
10. P. E. Eberly, Jr., *Ind. Eng. Chem. Process Des. Dev.* **8**, 25 (1969).
11. P. E. Eberly, Jr., C. N. Kimberlin, W. H. Miller, and H. V. Drushel, *Ind. Eng. Chem. Process Des. Dev.* **5**, 193 (1966).
12. F. Y. A. El-Kady and R. Mann, *J. Catal.* **69**, 147 (1980).
13. G. F. Froment, *Stud. Surf. Sci. Catal.* **6**, 1 (1980).
14. B. C. Gates, J. R. Katzer, and G. C. A. Schuit, "Chemistry of Catalytic Processes," p. 43. McGraw-Hill, New York, 1979.
15. R. G. Haldeman and M. C. Botty, *J. Phys. Chem.* **63**, 489 (1959).
16. H. M. Heinemann, *Ind. Eng. Chem.* **43**, 2098 (1951).
17. V. C. F. Holm and R. W. Blue, *Ind. Eng. Chem.* **43**, 501 (1951).
18. O. A. Hougen and K. M. Watson, "Chemical Process Principles," Vol. 3, p. 936. Wiley, New York, 1947.
19. B. H. Isaacs and E. E. Petersen, *J. Catal.* **85**, 1, 8 (1984).
20. L. D. Jossens and E. E. Petersen, *J. Catal.* **76**, 265 (1982).
21. H. E. Kluksdahl, U.S. Patent 3,415,737 (1968).
22. M. E. Levinter, G. M. Panchekov, and M. A. Tanatarov, *Int. Chem. Eng.* **7**, 23 (1967).
23. E. L. Mackor, A. Hofstra, and J. H. van der Waals, *Trans. Faraday Soc.* **54**, 66 186 (1958).
24. R. H. Mieville, *J. Catal.* **87**, 437 (1984).
25. G. A. Mills, E. R. Boedeker, and A. G. Oblad, *J. Am. Chem. Soc.* **72**, 1554 (1950).
26. C. G. Myers, W. H. Lang, and P. B. Weisz, *Ind. Ing. Chem.* **53**, 299 (1961).
27. Y. Ozawa and K. B. Bischoff, *Ind. Eng. Chem. Process Des. Dev.* **7**, 67 (1968).
28. M. A. Pacheco and E. E. Petersen, *J. Catal.* **86**, 75 (1984).
29. M. A. Pacheco and E. E. Petersen, *J. Catal.* **88**, 400 (1984).
30. M. A. Pacheco and E. E. Petersen, *J. Catal.* **96**, 499 (1985).
31. C. D. Plank and D. M. Nace, *Ind. Eng. Chem.* **47**, 2374 (1955).
32. C. D. Prater and R. M. Lago, *Adv. Catal.* **8**, 298 (1956).
33. I. H. Ramser and B. P. Hill, *Ind. Eng. Chem.* **50**, 117 (1958).
34. J. T. Richardson, *Ind. Eng. Chem. Process Des. Dev.* **11**, 8 (1972).
35. J. T. Richardson, *Ind. Eng. Chem. Process Des. Dev.* **11**, 12 (1972).
36. G. W. Roberts and K. E. Westfield, File No. P26 27 622.8 (1977).
37. C. G. Rudershausen and C. C. Watson, *Chem. Eng. Sci.* **3**, 110 (1954).
38. W. M. H. Sachtler and P. Biloen, *Div. Pet. Chem., Prepr., Am. Chem. Soc.*, Seattle p. 482 (1983).
39. V. K. Shum, J. B. Butt, and W. M. H. Sachtler, *J. Catal.* **96**, 371 (1985); **99**, 126 (1986).
40. K. Suga, Y. Morita, E. Kunugita, and T. Otaki, *Int. Chem. Eng.* **7**, 742 (1967).
41. H. H. Voge, J. M. Good, and B. J. Greensfelder, *Proc. World Pet. Congr., 3rd* **4**, 124 (1951).
42. A. Voorhies, Jr., *Ind. Eng. Chem.* **37**, 318 (1945).
43. N. Wagstaff and R. Prins, *J. Catal.* **59**, 434 (1979).
44. D. E. Walsh and L. D. Rollman, *J. Catal.* **49**, 369 (1977).
45. P. B. Weisz and R. B. Goodwin, *J. Catal.* **2**, 397 (1963).

CHAPTER 4

Deactivation by Poisoning

> Some rare disease ... or maybe it's fleas ... ?
> *West Side Story*

Perhaps the major factor one encounters in trying to draw a distinction between the general characteristics of deactivation by coke formation or fouling and deactivation by poisoning is that the latter is a much more well-defined event from a chemical point of view. At least part of the discussion of the preceding chapter suggests that because of some vagueness in the detailed chemical processes involved in coke formation—in fact, considerable variation in the mechanisms occurring in different cases—the literature has offered until recently only many versions of time-on-stream correlations. These reflect at least partial ignorance of the detailed chemistry of coke formation and imply conversion independence of the rate of deactivation, which certainly cannot be true in general. Now as we have said above, and in some detail in Chapter 1, poisoning is a much more well-defined chemical event and in most instances can be regarded as a chemisorption process. It is competitive in nature in that the poison is able to be preferentially chemisorbed on the catalytic surface in the presence of (often) considerable excess of reactant or product, and the chemisorption may be reversible or irreversible in nature, dependent on the particular elements (compounds) and the conditions of temperature and concentration involved. We remind the reader at this point to note that the classification of poisoning set forth in Chapter 1 excluded reaction rate inhibition terms appearing in the denominator of Longmuir-Hinshelwood-Hougen-Watson (LHHW) rate equations when these terms refer to reactants or products of the main reaction. Of course, modeling the effect of a reversible poison on reaction kinetics via inhibition terms in LHHW equations is a legitimate approach and this is not excluded from consideration.

Finally, we also remind the reader of the distinctions pointed out between selective and nonselective poisoning, the former often being associated with chemisorption site strength distributions on the surface. However, the description is not that simple in general, and we shall see later that it is possible for the same poison on the same surface to be selective for one reaction and nonselective for another.

I. SOME BEGINNING IDEAS: A GENERAL DISCUSSION

In an early review (67) a number of typical examples were selected to illustrate a general approach to the analysis of the kinetics of deactivation processes. Some of these were cited previously in Table 1-2, including processes in addition to poisoning. If we select from available reviews (16, 67) what might be considered classical examples of early work done in specific regard to poisoning, we might fashion a table such as Table 4-1 (which is partially redundant with Table 1-2). The list is naturally not comprehensive, but one sees in each case a well-defined chemical interaction between catalytic surface and poison.

TABLE 4-1

Some Early Examples of Catalyst Poisoning Investigations

System	References[a]
Liquid-phase hydrogenations with Pt poisoned either by metals such as As (from arsine) or by nonmetallic elements such as O or S contained within nonshielded compounds	A
Para-hydrogen conversion on W poisoned by O_2	B
Hydrogenation of ethylene on Cu poisoned by CO	C
Hydrogenolysis of cyclopropane on Pt poisoned by AsH_3	D
Cracking of cumene on SiO_2 or SiO_2/Al_2O_3 poisoned by basic nitrogen compounds, quinoline, quinaldine, and so forth	E
Isomerization of cyclohexane, isomerization of 3,3-dimethyl-1-butene, and dehydration of 1-butanol on Al_2O_3 poisoned by Na and K	F

[a] A, E. B. Maxted, *Adv. Catal.* **3**, 129 (1951); B, D. D. Eley and E. K. Rideal, *Proc. R. Soc. London, Ser. A* **178**, 429 (1941); C, R. N. Pease and L. Stewart, *J. Am. Chem. Soc.* **47**, 1235 (1925); D, R. D. Clay and E. E. Petersen, *J. Catal.* **16**, 32 (1970); E, G. A. Mills, E. R. Boedeker, and A. G. Oblad, *J. Am. Chem. Soc.* **72**, 1554 (1950); F, H. Pines and W. O. Haag, *J. Am. Chem. Soc.* **82**, 2471 (1960).

I. Some Beginning Ideas: A General Discussion

A. Metals

The susceptibility of various metals to poisoning was classified long ago by Maxted (46), and one may summarize the findings by a truism: all metals that are good catalysts are susceptible to poisoning. Table 4-2 gives a

TABLE 4-2

Some Reactions Catalyzed by Metals[a]

Reaction	Metals known to show catalytic activity	Examples of metals with high activity
Hydrogen–deuterium exchange	Most transition metals: some nontransition metals >600 K	W, Pt
Deuterium-saturated hydrocarbon exchange		
(a) involving σ bonding to surface	Most transition metals	W, Rh
(b) involving π allyl bonding to surface	Pd, Pt, Ni, Rh, W	Pd
Hydrogenation of alkenes	Most transition metals, Cu	Rh, Ru, Pd, Pt, Ni
Deuterium–alkene exchange	Most transition metals, Cu, Au	W, Rh, Pd
Double bond shift	Most group VIII metals	Pd
Hydrogenation of alkynes	Most group VIII metals, Cu	Pd
Hydrogenation of aromatics	Most group VIII metals, W, Ag	Pt, Rh, Ru, W, Ni
Deuterium–aromatic exchange	Most group VIII metals, W, Cu, Ag	W, Pt
Hydrogenolysis of C–C bonds	Most transition metals	Os, Ru, Ni
Skeletal isomerization of hydrocarbons	Pt, Ir, Pd, Au	Pt
Deuterium exchange with NH_3, $-NH_2$, H_2O, $-OH$	Pt, Rh, Pd, Ni, W, Fe, Ag	Pt
Hydrogenolysis of C–N bonds	Most transition metals, Cu	Ni, Pt, Pd
Hydrogenolysis of C–O bonds	Most transition metals, Cu	Pt, Pd
Hydrogenation of carbonyl group	Pt, Ni, Fe, W, Pd, Au	Pt
Hydrogenation of carbon monoxide	Most group VIII metals, Cu, Ag	Fe, Co, Ru—(Fischer–Tropsch); Ni—(methanation)
Hydrogenation of carbon dioxide	Co, Fe, Ni, Ru	Ru, Ni
Hydrogenation of nitrogen oxides	Most platinum group metals	Ru, Pd, Pt
Hydrogenation of nitro group	Most group VIII metals, Cu	Pt, Pd, Ni
Hydrogenation of nitriles	Co, Ni	Co, Ni

(*continues*)

TABLE 4-2 (*continued*)

Reaction	Metals known to show catalytic activity	Examples of metals with high activity
Hydrogenation of nitrogen (ammonia synthesis)	Fe, Ru, Os, Re, Pt, Rh (Mo, W, U, probably as nitrides)	Fe
Ammonia decomposition	Most transition metals, group IB	Pt
Dehydrogenation, cyclization, aromatization of hydrocarbons	Most group VIII metals	Pt
Decomposition of alcohols	Most transition metals, group IB	Cu, Ni
Decomposition of formic acid	Most transition metals, group IB	Pt, Ir
Oxidation of hydrogen	[b] Platinum group metals, Au	Pt
Oxidation of hydrocarbons		
(a) ethylene to ethylene oxide	Ag	Ag
(b) oxidation of other hydrocarbons	[b] Platinum group metals, Ag	Pd, Pt
Oxidation of carbon monoxide	[b] Platinum group metals	Pd, Pt
Oxidation of carbon monoxide with steam (water-gas shift)	Cu	Cu
Oxidation of ammonia	[b] Platinum group metals	Pt
Oxidation of sulfur dioxide	[b] Platinum group metals, Au	Pt
Oxidation of alcohols, aldehydes	[b] Platinum group metals, Ag, Au	Ag, Pt
Oxidation of methane with steam	Ni, Co, platinum group metals	Ni, Pt

[a] From J. R. Anderson, "Structure of Metallic Catalysts," Academic Press, New York, 1975. Reproduced by permission of Academic Press, Inc.

[b] Activity of metals outside the platinum group and noble metals is known, but under reaction conditions the metal oxide is the active catalyst.

classification of metal catalysts used for various reactions; for classification with respect to poisoning it will be more fruitful to concentrate on the poison classification than the catalyst classification. In part this is because *all* metals active in chemisorption and catalysis can be visualized as forming surface bonds via "dangling orbitals" of the surface metal atoms (10). If one assumes that the arrangement of orbitals is the same at the surface as in the bulk metal, then the surface metal atom, missing a certain number of nearest neighbors, will be coordinatively unsaturated and a number of both *dsp* orbitals (bonding orbitals in the bulk) and *d* orbitals (not involved in bulk bonding) will be available for surface bonding. Of course, the assumption that the orbital geometry of a surface atom is the same as that of a bulk atom is certainly incorrect in detail, but the qualitative picture

I. Some Beginning Ideas: A General Discussion

seems reasonable enough. Thus, if we change our classification from attention to the susceptibility of an individual metal to poisoning to the potential of various materials to be poisons, a rather simple picture emerges. Among nonmetallic poisons, molecules containing elements of groups Vb and VIb or the free elements (other than N_2) are potential poisons if they have the proper electronic configuration, as well as molecules containing multiple bonds such as CO, dienes, acetylenes, and aromatics. In the latter grouping, however, we must be attentive to our definition of poisoning, since many instances of strong adsorption of molecules with multiple bonds fall into the category of adsorption inhibition.

The "proper electron configuration" mentioned above also has a simple interpretation. For molecules containing group Vb or VIb elements, toxicity is related to the extent of bonding to other atoms in the molecule. If the potential poison atom has its normal valence orbitals saturated by bonding to other elements in the molecule, there is no toxic activity. However, if there are unshared electron pairs or unoccupied orbitals, then chemisorptive bonding via the dangling bonds of the metal is possible and the material is a poison. Maxted referred to a general class of such compounds as "shielded" or "unshielded" structures. For example, in hydrogenation reactions arsenic in the form of arsene

$$\begin{bmatrix} H : \ddot{A}s : H \\ \ddot{H} \end{bmatrix} \qquad \text{I}$$

is a strong poison for catalysts such as platinum. Yet no effect on catalytic activity is seen in the decomposition of hydrogen peroxide, presumably because the arsenate is formed under strong oxidizing conditions:

$$\begin{bmatrix} O \\ O : \ddot{A}s : O \\ \ddot{O} \end{bmatrix}^{3-} \qquad \text{II}$$

Arsenate, then, is a shielded compound and arsene an unshielded one in the terminology of Maxted. Table 4-3 gives a number of other examples demonstrating the influence of electronic configuration on toxicity. From these qualitative considerations, it is relatively easy to make some judgment as to whether a given compound has the potential to be a poison for a given catalyst. The question of how effective a poison may be is another matter. As is apparent from the discussion above, in somewhat roundabout manner, and as we explicitly stated in Chapter 1, catalyst poisoning is in fact a special case of parallel competitive chemisorption. The "strength" of a poison is thus directly related to the strength of the chemisorptive bond, and in turn the covalent chemisorptive bond should be describable in terms similar to those for bonds in covalent molecules. If we make the assumption

TABLE 4-3

Relationship of Electronic Structure to Toxicity

Toxic compounds—unshielded	Nontoxic compounds—shielded
Hydrogen sulfide, phosphine	Phosphate ion
Sulfite ion, selenite, tellurite	Sulfate ion, selenate, tellurate
Organic thiols	Sulfonic acid
Organic sulfides	Sulfone
Pyridine	Pyridinium ion
Piperidine	Piperidinium ion

of bonding to a *single* metal atom (not always the case, of course, but we have to start somewhere), some long-standing correlations have been used to estimate heats of chemisorption for the adsorption of small molecules such as hydrogen on metals. For example, the energy of a covalent bond between a surface metal atom and a hydrogen atom, $D(M-H)$, can be written in terms of individual bond energies $D(M-M)$ and $D(H-H)$ using a correlation proposed by Pauling:

$$D(M-H) = \tfrac{1}{2}[D(M-M) + D(H-H)] + (\chi_m - \chi_H)^2 \qquad (4-1)$$

where χ is the electronegativity. Unfortunately, the application of such methods to the more complicated molecules of Table 4-3 is not clear and alternative theoretical proposals rapidly transgress the boundary of our interests here. So we are left in a familiar position, relying on experience, intuition, and available data to make judgments on the strengths of various poisons. Fortunately, a relatively large body of information is available for heats of chemisorption of a number of molecules on metals, as summarized in Table 4-4. However, the adsorption behavior, and hence the poisoning behavior, of even simple molecules is often complicated by the existence of more than one binding state on the metal surface. A classical example is the chemisorption of CO on Pd, where both linear and bridged species can coexist (24). Multiple bonding states have now been reported for a very wide variety of systems, using diverse experimental techniques including temperature-programmed desorption, infrared spectroscopy, and low-energy electron diffraction. Thus, in addition to attempting classification of poisoning by the strength of the poison, it is also reasonable to classify according to the complexity of the phenomena involved (35). Such classification would extend to both metals and oxides and in ascending order of complexity would be categorized as poisoning of monofunctional catalysts with uniform sites, monofunctional catalysts with site strength distribution, multifunctional catalysts, and catalysts where the support surface induces behavior not seen on the unsupported material. We shall

TABLE 4-4

Heats of Adsorption of Gases on Metals[a]

Metals	Adsorbate[b,c]						
	H_2	O_2	N_2	CO	CO_2	CH_4, C_2H_6	C_2H_4
Group IA							
Li	—	a	—	—	—	—	—
Na	—	a	—	—	—	—	—
K	b	a	b	b	—	(b)	b
Rb	—	a	—	—	—	—	—
Cs	—	a	—	—	—	—	—
Group IIA							
Be	—	a	—	—	—	—	—
Ng	—	a	—	—	—	—	—
Ca	a	a	a	a	—	—	—
Sr	—	a	a	a	—	—	—
Ba	a	a	a	a	—	—	—
Group IIIA							
B	—	—	—	—	—	—	—
Al	b	525	b	—	—	—	—
Ga	—	a	—	—	—	—	—
Ir	b	a	b	b	—	—	b
Tl	—	a	—	—	—	—	—
Group IVA							
C	—	400	—	—	—	—	—
Si	—	870	—	—	—	—	—
Ge	a	550	—	—	—	—	—
Sn	b	a	b	b	—	(b)	b
Pb	b	a	b	b	—	b	b
Group IIIB							
Sc	—	(a)	—	—	—	—	—
Y	—	(a)	—	—	—	—	—
La	(a)	a	a	a	—	—	—
Group IVB							
Ti	a	990	a	640	787	(a)	a
Zr	a	a	630	620	a	a	a
Hf	(a)	(a)	(a)	(a)	(a)	(a)	(a)
Group VB							
V	a	a	(a)	(a)	(a)	a	(a)
Nb	a	870	a	553	626	(a)	(a)
Ta	188	890	585	560	750	a	580
Group VIB							
Cr	189	730	(a)	a	a	a	a
Mo	170	755	272	310	449	a	a
W	184	845	490	420	504	a	420
Group VIIB							
Mn	a	630	(a)	326	260	a	a
Tc	—	—	—	—	—	—	—
Re	a	a	>210	a	(a)	a	a

(*continues*)

TABLE 4-4 (*continued*)

Metals	Adsorbate[b,c]						
	H_2	O_2	N_2	CO	CO_2	CH_4, C_2H_6	C_2H_4
Group VIII(1)							
Fe	142	570	168	192	280	a	285
Ru	≈118	a	—	a	b	a	a
Os	a	a	(a)	a	—	(a)	(a)
Group VIII(2)							
Co	101	420	—	197	152	a	a
Rh	117	503	b	193	b	a	a
Ir	109	a	243	a	(b)	a	a
Group VIII(3)							
Ni	155	500	b	176	222	a	244
Pd	117	294	b,*	180	b	a	a
Pt	109	294	b	201	b	a	a
Group IB							
Cu	34;b	a	b	39	b	b	76
Ag	b	a	b	b	b	b	36
Au	b	b	b	36	(b)	—	87
Group IIB							
Zn	b	a	b	b	b	b	b
Cd	b	a	b	b	b	b	b
Hg	—	a	—	—	—	—	—

[a] From J. R. Anderson, "Structure of Metallic Catalysts," Academic Press, New York, 1975. Reproduced by permission of Academic Press, Inc.

[b] Notes: a, chemisorption occurs; b, no chemisorption at 273 K; (a), (b), reasonably reliable estimate; *, weak molecular chemisorption on extremely small nickel crystallites.

[c] Numerical values are heats of chemisorption (kilojoules per mole) at low coverage on polycrystalline surfaces at about room temperature.

in general follow this classification, since increasing degrees of complexity of poisoning systems lead us in the overall direction from nonselective to selective deactivation.

B. Oxides

If we turn now to the earlier work on poisoning of *oxides*, classical examples have already been given in Chapter 1 in the discussion of the titration methods for identification of surface chemical sites. As pointed out there, many oxides possess surface acidity and thus are active in acid-catalyzed reactions such as hydrocarbon isomerization. In consequence, any basic material will act as a poison to the acidic function of the surface, and we might expand the poison classification here to include

I. Some Beginning Ideas: A General Discussion

basic impurities in the oxide (alkali metals for example) as well as the chemisorption of basic molecules, since minor impurity levels are often encountered in materials such as alumina or silica/alumina.

A prototype study of poisoning of acidic oxides was conducted by Mills and co-workers (50, 52) on the poisoning of cracking catalysts with nitrogen-containing organic molecules. Even though this work was carried out many years ago, it established much of the basis even for present understanding of the poisoning of acidic surfaces and, further, the nature of the acidic sites on such surfaces. In one set of experiments comparing the uptake of quinoline on silica and silica/alumina catalysts of similar surface area, a reversible chemisorption was demonstrated to occur for the former, but a certain additional irreversible chemisorption occurred on the latter. Further observations of this sort over the years have led to the distinctions between Lewis acidity and Brønsted acidity discussed in Chapter 1 and the development of titration methods for identification of surface sites.

In an earlier part of this section, we attempted to identify the strength of a poison for a metal catalyst via its heat of chemisorption—with less than complete success. With the class of acidic oxide catalysts studied by Mills *et al.*, a slightly different classification of poisoning ability bears fruit. This is best illustrated in Fig. 4-1 for the poisoning of an SiO_2-Al_2O_3 catalyst by various basic organic nitrogen compounds for cumene cracking at 425°C and 1.5 liquid hourly space velocity (LHSV). Ranked in order of their effectiveness, they are quinaldine > quinoline > pyrrole > piperidine > decylamine > aniline. This is not quite the same order as their relative basicity; however, when correction is made for the extent of cracking of

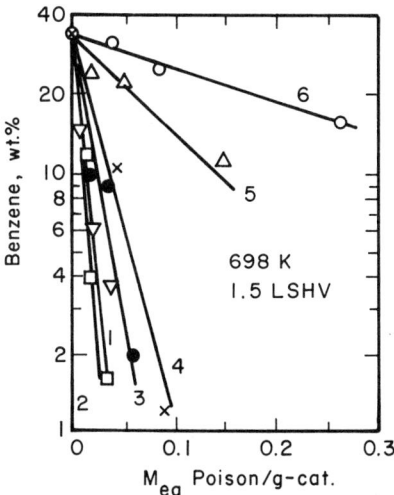

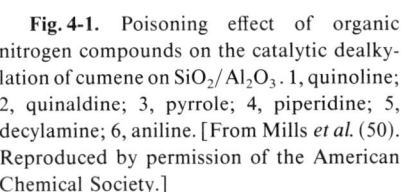

Fig. 4-1. Poisoning effect of organic nitrogen compounds on the catalytic dealkylation of cumene on SiO_2/Al_2O_3. 1, quinoline; 2, quinaldine; 3, pyrrole; 4, piperidine; 5, decylamine; 6, aniline. [From Mills *et al.* (50). Reproduced by permission of the American Chemical Society.]

the individual poison molecules under experimental conditions the correlation becomes direct. Hence basicity is a measure of poison strength in this type of system.

An interesting corollary to these results is that it was possible to develop a predictive correlation for the cracking activity of a number of different catalysts (representing the state of the art in 1950) simply by measuring the extent of quinoline chemisorption at set conditions. This we now know to be the effect of titration of surface acidity by quinoline, which remained in a stable configuration on the surface under the conditions of the experimental activity tests. One final nugget of information contained in the work of Mills *et al.* lends additional credence to the concept of individual acidic sites as set forth previously. Assigning an area of 36 Å^2 to the flat-lying quinoline molecule (and the area is probably less than this), one calculates that only about 4% of the total surface area is covered under conditions where the total cracking activity is decreased by a factor of seven. Thus the vast majority of surface area apparently has nothing to do with catalytic activity.

II. POISONING OF NONUNIFORM SURFACES: TRUE AND APPARENT

Lest we become too attached by the appealingly simple correlation of Mills *et al.*, let us step up the scale of complexity mentioned before. In this exercise we shall vary the intrinsic acidity of the catalyst by incorporation of different amounts of alkali metal in the oxide and study relative effects on reactions requiring different degrees of acidity for catalysis. This would in effect then be looking at monofunctional catalysts with site strength distributions. To the point is the work of Pines and Haag (57) on the isomerization of cyclohexane (CH) and 3,3-dimethyl-1-butene (3,3-DMB) and the dehydration of 1-butanol on a series of modified aluminas. The individual steps involved in these reactions are shown in Fig. 4-2, designated as I, II, and III, respectively. Carbonium ion mechanisms occur in all these reactions and the relative stability of these ions can be used as a measure of the acid strength required to effect catalysis of a given reaction. Hence the CH isomerization (2° → 1°) should be more difficult to carry out on a given surface than 3,3-DMB (IIa) isomerization, where the 2° → 3° rearrangement is involved. Subsequent steps in the 3,3-DMB sequence of II are also of interest. Step IIb, yielding 2-methylpentene from 2,3-DMB, is a 3° → 1° transformation and would proceed more slowly than the first step. Similar deductions can be made concerning IIc and IId as shown; the point is that it is not only the total reaction of 3,3-DMB which can be used to indicate

II. Poisoning of Nonuniform Surfaces: True and Apparent

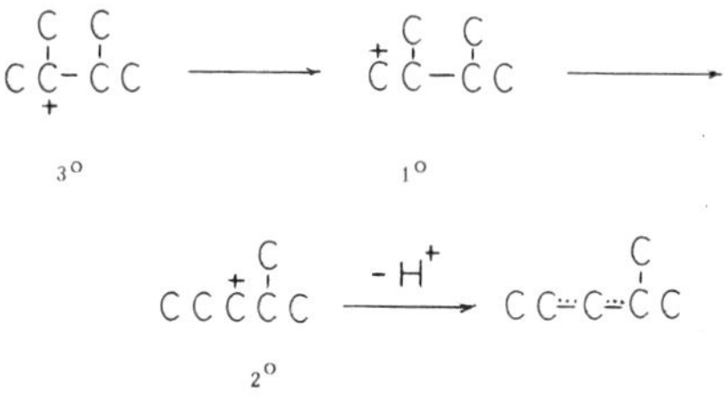

I

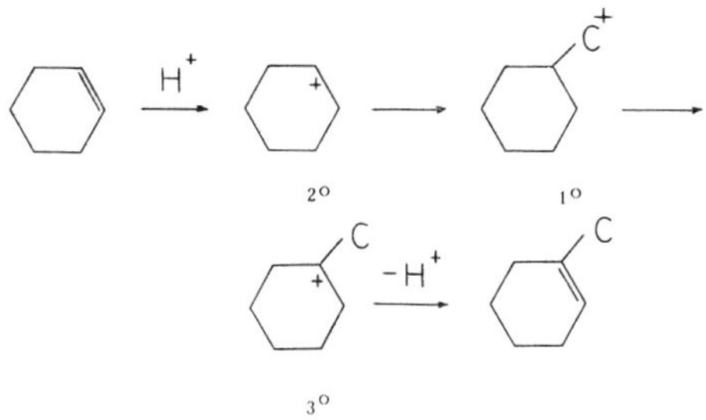

$$\text{cyclohexene} \xrightarrow{H^+} \text{(cyclohexyl cation, 2°)} \longrightarrow \text{(methylcyclohexyl cation, 1°)} \longrightarrow$$

$$\text{(methylcyclohexyl cation, 3°)} \xrightarrow{-H^+} \text{(methylcyclohexene)}$$

II (a)

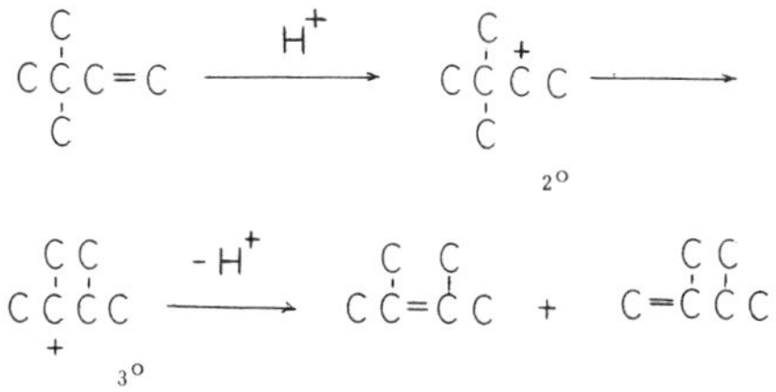

$$\underset{C}{\overset{C}{C\,C\,C=C}} \xrightarrow{H^+} \underset{C}{\overset{C}{C\,C\,\overset{+}{C}\,C}} \;\; (2°) \longrightarrow$$

$$\underset{+}{\overset{C\;C}{C\,C\,C\,C}} \;\; (3°) \xrightarrow{-H^+} \overset{C\;\;C}{C\,C=C\,C} + \overset{C\;C}{C=C\,C\,C}$$

(b)

$$\underset{+}{\overset{C\;\;C}{C\,C-C\,C}} \;\; (3°) \longrightarrow \overset{+\;C\;\;C}{C\,C-C\,C} \;\; (1°) \longrightarrow$$

$$\underset{}{\overset{C}{C\,C\,\overset{+}{C}\,C\,C}} \;\; (2°) \xrightarrow{-H^+} \overset{C}{C\,C=C=C\,C}$$

(c)

$$\overset{+\,C}{C\,C\,C\,C\,C} \;\; (2°) \longrightarrow \overset{C\,+}{C\,C\,C\,C\,C} \;\; (2°) \longrightarrow \overset{C}{C\,C\,C=C\,C}$$

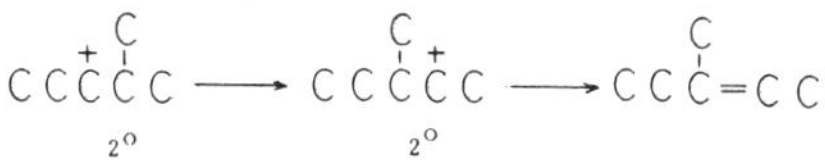

(d)

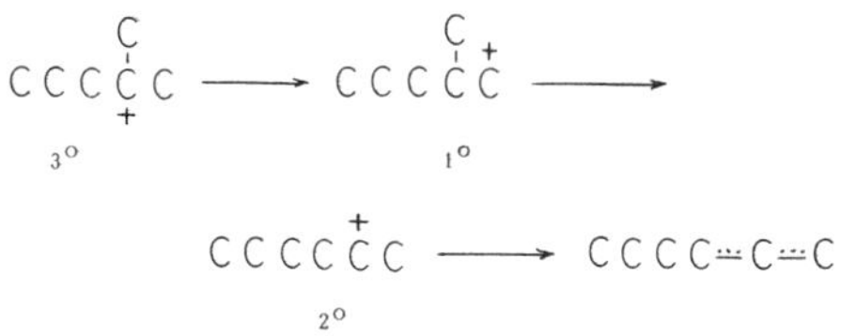

$$\underset{+}{\overset{C}{C\,C\,C\,C\,C}} \;\; (3°) \longrightarrow \overset{C\,+}{C\,C\,C\,C\,C} \;\; (1°) \longrightarrow$$

$$\overset{+}{C\,C\,C\,C\,C\,C} \;\; (2°) \longrightarrow C\,C\,C\,C=C=C$$

III

$$CCCC-OH \longrightarrow HOH + C=CCC$$
$$+ CC=CC$$
$$+ \underset{C}{CC=C}$$

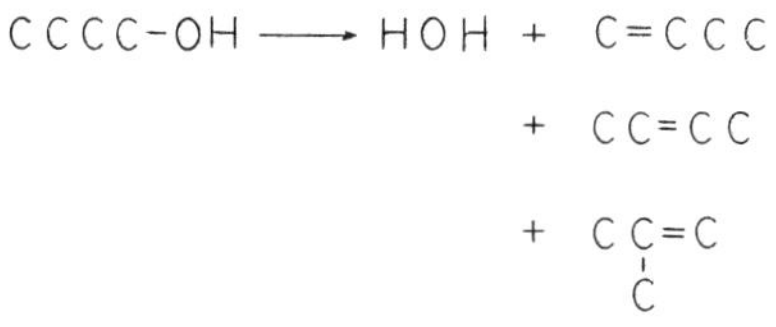

Fig. 4-2. Model reactions used to evaluate relative activities of aluminas.

some measure of total acidity but also the selectivity between the individual steps which can be used to give some indication of the distribution of acid strength. Rearrangements via 1° carbonium ions would occur only on relatively strong acid sites at measurable rates, 2° and 3° on both strong and weak sites. Finally, alcohol dehydration should also occur on both strong and weak acid sites.

The aluminas studied were prepared by varying techniques designed to produce different surface acidities. The major types were as follows (see Table 4-5).

(a) Pure—prepared from aluminum isopropoxide or from aluminum hydroxide and aluminum nitrate (method A or B).

(b) Impregnated—pure catalyst impregnated with NaCl or NaOH.

(c) Aluminate—alkali containing alumina precipitated from KOH solution (method C).

The incorporation of alkali in the impregnated or aluminate catalysts decreases the acidity of these materials relative to the isopropoxide preparations, hence the model reactions should be responsive to the different catalyst formulations.

TABLE 4-5

Methods of Preparing and Designating Alumina Catalysts[a]

Catalyst	Method of preparation	Temperature (°C)	Conditions [hr (atm)]	Na or (K) (wt. %)	Area (m^2/g)
1	A (pure)	400	4 (10 mm)	—	246
2	A	600	4 (10 mm)	—	147
3	A	700	4 (10 mm)	—	152
9-1	A impregnated with NaOH	600	4N$_2$	0.11	280
9-2	A impregnated with NaOH	600	4N$_2$	0.2	—
9-3	A impregnated with NaOH	600	4N$_2$	0.4	—
10-1	A impregnated with NaCl	600	4N$_2$	0.2	—
10-2	A impregnated with NaCl	600	4N$_2$	0.6	215
10-3	A impregnated with NaCl	600	4N$_2$	1.5	—
11(2)[b]	C (in NaOH)	600	4N$_2$	0.65	371
11(4)[b]	C (in NaOH)	600	4N$_2$	—	—
11(6)[b]	C (in NaOH)	600	4N$_2$	—	—
18	C (in KOH)	360	16N$_2$	—	—
19	C(7)[b] (in KOH)	360	16N$_2$	(0.09)	384
19[b]	C(7)[b] (in KOH)	700	4N$_2$	—	298

[a] From Pines and Haag (57). Reproduced by permission of the American Chemical Society.
[b] Number of washings after precipitation.

II. Poisoning of Nonuniform Surfaces: True and Apparent

The results (the relative catalytic activity here is based on conversion under set conditions rather than directly determined rates) for CH isomerization are in good agreement with expectations based on catalyst acidity. The pure isopropoxide catalysts demonstrated high activity for the isomerization, while the alkali-doped types had very little activity for CH isomerization. The 3,3-DMB isomerizations were also in line with expectation, although the results here were somewhat more complex. The overall degree of isomerization was higher than for CH for all catalysts, indicating that some acidity was effective in this reaction which did not participate in the CH isomerization, although the product distributions did vary considerably for the different materials. Butanol dehydration also occurs on all types at conversion levels above the CH isomerization, again with a wide variation in product distribution (between selective 1-butene or an n-butene mixture).

Generally, all the experimental results are in accord with the expectation that there exists a distribution of acid site strength on alumina, but this is far from all the information available to us from the study. What are the factors that affect this acid strength which so markedly affects the catalytic properties of an alumina? Surface hydration is one choice, since cracking catalysts require a small concentration of water for maximum activity. The CH isomerization was conducted with a series of isopropoxide catalysts heated to different levels before use, and the measured specific isomerization activity was determined as a function of calcination temperature, as shown in Fig. 4-3. Sintering resulting from the treatment at different temperatures is accounted for in the specific activities, so there appears to be a quite

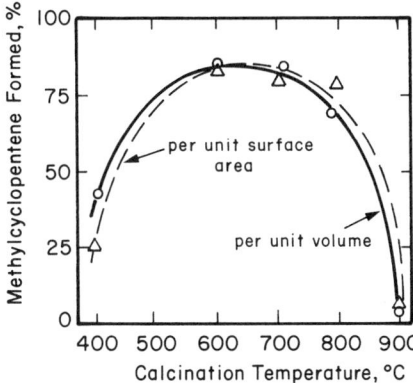

Fig. 4-3. Isomerization of cyclohexene [410°C, 2.0 liquid hourly space velocity (LHSV)] as a function of calcination temperature of alumina. Isomerization activity is shown per unit volume and per unit surface area. [From Pines and Haag (57). Reproduced by permission of the American Chemical Society.]

strong dependence of the catalytic properties (i.e., strongly acidic properties) on the calcination temperature. Since the latter would determine equilibrium surface water content, one infers that acidity, activity, and surface hydration are all related. Subsequent experiments of this sort gave similar results for the 3,3-DMB isomerization, hence the hydration effect appears related to total acidity. It is also interesting to note that very large amounts of water on the surface (lower calcination temperatures) act as a poison for the isomerization activity, while at the opposite end of the scale some water is required for catalytic activity. These notions of surface hydration are consistent with dissociative adsorption of water via hydroxylation of the surface aluminum and protonation of the oxide; the acidity of the surface arises from sites corresponding to partial hydration. Peri (56) has been able to use this as the basis for a Monte Carlo simulation of the surface hydration of alumina which results in a reasonable interpretation of observed infrared spectra in terms of particular groupings of partially hydrated sites.

A second factor affecting the acidity of the catalysts of Pines and Haag is certainly the amount of alkali incorporated into the impregnated or aluminate samples. Figure 4-4 gives some of the results of tests conducted on both CH and 3,3-DMB isomerizations and the butanol dehydration. In all cases the activity is diminished by the presence of the alkali, and the similarity in behavior for the various reactions again suggests that similar types of sites are responsible for the catalytic properties of the alumina. The aluminate catalysts are poisoned by alkali contents an order of magnitude smaller than those shown for the impregnated material; presumably this is related to the distribution of poison on the internal surface of the alumina as well as the mechanism of poisoning. The question of poison

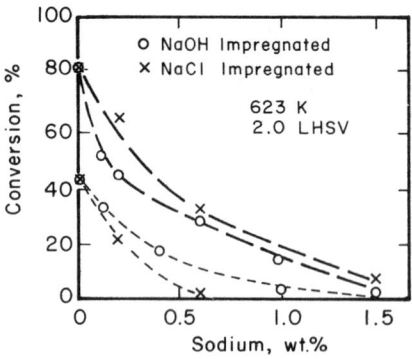

Fig. 4-4. Isomerization of cyclohexene and dehydration of 1-butanol over impregnated aluminas (350°C, 2.0 LHSV). (-----) Isomeration; (– – –) dehydration. [From Pines and Haag (57). Reproduced by permission of the American Chemical Society.]

distribution is one to which we shall return later in more detail with regard to simultaneous diffusion and deactivation phenomena.

In summary, high activity can result from either a large number of acid sites or strong acidity of the sites. The CH isomerization is a measure of strong acid sites, dehydration a measure of total acidity. Similarly, the extent and selectivity of the 3,3-DMB isomerization is a measure of total acidity (overall conversion), weak acidity (production of primary product 2,3-DMB), and strong acidity [further isomerizations to 2- and 3-methyl pentane (MP)]. Aluminas without alkali poisons have a strongly acidic surface, those prepared from the aluminate have no strong acid sites but are weakly acidic, while impregnation seems to deactivate both strong and weak sites indiscriminately. From further experiments on the adsorption of alkali, Pines and Haag obtain values of about 10^{13} strong acid sites per square centimeter (aluminate measurements) and about 7.5×10^{13} total sites involved in 1-butanol dehydration. Thus pure alumina has roughly 10^{14} acidic sites per square centimeter, with about 10% of this number effective in isomerization.

We recall that Mills *et al.* correlated the activity of alumina and silica-alumina with quinoline adsorption, and the amine index is another well-known means for determination of the acidity of solid surfaces. An interesting further result obtained by Pines and Haag is the relationship between amine index and catalytic activity. Apparently the index measures total acidity, and hence one can determine relative activities for a "family" of related aluminas, as shown in Figs. 4-5 to 4-7. However, if one attempts to compare activity for aluminas from different sources, no correlation is

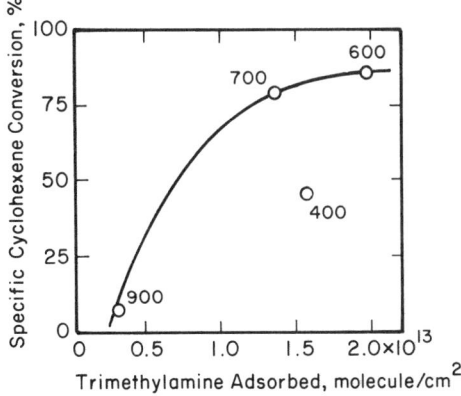

Fig. 4-5. Comparison of trimethylamine adsorption and isomerization of cyclohexene at 410°C, 0.5 LHSV. [From Pines and Haag (57). Reproduced by permission of the American Chemical Society.]

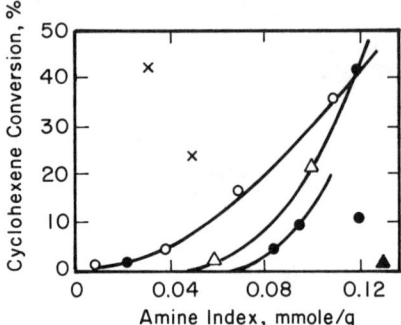

Fig. 4-6. Cyclohexene conversion at 350°C and 2.0 LHSV versus acidity. Aluminas: (○) NaOH-impregnated; (●) from potassium aluminate; (△) NaCl-impregnated; (▲) from sodium aluminate; (×) from isopropoxide. [From Pines and Haag (57). Reproduced by permission of the American Chemical Society.]

obtained, as shown by the differing curves in Fig. 4-6 for the three types of alumina in a strong acid reaction or the shotgun pattern of Fig. 4-8 for a total acid reaction. The fact that amine index is generally a satisfactory measure of cracking catalyst activity is at least indirect evidence of a relatively homogeneous acidity of the sites on silica–alumina.

The results with trimethylamine chemisorption on alumina are thus in good agreement with the reaction experiments. While elucidation of the detailed catalytic chemistry of alumina was not the object of the present discussion, it is interesting to note that much was learned in this regard and this was accomplished by specific studies of the poisoning behavior of the catalyst.

Another example, apparently involving site strength distributions, has a radically different origin. In this case we examine a surface that intrinsically has energetically equivalent sites and look at the interactions that occur as it is progressively poisoned. A classical discussion of this problem has been given by Herrington and Rideal (36) for multisite adsorption of reactant and poison. The poison is assumed to block a geometrically fixed number

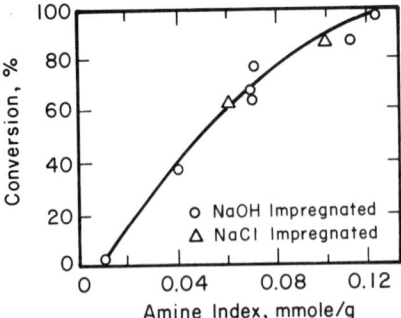

Fig. 4-7. Conversion of 3,3-dimethyl-1-butene at 350°C and 2.0 LHSV versus acidity. Aluminas: (○) NaOH-impregnated; (△) NaCl-impregnated. [From Pines and Haag (57). Reproduced by permission of the American Chemical Society.]

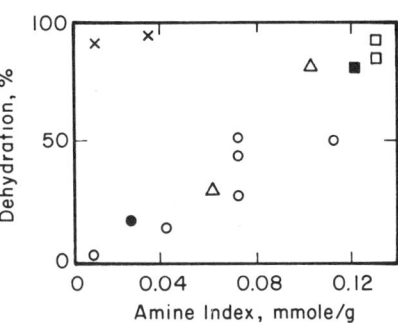

Fig. 4-8. Dehydration of 1-butanol at 350°C and 2.0 LHSV versus acidity. Aluminas: (×) from isopropoxide; (●) from potassium aluminate; (□) from sodium aluminate; (○) NaOH-impregnated; (△) NaCl-impregnated. [From Pines and Haag (57). Reproduced by permission of the American Chemical Society.]

of sites, in which case two types of poison–reactant interaction can be identified, shown as A and B on Fig. 4-9. In case A the sets of adsorption sites can overlap, while they are widely separated in case B. In the latter case no mutual interference is possible and as a result activity is directly proportional to the number of groups of active sites. For a surface coverage of $(1 - \theta_A)$ and poison occupying one active site, the number of unpoisoned groups will be proportional to θ_A^n if a group of n isolated centers is required for reaction.

In the more general case where adsorption sites are overlapping, the relationship between reaction rate and fraction of surface poisoned may become quite complicated. Herrington and Rideal investigated specific cases where the number of adsorption sites per reactant molecule can be $m = 1, 2,$ or 7.[1] Monte Carlo calculations were used to determine the percentage of surface covered by reactant as a function of percent coverage by poison for the three cases—specific to the (111) plane of a face-centered cubic (fcc) crystal. The basic results are in accord with expectation, as shown in Fig. 4-10; for multiple-site adsorption of reactant the surface is much more strongly deactivated by a given amount of poison than for single- or dual-site adsorption. The curves are reminiscent of nonselective versus selective poisoning discussed previously, except that here we are making the analysis from a geometric rather than a site strength distribution basis.

These computations were carried out assuming that the structure of the adsorbed molecules is rigid and no rearrangement occurs. In several instances, it has been documented that the adsorption site requirement for a poison changes with surface coverage. Most experimental data to this effect are reported in terms of changes in poisoning as a function of temperature.

---

[1] The approach is certainly derivative of the geometric multiplet theory of Balandin (3), much of which was based on data for cyclohexane dehydrogenation, where $m = 6$ for single-molecule adsorption excluding the central site rendered inaccessible by the adsorbed molecule. If geometry such as envisioned in case A is assumed, then $m = 7$.

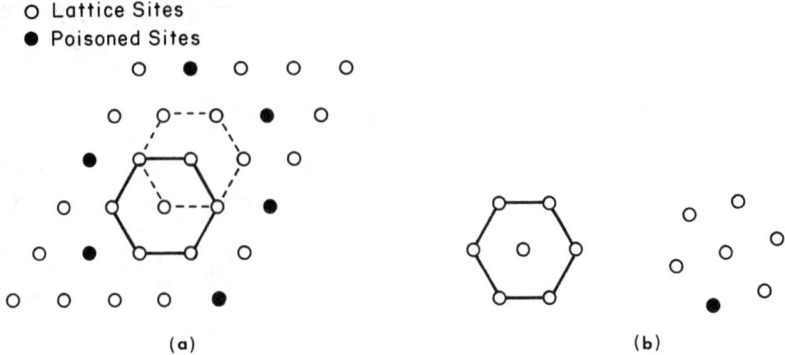

Fig. 4-9. (a) Interference between two possible modes of adsorption. (b) Adsorption on isolated sets of sites. [From Herrington and Rideal (36). Reproduced by permission of the Royal Society of Chemistry.]

A good example of this is the interaction of thiophene with nickel. As reported by Lyubarskii and co-workers (43), thiophene bonded to the Ni atom has a planar orientation, with bonding of the sulfur atom to the nickel and the unsaturated ring interacting with the Ni d orbitals. When acting as a poison in hydrogenation reactions, thiophene can itself be hydrogenated to tetrahydrothiophene, and when this occurs the planar adsorption

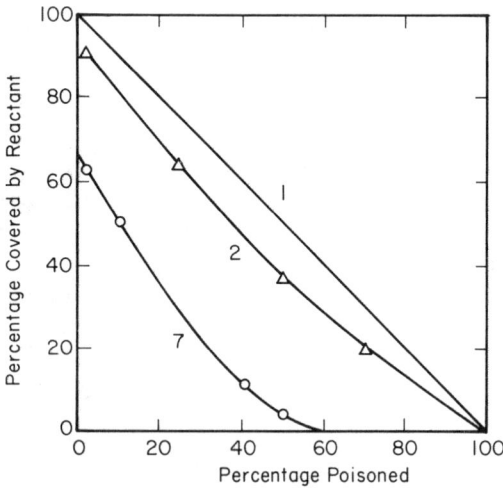

Fig. 4-10. Poisoning of a face-centered cubic (111) plane by a single-site poison, for 1-, 2-, and 7-site reactants. [From Herrington and Rideal (36). Reproduced by permission of the Royal Society of Chemistry.]

configuration disappears and is replaced by bonding solely via Ni–S and the molecule assumes an upright or "anchor" position. Thus the apparent adsorption capacity for thiophene can increase as the shielded sites are released on hydrogenation. This can manifest itself as a strong function of temperature. At low temperatures the main hydrogenation reaction may progress yet the thiophene itself may not be hydrogenated—hence a very strong and selective poisoning by the planar configuration. Inversely, at high temperatures the actual poisoning species is the anchor-bonded tetrahydrothiophene and poisoning is more nearly nonselective. Billimoria and Butt (8) report a fourfold increase in apparent thiophene adsorption capacity of a commercial Ni/kieselguhr catalyst in benzene hydrogenation as temperature is increased from about 65 to 150°C, so the effect is not small. From the magnitude of the change one would estimate that the planar configuration shields four or five Ni atoms, that is, $m = 4$–5 in the notation of Herrington and Rideal. Further calculations, but for dual-site poisoning mechanisms, have been reported more recently by Verma and Ruthven (72).

In summary, then, we see the effect of poisoning on a truly nonuniform surface site distribution (Pines and Haag) mimicked by that on a uniform surface where geometric factors are superimposed (Herrington and Rideal). We have not mentioned electronic interactions to this point because they are not easily uniquely identified independently of geometric effects, particularly for technological catalyst systems. To examine in more detail both geometric and electronic factors in catalyst poisoning we will now turn to more recent studies of some idealized systems, returning for the moment to homogeneous surfaces.

III. SOME STUDIES OF HOMOGENEOUS SURFACES

Over the past 15 years or so we have been witness to an onslaught of surface-sensitive techniques of experimentation involving primarily various types of electron spectroscopy under high-vacuum conditions. The relationship between such studies and the catalysts of technological application has been considered to be somewhat tenuous, since one is limited to high-vacuum conditions and, in most cases, to metal surfaces. Nonetheless, a wealth of information has been acquired concerning surface structure, the geometry of adsorbed layers, interaction in coadsorbed layers, and the like. Among the most useful of these techniques have been low-energy electron diffraction (LEED) and Auger electron spectroscopy (AES), although many other methods have found considerable use. We cannot at this point enter into a long discussion of these surface spectroscopies; fortunately, a good book in the area exists (33).

Probably the most extensively investigated atomic species as a poison is sulfur, and the most extensively investigated metal is platinum. The reader is referred to an excellent review by Oudar (55) on the topic of sulfur adsorption and poisoning of metallic catalysts. One general point that clearly emerges from studies of S-Pt, S-Ni, and others is the geometric complexity of events and the fact that a single metal atom is very rarely involved with a single sulfur atom, but rather with the ensembles of surface atoms envisioned by Herrington and Rideal. Figure 4-11 illustrates the point. On the Pt(100) surface, LEED studies indicate the existence of two structures of chemisorbed sulfur, as shown (7, 34). In Fig. 4-11A we see a structure involving five sulfur atoms, which in each case interact with four Pt atoms in the surface; note also that the central sulfur shares its interactive Pt atoms with each of its four neighbors. In the language of the LEED literature, this is termed a centered c(2×2) structure and corresponds to one-half monolayer coverage at saturation. In Fig. 4-11B the central S is not present and this primitive p(2×2) structure corresponds to one-fourth monolayer coverage at saturation. The c(2×2) represents the maximum surface coverage possible in any type of ordered structure on this surface, and even though saturation coverage is only half of a monolayer, one can see that the surface is completely deactivated. Similar LEED studies of S-Ni(110) indicate ensembles of Ni atoms interacting with a single S, with the ensemble size changing with changing exposure (51).

The influence of structures such as those of Fig. 4-11 on catalytic activity has been studied in detail by Fischer and Kelemen (27), especially for the reduction of NO by CO on the same Pt(100) surface:

$$2CO + 2NO \rightarrow 2CO_2 + N_2$$

In Fig. 4-12 are given the results of poisoning studies in terms of the relative rate of CO_2 formation with respect to that of an unpoisoned surface as a function of surface coverage of sulfur. The initial linear relationship is observed only for very low coverage, $\theta < 0.1$, where the ordered structures cannot form due to low surface concentration of S. Activity here is then governed by isolated sulfur atoms and, while ensembles of four Pt atoms may still be involved in sulfur chemisorption, the distances between such

(a)

(b)

Fig. 4-11. Sulfur overlayers on a Pt(100) surface. (a) Centered c(2×2) structure. (b) Primitive p(2×2) structure. [From Fischer and Kelemen (27). Reproduced by permission of Academic Press, Inc.]

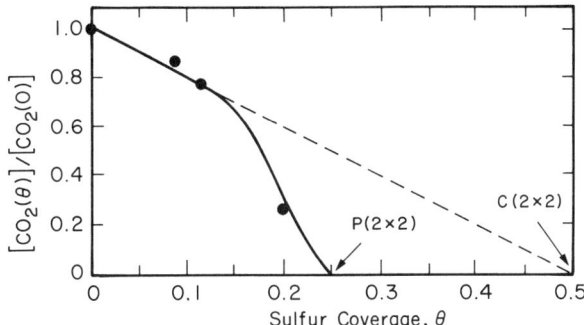

Fig. 4-12. Rate of the CO + NO reaction as a function of sulfur coverage on Pt(100). [From Fischer and Kelemen (27). Reproduced by permission of Academic Press, Inc.]

ensembles are still too large for significant interaction to be noted. However, when ordered structures are formed there is a catastrophic loss of activity and, as shown, the p(2×2) structure ($\theta = 0.25$) is sufficient to completely poison the surface. We note from the structure of Fig. 4-11 that a vacant adsorption site exists at the center of p(2×2), hence chemisorption of CO or NO at this position might still be possible. However, reactant molecules so adsorbed are isolated and the bimolecular surface reaction is blocked. This rough picture is supported by studies of CO adsorption on Pt(110) surfaces partially covered with S (11) using a flash desorption technique. These showed that CO would be chemisorbed on surfaces containing up to one-half monolayer of sulfur (indeed, even a bit more), though as we have seen the NO + CO reaction is extinguished at half this value.

In addition to the geometry we have been discussing, it is important to note that the presence of poison on a surface can affect the strength of bonding of other molecules on unpoisoned sites. In the Pt–S system, the very strong chemical bond between Pt and S weakens the interaction of Pt with other adsorbates; much the same may be said for Ni–S (60). For example, in the flash desorption studies of Bonzel and Ku (11) it was found that CO existed in two binding states on the fresh (110) surface, a high-energy state characterized by a sharp maximum in the desorption spectrum at temperatures ca. 250°C and a lower-energy state which appeared as a shoulder on the high-temperature peak at ca. 150°C. As the surface coverage of sulfur increased, the high-temperature peak was sharply reduced and eventually at higher coverage completely eliminated. Under these conditions the spectrum was characterized by a single broad peak, again in the vicinity of 150°C, much reduced in intensity. Another example of poison-induced changes in the strength of chemisorption bonding is provided by Rhodin and Brucker (59) for CO chemisorption on α-Fe(100) poisoned by sulfur.

In this work, employing the technique of UV photoemission spectroscopy [UPS in the alphabet soup of surface chemical physics (SCP)] it was found that the C–O bond over a fresh surface is distorted (elongated), which leads to dissociation. The presence of sulfur on the surface reduces the electron transfer (both forward from C to the metal and backward from the metal to the π bond of CO) and thus reduces the tendency for distortion in the molecule. This, of course, leads to much weaker net interactions between the metal and CO.

One sees to this point attempts made to explain various poisoning results on the basis of either/both geometric and electronic factors. The student of catalysis will certainly know that many attempts have been made with both geometric and electronic models to explain the observations of apparent structure sensitivity of various reactions on supported metal catalysts. However, contributions from those individual factors are difficult to separate, particularly when considering small crystallites of size order 1–10 nm. This pertains to unpoisoned reactions as well as those undergoing deactivation.

As an example, let us consider what has been proposed concerning geometric/electronic factors for hydrogenolysis reactions on supported metals. Most geometric models assume that the active sites for hydrogenolysis are atoms of low coordination (i.e., corners and edges, perhaps in some ensemble) on low-index surface planes which bound the particle. However, the underlying assumption that this can be correlated by assuming that the crystallites are bounded by an equal distribution of complete planes is questionable, to say the least, and deviations from this assumption can lead to large differences in correlation (25, 37, 67).

Earlier suggestions concerning the influence of electronic factors relate hydrogenolysis to percentage of d character (13, 64). However, in view of results obtained with bulk metals, Lornston (41) and Rorris *et al.* (61) have proposed that electronic and geometric factors are interrelated such that the d-band vacancies increase with increasing percentage exposed. For example, Konigsberger *et al.* (40) report evidence for the existence of some Rh^{n+} ions in highly dispersed Rh/Al_2O_3; these ions are in close contact with Rh^0 and may function as anchors for the crystallite to the support. From an electronic point of view, these ions would pull electron charge from the metal particle; presumably the smaller particles would lose proportionately more electron density than the larger ones, such that an increase in d-electron vacancies with particle size is observed. Now, the correlation between increasing activity and increasing d-band vacancies presumes that one is on the ascending portion of the volcano curve relating activity and strength of bonding. This is in accord with what is known concerning hydrogenolysis on Pt/SiO_2 and Rh/SiO_2.

A correlation of electronic factors associated with metal crystallite size, separable from geometric effects, may be provided by measurement of x-ray absorption edge data. The near-edge x-ray absorption spectrum is a function of the electronic structure of the absorbing atom. In particular, the L_{III} absorption resonance spectra can be interpreted as an electronic transition from a core $2p_{3/2}$ to vacant d states of the absorbing atom. Since the transition probability is related to empty d states, the threshold resonance area should be correlated with d-band vacancies and as well be reflective of the chemical state of the atom (14, 43). Further support for this view is provided by the work of Rorris et al. on propylene hydrogenation over 6.2 and 81% exposed Pt/SiO_2. In that work the area of the L_{III} absorption edge with reference to bulk Pt foil increased with increasing percentage exposed. This correlated with an increase in turnover frequency for the reaction. It appears that absorption edge measurements of poisoned catalysts in comparison with those of corresponding fresh catalysts could be useful in untangling combined geometric and electronic contributions to catalyst poisoning.

For some cases of reactions on fresh catalysts this approach to electronic interpretation seems quite consistent. The observed order of activity for some noble metal catalysts (on SiO_2) for methylcyclopropane hydrogenolysis is Rh > Pt > Pd. This can be correlated if the d-band vacancies may be related back to the number of d electrons in the constituent metal atom. Thus, Rh ($[Kr]4d^85s^1$), Pt ($[Xe]4f^{14}5d^96s^1$), and Pd ($[Kr]4d^{10}$) have an increasing number of d electrons in the order of decreasing activity. If, for a given metal particle size, the amount of electron density pulled from the crystallite to anchor it to the support is independent of the metal type, then Rh will always have less electron density or a larger number of d vacancies than Pt and Pd. It would be of interest to pursue these concepts further with data on poisoning.

To carry the concept of changes in chemisorptive behavior on poisoning one step further we may also cite some examples where the mechanism of poisoning involves not only competitive chemisorption but also subsequent chemical reaction. Many of these cases arise from the interaction of oxygen with metallic surfaces, although sulfur is also important in this respect. Amirnazmi and Boudart (1) have shown that when oxygen is added to the NO feed for the decomposition reaction of NO over Pt foils at 600–1000°C a substantial decrease in reaction rate is observed. This is due not only to the competitive chemisorption of oxygen on the Pt surface but also to the formation of an oxidized surface, PtO_2, produced by slow rearrangement of the reduced Pt surface after oxygen addition. In this instance both modes of deactivation were reversible; full activity was regained over a period of several hours after removal of O_2 from the feed. However, it may be that oxygen does not need to be present as a feed contaminant for such oxide formation to occur. The NO decomposition and NO+CO reaction were studied over Fe and Ni films by Baker and Peterson (2). Under the conditions of the investigation the interaction of NO with the surface was dissociative (some N_2O also produced) and the oxygen so formed reacted with the metal to form an inactive oxide layer. Examples of metal sulfide formation with consequent diminution or complete loss of activity for various metals and various reactions abound in the literature, but these seem mostly to be concerned with supported metals rather than the ideal homogeneous surfaces we are discussing in this section. Classical examples are hydrodesulfurization/hydrotreating catalysts such as $Co-Mo/Al_2O_3$, in which the metallic components exist as sulfides under reaction conditions. In this case, the metal sulfides are not without activity for the reactions of interest, but come to an equilibrium composition and activity level dependent on the process operating conditions.

A final type of deactivation that has been investigated on homogeneous surfaces leaves us in something of a quandary as to how it should be classified. Hegedus (35) refers to it as "chemically induced surface reconstruction," although an alternative term would be "chemically induced sintering," which would land the topic in another section of this book. We discuss the matter here although it stretches our careful definition of poisoning. Much of the work has been concerned with the restructuring of Pt-Rh gauzes or Pt single crystals used for ammonia oxidation. The temperatures involved in this reaction are high (800–1400°C), and one observes extensive restructuring of the surface including the formation of pits, evolution of various index planes, or formation of curved surfaces under reaction conditions. It has also been shown (63) that changes in surface morphology are enhanced in the presence of typical poisons such as H_2S, even though the surface is devoid of sulfur at such high temperatures. Restructuring is also observed, however, in the absence of impurity species (28, 48) and eventually leads to degradation of the physical integrity of the surface. For gauze catalysts, this leads to macroscopic collapse of the network. Another type of chemically induced sintering, exhibiting almost classical nonselective behavior, was reported by Clay and Petersen (cited in the previous chapter) for the poisoning of the Pt-film-catalyzed cyclopropane hydrogenolysis by arsine. Typical results are shown in Fig. 4-13 for experiments over a range of temperatures. Since both the adsorption and the reaction rate constants decreased linearly with increasing poison coverage, this was taken as evidence that the area available for reaction was the same (or at least proportional to) that for the adsorption of AsH_3.

In this section we have tried to illustrate some very fundamental features of poisoning processes from studies on very well characterized systems. Our

III. Some Studies of Homogeneous Surfaces

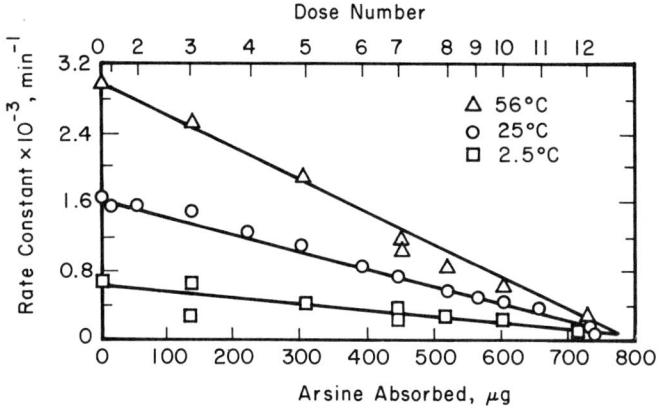

Fig. 4-13. Rate constants of cyclopropane hydrogenation over a Pt film as a function of the amount of AsH_3 adsorbed. [From R. D. Clay and E. E. Petersen, *J. Catal.* **16**, 32 (1970). Reproduced by permission of Academic Press, Inc.]

list of illustrations is very small compared to the total amount of information available; however, we feel they are representative. Certain qualitative features should stand out. The first is that poisoning processes, even on the chemisorption of simple atoms or molecules, have a high likelihood of involving multiple atoms in the surface. Hence, if even for only geometric reasons, the poisoning of homogeneous surfaces can assume a very selective nature. The second feature one might envision as a reinforcement of the strength of the poisoning effect as the level of poison on the surface is increased. This is the alteration of chemisorption bonding strength for reactant molecules—decreasing as the surface coverage of poison increases. Probably this can be attributed to electronic effects induced in the surface layers as the extent of poisoning increases, and there is an increasing body of literature, both experimental and theoretical, to support this view. From a more practical point of view, it is probably worthwhile to think of poisoning effects on chemisorptive bonding in terms of the classical "volcano" correlation of catalytic activity. We recall that, over a range of surfaces for a given reaction, activity will pass through a maximum for certain materials, and this has successfully been correlated with the heat of chemisorption of key species involved in the reaction (6) both in the older literature or in the more recent offering of Vannice (71). The classic example of this approach is probably that for formic acid decomposition (26). In this sense, then, we are just sliding down the left-hand side (LHS) of the volcano (decreasing activity with decreasing heat of chemisorption) as the influence of the poison becomes more predominant. By the same token, the presence of impurity adlayers on the surface should not necessarily *always* be deleterious to activity if we happen to be dealing with a catalytic surface that lives on the right-hand side (RHS) of the volcano. Such examples exist: lead is a promoter of the reduction of NO by CO on CuO (ascent of the RHS) but poisons the oxidation of ethylene over the same catalyst (descent of the LHS) (65). In the opposite sense, the formation of surface-inactive compounds represents a descent on the RHS to very stable materials that might be characterized by some pseudo equivalent high heat of chemisorption in formation and consequent stability.

IV. HETEROGENEOUS SURFACES: SITE STRENGTH DISTRIBUTIONS

As we have by now pointed out several times, the existence of a distribution of active site strengths typically leads to selective poisoning in which the poison preferentially interacts with the most active sites first. In fact, we must be careful with this view, since the very "strongest" sites on a surface (say in terms of the magnitude of the heat of chemisorption for reactants) may indeed interact so strongly with the reactants that they do not contribute to catalytic activity; that is, we are on the right-hand side of the volcano again. In other cases, indeed, the poison may preferentially bind with surface sites that are less active for the main reaction.

We have already used what is probably our best example of the detailed analysis of poisoning of a heterogeneous surface in discussing the work of Pines and Haag. However, an important point remains that was perhaps not discussed with present objectives in mind. This has to do with the interaction of the selectivity of the chemical reactions with the selectivity of the deactivation of the surface. In particular, for the isomerization of 3,3-dimethyl-1-butene, shown in detail in Fig. 4-2, the different steps in the overall reaction involve various types of carbonium ions, as indicated. Hence the poisoning of a given portion of the surface will affect these individual steps in different ways and thus will induce changes in the reaction selectivity as well as the overall catalyst activity. For example, step II(b) of Fig. 4-2, proceeding from tertiary to secondary via a primary carbonium ion intermediate, is much more severely affected by the poisoning of strong acid sites than step II(a), going from secondary to tertiary. Another example of product selectivity alteration on poisoning has been reported by Dalla Betta *et al.* (21) for a vastly different reaction and catalyst—methanation of CO on supported Ni, Rh, and Ru. In this case, the presence of H_2S significantly reduced overall methanation rates but at the same time more severely affected the hydrogenation function of the catalysts than C–C bond formation. Hence the overall decrease in net CO conversion to methane was

IV. Heterogeneous Surfaces: Site Strength Distributions

accompanied by an increase in the fraction of higher molecular weight products. The "toning" of fresh supported noble metal reforming catalysts with sulfur compounds is also a well-known example of an industrial procedure in which both activity and selectivity are affected. These are, of course, bifunctional catalysts and we shall discuss further some aspects of their poisoning behavior in a subsequent section.

Complex oxide surfaces, complicated reactions, or bifunctional catalysts are not required for heterogeneous/selective poisoning, of course. A model study has been reported by Völter and Hermann (73) for a very simple reaction, ortho-para H_2 conversion, over Pt foil catalysts with CO as the catalyst poison. In this case, actually, one is confronted with a combination of selective and nonselective poisoning, as indicated in Fig. 4-14, in which the areal surface rate constant is plotted as a function of surface coverage of carbon monoxide. Accepting 10^{15} as a nominal surface atom density per square centimeter for the Pt, we observe that complete deactivation occurs essentially at a surface coverage of unity. However, a surface coverage of

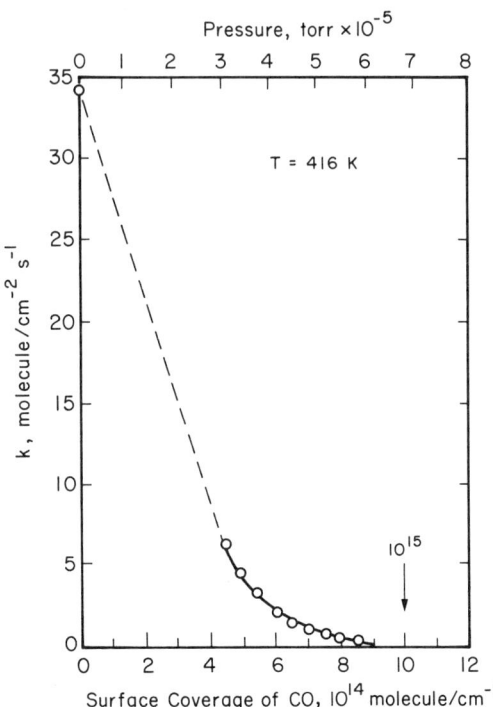

Fig. 4-14. CO poisoning of para-H_2 conversion over a Pt foil. [From Völter and Hermann (73). Reproduced by permission of Johann Barth, Publishers.]

about 40% is sufficient to reduce activity by almost 80% from the initial level (and the change in this range is nonselective), while the remaining 60% of the surface possesses only 20% of the activity. In this case it was found that the decrease in rate constant with increasing CO coverage was accompanied by an increase in the activation energy of the reaction, which was also correlated with an increase in the heat of H_2 chemisorption. Hence, the selective portion of the poisoning correlation is associated with CO chemisorption on high-energy sites with strong bonding and lower intrinsic activity (again, the right-hand side of the volcano).

This combination of a region of nonselective poisoning at low loadings and selective poisoning at high loadings has been observed in a very wide variety of catalyst and poison systems and for a wide variety of reactions. Figure 4-15 illustrates the similar behavior of arsine on a Pt hydrogenation catalyst, where, again, a very large fraction of total surface activity resides in the region of nonselective poisoning.

A final example of poisoning of heterogeneous surfaces is given in the work of Baron (5) for the effect of lead poisoning of Pt films in the CO oxidation reaction. Major experimental results are given in Fig. 4-16. Here we note three distinct regions of poisoning behavior, with the familiar combination of nonselective/selective poisoning for Pb coverages greater than about 20% surface coverage. However, at very low coverages the poisoning curve exhibits a shape opposite to what one expects from selective poisoning on a heterogeneous surface. In this event of "antiselective" poisoning, Baron attributed the peculiar behavior at coverages <20% to preferential binding of Pb on surface sites which are less active for CO oxidation.

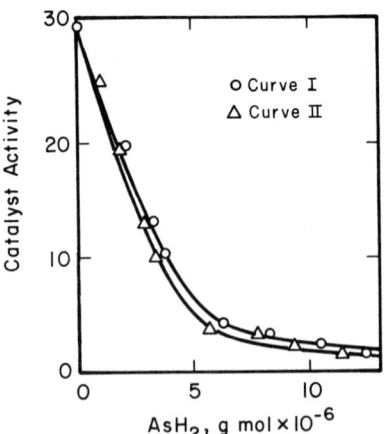

Fig. 4-15. Effective and true poisoning curves for a Pt catalyst (0.05 g) for hydrogenation poisoned with AsH_3. Curve I, effective toxicity, is based on total poison present in the system; curve II, true toxicity, is based on the amount of poison actually adsorbed on the catalyst. [From Maxted (46). Reproduced by permission of Academic Press, Inc.]

V. Particle Size Dependence

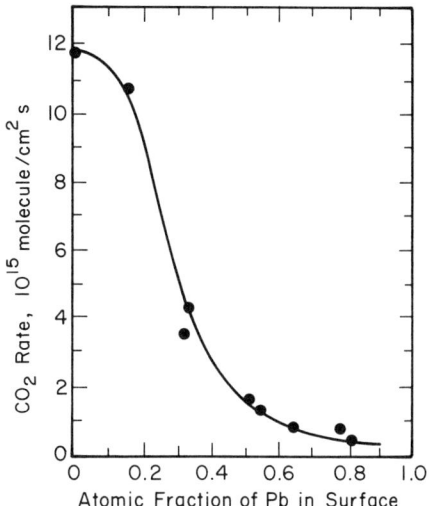

Fig. 4-16. Effect of lead coverage on the rate of CO oxidation over a Pt film. [From Baron (5). Reproduced by permission of Elsevier-Sequoia, S.A.]

V. PARTICLE SIZE DEPENDENCE

The concept of "structure-sensitive" catalytic reactions has now been with us for many years, particularly with respect to reactions catalyzed by supported metals. In general, the exact topography of the crystallite particles of supported metal catalysts is not known; however, general characterization methods such as hydrogen chemisorption or x-ray diffraction emphasize determination of the fraction of surface metal atoms exposed, as we have already pointed out in Chapter 1. From these characterization studies a turnover frequency defined in terms of rate of reaction per surface metal atom can be determined. The variation in the turnover number with the percentage of the catalyst exposed can yield information which relates the surface structure of a catalyst to its activity. Boudart and co-workers (12) provided an early realization of this concept, measuring the turnover frequency for the hydrogenation of cyclopropane on a series of platinum-on-alumina and platinum-on-silica catalysts. For this reaction, the turnover frequency varied by a factor of two as the specific surface area was changed by four orders of magnitude. Since in this case the turnover frequency was within the errors of their experiments, essentially independent of the percentage exposed, the reaction was structure-insensitive. Some studies have suggested that, within experimental error, a number of other reactions exhibit structure-insensitive behavior over platinum catalysts. Table 4-6 lists

TABLE 4-6
Summary of Some Structure-Insensitive Reactions

Reaction	Catalyst	Variation in activity (%)	References[a]
Cyclopropane hydrogenolysis	Pt/SiO_2 Pt/Al_2O_3	±50	A
Ethylene hydrogenation	Pt/SiO_2 (reduced at 500°C)	±30	B
Ethylene hydrogenation	Pt/SiO_2 (O_2, 300°C; H_2, 400°C)	±20	C
Benzene hydrogenation	Pt/SiO_2	±30	D
Benzene hydrogenation	Pt/Al_2O_3	±15	E
Benzene hydrogenation	Pt/Al_2O_3	±20	F
Isomerization of n-pentane	Pt/SiO_2	±25	G
Epimerization of cis-1,4-dimethylcyclohexane	Pd/charcoal	—	H

[a] A, M. Boudard, A. M. Aldag, J. E. Benson, N. Dougharty, and G. Harkins, *J. Catal.* **6**, 92 (1966); B, T. A. Dorling, M. J. Eastlake, and R. L. Moss, *J. Catal.* **14**, 23 (1969); C, J. Schlatter and M. Boudart, *J. Catal.* **24**, 482 (1972); D, T. A. Dorling and R. L. Moss, *J. Catal.* **5**, 111 (1966); E, J. Aben, H. Vander Eijk, and J. M. Olderik, *Catal., Proc. Int. Congr. 5th, 1972*, Vol. 1, p. 717 (1973); F, J. P. Brunelle, A. Siegner, and J. LePage, *J. Catal.* **43**, 273 (1976); G, F. M. Dautzenberg and H. Platteuw, *J. Catal.* **19**, 41 (1970); H, J. K. A. Clarke, E. McMahon, and A. D. Cinneide, *Catal., Proc. Int. Congr., 5th, 1972*, Vol. 1, p. 685 (1973).

some of these studies, which include ethylene hydrogenation, benzene hydrogenation, and the isomerization of n-pentane and n-hexane. The range of specific surface in these studies is, however, far less than in the cyclopropane work.

The variance in activity is quite large in some of these reported structure-insensitive reactions. In early experiments, large errors in the characterization of the percentage exposed and in the determination of reaction rates contributed to this variation in turnover frequency. However, in recent years the techniques for evaluating the percentage exposed of platinum catalysts has greatly improved, reducing the errors in the turnover frequencies to within ±10–15%. Therefore, any variation in carefully measured turnover frequencies as a function of percentage exposed greater than 15% should be attributed to the influence of the surface structure on the catalytic activity.

If the turnover frequency is a strong function of the percentage exposed of a catalyst, then the reaction can be classified as a primary structure-sensitive reaction. In this type of reaction, the environment around each active site greatly influences the catalytic activity of that site; therefore, it may be possible to infer the likely catalytic sites for reaction.

Primary structure-sensitive reactions are classified into two types. Type one is a reaction in which the turnover frequency increases with increasing percentage exposed, while type two is one in which the turnover frequency decreases with increasing percentage exposed. Table 4-7 lists examples of both types of structure-sensitive reactions. This table shows that the dehydrogenation and hydrogenolysis reactions adhere to type one structure sensitivity while the isomerization reactions adhere to type two structure sensitivity.

Simple geometric models of the metal crystallites can be used to indicate the possible active sites for these reactions. If all the metal crystallites are assumed to have a cubic shape, as discussed by Dorling and Moss (23),

TABLE 4-7

Summary of Some Primary Structure-Sensitive Reactions

Reaction	Catalyst	Assumed active site	References[a]
Type I			
Dehydrogenation of 2,3-dimethylbutane	Pt/charcoal	Corner atoms of a cube	A
Hydrogenolysis of *n*-hexane and methylcyclopentane	Pt films	Low-coordination atoms	B
Hydrogenolysis of *n*-pentane	Pt/SiO$_2$	Low-coordination atoms at edges and corners of a cubo-octahedron	C
Hydrogenolysis of cyclopentane	Pt/Al$_2$O$_3$	Low-coordination sites near defects in crystal	D
Hydrogenolysis of *n*-heptane	Pt/Al$_2$O$_3$	No model	E
Type II			
Isomerization of neopentane	Pt/SiO$_2$ Pt/Al$_2$O$_3$	Triplet sites occurring on (111) planes	F
Isomerization of butanes	Pt films	(111) oriented platinum surface	G
Isomerization of labeled hexanes	Pt/Al$_2$O$_3$ Pt/SiO$_2$	Multiple atom site	H
Isomerization of *n*-heptane	Pt/Al$_2$O$_3$	No model	E

[a] A, M. Nakamura, M. Yamada, and A. Amano, *J. Catal.* **39**, 125 (1975); B, J. R. Anderson and M. Shimoyama, *Catal., Proc. Int. Congr. 5th, 1972*, Vol. 1, p. 695 (1973); C, J. P. Brunelle, A. Siegner, and J. LePage, *J. Catal.* **43**, 273 (1976); D, R. Maurel, G. Leclercq, and L. Leclercq, *Bull. Soc. Chim. Fr.*, p. 481 (1972); E, H. J. Maat and L. Moscou, *Proc. Int. Congr. Catal., 3rd, 1964* p. 1277 (1965); F, M. Boudart, A. M. Aldag, L. Ptak, and J. E. Benson, *J. Catal.* **11**, 35 (1968); G, J. R. Anderson and N. R. Avery, *J. Catal.* **5**, 446 (1966); H, G. Maire, G. Corroleur, D. Juttard, and F. Gault, *J. Catal.* **21**, 250 (1971).

then a type one structure-sensitive reaction would be defined as one in which the turnover frequency increases as the mean size of the cube decreases. The fraction of surface metal atoms occurring in corners and on edges also increases with decreasing cube diameter. Therefore, there is a direct correlation between an increase in a turnover frequency and an increase in the fraction of surface corner and edge atoms. This conclusion led Nakamura *et al.*, Anderson and Shimoyama, and Brunelle *et al.* to assume that the active sites for the dehydrogenation and hydrogenolysis reactions are the low coordination number atoms in the crystallites. This result is consistent with the results of Blakely and Somorjai (9), who correlated the number of kinks, low-coordination atoms, with the turnover number for the hydrogenolysis of cyclohexane.

If the cubic crystal model is used to determine the active sites for a type two structure-sensitive reaction, the resulting correlation relates the fraction of surface metal atoms in the faces of the cubic crystallite to the turnover frequency. This assumption led Anderson and Avery and Boudart *et al.* to conclude that isomerization reactions involve a multiple-bonded intermediate which is more likely to occur on flat planar surfaces. As an example, Boudart *et al.* showed that the isomerization of neopentane involved a triadsorbed intermediate which would be found most frequently on the (111) planes of the metal. These authors claimed that firing a catalyst in vacuum at 900°C would cause rearrangement of the crystallites so that a larger fraction of the surface contained (111) planes. This claim was supported by a large increase in the selectivity for the isomerization reaction on the fired catalysts. Further very detailed analysis of the geometry of small metal particles is given by van Hardeveld and Hartog (70).

Returning for the moment to the structure-insensitive examples shown in Table 4-6, it has been found in several cases that such reactions are structure-insensitive only under certain pretreatment conditions. In the ethylene hydrogenation study of Dorling *et al.*, for example, the authors found a correlation between the number of chloride ions on the catalyst and its activity. After the catalyst was heated in hydrogen at 400°C to desorb all the chloride ions, its activity was similar to that of the supported platinum catalysts prepared without chloride compounds. In that study, the presence of chloride on the catalyst affected the catalytic activity so that the reaction appeared to be structure-sensitive.

In the same manner, Schlatter and Boudart observed that carbonaceous impurities deposited on the support caused the ethylene hydrogenation reaction to be falsely labeled as a structure-sensitive reaction. After the carbon was burned off with oxygen, the reaction, as measured on the metallic platinum, was structure-insensitive.

V. Particle Size Dependence

Manogue and Katzer (44) have described the effects of chemical impurities (we can think of them as predeposited poisons) on catalytic activity as secondary structure-sensitive reactions. In this case, the chemical substituent which is adsorbed on the support or the metal influences the catalytic activity so that the effects of the surface structure cannot be determined. Table 4-8 lists examples of chemical impurities influencing the catalytic activity of supported platinum catalysts. As illustrated by this table, the most common chemicals that influence catalytic activity are chlorine, oxygen, and carbon. We present here some details on the first two.

Chloride is present on all supported platinum catalysts that are prepared with H_2PtCl_6. Aside from the ethylene hydrogenation work discussed above, the effects of chloride on supported Pt have been discussed for a number of other reactions. McHenry et al. (49) showed that the chemical nature of a Pt/Al_2O_3 catalyst for the isomerization and dehydrocyclization of n-heptane is greatly altered in the presence of chloride compounds. These authors stated that the chloride does not only increase the acidity of the alumina support. Rather, they reported that there is a synergistic effect with a chlorinated platinum-on-alumina catalyst for the isomerization and dehydrocyclization of n-heptane such that the combined catalyst yields a higher

TABLE 4-8

Summary of Some Secondary Structure-Sensitive Reactions

Reaction	Catalyst	Chemical affecting rate	References[a]
Ethylene hydrogenation	Pt/SiO_2	Chloride compounds	A
Ethylene hydrogenation	Pt/SiO_2	Carbon	B
Nitric oxide reduction with NH_3	Pt/Al_2O_3	Oxygen	C
Ammonia oxidation	Pt/Al_2O_3	Molecular oxygen	D
Dehydrogenation of cyclohexane	Pt/SiO_2	Oxygen	E
Dehydrogenation of cyclohexane and cyclohexene	Pt crystals	Carbon	F
Hydrogenolysis of cyclopentane and benzene hydrogenation	Pt/Al_2O_3	Sulfur H_2S/SO_2	G

[a] A, T. A. Dorling, M. J. Eastlake, and R. L. Moss, J. Catal. **14**, 23 (1969); B, J. Schlatter and M. Boudart, J. Catal. **24**, 482 (1972); C, R. J. Pusateri, J. R. Katzer, and W. H. Manogue, AIChE J. **20**, 29 (1974); D, J. J. Ostermaier, J. R. Katzer, and W. H. Manogue, J. Catal. **41**, 277 (1976); E, A. N. Mitofanova, V. S. Boronin, and G. M. Poltorak, Zh. Fiz. Khim. **46**, 32 (1972); F, D. W. Blakely and G. A. Somorjai, J. Catal. **35**, 200 (1976); G, R. Maurel, G. Leclercq, and J. Barbier, J. Catal. **37**, 324 (1975).

activity for these reactions than the addition of a chlorinated alumina catalyst mixed with a platinum-on-alumina catalyst. This synergistic effect has been attributed to the presence of localized surface complexes containing platinum, alumina, and chloride compounds. These surface complexes are assumed to be the active sites for the reaction. The source of the chloride may also be important in determining the magnitude of such effects.

The presence of oxygen can also greatly affect the activity of supported metal catalysts. Some of the anomalies of the oxygen–hydrogen titration as discussed in Chapter 1 are caused by unusual adsorption of oxygen on the metal surface. In these titrations, the adsorption of oxygen occurred at 25°C. However, sometimes in the pretreatment of supported catalysts, oxygen is contacted at elevated temperatures to remove all carbonaceous materials. Gruber (31) reported that oxygen at 350°C can penetrate platinum crystallites in more than one monolayer. Ostermaier *et al.* also also concluded that oxygen at high temperatures can penetrate the lattice at least three or four monolayers deep on large platinum crystallites but can oxidize small platinum crystallites completely to a bulk platinum oxide. Uchijima *et al.* (69) concluded that small platinum crystallites (approximately 2 nm) become a total oxide after contact with oxygen at 300°C.

In gas-phase hydrogenation reactions, the catalyst is usually pretreated with hydrogen at various temperatures after the oxygen treatment. Uchijima *et al.* found that a pulse of hydrogen at room temperature requires at least 450 min to completely displace the oxygen from platinum-on-silica catalysts with crystallite sizes less than 3 nm. Thus, there is a required time for the complete reduction of the platinum crystallites. Also, Ratnasamy *et al.* (58) indicated, using the radial electron density distribution technique, that the surface of a platinum catalyst after exposure to oxygen and pretreatment in hydrogen contains many more low coordination number sites than exist in the bulk platinum surface. Therefore, the oxygen can affect catalytic activity in two ways: by the oxygen itself present after the catalyst has been reduced, and by the creation of more low-coordination sites than on an equilibrium surface structure.

The examples of structure sensitivity, structure insensitivity, and secondary structure sensitivity presented here for supported Pt have by now many counterparts for other supported metal catalysts. In what way does this fit into a discussion of poisoning? It is apparent that the secondary structure sensitivities induced either by the method of catalyst pretreatment or by preparation can be regarded as a special type of poisoning involving predeposition of the poison rather than competitive parallel chemisorption. Again, this involves us in a delicate extension of the definition of poisoning made so carefully in Chapter 1. The fact that small quantities of simple molecules are able to alter something as fundamental as the structure

V. Particle Size Dependence

sensitivity of a reaction, however, leads to an even more intriguing concept: that of structure-sensitive poisoning.

The question of structure-sensitive poisoning was, in fact, touched on in the work of Maurel et al. (45) on the sulfur poisoning of a series of Pt/Al_2O_3 catalysts for the reactions shown in Table 4-8, although the results of that study were not directly reported in terms of activity versus percent exposed metal. Also, Fuentes and Figueras (29) investigated a similar problem for a series of Pd/Al_2O_3 and Pd/SiO_2 catalysts as affected by metallic poisons (or modifiers) such as Fe. However, again the catalysts were preloaded with the impurities under examination in these studies and only reference activities were reported.

Closer to the point are the data of Barbier et al. (4), who prepared a series of Pt/Al_2O_3 catalysts with metal exposed ranging from 5 to 80% and studied a wide variety of reaction and poisons on them. The reactions included, among others, benzene hydrogenation, cyclopentane hydrogenolysis, and deuterium exchange with benzene and cyclopentane, with a wide variety of reversible and irreversible poisons. A primary series of results obtained for benzene hydrogenation with ammonia poisoning of the catalyst series is shown in Fig. 4-17. This reaction over the unpoisoned catalyst was structure-insensitive when carried out at 85°C, yet it is seen from the figure that a relatively large degree of structure sensitivity appears as the surface concentration of ammonia is increased. Quite clearly, the smaller Pt particles are poisoned more severely than the larger ones, and the resistance to deactivation increases monotonically with increasing particle size. Since the reaction on fresh catalyst appears to be structure-

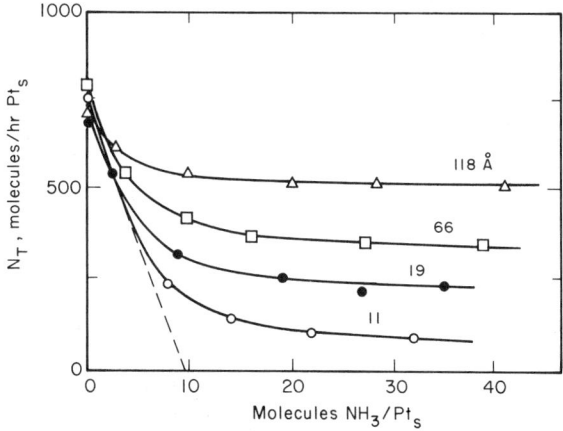

Fig. 4-17. Effect of Pt particle size on the ammonia inhibition of benzene hydrogenation at 85°C. [From Barbier et al. (4). Reproduced by permission of Société de Chimie Belge.]

insensitive, this is a very clear example of *both* poison-induced structure sensitivity and structure-sensitive resistance to poisoning. It is not clear what the origins of the development of such structure sensitivity are, but one possibility is that if an ensemble of surface metal atoms is involved in the hydrogenation reaction, then a single NH_3 molecule would be more effective in hindering chemisorption/hydrogenation on small particles than on large ones.

The work on ammonia oxidation by Ostermaeir *et al.* (54) also gives an example of structure-sensitive poisoning. Here the catalyst was 2-15-nm Pt/Al_2O_3, or Pt black, and the reaction was conducted at 368-473 K. It was found that the extent of poisoning increased with decreasing temperature and the temperature dependence of the poisoning effect was a function of metal particle size. As in the case of Barbier *et al.*, the poisoning was more severe with smaller crystallites, but it was always possible to reactivate the surface with H_2 at 673 K, suggesting that oxygen was the poison and PtO (or even some bulk oxide) the deactivated surface. These conditions of reduction are well within the range found by Uchijima *et al.* to be sufficient for reduction of even the highest percent exposed Pt catalysts.

Another example of structure-sensitive poisoning for a structure-insensitive reaction is provided by Boudart *et al.* (12) for the hydrogenolysis of cyclopropane. Here there was again an enhanced deactivation of more highly dispersed catalysts, in this case by oxygen, although the poison-free reaction rates varied by approximately a factor of only two over the series investigated. However, in contrast to this, Önal (53) found the pretreatment in oxygen at elevated temperatures of the Pt/SiO_2 series previously studied by Uchijima *et al.* to result in substantial increases in activity of the catalysts for methylcyclopropane hydrogenolysis at 0°C. The effect is structure-sensitive, as shown in Fig. 4-18, where a modified turnover frequency for the hydrogenolysis reaction is plotted against the percentage exposed measured via hydrogen chemisorption, D_h. The only difference between the two sets of experiments is that the second series (circles in the figure) were subject to a pulse of oxygen (of total size $\gg Pt_s$) at 300°C after reduction by H_2 at 300°C for 1 hr. The enhancement in activity is also sensitive to the temperature of O_2 pretreatment, shown in Fig. 4-19, although the relative variation is not as great as the distinction between oxygen pretreatment and no oxygen pretreatment. In addition to the large changes in activity noted, it was found that the selectivity of the hydrogenolysis, expressed as the ratio of isobutane to *n*-butane in the product, decreased from about 20 to 5 after the oxygen treatment, independent of percent exposed metal.

The role that oxygen plays in changing the activity and selectivity of these Pt catalysts is complex and certainly not well understood. Three models to explain the effects of strongly adsorbed oxygen on Pt have been

V. Particle Size Dependence

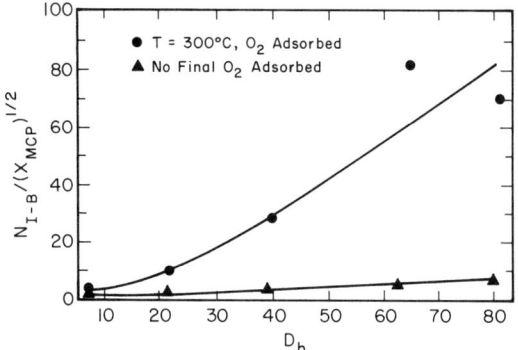

Fig. 4-18. Isobutane turnover frequency versus percentage exposed Pt with oxygen pulse pretreatment. [From Önal and Butt (53). Reproduced by permission of the Royal Society of Chemistry.]

proposed by McCabe and Schmidt (47). These are: (i) the formation of a surface layer of oxide results in a change in the electronic structure of the surface Pt atoms, (ii) strongly adsorbed oxygen atoms are active in compound formation with other adsorbates, and (iii) oxidation of the Pt results in reconstruction of the surface. Although definitive evidence is lacking, it seems most plausible that cases of oxygen-induced enhancement of activity are best explained by induced electronic changes in the Pt surface, while cases in which oxygen acts as a poison are the result of the formation of a surface layer of inactive oxide. In the former case, we might envision subsurface oxygen atoms in the Pt crystallite that would tend to remove Pt valence electrons, resulting in a positive charge on the surface Pt atoms. This amounts to inducing a certain acidic character of the metal that could

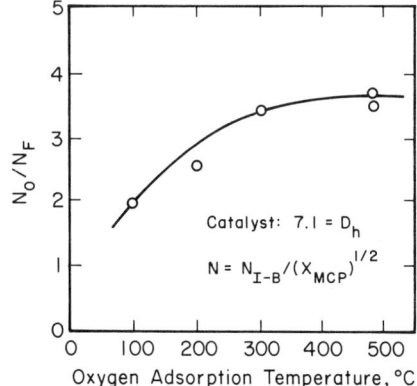

Fig. 4-19. Enhancement in turnover frequency as a function of oxygen adsorption temperature. [From Önal and Butt (53). Reproduced by permission of the Royal Society of Chemistry.]

affect the binding of hydrogen and hydrocarbon reactants, thus affecting the activity and selectivity.

Quite a different picture emerges if we look at results obtained for the same series of Pt/SiO$_2$ catalysts, the same reaction at the same conditions, but with preexposure to CO instead of O$_2$. The stoichiometry of CO chemisorption on these catalysts was verified to be 1:1 with respect to surface Pt at 25°C, so that surface coverage could be varied by controlling the size and/or number of adsorption pulses. In this case, CO is an effective poison for the hydrogenolysis reaction at 0°C, as shown in Fig. 4-20. A rather complex structure appears, since the results are not monotonic with respect to particle size as were those of Barbier *et al.* This is revealed more explicitly in the cross-plot of Fig. 4-21, which is directly comparable to Fig. 4-17. The structure sensitivity of the reaction is affected by the extent of CO coverage on the surface, as was the case with Barbier *et al.*, but now there is also a change in the structure sensitivity of the poisoning. For example, the catalyst of D_h (percentage exposed determined by hydrogen chemisorption) = 63 is optimum for coverage of 50%, while we would want to select $D_h = 81$ for coverage of 20%. Such results offer the possibility that for certain systems one may be able to tailor metal particle sizes and/or distributions not only for maximum activity but also for maximum resistance to poisoning.

The pattern of structure sensitivity of poisoning can be different, however, even for the same reaction and active metal component, the same pretreatment, and the same reaction conditions, but a different support. This is illustrated by the results of Damiani and Butt (22) for the CO-poisoned

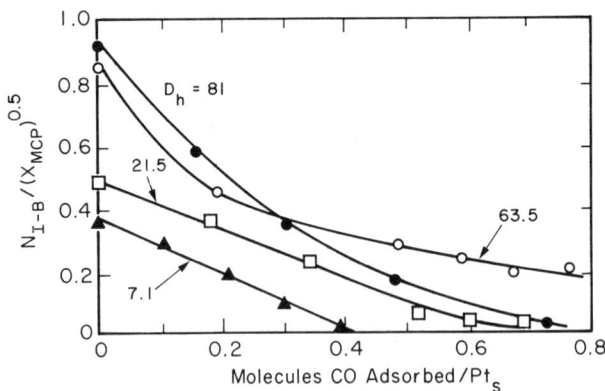

Fig. 4-20. Isobutane turnover frequency versus CO surface coverage for methylcyclopropane hydrogenolysis on a series of Pt/SiO$_2$ catalysts. [From Önal and Butt (53). Reproduced by permission of the Royal Society of Chemistry.]

V. Particle Size Dependence

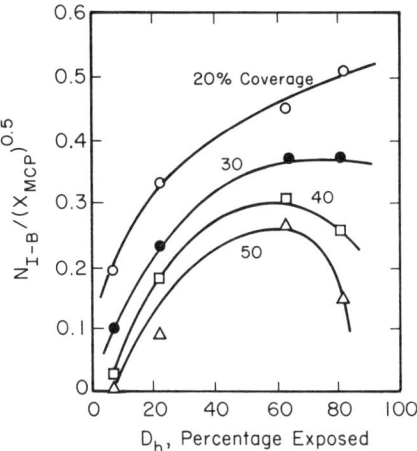

Fig. 4-21. Structure sensitivity as a function of CO coverage; methylcyclopropane hydrogenolysis on Pt/SiO$_2$. [From Önal and Butt (53). Reproduced by permission of the Royal Society of Chemistry.]

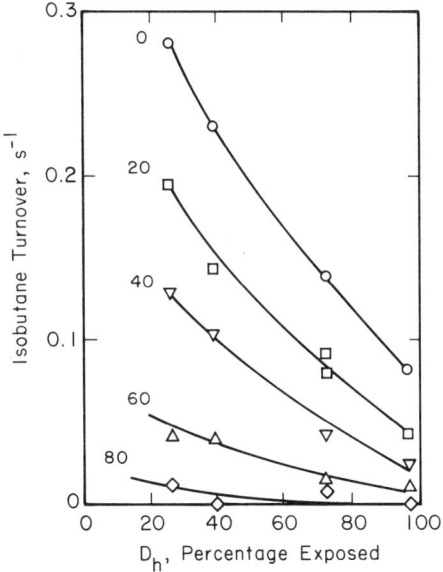

Fig. 4-22. Structure sensitivity as a function of CO coverage; methylcyclopropane hydrogenolysis on Pt/Al$_2$O$_3$. [From Damiani and Butt (22). Reproduced by permission of Academic Press, Inc.]

methylcyclopropane hydrogenolysis on a Pt/Al_2O_3 series. The unpoisoned reaction in this series is somewhat structure-sensitive, but the general pattern of structure sensitivity, unlike that of Pt/SiO_2, does not change on progressive poisoning, as shown in Fig. 4-22. The change in absolute magnitude of activity is greater for the larger particle sizes, although these lower percentage exposed catalysts still maintain slightly higher activity on more severe poisoning. Because of the similarities in preparation, pretreatment, reaction, and reaction conditions for these two series of catalysts, the different results can only be ascribed to different metal–support interactions. Generally, alumina interacts more strongly with the supported metal than silica, as will become evident in Chapter 5. Thus, particles supported on alumina may be more uniform due to stabilization by the support interaction(s). Particles supported on silica are less influenced by the support and would tend to retain characteristics acquired during preparation. It should be noted that the term "support interaction" here includes the possible influence of different levels of impurities in the two supports, although neither of these supports is included among those exhibiting strong metal–support interactions (SMSI).

VI. MULTIFUNCTIONAL CATALYSTS

In the discussion of heterogeneous surfaces we concentrated on catalytic sites which had differing activities or strengths but were of the same chemical nature. There are, of course, important types of catalysts that have active sites of differing chemical nature and that promote, simultaneously, different chemical transformations. The Pt/Al_2O_3 family of reforming catalysts (and related alloy formulations) is the prototype of such multifunctionality, in which the metal participates in hydrogenation/dehydrogenation-type reactions and the alumina series acts as both support and acidic function catalyst for reactions such as cracking and isomerization. Many other examples can be cited. Some metal oxide catalysts can have both acidic and basic surface sites, and a number of mechanistic schemes for hydrocarbon reactions on such surfaces postulate the concerted action of both acidic and basic sites on the hydrocarbon molecule.

A factor of major importance in the poisoning of bi- or multifunctional catalysts is that if one of the functions is preferentially poisoned, then the catalyst selectivity is altered in addition to the activity. This is by no means a feature unique to multifunctional catalysts, as we have seen previously, but it assumes particular significance here. It is hard to avoid using the term "selective" to describe the preferential poisoning of one function, and it will be understood that this term does not mean the same thing in the

VI. Multifunctional Catalysts

sense in which we used it before. If the poison is nonpreferential and deactivates all functions, by the same token we may use the term "nonselective."

Nonselective poisoning was analyzed mathematically by one of us a number of years ago (17), and it was shown for a bifunctional catalyst that the composition (balance between functions) which optimizes initial performance (minimum temperature for specified selectivity and conversion) is not the same as that which optimizes long-term performance when temperature-variant, constant-conversion operation is considered. Indeed, when there is an imbalance between the activation energies of the two functions, even nonpreferential poisoning of the two functions will drive the reaction selectivity to limiting values involving only one product. This type of behavior is discussed at some length in Chapter 10.

An extensive review of the poisoning of multifunctional oxide catalysts was published by Knözinger in 1976 (39). As mentioned above, the functionalities involved are likely to be as different as acid and base, so generally the poisoning of such oxides is strongly selective. On alumina, for example, we have already given as an example in several places the development of both Brønsted and Lewis acidity associated with the degree of hydroxylation of the surface and the net coordination of surface Al ions. Hightower and co-workers (38, 62) have provided a good example of the multifunctional nature of alumina in their studies of deuterium exchange with benzene and olefins and with olefin isomerization. The selective poisoning effects of a number of simple molecules such as NO, CO, NH_3, and CO_2 were investigated, and the strongest poison was identified as CO_2. On the basis of the CO_2 poisoning results and temperature-programmed desorption, four types of sites, denoted A, B, I, and E, were identified. Sites A and B were strong chemisorption sites for hydrocarbon and were not poisoned by CO_2, while sites E were associated with exchange activity and were strongly poisoned by CO_2. Thus, CO_2 chemisorption on the alumina would selectively poison isomerization without an effect on exchange activity, and we have four types of sites, chemically identifiable, on the single alumina surface that differ substantially in their response to poisoning.

Another interesting study, also involving alumina and a four-site conception, is that of Chorbel *et al.* (18) on the isomerization of 1-butene on amorphous alumina. The four types of sites were postulated to be acidic, basic, electron accepting, and electron donating, for which selective poisons can be identified as ammonia, acetic acid, phenothiazine, and tetracyanoethylene, respectively. Unfortunately, any one or any combination of these molecules was capable of poisoning the isomerization reaction, hence one could only conclude that a concerted mechanism involving oxidative acidic sites and reductive basic sites is occurring. Many other selective poisoning studies exploring multifunctional oxides have been reported; a number of these are summarized in the previously cited review by Hegedus.

As mentioned earlier, bifunctional supported metal catalysts, particularly those employed in reforming, have been widely investigated and a relatively large amount of information is available on selective poisoning of these materials. Sterba and Haensel (66) have given an interesting review of the development of reforming catalysts and give examples of selective poisoning of Pt via arsenic-containing compounds in feedstock materials and the toning of fresh catalyst with sulfur to modify selectivity properties. In Chapter 1 we discussed two types of general reaction schemes as being at least qualitatively descriptive of deactivation by poisoning and coking. These were of either parallel or series nature, and by our definition of poisoning, the parallel description would ordinarily be applicable for this mechanism of deactivation (68). For bifunctional supported metal catalysts, somewhat similar schemes apply, but with more complexity. Consider the isomerization of n-butane on a Pt/Al_2O_3 catalyst. This reaction is generally thought to proceed via the following series of steps:

$$n\text{-}C_4(g) \xrightleftharpoons{Pt} n\text{-}C_4^=(a) + H_2$$

$$n\text{-}C_4^=(a) \xrightarrow{Al_2O_3} i\text{-}C_4^=(a)$$

$$H_2 + i\text{-}C_4^=(a) \xrightleftharpoons{Pt} i\text{-}C_4(g)$$

In the first step there is a dehydrogenative chemisorption of n-butane on the Pt function to yield an adsorbed olefin. By some mechanism (the nature of which is still a topic of discussion) this adsorbed olefin is transferred to the alumina (acidic) function, where it is isomerized; thence the adsorbed isobutene is transferred back to the Pt function, rehydrogenated, and desorbed into the gas phase as i-butane. Obviously this is a series sequence, and one can interrupt the reaction in several ways via selective poisoning. A poison selective for the Pt function, for example, would greatly decrease the overall activity for isomerization; however, one would still observe i-butane as the reaction product. In the limit of complete poisoning of the Pt function, the reaction would simply be extinguished because the normal paraffin would not isomerize on the Al_2O_3. On the other hand, if the poison is selective for Al_2O_3 and does not affect Pt, then in the limit we would see some mixture of n-butane and n-butene but no isomerized products. In simple terms (perhaps too simple), then, one can think of the selective poisoning of Pt as changing the activity but not the selectivity of the catalyst, while selective poisoning of Al_2O_3 alters both activity and selectivity.

VI. Multifunctional Catalysts

A particularly nice example of selective poisoning of a metal on oxide catalyst exists in the literature (74). This is for the hydroisomerization of 1-butene on an Rh/SiO_2 catalyst, poisoned by mercury. This is an example of parallel bifunctionality, where

$$n\text{-}C_4^= + H_2 \rightleftarrows n\text{-}C_4 \quad (Rh)$$

$$n\text{-}C_4^= \rightleftarrows i\text{-}C_4^= \quad (SiO_2)$$

The mercury is a preferential poison for the metallic function, so one can ultimately terminate the hydrogenation reaction while leaving the isomerization relatively unscathed. Experimental results are demonstrated in Figs. 4-23 and 4-24 for the hydrogenation and isomerization activity, respectively. It is clear that there is a progressive decrease in hydrogenation activity with increasing mercury surface coverage but that the isomerization activity is unaffected for coverages as high as 0.8. Deuterium exchange with 1-butene was also investigated with this catalyst, Fig. 4-25; the parallelism in poisoning behavior to the hydrogenation function strongly suggests that the metal function is also active for exchange.

An example of series bifunctionality was reported by Burnett and Hughes (15) for the disproportionation of butane over a mixture of Pt/Al_2O_3 and

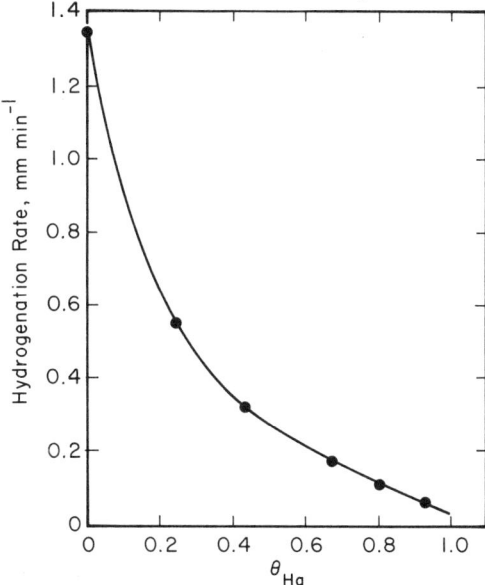

Fig. 4-23. 1-Butene hydrogenation rate as a function of Hg coverage on supported Rh. [From Webb and Macnab (74). Reproduced by permission of Academic Press, Inc.]

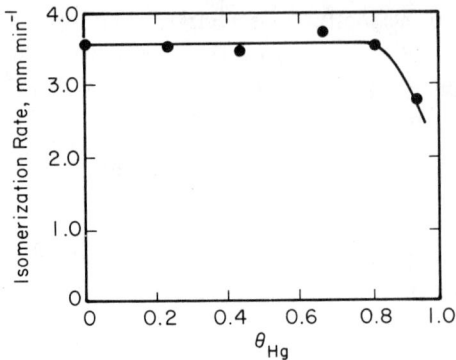

Fig. 4-24. 1-Butene isomerization rate as a function of Hg coverage on supported Rh. [From Webb and Macnab (74). Reproduced by permission of Academic Press, Inc.]

WO_3, with water as a poison for the Pt function. Here we have

$$2C_4 \xrightarrow[-H_2]{Pt} 2C_4^= \xrightarrow{WO_3} C_2^= + C_6^= \xrightarrow[+H_2]{Pt} C_2 + C_6$$

The preferential poisoning of Pt/Al_2O_3 by water vapor suppresses all product formation in this case. However, the most important practical example of the selective deactivation of bifunctional catalysts remains the long-term sintering of the metallic function in reforming catalysts, which is, of course, not poisoning.

Metals deposition on oxide catalysts is a final important example of selective poisoning leading to alteration in product distribution as well as

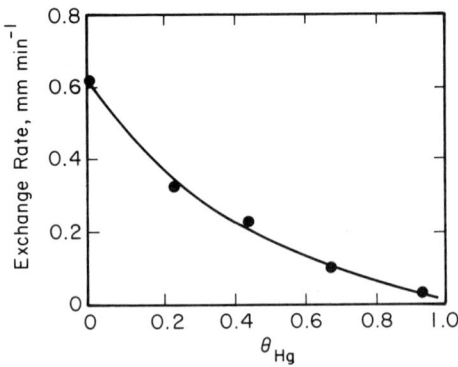

Fig. 4-25. 1-Butene exchange rate with deuterium as a function of Hg coverage on supported Rh. [From Webb and Macnab (74). Reproduced by permission of Academic Press, Inc.]

VI. Multifunctional Catalysts

depressed activity. Ordinarily one thinks of metals deposition as a problem primarily in various hydrotreating processes, attended by simultaneous coke deposition, and the net deactivation as the result of metals poisoning, coke deposition, and eventual pore plugging. In such cases it is difficult to identify the role of metals as a poison separately from all the other things going on simultaneously. However, there is an incentive to use hydrotreated, demetallized residua for fluid catalytic cracking (FCC) units. Such feedstocks will still contain metals, primarily nickel and vanadium compounds, but at greatly reduced levels after hydrotreating. The action of metals on FCC catalysts is more readily identifiable in terms of poisoning action, and early studies of this were reported by Connor and co-workers (20, 30). Subsequent to this work, zeolite cracking catalysts were brought into extensive commercial use, and a comparison of metals poisoning on zeolite and amorphous silica/alumina has been reported by Cimbalo et al. (19). A good example of the poisoning by metals in hydrotreated resid cracking is found in the work of Habib et al. (32). In this work equilibrium catalysts were

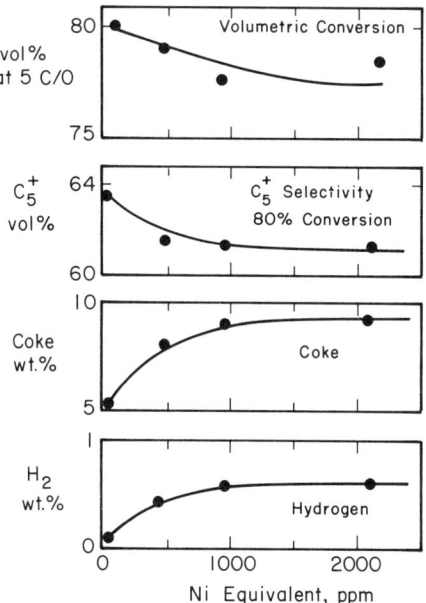

Fig. 4-26. Metals poisoning in fluid catalytic cracking of hydrotreated resid. Feedstock properties: density (60°F), 0.9065 g/cm^3; molecular weight, 524; Conradson carbon, 2.6%; basic N_2, 179 ppm; Ni+V+Fe, 2.3 ppm; weight percent composition, 23.8 paraffins, 26.2 naphthenes, 50.0 aromatics. [From Habib et al. (32). Reproduced by permission of the American Chemical Society.]

preloaded by impregnation with a solution of nickel and vanadium naphthenate to 510, 870, and 2080 ppm Ni equivalent. Cracking was carried out with hydrotreated resid of high reactivity at 1000°F and 3-5 sec contact time. It was found that total conversion dropped only slightly while C_5^+ gasoline yield was about 3% lower but essentially independent of metals loading, as shown in Fig. 4-26. Coke formation was nearly doubled up to the 900 ppm level but thereafter remained constant. Gasoline fraction octane numbers were about 1.5 units *higher* for cracking over the poisoned catalysts, which was attributed to the slightly higher olefin and slightly lower paraffin content of this fraction. Most of these results can be interpreted rather simply in the sense, suggested previously, that metals deposited on the surface do not act so much to poison cracking reactions (other than as the result of physically blocking some active surface) as they do to alter selectivity by superimposing their dehydrogenation activity on the cracking function. Hence, incremental amounts of feedstock are converted to gaseous products and coke that ordinarily would be found in C_5^+ gasoline yield. Nickel, of course, is a particularly effective dehydrogenation catalyst. The increase in H_2 yield shown is significant in operational problems, since this represents a large increase on a volumetric basis and much more stringent gas-handling requirements in commercial operation.

VII. SUMMARY AND EVALUATION

The general discussion of poisoning presented in this chapter has given a large amount of experimental information with some suggestions as to organization via the type of poison, the nature of the catalytic surface (homogeneous, multifunctional), and catalyst morphology. To the extent that such classifications are important in understanding in an overall way the mechanisms of poisoning, they are useful. However, to the practitioner of the art, classifications and correlations are not the whole story, since it is necessary to cope with poisoning on more than an intellectual level. We have emphasized the view of the poisoning process as a competitive chemisorption that produces, in most cases, undesirable and irreversible alteration of the active surface. In this sense, then, one must try to circumvent one type of chemistry with another; this is not an easy task in any event since one is trying to defeat the basic chemical tendencies of a given reaction/poison/catalyst system that do not wish to be altered.

How is this to be done? Generalization is not possible. As we pointed out in the first paragraphs of this chapter, poisoning is a well-defined chemical event, thus individual strategies are defined by individual chemical reactions. To the extent that one can be general, it seems that procedures designed to alter the strength of the chemisorption bond between poison

VII. Summary and Evaluation

and catalytic surface would be most fruitful. Yet the difficulty here is to obtain specificity with respect to the poison without altering appreciably the catalysis of the major, desired reaction. After all is said and done, this really implies altering the electronic nature of the catalyst in a very specific manner, which in turn implies detailed knowledge of both the mechanism of the main reaction and the interaction of the poison with the surface. This requires that one must know a considerable amount of detail concerning the chemistry involved—which, in reality, is not always available.

Let us pursue the point a little further in terms of what might be required. It is simple enough, maybe, to identify the impurity components in a reaction mixture that might be poisons. All that is required is sufficient experimental effort and analytical capability. The problem here is that identification of potential poisons in the reaction mixture does not describe the nature of the *surface* interaction. We have discussed the chemisorption of thiophene on nickel to some extent in this chapter and found a rather complex picture emerging concerned with the geometric configuration as a function of temperature. Yet, that is not the whole picture by any means. At sufficiently high temperatures thiophene will decompose on a nickel surface, leaving a sulfur atom bonded to the chemisorption site. Now the picture of thiophene poisoning previously depicted was stated largely in geometric terms with deactivation by site blockage or screening. On decomposition, this geometric picture is changed, since studies of sulfur interactions in the chemisorption of various molecules on single metal crystal surfaces give evidence of more than short-range electronic interactions. Hence, for thiophene-nickel, we have a perhaps non-atypical example of combined geometric–electronic interactions dependent on the nature of the poison as it exists on the surface. Studies of, and controversies concerning, the nature of the "electronic factor in catalysis" vis-à-vis the "geometric factor in catalysis" have been with us for most of this century; the situation is certainly no different for many cases of chemisorptive poisoning.

In addition to the geometric–electronic factor concerns, it is fruitful in some cases to be aware of subsequent chemical reactions of poisons with the surface. In a sense, the decomposition of thiophene discussed above can also be thought of as an example of this. In such instances one might wish to circumvent a specific deactivation by intercepting a surface reaction pathway of a poison (or poison precursor). In some oxidation reactions on metals, for example, inhibition of the oxidation of the surface can result in enhanced activity maintenance and selectivity. Alteration of metallic oxides by sulfur compounds to form sulfides is also well known and widely encountered.

One of the interesting concepts that comes forward from some of the material in this chapter is control of poisoning via control of catalyst morphology. The best examples of this so far known pertain to supported metal catalysts, and the phenomenon seems related to the structure sensitivity of many reactions on these materials. Simply stated, if there are such things as structure-sensitive reactions, is it not possible that there also exists structure-sensitive deactivation? The answer in the case of deactivation by poisoning would appear at least to be a qualified yes.[2] With respect to the particular case of supported Pt, such effects might have to do with preferential retention of small amounts of oxygen (the chemisorption of which is structure-sensitive) for certain reactions that can be catalyzed by either a metallic function or an acidic function (albeit via different mechanistic pathways). The example of methylcyclopropane hydrogenolysis given in this chapter is one such case. Another example is the effect of β-hydride formation on supported Pd in hydrogenation reactions. The hydride formation is structure-sensitive, and hydride is not formed on high-dispersion materials, resulting in a depression of hydrogenation activity for the larger crystallites on which it is formed. Formation of the hydride, however, may also enhance the resistance to poisoning, resulting in apparent optimal resistance to deactivation for catalysts in a certain range of metal particle sizes. Isolated reports of such behavior have appeared in the literature; research on this is not sufficiently well developed for us to give it more than passing mention here, but the prospects and implications of morphological modification of poisoning behavior are sufficiently important to warrant considerable further study.

Finally, we turn to the control of catalytic selectivity via poisoning. As illustrated in this chapter, this has particular application in bi- or multifunctional catalysts and operates to intercept one reaction pathway to the benefit (or hindrance) of another. The "toning" of reforming catalysts via low-concentration sulfur pretreatment to decrease hydrocracking activity is another. As the importance of new formulations for various applications involving mixed metal oxides or multifunctional metal increases, there is very likely to be highly increased interest in the use of specific poisons to control details of selectivity. Ultimately, in words used above, this means one has sufficient understanding to supplant one type of chemistry with another. Indeed, in very general terms this appears to be the philosophy one should employ in attacking a given problem in any sense more than just to "cope" with catalyst poisoning.

REFERENCES

1. A. Amirnazmi and M. Boudart, *J. Catal.* **39**, 383 (1975).
2. B. G. Baker and R. Peterson, *Proc. Int. Congr. Catal., 6th, 1976* p. 988 (1977).

[2] We will see in the next chapter that this is definitely the situation for sintering deactivation, where rates of sintering are often correlated as a function of active surface area.

References

3. A. A. Balandin, *Z. Phys. Chem., Abt. B* **2**, 289 (1929).
4. J. Barbier, A. Morales, P. Marecot, and R. Maurel, *Bull. Soc. Chem. Belg.* **88**, 569 (1979).
5. K. Baron, *Thin Solid Films* **55**, 449 (1978).
6. O. Beeck, *Discuss. Faraday Soc.* **8**, 118 (1950).
7. Y. Berthier, M. Perderaux, and J. Oudar, *Surf. Sci.* **36**, 225 (1973).
8. R. M. Billimoria and J. B. Butt, *Chem. Eng. J.* **22**, 71 (1981).
9. D. W. Blakely and G. A. Somorjai, *J. Catal.* **35**, 200 (1976).
10. G. C. Bond, *Discuss. Faraday Soc.* **14**, 200 (1966).
11. H. P. Bonzel and R. Ku, *J. Chem. Phys.* **58**, 4617 (1973).
12. M. Boudart, A. M. Aldag, J. E. Benson, N. Dougharty, and G. Harkins, *J. Catal.* **6**, 92 (1966).
13. M. Boudart and L. D. Ptak, *J. Catal.* **16**, 90 (1970).
14. M. Brown, R. E. Peierls, and E. A. Stern, *Phys. Rev. B* **15**, 738 (1977).
15. R. L. Burnett and T. R. Hughes, *J. Catal.* **31**, 55 (1973).
16. J. B. Butt, *Adv. Chem. Ser.* No. 109, 259 (1972); in "Catalysis—Science and Technology" (J. R. Anderson and M. Boudart, eds.), Vol. 6, Chap. 1. Springer-Verlag, Berlin and New York, 1985.
17. J. B. Butt, *Chem. React. Eng., Proc. Eur. Symp., 4th, 1968*, p. 255 (1971).
18. A. Chorbel, C. Hoang-Van, and S. J. Teichner, *J. Catal.* **33**, 123 (1974).
19. R. N. Cimbalo, R. L. Foster, and S. J. Wachtel, *Oil Gas J.* **70**, 122 (1972).
20. J. E. Connor, Jr., J. J. Rothrock, E. R. Birkheimer, and L. N. Leun, *Ind. Eng. Chem.* **49**, 276 (1957).
21. R. A. Dalla Betta, A. G. Piken, and M. Shelef, *J. Catal.* **40**, 173 (1975).
22. D. E. Damiani and J. B. Butt, *J. Catal.* **94**, 203 (1985).
23. T. A. Dorling and R. L. Moss, *J. Catal.* **7**, 378 (1967).
24. R. P. Eischens and W. A. Pliskin, *Adv. Catal.* **10**, 1 (1958).
25. V. Eskinazi and R. L. Burwell, Jr., *J. Catal.* **75**, 118 (1975).
26. J. Fahrenfort, L. L. von Reyen, and W. M. J. Sachtler, in "The Mechanism of Heterogeneous Catalysis" (J. H. de Boer, ed.), p. 23. Elsevier, Amsterdam, 1960.
27. T. E. Fischer and S. R. Kelemen, *J. Catal.* **53**, 24 (1978).
28. M. Flytzani-Stephanopoulos, S. S. Wong, and L. D. Schmidt, *J. Catal.* **49**, 51 (1977).
29. S. Fuentes and F. Figueras, *J. Catal.* **54**, 397 (1978).
30. H. R. Grane, J. F. Connor, Jr., and G. P. Mosobgites, *Pet. Refiner* **40**, 168 (1961).
31. H. Gruber, *J. Phys. Chem.* **66**, 48 (1962).
32. E. T. Habib, Jr., H. Owen, P. W. Snyder, C. W. Streed, and P. B. Venuto, *Ind. Eng. Chem. Prod. Res. Dev.* **16**, 291 (1977).
33. G. L. Haller, W. N. Delgass, R. Kellerman, and J. H. Lunsford, "Spectroscopy in Heterogeneous Catalysis." Academic Press, New York, 1979.
34. W. Heegemann, K. H. Meister, E. Berthold, and K. Hayek, *Surf. Sci.* **49**, 161 (1975).
35. L. L. Hegedus, *Catal. Rev.—Sci. Eng.* **23**, 377 (1981).
36. E. F. G. Herrington and E. K. Rideal, *Trans. Faraday Soc.* **40**, 505 (1944).
37. R. F. Hicks and A. T. Bell, *J. Catal.* **91**, 104 (1985).
38. M. P. Rosynek and J. W. Hightower, *Catal., Proc. Int. Congr., 5th, 1972*, p. 851 (1973).
39. H. Knözinger, *Adv. Catal.* **25**, 184 (1976).
40. D. C. Konigsberger, J. B. A. D. van Zon, H. F. J. van't Blik, J. G. Visser, R. Prins, A. N. Mansour, D. E. Sayers, D. R. Short, and J. R. Katzer, *J. Phys. Chem.* **89**, 4075 (1985).
41. J. M. Lornston, Ph.D. Thesis, Univ. of Delaware, Newark, 1981.
42. F. W. Lytle, *J. Catal.* **43**, 376 (1976).
43. G. D. Lyubarskii, L. B. Andreeva, and H. Y. Kulkova, *Kinet. Katal.* **3**, 102 (1962).
44. W. H. Manogue and J. R. Katzer, *J. Catal.* **32**, 166 (1975).
45. R. Maurel, G. Leclercq, and J. Barbier, *J. Catal.* **37**, 324 (1975).
46. E. B. Maxted, *Adv. Catal.* **3**, 129 (1951).

47. R. W. McCabe and L. D. Schmidt, *Surf. Sci.* **60**, 85 (1976); **65**, 189 (1977).
48. R. W. McCabe, T. Pignet, and L. D. Schmidt, *J. Catal.* **32**, 114 (1974).
49. K. W. McHenry, R. J. Bertolacini, H. M. Brennan, J. L. Wilson, and H. S. Seelig, *Proc. Int. Congr. Catal., 2nd* p. 117 (1960).
50. G. A. Mills, E. R. Boedeker, and A. G. Oblad, *J. Am. Chem. Soc.* **72**, 1554 (1950).
51. S. Mroz, *Surf. Sci.* **83**, 1625 (1979).
52. A. G. Oblad, T. H. Milliken, Jr., and G. A. Mills, *Adv. Catal.* **3**, 199 (1951).
53. I. Önal and J. B. Butt, *J.C.S. Faraday I* **78**, 1887 (1982).
54. J. J. Ostermaeir, J. R. Katzer, and W. H. Manogue, *J. Catal.* **41**, 277 (1976).
55. J. Oudar, *Catal. Rev.—Sci. Eng.* **22**, 171 (1980).
56. J. B. Peri, *J. Phys. Chem.* **69**, 231 (1965).
57. H. Pines and W. O. Haag, *J. Am. Chem. Soc.* **82**, 2471 (1960); see also H. Pines and J. Manassen, *Adv. Catal.* **16**, 49 (1966).
58. P. Ratnasamy, A. J. Leonard, L. Rodrique, and J. J. Fripiat, *J. Catal.* **29**, 314 (1973).
59. T. N. Rhodin and C. F. Brucker, *Solid State Commun.* **23**, 275 (1977).
60. C. H. Rochester and R. J. Terrell, *J.C.S. Faraday I* **73**, 609 (1977).
61. E. Rorris, J. B. Butt, R. L. Burwell, Jr., and J. B. Cohen, *Proc. Int. Congr. Catal., 8th* **4**, 321 (1984).
62. P. C. Saunders and J. W. Hightower, *Prepr., Div. Pet. Chem., Am. Chem. Soc.* **15**, A79 (1970).
63. L. D. Schmidt and D. Luss, *J. Catal.* **22**, 269 (1971).
64. J. H. Sinfelt, *Adv. Catal.* **23**, 91 (1970).
65. L. L. C. Sorensenand and K. Nobe, *Ind. Eng. Chem. Prod. Res. Dev.* **11**, 423 (1972); *Environ. Sci. Technol.* **6**, 239 (1972).
66. M. J. Sterba and V. Haensel, *Ind. Eng. Chem. Prod. Res. Dev.* **15**, 2 (1976).
67. B. E. Sundquist, *Acta Metall.* **12**, 67 (1964).
68. S. Szépe and O. Levenspiel, *Chem. React. Eng., Proc. Eur. Symp., 4th, 1968*, p. 265 (1971).
69. T. Uchijima, J. M. Herrmann, Y. Inoue, R. L. Burwell, Jr., J. B. Butt, and J. B. Cohen, *J. Catal.* **50**, 464 (1977).
70. R. van Hardeveld and F. Hartog, *Surf. Sci.* **15**, 189 (1969); *Adv. Catal.* **22**, 76 (1972).
71. M. A. Vannice, *J. Catal.* **37**, 462 (1975).
72. A. Verma and D. M. Ruthven, *J. Catal.* **46**, 160 (1977).
73. J. Völter and M. Hermann, *Z. Anorg. Allg. Chem.* **405**, 315 (1974).
74. G. Webb and J. I. Macnab, *J. Catal.* **26**, 226 (1972).

CHAPTER 5

Deactivation by Sintering

> Less is more
> *Mies van der Rohe*

> More or less
> *JBB & EEP*

In Chapter 1 we defined sintering as the loss of active surface area by the agglomeration of small crystallites into larger ones in the case of supported metal catalysts, or the collapse of pore structure and loss of internal surface area in the case of supports or various unsupported (normally oxide) catalysts. Sintering has long been of interest in various aspects of powder metallurgy, and detailed theories have been worked out for the sintering of supported metal powders. However, the mobility of metal atoms over metal surfaces is relatively large; in supported metal catalysts the individual metal particles are separated from each other by relatively large distances, as small islands in a sea of the support which itself is generally not a metal. Hence one would expect to find, and does find, significant differences in the sintering behavior of supported and unsupported metals. Figure 5-1 shows a typical transmission electron micrograph (TEM) of Pt crystallites supported on a γ-Al_2O_3 film; in this specimen the crystallites are approximately 110 Å in size, and while the spacing is somewhat irregular, they are on average separated from each other by four to five times this average dimension.

In this chapter we shall be primarily concerned with phenomena responsible for and accompanying the sintering of supported metals. The sintering of oxides or other porous unsupported catalysts has been treated to some extent in the literature; however, probably the most important single area where deactivation by sintering is encountered is in applications of suppor-

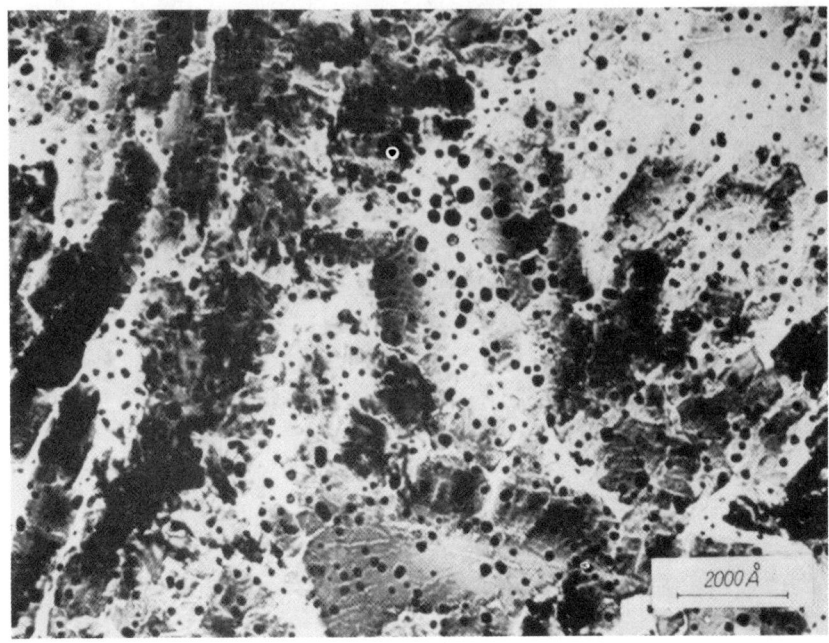

Fig. 5-1. TEM of Pt crystallites on a γ-Al_2O_3 film support. [From Ruckenstein and Malhotra (55), Fig. 5. Reproduced by permission of Academic Press, Inc.]

ted metals. Even in the latter case, though, we will see that our knowledge is restricted and our understanding even more so. Sintering, either of supported metals or other materials, is a complicated phenomenon where a variety of mechanisms can be envisioned, all of which probably occur under one set of circumstances or another.

I. SINTERING IN SUPPORTED METAL SYSTEMS

A. General Aspects

One may argue with some generality that, if metal–metal bond energies are greater than metal–support interaction energies, the larger a metal crystallite is the more stable it will be. Two differing but quite general pictures have been set forth for the mechanism of crystallite growth. In one, termed the *atomic migration model*, sintering occurs via the sequence of escape of metal atoms from a crystallite, transport of these atoms across the surface of the support, and subsequent capture of the migrating atoms on collision with another metal crystallite. Since larger crystallites are more stable than smaller ones, one envisions as a consequence that small crystallites diminish in size and larger ones increase. This mechanism is also sometimes termed *Ostwald ripening* by analogy to the corresponding process of particle growth in gaseous or liquid environments resulting from the motion of individual atoms or molecules. The theoretical development of this model has been presented in detail by Flynn and Wanke (18), among others, and further discussion of its quantitative development will be given subsequently. A second model, termed the *crystallite migration model*, visualizes sintering to occur via migration of the crystallites themselves along the surface of the support. This also has another name; it is sometimes referred to as the *Smoluchowski coagulation model*. The motion of metal atoms on the surface of the crystallite in a random manner produces the two-dimensional analog of Brownian motion of the particles on the support, and crystallite growth occurs on the collision and coalescence of two crystallites. The theory of this model has been developed by Ruckenstein and co-workers (54, 56); Wynblatt and Gjostein (77) have also derived expressions for particle growth via either surface diffusion or transport through the vapor phase. Again, more detail concerning this model will be given later.

It is apparent that a number of different rate-limiting steps can potentially be identified in either model. In atomic migration these could be the rate of detachment of atoms from a crystallite, the rate of migration across the surface, or possibly the rate of capture of atoms by the growing crystallite. For crystallite migration the rate could be determined by the rate of collision or coalescence of migrating crystallites. Regardless of mechanism, metal–metal and metal–support interaction energies (and all other factors determining the free energy of small crystallites) are of central importance, since they will dictate in large measure the rates of atom detachment, rates of migration of atoms or crystallites, and coalescence rates. Unfortunately, these are, for the most part, unmeasured and unknown. However, experimental observations of a more macroscopic nature have been made on sintering, and it will be instructive at this point to turn our attention to some of these and the resulting correlations.

B. Experimental Studies—Characterization Techniques

In 1975 Wanke and Flynn (70) presented a thorough review of the literature on the sintering of supported metals up to that date. Much of the information appearing since then is similar, although further detail and refinement have been added (71). A prime factor to be considered is the experimental methodology employed in study of the sintering phenomena, that is, how the state of a supported metal, in the form of extremely small discrete crystallites, is characterized. Some discussion of this is in order here.

Most modern literature expresses primary data on sintering as the change in percentage exposed, or dispersion, with time-temperature-atmosphere. Percentage exposed (cf. Chapter 4) is defined as the percent of total metal atoms which are surface atoms, and this measure may be translated into an equivalent crystallite dimension if some characteristic geometry (i.e., hemisphere, sphere, cube) is assumed for the crystallite. In practice, several experimental techniques may be employed to determine percent exposed, the most widely used of which appear to be transmission electron microscopy (TEM), selective gas adsorption, and x-ray diffraction (XRD). The order here may vary with one's prejudices and the laboratory equipment available; it is, in fact, not particularly important since each method has advantages and disadvantages.

Transmission microscopy is, of course, the most direct method of examination, and in recent years the resolution claimed has increased to include individual particles (clusters) smaller than 1 nm. Since individual particles are detected via TEM, particle size distributions can be directly determined. As pointed out by Flynn and Wanke (19), though, such distributions are based on three implicit assumptions:

(i) The size of a particle is the same as its image recorded on the micrograph, corrected for magnification.

(ii) Detection of a given size of particle implies that all particles of that size and larger are being detected.

(iii) Contrast of the metal particles is distinguishable from the contrast arising from the support material.

Each of these assumptions is nontrivial. Assumption (i) deals with the fact that the photographic image is two-dimensional; crystallites which are not equiaxed will have apparent different dimensions depending on their orientation. Assumption (ii) is essentially statistical in nature; if particle size distributions are being determined, is the sampling a valid representation of the true distribution in both size range and size frequency? Another way to phrase this is, do not worry about what you see in TEM, but what you do not see. Assumption (iii) is somewhat related to (ii), since if insufficient contrast exists between metal and support it may be possible to miss the identification of some images. Indeed, these three assumptions are in general not correct for small particles (ca. 2 nm) and it is necessary to establish limits for both the smallest reliable particle size detected and the smallest reliable difference between two particle sizes. The procedures involved in such studies have to do with details of TEM theory and operation with which we assume many readers, like ourselves, are not expert. However, a

I. Sintering in Supported Metal Systems

rather definite conclusion is produced: particle size distributions determined by TEM become increasingly unreliable as the particle size extends below 2.0 nm, and accurate detection and identification of particles below this size are "extremely unlikely." Further, "variation in apparent particle size up to 1 nm suggests this value as the lowest meaningful division of diameters in a particle size distribution." We should also point out that by far the bulk of TEM data published have to do with supported Pt or Pd; as one goes to supported metals of lower atomic number (e.g., Fe) these problems—particularly those associated with contrast—are exacerbated. Additional interesting studies of supported metals via TEM are provided by Freel (23), Wilson and Hall (75), and Baker et al. (4, 5).

Selective chemisorption is probably the most widely employed method for determination of percentage exposed. For group VIII metals the method is based on the well-known fact that hydrogen chemisorption on most of them is dissociative with hydrogen atom bonding to individual surface metal atoms. Similarly, carbon monoxide chemisorption is generally nondissociative, also with bonding to surface metal atoms. These systems typically exhibit Langmuir isotherms, so determination of monolayer coverage gives directly the number of surface metal atoms—given the appropriate chemisorption stoichiometry—and hence the percentage exposed. Figure 5-2 illustrates some typical data of this sort for the chemisorption of hydrogen and carbon monoxide on several silica-supported group VIII metals. Additional information may be derived from these measurements, such as metal surface area or average crystalline size, as shown in Table 5-1. It is important, however, to distinguish between catalysts exhibiting normal chemisorption behavior and those with so-called strong metal-support interactions (SMSI), as will be discussed later.

It is probably worthwhile to spell out in more detail the role of stoichiometry in such measurements. For unit stoichiometry the ratio H/M or CO/M is, of course, a direct measure of percentage exposed; however, more formally

$$\phi = (G/M_T)/(G/M_s) \tag{5-1}$$

where ϕ is the fraction exposed, (G/M_T) the total uptake of gas on the metal, and (G/M_s) the stoichiometry of the chemisorption. For these systems it was shown by independent means that (G/M_s) was unity.

The metal surface areas are determined from the value of fraction exposed by

$$S_M = M_T A_M \phi \tag{5-2}$$

where S_M is the metal surface area, M_T the total number of metal atoms in the sample, and A_M the area occupied by a metal atom in the surface. The

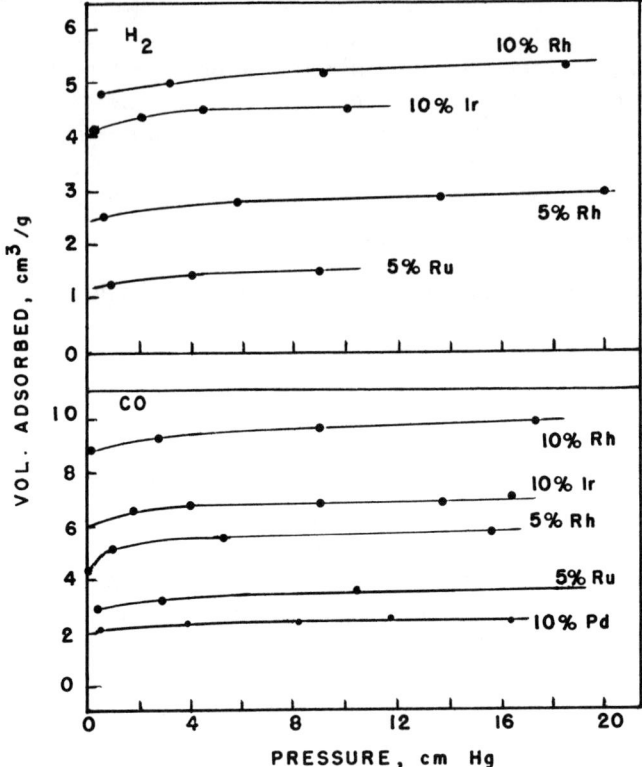

Fig. 5-2. Chemisorption of hydrogen and carbon monoxide on silica-supported group VIII metals. [From Sinfelt and Yates, *J. Catal.* **8**, 82 (1967). Reproduced by permission of Academic Press, Inc.]

value of A_M may be estimated from the reported value of 6.5 Å² determined for nickel (36) by adjusting for the difference in atomic radius between nickel and the metal under consideration. Hence:

$$A_M = A_{Ni}(\tau_M/\tau_{Ni})^2 = 6.5(\tau_M/\tau_{Ni})^2 \qquad (5\text{-}3)$$

An average particle dimension may also be evaluated from metal surface areas by assuming the particles are cubes of dimension *l* with one side bound to the support according to

$$l = 5/S_M\rho \qquad (5\text{-}4)$$

where ρ is the density of the metal and S_M the metal surface area per gram of metal. Of course, other geometries may be assumed which will yield somewhat different values of the dimension from the same measurement of adsorbate–metal ratio.

TABLE 5-1

Surface Areas and Crystallite Sizes of Silica-Supported Metals[a]

Catalyst[b]	H/M[c]	CO/M[d]	Area (m^2)/g M[e]	Size (nm)
10% Pd	—	0.11	47	10.6
10% Rh	0.48	0.44	214	2.3
5% Rh	0.53	0.52	237	2.0
5% Ru	0.27	0.31	121	4.2
10% Ir	0.78	0.58	186	1.4

[a] From J. H. Sinfelt and D. J. C. Yates, *J. Catal.* **8**, 82 (1967). Reproduced by permission of Academic Press, Inc.
[b] All metals supported on Cabosil.
[c] Atoms of H adsorbed (at 10 cm H_2 pressure) per metal atom.
[d] Molecules of CO adsorbed (at 10 cm pressure) per metal atom.
[e] Calculated from the hydrogen adsorption data.

An alternative method for surface characterization of supported metals has been developed that proposes titration of the oxidized surface rather than chemisorption on the metal surface after reduction and desorption (6). This has been employed widely for supported platinum; oxygen is chemisorbed on the metal surface to form PtO_x, and the titration proceeds according to

$$PtO_x + (x + y/2)H_2 \to Pt + xH_2O \tag{5-5}$$

If $x = y = 1$, then three hydrogen atoms are consumed for each platinum atom, so there is a considerable increase in sensitivity compared to conventional hydrogen chemisorption.

As is the case for transmission microscopy, these chemisorption/titration methods are not without their difficulties. First, it is apparent that only total uptake is measured and thus average values are determined. There is no way to measure a size distribution from chemisorption or titration. Second, evaluation of percentage exposed requires an assumption or an independent determination of the stoichiometry of chemisorption/titration. This has been a particular thorn in the side of the titration procedure, where the stoichiometry of Eq. (5-5) has been reported to vary from $x = y = 1$ to $x = 1$, $y = 2$ as the particle size decreases (59). This may be the result of changes in the intrinsic stoichiometry of the H–O reaction or, as has been shown by Uchijima *et al.* (68), to a change in the nature of what is termed PtO_x with particle size. For measurements via chemisorption rather than titration, stoichiometric uncertainty also exists for CO on some metals. For example, CO chemisorbs on palladium in either a linear or bridge-bonded form, leading to stoichiometries of 1:1 or 1:2 CO:M (surface), respectively. Additional aspects of these chemisorption and titration methods have been discussed in Chapters 1 and 2. Measurements with hydrogen seem the most reliable for most group VIII metals with the stoichiometry well established as 1H:1M (surface), although even here we must beware of the formation of hydride phases in some cases or of kinetic limitations in others. There has been considerable discussion in the literature that migration of hydrogen to the support surface after initial chemisorption on the metal ("spillover") might lead to erroneously high values of H:M, and reviews have been written on the topic (59). However, the accumulation of evidence over the years, and in particular in cases where the results of hydrogen chemisorption have been compared with other techniques (53, 57), indicates that spillover is not an overriding factor, at least for prototype catalysts such as Pt/SiO_2 or Pt/Al_2O_3.

We stated above that in general for group VIII metals the stoichiometry of hydrogen chemisorption seems fairly well established as unity on an atomic basis. However, the past few years have seen a number of reports of reliable work that show rather definitely that there are certain conditions in certain systems where large deviations from this are observed, stoichiometries can become variable (structure-sensitive), and certainly hydrogen uptake cannot be used to determine metal dispersions. One can identify in general two problem areas: (i) measurement of hydrogen uptake after a catalyst on a typical support such as alumina has been treated in hydrogen at high temperatures (>500°C), and (ii) measurement of hydrogen uptake on Pt group metals supported on reducible oxides such as TiO_2, V_2O_3, and Nb_2O_5.

The first effect, normally referred to as that of "strongly bound hydrogen," seems first to have been identified as such by Dautzenberg and co-workers (13, 14) for Pt/Al_2O_3, but on the basis of a number of studies in recent years this would appear alternatively to be a phenomenon associated with the high-temperature reduction in hydrogen of Pt group metals on conventional supports. The area has been summarized in an excellent review by Paál and Menon (50). The means of avoiding problems with strongly bound hydrogen when measuring metal dispersion are very simple: avoid if possible any pretreatment procedure that employs hydrogen at temperatures greater than about 500°C. From the available data it appears that temperature rather than time is the overriding factor; thus even fairly short exposure of Pt/Al_2O_3 or Pt/SiO_2 to hydrogen at, say, 550°C can lead to significant strongly bound hydrogen, whereas prolonged exposure of the same materials at 350°C has no effect. The general result of strongly bound hydrogen is to reduce the uptake of hydrogen compared to that of the normal metal in subsequent chemisorption experiments used to determine the dispersion (or percentage exposed), so that one tends to underestimate metal exposed or, if making extensions of such data to crystallite size, overestimate the average particle dimension.

The second effect, also associated with high-temperature reduction but on reducible oxides, is now commonly termed the strong metal–support interaction (SMSI). Again, there has been much work in this area, but the fundamentals were well outlined in the early work of Tauster and co-workers (66, 67). The distinction here also seems to be between reduction temperatures of the order of 300–400°C and those greater than 500°C, but unlike SiO_2 or Al_2O_3 the oxides associated with SMSI are themselves reduced at the higher temperatures. The origin of the metal–support interaction has been widely discussed, with a predictable divergence of opinion. One thought ascribes the interaction to an electron transfer mechanism with the reduced oxide, another to possible encapsulation of the metal by the oxide, and a number of other possibilities have been suggested. For those interested, the SMSI menu is generally outlined in two representative symposia (3, 63). Our interest here, however, is not so much in the mechanism as in the effect—again a suppression of apparent chemisorption capacity that can lead to large errors in the estimates of metal dispersion based on measurements of the uptake of hydrogen or carbon monoxide.

As Wanke *et al.* (71) point out, in either case the influence of reduction temperature on hydrogen chemisorption has a serious influence on sintering studies, and it really does not make much difference whether the effect is due to strongly bound hydrogen or SMSI. The apparent depression in uptake of calibrating gases leads to false information that can completely disguise the interpretation of particle size, which is in turn, of course, the key to understanding sintering. This is particularly applicable to the interpretation of results reported for sintering in reducing atmospheres (e.g., H_2), and some results of this sort are discussed subsequently.

The third method of characterization, x-ray diffraction, recently reviewed by Gallezot (24), has most commonly employed the wide-angle diffraction pattern in which the half-width or the integral breadth of a peak is used to obtain a size. In general, the width of an x-ray diffraction peak is a function of both particle size and strain, but in the absence of strain the particle size is given by the Scherrer equation:

$$t = 0.9\lambda / B \cos \theta_B \qquad (5\text{-}6)$$

where B is the peak width at half-maximum, λ the x-ray wavelength, θ_B the diffraction angle, and t the particle diameter, in angstroms. If a number of crystallographic directions are analyzed, the shape of the particles can be ascertained. In actual measurement, the instrument itself makes a contribution to the measured width (instrumental broadening). It is normally assumed that the shapes of the particle size distribution and of the instrumental broadening are Gaussian and therefore can be deconvoluted by the simple relationship

$$B^2 \text{ (actual)} = B^2 \text{ (measured)} - B^2 \text{ (instrumental)} \qquad (5\text{-}7)$$

Instrumental broadening, B^2 (instrumental), is determined by running a diffraction pattern on large particles (>100 nm) of the metal being investigated. Almost all reported x-ray measurements of supported metal crystallite size in the catalytic literature have been obtained by this method or variants on it, and the topic has been reviewed by Whyte (74) in addition to Gallezot. Difficulties with the x-ray method arise from a number of sources. The use of Eq. (5-6) presumes the absence of strain, yet for large crystallites possibly approaching the dimensions of the micropore structure of a support there may exist strain resulting from orientated growth; in this event the estimated particle size will be greater than the true particle size. Particle size distributions may not be Gaussian, as required by the instrumental broadening correction of Eq. (5-7). In a number of metal–support systems there are interferences between the diffraction pattern of the metal and of the support, and it is difficult to obtain unobscured peaks of sufficient intensity to make reliable measurements of B; Pt/Al_2O_3 is a very important example of this. Wanke et al. (71), however, have discussed some methods for avoiding this problem via subtraction of blank and active spectra. Finally, but most importantly for supported metal catalysts, there is a lower limit of particle sizes detectable by x ray. This limit is determined by the coincidence of the wavelength λ and particle size (or, more precisely, as the two approach each other in dimension), at which point the sample appears nearly amorphous to x rays and the diffraction "peaks" become very broad and of low intensity. Controversy appears in the literature from time to time as to what this limit is, but a figure of 4 to 5 nm is often quoted. This is singularly unfortunate for catalysis; for example, a quick glance at the representative catalysts in Table 5-1 shows that four out of five have crystallite sizes via chemisorption below this range and hence would be inaccessible to reliable characterization via x-ray line broadening. However, more modern instrumentation and analysis has permitted the elimination of a number of these limitations of x-ray analysis. Fourier analysis of the diffraction profile, called the Warren–Averbach method (72), permits the determination of average crystallite size, crystallite shape, microstrain, and (in the absence of strain) crystallite size distribution. This technique is a very powerful one and has been employed using conventional diffraction equipment to characterize supported platinum particles to sizes as small as 2.5 nm. Figure 5-3, for example, shows some particle size distributions determined by Sashital et al. (57) for Pt/SiO_2 of this average dimension. Note that there is a large information content in these distributions. The average sizes $\langle L \rangle$ in three different directions, (111), (100), and (311), are 3.0, 2.6, and 2.0 nm, respec-

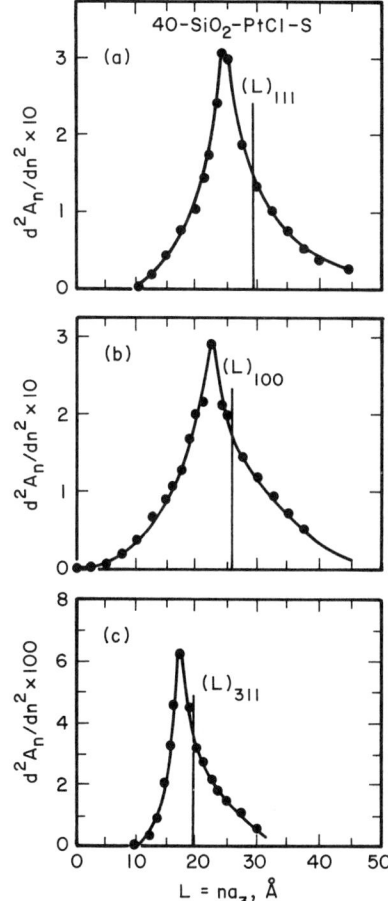

Fig. 5-3. Particle size distribution of a 40% exposed Pt/SiO$_2$ catalyst determined by Warren–Averbach analysis of x-ray diffraction pattern. [From Sashital et al. (57). Reproduced by permission of Academic Press, Inc.]

tively, so the crystallites are nearly equiaxed, or roughly spherical. The peak widths at half-maximum are ca. 1 nm so the distributions are relatively narrow. Finally, the distributions appear somewhat skewed toward larger crystallite sizes, which, as we shall see, some of the theoretical development regards as evidence for the mechanism of sintering via coalescence growth (Smoluchowski model). It should be pointed out that the material in Fig. 5-3 was not sintered; however, the shape of the particle size distribution is still evidence of the mechanism of particle growth during preparation.

Percentage exposed may also be determined from x-ray data, given a value (or values) for $\langle L \rangle$. This is calculated via

$$D_x = 100 \frac{\text{(surface)}}{\text{(volume)}} \left(\frac{\rho_s}{\rho_B} \right) = \frac{600}{\langle L \rangle} \frac{\rho_s}{\rho_B} \tag{5-8}$$

where D_x is percentage exposed, x-ray, and ρ_s and ρ_B are surface and bulk densities. For Pt, ρ_s would normally be taken as the average for (111), (100), and (110) planes. For the example of Fig. 5-3, D_x was determined as 45% (average of two measurements) versus 40% determined via hydrogen chemisorption (D_h).

The technique of small-angle x-ray scattering offers promise for the study of very small particles; x rays are scattered because of the discontinuity in electron density between the particles and the surrounding medium, and it is not necessary to know the crystal structure of the diffracting particles to determine their size. Interference by scattering by the porous support is a serious problem, but the use of pore maskant material of electron density comparable to that of the support can suppress such interference (53). For many silica gels, for example, a mixture of methyl iodide and methylene iodide can be used for this purpose. Sashital et al. have shown that small-angle scattering may also be used to measure the internal surface area of porous supports. A comparison of chemisorption, microscopic, and diffraction techniques has recently been given by Matyi et al. (42).

C. Some Experimental Observations—General

Wanke and Flynn divide experimental studies of sintering of supported metals into three general categories: (i) variable time–variable temperature experiments, (ii) constant time–variable temperature experiments, and (iii) constant temperature–variable time experiments. It is convenient in many instances to maintain this distinction here, and it will be seen that the vast majority of such data are for supported Pt.

Some early data on the sintering of supported Pt/Al$_2$O$_3$ in variable temperature–variable time experiments were reported by Herrmann et al. (32), Somorjai (61), and Hughes et al. (33). More recently, experiments have been reported by Bett et al. (7) for Pt on graphitized carbon black. We will not attempt to give a comprehensive survey of all results here (more details are given in the review by Wanke and Flynn previously cited), but some of the essential points from these experiments are surveyed in Table 5-2.

Two facts are indicated from these data; sintering rates are strongly affected by temperature, and there appears to be an effect of the atmosphere on the rate of sintering. The approximately 100°C difference in sintering temperatures employed by Herrmann et al., for example, results in approximately an order of magnitude increase in the rate of loss of dispersion, while the change in atmosphere from oxidizing to reducing employed by Somorjai results in some retardation in the rate of sintering, although the effect is not large. Indeed, the data of Bett et al. are inconclusive in this

I. Sintering in Supported Metal Systems

TABLE 5-2

Typical Variable Time–Variable Temperature Sintering Data

(1) Investigator: R. A. Herrmann et al. (32)
    Catalyst: Commercial reforming, 0.375% Pt/Al_2O_3, BET = 176 m^2/g

Atmosphere	Temperature (°C)	Time (hr)	D/D_0
N_2	564	44	0.45
		167	0.17
		353	0.10
N_2	625	4	0.29
		8	0.12
		18	0.10
		40	0.08

(2) Investigator: G. A. Somorjai (61)
    Catalyst: 5% $Pt/\eta\text{-}Al_2O_3$

Atmosphere	Temperature (°C)	Time (hr)	D	D_p (nm)
Air	600	1	0.082	12.4
		6	0.051	20.0
		24	0.043	23.6
		48	0.040	25.4
H_2	600	1	0.094	10.8
		6	0.055	18.4
		24	0.050	20.4
		96	0.046	22.1

(3) Investigator: T. R. Hughes et al. (33)
    Catalyst: 0.4% Pt/Al_2O_3
    Atmosphere: H_2 at T = 900 and 1000°F, $0 < t < 1000$ hr
    Correlation: $D = at^b$ $a = 0.73\ (900),\quad 0.67\ (1000)$
                              $b = -0.13\ (900),\quad -0.14\ (1000)$

(4) Investigator: J. A. Bett et al. (7)
    Catalyst: 20% Pt on graphitized C (Vulcan XC-72, Cabot Corp., 80 m^2/g) via impregnation from $Pt(NH_3)_2(NO_3)_2$

Atmosphere	Temperature (°C)	Time (hr)	D
N_2	600	0	0.26
		1.5	0.16
		6	0.16
		16	0.14
		65	0.13
H_2		0.5	0.16
		2	0.15
		6	0.16
		16	0.13

regard, though it should be noted that comparison is only provided between inert and reducing atmospheres.

The correlation of Hughes *et al.* is reminiscent of time-on-stream correlations for coke deposition, and, indeed, correlations of this sort for sintering have been presented periodically in the literature. There may even be a bit more rationale for a time-on-stream correlation for sintering than coking, since the former process may very well be independent of conversion level (an implicit assumption of time-on-stream approaches), while there is little chemical reason that the latter should be. Thus, such correlations may have use as interpolation models but one should not ascribe particular physical significance to the parameters involved, nor expect too much by way of correlation between parametric values and the particular conditions of sintering or the properties of the catalyst investigated.

A second type of sintering data is obtained from experiments run at varying temperatures for the same amount of time. Examples of this, again abstracted from the review of Wanke and Flynn (70), are given in Table 5-3 for two types of Pt/SiO_2 catalysts. Comparison of the air-sintered samples of Sagert and Pouteau at 500 and 600°C, for example, reveals again the activated nature of the sintering process. Also, an interesting comparison is provided by considering series a and b from Benesi *et al.* It appears that ion exchange provides a means of obtaining higher initial dispersions; however, it is not clear from these data whether the ion-exchanged materials are more or less resistant to sintering. Interpretation of the kinetics of sintering from such variable temperature–constant time experiments is a dangerous sport, since only the behavior at a single point is considered. If there is a crossover in the continuous curves of percentage exposed versus time, then comparison of two samples on one side of the crossover point would yield opposite results from comparison on the other side of this point. Yet, one cannot know this unless the complete time histories of both samples are determined. Conclusions based on single-point comparisons, thus, must in general be regarded as tentative ("suspect" is a more harsh but perhaps more accurate term). It may seem trivial to belabor a point such as this, but the temptation to make such point comparisons has not always been avoided in the literature.

Perhaps the most common type of experimental data on sintering comes from constant temperature–variable time studies. Two examples of these data are given in Table 5-4a, also abstracted from the review of Wanke and Flynn. Here somewhat less ambiguous comparisons may be made since absolute dispersions are reported and experiments are run for similar catalysts in differing atmospheres and at different temperature levels. The much slower sintering of Pt/Al_2O_3 at the lower temperature in the reducing atmosphere is strong evidence that both factors act to retard this process.

I. Sintering in Supported Metal Systems

TABLE 5-3

Constant Time–Variable Temperature Sintering Data

(1) Investigator: N. H. Sagert and R. M. L. Pouteau, *Can. J. Chem.* **49**, 3411 (1971)
Catalyst: 1% Pt/SiO$_2$, impregnated (a) or ion-exchanged (b), BET = 285 m^2/g

Sample[a]	Temperature (°C)	Time (hr)	Atmosphere	D/D_0[b]
a	500	5	Air	0.27
b	500	5	Air	1.0
a	500	5	Air	0.048
a	600	5	Air	0.15
b	600	5	Air	0.70
a	600	5	Air	0.041

(2) Investigator: H. A. Benesi, R. M. Curtis, and H. P. Studer, *J. Catal.* **10**, 328 (1968)
Catalyst: 1.6% Pt/SiO$_2$, impregnated (a) or ion-exchanged (b), BET = 370 m^2/g

Sample[a]	Temperature (°C)	Time (hr)	Atmosphere	D/D_0[b]
a	500	2	H$_2$	0.66
a	700	2	H$_2$	0.51
a	800	2	H$_2$	0.40
b	500	2	H$_2$	1.0
b	700	2	H$_2$	0.87
b	800	2	H$_2$	0.69

[a] a, Impregnation with H$_2$PtCl$_6$ to incipient wetness; b, ion exchange of support with Pt(NH$_3$)$_4^{2+}$.

[b] In each series relative dispersions are reported with reference to the highest-dispersion sample, $D/D_0 = 1$.

Constant temperature–variable time data are, of course, more amenable to kinetic interpretation.

A quick summary of the data in Tables 5-2, 5-3, and 5-4a reveals primarily that temperature level is the single most important factor in determining sintering rates. This is no surprise in the face of some remarks we made earlier in this chapter. The type of atmosphere—for instance, reducing, oxidizing, or neutral—is also of importance. The severe sintering of platinum group metals supported on silica or alumina in oxidizing atmospheres appears to be most pronounced for temperatures >600°C. However, for supported Ru or Ir (see Section IV,B) sintering may occur at temperatures as low as 400°C (26, 60).

The data of Tables 5-2 to 5-4a, largely abstracted from the 1975 review of Wanke and Flynn, are in some cases somewhat elderly. However, the general conclusions regarding effects of temperature and atmosphere given before remain valid, as more recent results for Pt/Al$_2$O$_3$ reported in Table

TABLE 5-4a

Constant Temperature–Variable Time Sintering Data

(1) Investigator: H. J. Maat and L. Moscou (41)
    Catalyst: 0.6% Pt/γ-Al_2O_3, BET = 179 m^2/g

Temperature (°C)	Atmosphere	Time (hr)	D
780	Air	0	0.85
		2	0.73
		4	0.26
		10	0.12
		17	0.055
		72	0.018

(2) Investigator: H. L. Gruber, *J. Phys. Chem.* **66**, 48 (1962)
    Catalyst: 1.1% Pt/η-Al_2O_3, BET = 210 m^2/g

Temperature (°C)	Atmosphere	Time (hr)	D
500	H_2	0	0.82
		72	0.52
		200	0.43
		1200	0.32

5-4b will verify. The exact importance of factors such as support, impurities in the support, and level of metal loading is less clear from available data, but they might reasonably be expected to be of secondary importance compared to temperature and atmosphere. Unfortunately, quite a bit of disparity still exists in the literature on such presumably secondary factors.

Earlier we gave some short attention to the factors of strong hydrogen adsorption and strong metal–support interactions as being potentially important effects in the determination of metal exposure or crystallite size as related to studies of sintering. In what immediately follows, we will give a summary of an investigation of this particular point; however, in ensuing parts of this chapter, when metal dispersion results are discussed, it will be understood that possible influences of strong chemisorption or SMSI have not been considered.

The question of disguise via high-temperature hydrogen pretreatment has been addressed for reducing and inert atmospheres by Wanke *et al.* (71). Just because of the high-temperature pretreatment in hydrogen used by many previous investigators, results can be suspect because interpretation of sintering was based on dispersion measurements following such treatments. Although there is some evidence that high-temperature treatment in hydrogen does not promote sintering, the issue is beclouded by the effect

TABLE 5-4b

Influence of Treatment in Oxidizing Atmospheres on Dispersion of Pt Supported on Alumina[a,b]

Catalyst	Treatment conditions[c] [atmosphere, temperature (°C), time (hr)]	Platinum dispersion	Method	References[d]
6.5% Pt/η-Al$_2$O$_3$	Fresh; H$_2$, 500, 1	0.15	TEM and H$_2$ adsorption	A
	Air, 600, 2	0.10		
	Air, 600, 8	0.09		
	Air, 600, 24	0.06		
2.75% Pt/γ-Al$_2$O$_3$	Fresh; H$_2$, 500, 8	0.67	H$_2$ adsorption	B
	Air, 590, 0.5	0.35		
	Air, 600, 0.7	0.30		
0.93% Pt/γ-Al$_2$O$_3$	Fresh; H$_2$, 500, 8	0.65	H$_2$ adsorption	B
	Air, 600, 5	0.48		
0.5% Pt/γ-Al$_2$O$_3$	Fresh; H$_2$, 500, 1	1.1	H$_2$ adsorption	C
	O$_2$, 800, —: seq$^+$	0.41		
	O$_2$, 550, 11: seq	0.60		
2.0% Pt/γ-Al$_2$O$_3$	Fresh; H$_2$, 400, 12	0.90	H$_2$ adsorption	D
	H$_2$, 700, 3: seq$^+$	0.39*		
	O$_2$, 520, 2: seq	0.74		
	O$_2$+HCl, 520, 2: seq	1.08		
	O$_2$+HCl, 600, 2: seq	0.95		
	H$_2$+HCl, 600, 2: seq	0.29*		
	O$_2$, 300, 2: seq	0.69		
	O$_2$+HCl, 520, 2: seq	0.92		
	O$_2$, 800, 2: seq	0.07		
	O$_2$, 520, 2: seq	0.09		

[a] From Wanke et al. (71). Reprinted by courtesy of Marcel Dekker, Inc.

[b] All catalysts prepared with chlorine-containing Pt precursors.

[c] All samples reduced before adsorption measurements; +, seq indicates that runs were done sequentially on the same sample; *, H/Pt ratios for these runs probably do not correspond to Pt dispersions.

[d] A, P. J. F. Harris, E. D. Boyes, and J. A. Cairns, J. Catal. 82, 127 (1983); B, C. R. Apesteguia, C. E. Brema, T. F. Garetto, A. Borgna, and J. M. Parera, J. Catal. 89, 52 (1984); C, H. Lieske, G. Lietz, H. Spindler, and J. Völter, J. Catal. 81, 8 (1983); D, T. J. Lee and Y. G. Kim, J. Catal. 90, 279 (1984).

of strongly bound hydrogen. Szymura and Wanke used a combination of x-ray diffraction and hydrogen chemisorption studies on Pt/Al$_2$O$_3$ to examine these effects on the stability of the material. Considered were comparisons of sintering in H$_2$, He, and N$_2$ at 800°C. The basic results from chemisorption experiments are shown in Table 5-4c. These measurements,

TABLE 5-4c

Influence of Treatment in Different Atmospheres on Hydrogen Adsorption by a 1.16 wt. % Pt/γ-Al$_2$O$_3$ Catalyst[a]

Run	Treatment conditions [atmosphere, temperature (°C), time (hr)]	Chlorine (wt. %)	H/Pt	Pt detected by XRD (wt. %)
27	H$_2$, 500, 1[b]	1.2	0.93	20
28	H$_2$, 800, 16	0.50	0.36	60
29	He, 800, 16; H$_2$, 500, 1	0.51	0.23	100
30	N$_2$, 800, 16; H$_2$, 500, 1	0.27	0.24	90
31	O$_2$, 700, 1; H$_2$, 500, 1	1.1	0.63	25

[a] From Wanke et al. (71). Reprinted by courtesy of Marcel Dekker, Inc.
[b] Treatments not done sequentially; before each treatment Pt was well dispersed.

which are probably as reliable as any to date, do show that sintering occurs in these atmospheres, but at a rate much lower than in oxygen. There are some differences in dispersion, however, since the sintering in an inert atmosphere compared to oxygen results in a larger fraction of Pt detectable by x ray, but with an average particle size much smaller, about 4 versus 25 nm. Various workers have reported higher sintering rates in H$_2$ than in He or N$_2$, but this may have been the result of oxygen impurities in H$_2$. The work of Wanke et al. was designed carefully to avoid this possibility; thus indicated differences in Table 5-4c are not due to oxygen contamination. These results and prior studies on Rh, Ir, and Ru indicate excellent thermal stability in reducing atmospheres (but watch out for Section IV,B).

Why is it, then, that many important sintering processes (e.g., reforming) occur in reducing atmospheres, and yet the regeneration procedures are carried out under oxidizing conditions? In the following we shall attempt to point out some of the particular aspects of all the various factors influencing sintering and redispersion in some detail, but the reader is warned that a truly general picture is not available.

1. Order and Activation Energy

In addition to the time-on-stream approach mentioned in the previous section, a number of workers have attempted correlation of sintering kinetics via power law forms:

$$dD/dt = -kD^n \tag{5-9}$$

where D is percentage exposed or dispersion and $k = k_0 e^{-E/RT}$. An alternative form of correlation is sometimes used which employs metal surface area, S, instead of dispersion. Values of the parameter n, which roughly corresponds to a reaction order, can be determined from constant temperature–variable time data. Integration of Eq. (5-9) yields

$$k = \frac{1}{t}\ln\left(\frac{D_0}{D}\right), \qquad n = 1 \qquad (5\text{-}10)$$

$$k = \frac{D_0^{1-n}}{t(n-1)}\left[\left(\frac{D_0}{D}\right)^{n-1} - 1\right], \qquad n \neq 1 \qquad (5\text{-}11)$$

where D_0 is the initial dispersion and D the dispersion at time t. Table 5-5 presents a summary of interpretations of reaction order for various examples of constant temperature–variable time data presented by Wanke and Flynn. The range of the n values shown reflects the fact that k turns out not to be a constant in many cases, varying with sintering time and hence with dispersion. Low values of n are associated with cases in which k at short times is greater than k at long times; large n values pertain to the opposite situation. All this is simply telling us that the kinetic model is not adequate to the task; however, to pursue the strong-arm comparison a bit further for the moment, some trends may be identified. The apparent order for reducing atmospheres appears to be somewhat larger than for oxidizing atmospheres, the order for nitrogen and air atmospheres is about the same, and higher initial dispersions result in higher values of n.

If we now turn to a similar analysis for activation energy, we can derive that E may be determined from variable time–variable temperature data as

$$E = R\left(\frac{T_1 T_2}{T_2 - T_1}\right)\ln\left(\frac{\Delta t_1}{\Delta t_2}\right) \qquad (5\text{-}12)$$

TABLE 5-5

Order Interpretations for Sintering[a]

Catalyst	Atmosphere	Temperature (°C)	D_0	n (range)
5% Pt/Al$_2$O$_3$	H$_2$	600	0.094	12–16
1% Pt/η-Al$_2$O$_3$	H$_2$	500	0.82	6–7
0.75% Pt/γ-Al$_2$O$_3$	N$_2$	564	0.70	2–3
		625	0.70	2
5% Pt/C	N$_2$	600	0.31	5–12
		700	0.31	5–14
5% Pt/Al$_2$O$_3$	Air	600	0.082	9–10
		700	0.082	8–14

[a] Abstracted from tabulations given in the review of Wanke and Flynn (70). Further data and original references are available in that source.

TABLE 5-6

Activation Energy Results for Sintering Kinetics[a]

Catalyst	Atmosphere	T (range (°C)	E (kcal/mol)
5% Pt/Al$_2$O$_3$	H$_2$	600–700	25
5% Pt/C	N$_2$	600–800	40
0.75% Pt/γ-Al$_2$O$_3$	N$_2$	564–625	90
5% Pt/Al$_2$O$_3$	Air	600–700	60

[a] Abstracted from tabulations given in the review of Wanke and Flynn (70). Further data and original references are available in that source.

where Δt_1 is the time required to obtain a change in dispersion from D_1 to D_2 at temperature T_1 and Δt_2 is the time required for the same change D_1 to D_2 at T_2. Some results from this analysis are given in Table 5-6. Activation energies may also be derived from constant time–variable temperature data if one is willing to make some commitment regarding the order n. In this case the determining relationship is

$$E = R\left(\frac{T_1 T_2}{T_1 - T_2}\right) \ln\left(\frac{D_0^{1-n} - D_1^{1-n}}{D_0^{1-n} - D_2^{1-n}}\right) \tag{5-13}$$

where D_0 is the initial dispersion, D_1 the dispersion at temperature T_1 after time t', and D_2 the dispersion at temperature T_2 after the same time t'. In general, higher values of n lead to higher values of E, as shown in Table 5-7.

The great variations in order and in activation energy shown in these tables are, as mentioned above, indicative of the inadequacy of a single, simple correlation such as Eq. (5-9) to cope with the complexities of sintering kinetics. However, even the results of Table 5-6, which do not require particular assumptions concerning order, reflect the fact that a wide range

TABLE 5-7

Activation Energy–Reaction Order Relationships for Sintering[a]

			E (kcal/mol) for $n =$			
Catalyst	Atmosphere	T (range) (°C)	2	6	10	14
1.6% Pt/SiO$_2$	H$_2$	700–800	—	30	50	—
1% Pt/SiO$_2$	Air	400–700	13	—	—	—
0.5% Pt/Al$_2$O$_3$	Air	600–800	35	45	—	—
0.4% Pt/Al$_2$O$_3$	Air	650–700	60	170	—	—
1% Ni/SiO$_2$-Al$_2$O$_3$	H$_2$	500–700	7	20	40	60

[a] Abstracted from tabulations given in the review of Wanke and Flynn (70). Further data and original references are available in that source.

of E is possible even for similar catalysts under varying conditions. As we shall see, this is possibly the result of sintering occurring by different mechanisms under different conditions; the postulation of a single "mechanism" of sintering, even for the same material at different times in a single experiment, may be hopelessly oversimplified.

2. The Influence of Atmosphere

Several of the previous examples pointed to the influence of atmosphere on the kinetics (and possibly the mechanism) of sintering. In recent years experimental data, at least on this point, have become much more well focused, and we shall pursue a few examples in this section. One aspect of the matter which is specifically excluded for the time being is the effect of atmosphere on the process opposite to sintering, that of "splitting" or redispersion. Such phenomena are sufficiently important and sufficiently different from sintering that they will be treated separately.

Most recent studies of sintering of supported metals have included atmosphere as one of the variables investigated. Among these one would include the later studies of Flynn and Wanke (20) on Pt/Al_2O_3. These involved both variable time-constant temperature and variable temperature-constant time experiments on two catalysts, a 0.5% Pt/Al_2O_3 and a 4.7% $Pt/Alon-Al_2O_3$. Both oxidizing and reducing atmospheres were investigated, and the general trends with respect to atmosphere and metals loading were in agreement with those pointed out in the previous section. One interesting result observed was that the inclusion of a portion of presintered catalyst in a fresh batch to be sintered increased the rate of sintering. This behavior was attributed to the action of larger Pt particles as a "sink" for the migration of metal from the smaller particles of unsintered material. If indeed this is true, it implies the possibility of migration over rather long distances in these experiments. Chu and Ruckenstein (10) also claim evidence for crystallite migration over "long" distances (>10 nm) via TEM observations of Pt on thin films of α-Al_2O_3. Again, sintering rates were found to be strong functions of temperature and atmosphere when oxidation-reduction cycling was employed, but there was no effect of nitrogen atmosphere. Chen and Schmidt (9) also used TEM to examine the behavior of Pt films supported on SiO_2, 400–700°C, in variable temperature-variable time experiments in O_2, N_2, H_2O, and Cl_2 atmospheres. While sintering rates differed, they reported no evidence of crystallite migration in any gas at any temperature, contrary to the observations of Chu and Ruckenstein.

Baker and France (4) have given some interesting data for the sintering of palladium evaporated onto graphite supports. Observations were made in controlled-atmosphere TEM in argon, oxygen, hydrogen, ethylene, and acetylene. The initial films, about one atomic layer in thickness, were first investigated for the conditions required for nucleation into discrete particles (ca. 10 nm). The sequence is quite interesting: at 5 torr pressure the temperature was 230°C for acetylene, 260°C for oxygen, 400°C for ethylene, and 535°C for argon. For only 1 torr H_2 pressure, nucleation occurred at 230°C. Growth rates subsequent to nucleation are shown in Fig. 5-4. The trends of growth are in general conformity to the nucleation temperature; growth in hydrogen was so rapid that it could not be observed with this method. The rapidly accelerated growth in O_2 at about 850°C was attributed to the agglomeration of very large particles (>50 nm), where considerable particle mobility was observed. It is apparent that there is a vast difference in sintering behavior at any given temperature in these different atmospheres.

The work of Dautzenberg and Wolters (13) gives a nice comparison between O_2 and H_2 on Pt/Al_2O_3. Heat treatments in both H_2 and air in their experiments led to a decrease in the H/M ratio determined by chemisorption, but there were apparently different mechanisms in the two cases. A range of catalysts containing 0.4 to 2 wt. % Pt on three different types of alumina supports was studied, and the primary experimental information was particle size distribution (via TEM) as a function of the heat treatment and average particle size from x-ray line-broadening analysis. For a 2% Pt supported on pseudoboehmite, calcining in air at 550°C for 3 hr resulted in a material with H/Pt = 0.23, while a staged calcination at 280°C, 1 hr; 500°C, 3 hr gave a material with H/Pt = 0.98. If this latter sample was then subjected to additional calcination at 550°C for 6 hr, the H/Pt was reduced from 0.98 to 0.62. Now, for the sample with H/Pt = 0.98, no crystallites were detected by either TEM or x-ray, while TEM of the high-temperature calcination (550°C) samples revealed nearly uniform crystallites of about 10–15 nm. The amounts of Pt detected via x-ray intensity

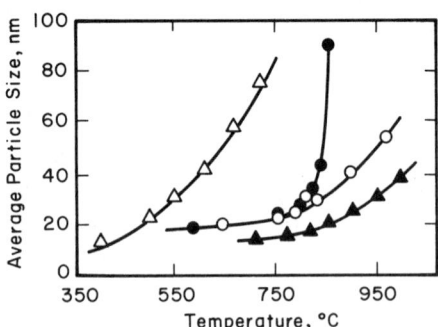

Fig. 5-4. Palladium particle size dependence on temperature in the presence of various gases. (△) Acetylene; (●) oxygen; (○) ethylene; (▲) argon. [From Baker and France (4). Reproduced by permission of Academic Press, Inc.]

I. Sintering in Supported Metal Systems

for the high-temperature samples were 82% (H/Pt = 0.28) and 42% (H/Pt = 0.62) of the initial loading, and it was concluded that the undetected portion consisted of crystallites of <3 nm, the limit of detection by x-ray analysis. Hence high-temperature treatment in an oxidizing atmosphere produced a *very distinct bimodal distribution* of Pt crystallite sizes consisting of uniform large particles of 10–15 nm and a significant fraction of <3 nm, with little in between. From the measured particle sizes 10–15 nm, the fraction <3 nm, and the overall H/Pt values, it was concluded that the two peaks of the distribution could be characterized separately by H/Pt = 0.1 for the large particles and 1.0 for the small particles.

Corresponding heat treatments in hydrogen apparently did not lead to the formation of a bimodal distribution. Figure 5-5 gives the results of a number of constant temperature–variable time experiments in terms of H/Pt. The decrease of this ratio is seen to be independent of the metal loading and the origin of the alumina support. In each case, the H/Pt ratio attains a final steady-state value dependent only on temperature, and the kinetics of sintering in H_2 were well represented by an equation of the form

$$-d(H/Pt)/dt = k[(H/Pt)_t - (H/Pt)_{ss}] \qquad (5\text{-}14)$$

where $(H/Pt)_t$ is the ratio at time t and $(H/Pt)_{ss}$ the steady-state value. Assuming that the rate constant in the above expression follows an Arrhenius form (a large assumption), Dautzenberg and Wolters report an activation energy for sintering of Pt/Al_2O_3 in H_2 of 20.9 kcal/g-at. Pt.

A very interesting effect was observed for these sintering experiments in H_2. For samples sintered at the higher temperatures (850°C), H/Pt was measured to be <0.2, which would correspond to particles >4 nm; however, the x-ray peak [(311) reflection] was very diffuse, indicative of very small

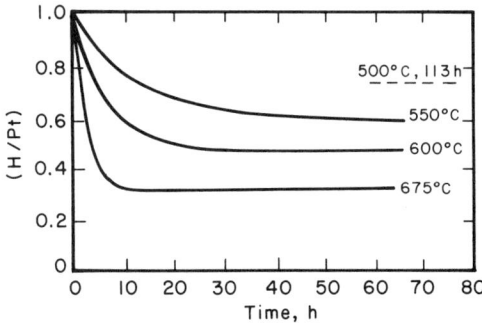

Fig. 5-5. Sintering of Pt/Al_2O_3 in H_2 atmospheres. Lines are for three different types of Al_2O_3 and Pt from 0.4 to 2.0%. [From Dautzenberg and Wolters (13). Reproduced by permission of Academic Press, Inc.]

particles. The TEM results corroborated the x-ray analysis; average particle sizes so determined were too small to explain measured H/Pt, although the amount detected was only a small fraction of Pt present in the samples. The H/Pt calculated from such data for all samples was uniformly much higher than the H/Pt measured. This result can be explained only by the possibility that after heat treatment in hydrogen a certain amount of Pt is no longer accessible for hydrogen chemisorption. Such fractions of "inaccessible" Pt varied from 0.3 to 0.6 from the low to the high end of the temperature range investigated. Oxidation-reduction cycling restores the lost Pt; the increase of H/Pt to the original value after five oxidation-reduction cycles was demonstrated. Restoration was independent of cycle length or O_2 partial pressure but strongly dependent on temperature in the range 400-500°C, depending on the alumina used. Inaccessible Pt was postulated to be formed via

$$Pt + Al_2O_3 + xH_2 \rightarrow Pt * Al_2O_{3-x} + xH_2O$$

and its recovery by

$$Pt * Al_2O_{3-x} + \tfrac{1}{2}xO_2 \rightarrow Pt + Al_2O_3$$

where the * complex denotes inaccessible Pt. The implications of the discovery of this inaccessible Pt go farther than the work of Dautzenberg and Wolters, of course, since one might legitimately wonder if all data obtained on the sintering of supported Pt (at least on Al_2O_3) using hydrogen chemisorption measurements under such conditions are not suspect. More recently we have learned a great deal more, although perhaps not understanding a great deal more, about such variations, as discussed later.

Den Otter and Dautzenberg (14) in a subsequent publication further investigated the changes in the dispersion of Pt/Al_2O_3 catalysts, treated in hydrogen, by means of oxygen-hydrogen titration and chemisorption. They observed a marked decrease in hydrogen chemisorption, while subsequent oxygen and hydrogen titration measurements indicated only a moderate decrease in dispersion. This behavior was attributed to the formation of a Pt-Al alloy. However, Straguzzi *et al.* (62) have argued that the electron microscopy experiments do not provide unequivocal evidence of the absence of metal sintering. They also performed an experiment on a 4.6% Pt/Al_2O_3 catalyst where CO and H_2 chemisorption was used to monitor the decrease in dispersion. Following treatment in H_2 at 750°C for 48 hr, the dispersion decreased to 34% but the CO/H ratio was not changed. Under the hypothesis of alloy formation, one would expect the CO chemisorption to be less affected than hydrogen chemisorption, since CO is not dissociatively adsorbed and is mostly linearly bonded. Thus the possibility of Pt-Al alloy formation and resulting partial reduction in hydrogen chemisorption capacity is still controversial.

An interesting comparison of the stability of various metals in oxidizing and reducing atmospheres has been given by Fiedorow et al. (16) for several Pt, Rh, and Ir catalysts on a typical alumina support (Kaiser KA 201). Sintering rates were compared in both hydrogen and oxygen atmospheres in the range 250-800°C. Three Pt/Al_2O_3 catalysts, 0.5% and two 2% loading, with initial dispersions of 0.20, 0.28, and 0.35, respectively; two Rh/Al_2O_3 samples, 0.5% and 1% loading, with initial dispersions of 0.37 and 0.32; and a single 2% Ir/Al_2O_3, initial dispersion 0.42, were investigated. (Although these initial dispersions were not exactly the same for the various samples, they are probably close enough to remove any controlling influence of that variable on the results obtained.) In oxygen it was concluded that the stability in decreasing order was Rh > Pt > Ir (>Rh in later work) and in hydrogen it was Ir > Rh > Pt. Under certain conditions redispersion (see later) was observed for Pt and Ir in oxygen, but not for Rh. The observed stabilities in oxygen are in general agreement with what might be predicted on the basis of the heat of formation of the corresponding oxides—in turn suggestive that a possible important step in the sintering process is the formation of MO_x molecules, which then become mobile. For hydrogen a corresponding correlation based on hydride formation is untenable; rather the stability seems to increase with increasing heat of sublimation of the metal. Fiedorow et al. also report data on the concurrent sintering of the support, an area of neglect in many published studies. There was no influence of atmosphere on the alumina, but time and temperature produced significant variations. Some area measurements were fresh, 213 m^2/g; after 700°C, 16 hr, 180 m^2/g; and after 800°C, 16 hr, 135 m^2/g. With such changes in the support, it is hard to envision that accompanying changes in pore structure would not have a significant influence on sintering kinetics.

Some further studies of stability of Ir on η-Al_2O_3 in oxygen are reported by McVicker et al. (44). Here the influence of various possible stabilizing agents was explored for sintering at temperatures above 450°C for 0.3, 1, and 2 wt. % Ir loadings. Group IIA oxides such as BaO, SrO, or CaO were found to completely inhibit sintering up to 650°C, but for this effect to be obtained the oxide concentration must be greater than the concentration of acidic sites on the support. This might seem a rather odd requirement since one does not immediately connect surface acidity with surface mobility. It was postulated that mobile iridium oxide molecules could be captured and stabilized by the IIA oxides occupying acidic sites. An activation energy for Ir sintering in O_2 was determined to be 16.5 ± 2 kcal/mol, which compares to the value of 16 reported for evaporation of Ir into O_2; hence it was concluded that sintering proceeded via formation of IrO_3, consistent with the findings of Fiedorow et al.

Further information on $Pt/\gamma\text{-}Al_2O_3$ is provided in two reports of Hassan et al. (30, 31). In the former study it was found that, *in vacuo*, metallic surface area decreased rapidly with time of sintering (300-800°C); however, stationary values of surface area were found to *decrease* with temperatures up to 400°C but thereafter showed an increase. Corresponding chemical reactivity measurements were made, which corroborated these findings (these will be discussed subsequently). In the second study, the investigation was extended to a series of 0.2, 0.4, and 2.0 wt. % $Pt/\gamma\text{-}Al_2O_3$ sintered in H_2, O_2, and N_2 atmospheres. A type of constant time-variable temperature experiment was used in which the samples were exposed for 2 hr to the given atmosphere at various temperatures in the range 300-800°C. The surface areas measured at 2 hr were claimed to be "stabilized," which we interpret to mean no further changes at longer times of exposure (a surprising observation). The alumina support was precalculated at 850°C before catalyst preparation, hence support sintering was not a factor (BET area was, in fact, rather moderate at 113.5 m^2/g). The basic series of results obtained is shown in Fig. 5-6 for sintering in the three atmospheres, as well as the vacuum results. It is clear that the metal surface area decreases in all cases at sintering temperatures below 400°C; however, for $T > 400°C$, there is an increase in surface area in vacuum, N_2, and H_2, which the authors attribute to a redispersion process which is rather temperature-dependent. Two aspects of the data in Fig. 5-6 are in sharp contrast to much of the information we discussed previously in this section. First, over large ranges below 400°C the retention of metal surface area in the oxygen atmosphere is greater than in the other environments, and it was concluded that the rate of sintering may be impeded by the formation of an oxide layer on the metal; this is essentially the opposite of what most other studies have shown. Second, the oxygen data show no redispersion occurring at $T > 400°C$, but this is measured for all other cases. Again, this result stands against what most other observers have reported. We shall return to the discussion of redispersion in a subsequent section devoted to that topic.

Another type of atmosphere effect on sintering behavior has been reported by Clay and Petersen (11). In this case already alluded to in Chapter 4, the chemisorption of a poison on the metal surface causes crystallite growth and we view this as a type of "chemically assisted" sintering. These workers observed the sintering of Pt films on chemisorption of arsine; measurements of the activity of the poisoned-sintered film for cyclopropane hydrogenolysis showed a nonselective deactivation which was correlated directly to the decrease in the surface area of the film. The net sintering effect was probably the primary result of the formation of $PtAs_2$ on the surface on chemisorption

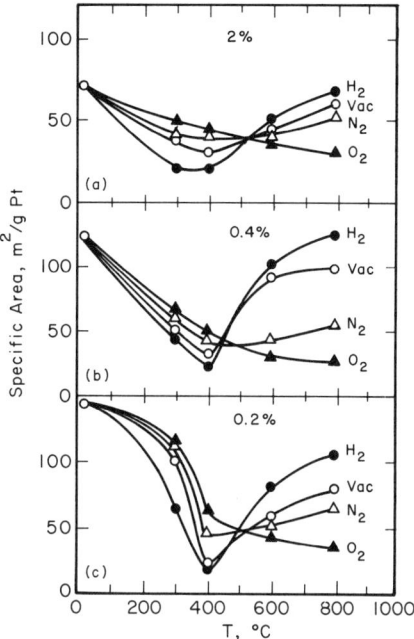

Fig. 5-6. Specific surface area of supported Pt (from H_2 chemisorption) as a function of temperature of sintering in various atmospheres. [From Hassan *et al.* (31). Reproduced by permission of Academic Press, Inc.]

of AsH_3 according to

$$Pt + 2AsH_3 \rightarrow PtAs_2 + 3H_2$$

$PtAs_2$ has a much lower melting point than Pt, hence its formation would tend to decrease considerably the thermostability of the catalyst at a given temperature level. Moreover, there is approximately a 35% increase in the lattice parameter associated with $Pt \rightarrow PtAs_2$, and the heat of chemisorption involved in the reaction above was estimated to be 90–120 kcal/mol. Thus, the combination of altered thermal and geometric properties of the surface, plus significant heat effects on chemisorption, was postulated to induce the loss of surface area and corresponding activity.

In their 1975 review, Wanke and Flynn concluded that in nonoxidizing atmospheres the general order of resistance to sintering increased with increased melting temperature; hence from data they summarize we have the stabilities $Rh > Pt > Pd < Ni$. The more recent data discussed above would also place $Ir > Rh$ (Fiedorow *et al.*) but suggest that heat of sublimation might be a better correlating variable than melting temperature. All metals, with the possible exception of Pd on hydride formation, are more stable in reducing than in oxidizing atmospheres (71).

3. Correlation of Sintering with Catalytic Activity

It is interesting that relatively few workers have reported data correlating the extent of sintering with concomitant changes in catalytic activity. An early but still quite timely example is provided by the work of Maat and Moscou (41), who correlated the sintering of a commercial reforming catalyst, 0.6% Pt/Al_2O_3 with 0.5% chloride, with its activity for n-heptane reforming. Variable time–constant temperature sintering experiments were carried out at 780°C in an inert atmosphere and the resulting Pt surface areas determined by hydrogen chemisorption and electron micrography. Corresponding crystallite sizes (average) ranged from about 1 nm for the fresh catalyst to 45 nm for the most severely sintered case (780°C, 72 hr) and good correlation of sintering kinetics was obtained via Eq. (5-9) with $n = 2$. Experimental results and correlations are shown in Fig. 5-7a. The corresponding effects of sintering on the n-heptane reforming reaction are shown in Fig. 5-7b for the conditions indicated. A significant change in activity is noted, although it is not as large as one might have expected. The total change in surface area is from 233 to 5 m^2/g Pt, a 98% decrease, yet total conversion decreases only from about 95 to 75%, roughly 25%. What is probably of more ultimate importance in these data is the change in product distribution. The selectivity for the various products changes markedly on sintering; as shown in the figure, the dehydrocyclization activity is severely affected, while the sum of isomerization plus cyclization remains approximately constant. (Isomerization activity actually increases, from 9 to 25 mol % product, with increasing sintering.) Since the aromatics produced in dehydrocyclization reactions would be prominent in determining product octane, this change represents a severe decline in a desired selectivity. The reason for the alteration in selectivity is found in the bifunctional nature of the catalyst; aromatization reactions are strongly dependent on the metallic function, while the isomerization activity is predominantly a function of the acidic Al_2O_3. Maat and Moscou do not report any notable sintering of the support in these experiments, so presumably the acidic function is relatively unchanged and the apparent increase in isomerization activity on sintering is the result of increased reactant availability in competition with the metallic function.

In the previous section we commented on the study by Hassan *et al.* of sintering of Pt/Al_2O_3 in various atmospheres, in which it was observed that oxygen exerted an apparent retarding effect on the observed rates of sintering. Since this is at odds with the bulk of observation and conventional wisdom on the matter, one might be skeptical; however, consider the

I. Sintering in Supported Metal Systems

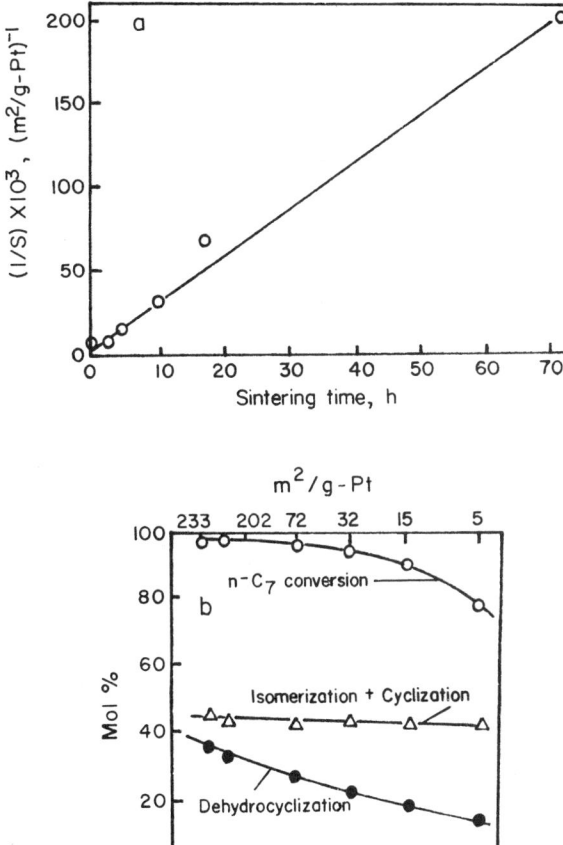

Fig. 5-7. (a) Second-order sintering of Pt/Al_2O_3 reforming catalyst. (b) Reforming of *n*-heptane as a function of the extent of sintering. 0.6% Pt/Al_2O_3, 200 psia, 500°C, 2.44 g/hr-g, $H_2/HC = 5.3$. [From Maat and Moscou (41). Reproduced by permission of North-Holland Physics Publishing.]

corresponding activity data for hydrogen peroxide decomposition given in Fig. 5-8a. The similarity to the sintering results shown in Fig. 5-6a is striking. Minima in activity with temperature of sintering are noted in exactly the same place as minima in surface area, the ordering with respect to atmosphere is the same, and the changes in relative activity are roughly proportional to the corresponding changes in surface area. Further, very similar patterns were observed for another test reaction, the hydrogenation of

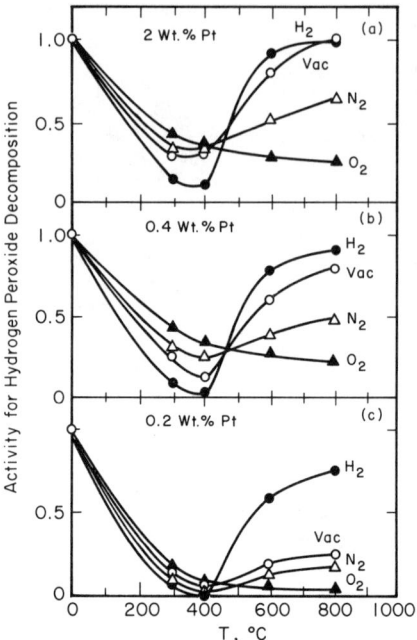

Fig. 5-8. Catalytic activity of Pt/Al_2O_3 catalysts of different weight percent Pt as a function of temperature of sintering in different atmospheres. [From Hassan et al. (31). Reproduced by permission of Academic Press, Inc.]

cyclohexene, with minima in relative rates occurring in the neighborhood of 400°C for N_2 and H_2 sintering atmospheres and no minimum noted for O_2 sintering. Hence, it appears that the results of Hassan et al. must stand as a rather well-documented exception to the general rule on Pt sintering in oxidizing atmospheres.

Correlation between sintering of charcoal-supported Pt and activity for a more complex reaction, dehydrogenation of 2,3-dimethylbutane, has been reported by Nakamura et al. (45). A series of five catalysts containing 0.12 to 9.33 wt. % Pt was prepared by conventional impregnation on 1200 m^2/g charcoal carrier. These samples were then sintered in hydrogen at 650°C for periods ranging from 2 to 72 hr and examined by TEM for particle size and size distribution. Corresponding activity and selectivity measurements were carried out for the dehydrogenation reaction at 460°C, 1 atm, $H_2/HC =$ 2.2 mol/mol, and ratio of weight of catalyst to feed rate (W/F) of 0.004– 0.5 g cat-hr/mol. Initial rates were determined from conversion versus W/F in the conventional manner and normalized with respect to total metal surface area determined from TEM.

I. Sintering in Supported Metal Systems

Investigation of the fresh samples is informative concerning the influence of metals loading on dispersion (percentage exposed). Table 5-8a gives the results of measurement of surface area and average particle size (the distributions will be discussed in a later section) as a function of loading. One would conclude from these data that conventional impregnation of Pt beyond about 0.5 wt. % results only in growth of particle size and a net reduction in exposed surface per unit weight of catalyst. While this result pertains to a rather idealized catalyst with very little support interaction, it is interesting to note that most commercial reforming catalysts, indeed, contain ca. 0.5% or less Pt.

Sintering data for one of the catalysts are given in Table 5-8b. There is about a 60% decrease in surface area for H_2 at 650°C, 72 hr; the corresponding reaction data are shown in Fig. 5-9a in terms of initial rate versus average diameter. In this case the specific rate decreases by about 75% for the 60% change in total metal surface, or as plotted versus average diameter one would conclude that the reaction is strongly structure-sensitive. In fact, the data of Fig. 5-9a are a very good example of what we might term the "double-edged sword" of sintering processes. In cases where there is structure sensitivity of a reaction in the sense that specific turnover frequencies decrease with increasing particle size (type one structure sensitivity in the terminology of Chapter 4), then one is defeated in two ways. First, overall activity decreases because of the loss of metal surface area, and second, each surface metal atom also becomes less active as the individual crystallite grows. Mother Nature really can be mean.

The 2,3-dimethylbutane dehydrogenation also exhibits a change in selectivity on sintering, shown in Fig. 5-9b. Selectivity, defined as percent 2,3-dimethyl-1-butene in the product, is shown as a function of particle diameter for 1% total conversion of reactant. The isolated point is that for a Pt wire,

TABLE 5-8

Diameter and Surface Area for Pt/C[a]

(a)			(b)			
Sample	$\bar{d}$ (nm)	S (m^2/g Pt)	Sample	Time (hr)	$\bar{d}$ (nm)	S (m^2/g Pt)
0.12%	2.0	140	0.50%	0	2.0	140
0.50%	2.0	140	0.50%	2	3.2	87
0.99%	2.6	108	0.50%	14	3.5	79
2.79%	4.5	63	0.50%	24	4.1	69
9.33%	4.8	58	0.50%	72	4.8	58

[a] From Nakamura et al. (45).

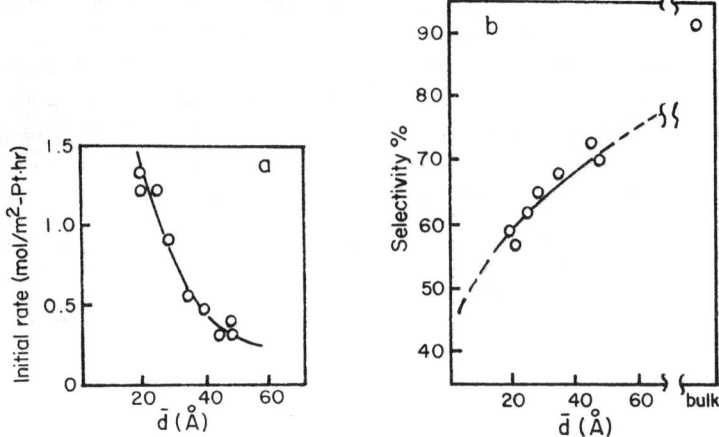

Fig. 5-9. Sintering effects on (a) activity and (b) selectivity for the Pt/C-catalyzed dehydrogenation of 2,3-dimethylbutane. [From Nakamura *et al.* (45). Reproduced by permission of Academic Press, Inc.]

which can be represented as the result for infinite particle size. The reasons for changes in selectivity here, however, are quite different from those pertaining to the results of Maat and Moscou, since this is not a bifunctional catalyst. The observed trend can be explained by assuming the reaction intermediate is either π-adsorbed or α,β-diadsorbed, depending on the size of the Pt particle. Lower-coordination Pt (small particles) favors formation of π-olefin or π-allyl intermediates in hydrogenolysis reactions; if this carries over to dehydrogenation, π-allyl intermediates produce equilibrium mixtures of butenes very rapidly on the smaller particles and the α,β intermediate, which is responsible for 2,3-dimethyl-1-butene production, becomes favored as the particles become larger.

Model supported metal systems have often been used in sintering studies in which the metal is deposited on a substrate of some type suitable for examination via TEM. Such efforts have been criticized as being too far removed from the situation of a metal crystallite on the interior pore surface of typical supports, but they have been valuable in yielding direct information on sintering and redispersion, albeit in some cases information which is contradictory [an interesting comparison of this sort, to which we return later, is provided by studies of Pt redispersion on γ-Al_2O_3 by Ruckenstein and Malhotra (55) and Stugla *et al.* (64)]. As in studies of supported metals, however, sintering results are rarely accompanied by reaction data. In this respect the data of Presland *et al.* (52) are of particular interest. They studied

I. Sintering in Supported Metal Systems

the sintering of silver films employed to catalyze the oxidation of ethylene. It was found that not only the total surface area of the metal but also the detailed morphology of the silver crystallites in the films was important in determining changes in activity and selectivity. Figure 5-10 and Table 5-9 give the data on changes in surface area on treatment with oxygen, showing about a threefold decrease with annealing time up to 60 min. The shadowed catalyst, prepared by evaporation of silver onto a Pyrex plate through a copper gauze, 0.21-cm mesh of 0.04-cm-diameter wire, is the one of interest here. Presland et al. show that this oxygen treatment has the same effect on the morphology of the silver as equivalent time and temperature conditions for the ethylene oxidation (i.e., that oxygen is effective in promoting the mobility of the silver crystallites on the surface), so the annealing data given in the table may be compared with the activity and selectivity results given in Fig. 5-10. The activity continuously decreases until a final plateau level,

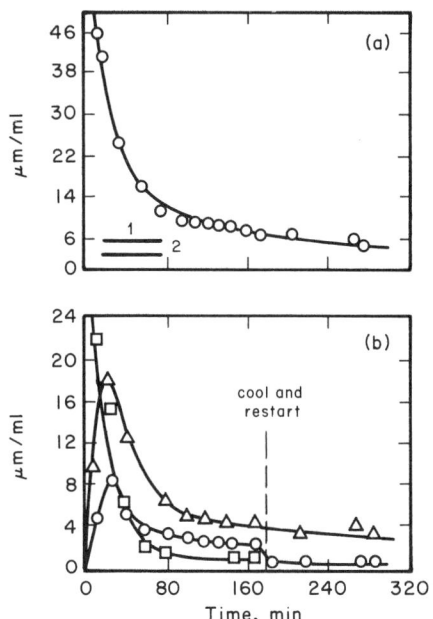

Fig. 5-10. (a) Decrease of activity with time for the oxidation of ethylene at 228°C, residence time 5 sec, $HC/O_2 = 2$. Circles correspond to activity of the shadowed catalyst, line 1 to activity of the continuous film, line 2 to activity of the film annealed to islands. (b) Product distribution as a function of time; conditions as in (a). (△) Ethylene oxide; (○) carbon dioxide; (□) water. [From Presland et al. (52). Reproduced by permission of Academic Press, Inc.]

TABLE 5-9

Surface and Interfacial Areas of the Silver Catalysts[a]

Annealing time, O_2, 266°C (min)	Surface area (cm^2)[c]		Interfacial area (%), continuous catalyst
	Continuous catalyst	Shadowed catalyst	
0	805	776	56
6	1020	—	53
60	1090	1050	19

[a] From Presland et al. (52).
[b] Geometric area = 878 cm^2. Area covered by one krypton molecule taken to be 19.5 Å^2.
[c] Accuracy ±5%.

almost eightfold lower than initial, is attained; yet, if the data in the table are correct, the total surface area has not decreased proportionately. Indeed, the selectivity data indicate a maximum for ethylene oxide formation after about 30 min, which is more easily interpreted in terms of an optimum crystallite size for the partial oxidation reaction rather than some total surface area effect. The changes in silver crystallite size on reaction were monitored at least qualitatively by electron microscopy (both scanning and transmission), and the following steps were noted:

(1) Breakup of the continuous films in the nonshadowed areas and migration of crystallites in the size range 10–200 nm into the shadowed area.

(2) Gradual disappearance of all crystallite sizes on the low end of this range (<50 nm) by agglomeration into larger particles.

(3) Formation of the final sintered form with large crystallites more or less uniformly dispersed on the surface.

The large changes in activity seem to be due to the loss of small crystallites and the maximum in selectivity with partial agglomeration (~100 nm) of the crystallites. Again, one observes here the interaction between sintering and structure sensitivity.

While the examples given in this section are by no means comprehensive, we hope to convince the reader that significant changes in activity and selectivity are normally associated with the loss of surface area on sintering, but that such changes may also not be in direct proportion to the loss of active surface. Particularly in instances where the reaction of interest is structure-sensitive in the sense of being preferentially catalyzed on the smaller crystallites, the effects of sintering are much aggravated.

II. PARTICLE GROWTH MODELS[1]

A. Crystallite Migration, Collision, and Coalescence

The basic theory of this view of the sintering process has been developed in most detail by Ruckenstein and Pulvermacher (56), with subsequent modification by Ruckenstein and Dadyburjor (54). The basis for the model is essentially a population balance. If n_k is the number of particles per unit area containing k units, then collisions between particles which have i and $k-i$ units will result in an increase of n_k, while collisions between those containing k units and any other size will decrease n_k. The rate of change is

$$\frac{dn_k}{dt} = \frac{1}{2}\left[\sum_{i+j=k} K_{ij} n_i n_j - n_k \sum_{i=1}^{\infty} (K_{i_k})(n_i)\right] \quad (5\text{-}15)$$

where K_{ij} are the rate constants and will depend in detail on the rate-determining step. Incorporation of this into the diffusion equation in cylindrical coordinates leads to the result

$$(1/S^n) - (1/S_0^n) = (C_1/S_0^n)(t) \quad (5\text{-}16)$$

where S is the area of particles per unit of support and C_1 is a constant. Equation (5-16) holds in both of the two limiting cases of sintering mechanisms according to this view: (i) when crystallite migration is rate-determining and (ii) when coalescence is rate-determining. Since $\bar{R}$, the average radius, is inversely proportional to S, a corresponding expression for $(\bar{R}^n - \bar{R}_0^n)$ can also be derived from Eq. (5-16). The values of n and C_1 depend on the controlling mechanism and the correlations invoked for the parameters involved. For crystallite migration rate-determining, for example,

$$n = 3 + m \quad (5\text{-}17)$$

where m is a parameter defining the diffusion coefficient of the particle via a correlation such as

$$D_{p_i} = C_2(1/R_i)^m \quad (5\text{-}18)$$

and D_{p_i} is the diffusion coefficient for radius R_i. Although the form of Eq. (5-16) is valid for either rate-determining process, Ruckenstein and Pulvermacher demonstrated that there were differences in the evolution of crystallite size distributions for the two. For crystallite migration control the distributions tend to become time-invariant after an initial transient, while for coalescence control the distributions continue to evolve although the

---

[1] The material of this section is based in part on Wynblatt and Ahn (76).

TABLE 5-10

Comparison of Coalescence and Migration Rates[a,b]

R_0 (Å)	τ_c (sec)[c]	D_p (cm^2/sec)[d]	t_p (sec)
25	3.5×10^{-4}	1.3×10^{-13}	12
100	8.9×10^{-2}	5.0×10^{-16}	5.0×10^4
250	3.5	1.3×10^{-17}	1.2×10^7
2500	3.5×10^4	1.3×10^{-21}	1.2×10^{13}

[a] From Wynblatt and Gjostein (77).
[b] Parameters: $T = 700°C$, D_s = surface diffusion coefficient = 2.8×10^{-9} cm^2/sec, N_0 = surface site density on particle = 10^{15} sites/cm^2, γ_m = surface energy = 2100 erg/cm^2, Ω = atomic volume of particle constituent = 1.5×10^{-23} cm^3, a_0 = atomic diameter = 0.277 nm.
[c] $\tau_c = 0.89(R_0^4/B)$, $B = D_s N_0 \gamma_m \Omega^2 / kT$.
[d] $D_p = 0.3 D_s (a_0/R_0)^4$, $t_p = (10 R_0)^2 / 4 D_p$.

changes are very slow at long times. Wynblatt and Gjostein, in view of the near indistinguishability of the two mechanisms of rate control, calculated some order-of-magnitude time constants that could be expected. The theory for coalescence of particles on a surface (48, 49) predicts that the relaxation time τ_c is proportional to the fourth power of the particle radius (identical spheres are assumed) and inversely proportional to the surface diffusion coefficient and the surface energy. On the other hand, particle migration rates considered as a type of Brownian motion over the surface can be modeled in terms of diffusion-random walk theory. The particle diffusion coefficient D_p used by Wynblatt and Gjostein for this process was proportional to the product of the surface diffusivity and the fourth power of the ratio of atomic to crystallite radii, and the characteristic time was chosen to be that required to traverse a distance 10 times that of the crystallite radius R_0. The results of these calculations are compared in Table 5-10. These show quite demonstratively that the coalescence times are much smaller than migration times when significant distances are involved, and obviously in the migration–coalescence sequence the former will be rate-determining. Note, however, that t_p becomes a very large number as crystallite size increases. If we accept $10R_0$ as a norm for the separation of crystallites in a typical catalyst (cf. Fig. 5-2), then the experimentally measured time scales for sintering that most observers have reported would rule out the particle migration–coalescence process as a mechanism for sintering of crystallites larger than about 5 nm.

B. Intraparticle Transport

The theory for sintering via the transport of individual entities—atoms in reducing atmospheres or oxide molecules in oxidizing atmospheres—has been developed by Wynblatt and Gjostein (77) and Flynn and Wanke (18, 20). The former consider growth to be possible by either of two modes of transport: diffusion over the support surface or diffusion through the vapor phase. The development by both sets of authors is an extension of the classical theory of Ostwald ripening; qualitatively one can show from the surface energies involved that the larger particles in a population will have lower chemical potential than smaller ones, hence net growth of the larger particles occurs at the expense of the smaller ones.

Consider the picture of two particles in the form of spherical segments partially wetting the support surface as shown in Fig. 5-11. In the case of *growth by diffusion over the support surface*, the net flux of diffusing entities is a complicated function of the contact angle θ, the value of R for the given particle, atomic (or molecular) dimension a, and, of course, concentration and diffusivities of the migrating entities. Again there can be differing regimes of rate control corresponding to either "diffusion control" (rate of migration of entities over the surface) or "interface control" (rate of attachment or detachment of entities from the particles). Wynblatt and Gjostein show that as long as $R \geq 10a$ (i.e., 2.5 nm or so for Pt) growth will be diffusion-controlled. Only in the case of very small particles would interface control play any significant role. In the case of diffusion control for Pt sintering, the ripening model eventually yields the simple form

$$(\bar{R}/\bar{R}_0)^4 = 1 + (1.03/\bar{R}_0)^4 K_D t \tag{5-19}$$

where $\bar{R}$ is the average particle radius of the distribution, $\bar{R}_0$ its initial value, and K_D a characteristic rate constant which is primarily dependent on the product of the surface diffusion coefficient and the concentration of diffusing species. It is important to note that in practice the form of Eq. (5-19) is indistinguishable from the crystallite migration result of Eq. (5-16), although

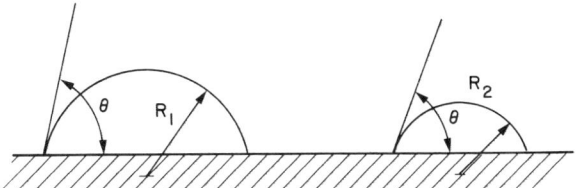

Fig. 5-11. Partially wetting metal particles on a support surface.

we have in principle fixed a value of $n = 4$ in the present case. The theoretical capability exists to compute values for K_D on an *a priori* basis; however, such computation requires parameters which are imperfectly known or not known at all—particularly the activation energy for diffusion across the support surface. Order-of-magnitude estimates of K_D from theory by Wynblatt and co-workers indicate that K_D for oxidizing atmospheres is much larger than for reducing atmospheres for platinum, a result that is in agreement with the bulk of experimental data presented previously.

The model of Flynn and Wanke for growth by surface diffusion was analyzed for two cases: (i) the rate of capture of migrating atoms is large, giving very small concentrations of surface diffusing species, and (ii) the rate of capture is small, yielding appreciable concentrations of diffusing species. In both of these cases the migration across the surface was assumed to be rapid, hence the two cases represent subsets of the interface control model of Wynblatt and Gjostein. Indeed, the results are very similar to the interface control model, except that Flynn and Wanke assume the rate constant for gas atom loss to be independent of crystallite size. This has the net effect, when their results are cast into the formalism of Eq. (5-16) or (5-19), of making the exponent n time-dependent in some situations; however, the basic form of the correlation is the same.

For particle *growth by vapor-phase diffusion* the transport process is very rapid and net growth rates are controlled by events at the interface. In this case a second-power growth law is obtained:

$$(\bar{R}/\bar{R}_0)^2 = 1 + (K_v/\bar{R}_0^2)^t \tag{5-20}$$

where K_v is a characteristic rate constant for the interfacial process. In reducing atmospheres, $K_v \sim 0$ for platinum; however, theoretical estimates are possible for oxidizing atmospheres.

C. Other Aspects of the Problem

It should be apparent from the foregoing that the kinetics of sintering processes are nonunique, similar to the situation in catalytic kinetics, where wide varieties of mechanistic proposals can lead to identical or very similar (i.e., experimentally indistinguishable) equations for the rate of reaction. One must then look to other aspects of the problem than simple time-dispersion or surface area relationships. Further, one might ask whether the crystallite migration and Ostwald ripening mechanisms, in their various forms of rate-controlling steps, encompass the total range of possibilities. At issue in this last question is the fact that the two general models discussed above yield growth laws with exponents n which have well-defined values, varying with the precise mechanism assumed, but which according to theory should not exceed the general range 5-7. Yet, we have shown data in a previous section, interpreted in terms of power law models, with values of n up to 12-16. In physical terms, such a large value of n reflects a very pronounced decrease in the rate of growth as the crystallite becomes larger. Wynblatt and co-workers suggested that this is the result of particle growth being inhibited by a nucleation process which occurs simultaneously with sintering. This is a concept adapted from crystal growth theory (8), proposing that faceted crystals are grown via a process of continued nucleation whereas unfaceted crystals do not involve this process. Rates of growth for faceted crystals are generally quite a bit lower than those for unfaceted crystals. Since there is considerable evidence from published work that faceting does occur in supported metals, nucleation processes may indeed play an important role in the rate of sintering. Detailed discussion of the nucleation-inhibited growth theory is beyond the scope of our present interest; however, in comparison to the classical ripening theory it does lead to time-dependent exponents n which can attain large values. Both general numerical solutions and an asymptotic solution have been obtained (76).

A number of workers have made the point that the various theories lead to different predictions of particle size distributions; both Ruckenstein and Pulvermacher and Flynn and Wanke give extensive discussions of the matter. Indeed, a lively controversy developed in the mid-1970s following a note by Granqvist and Burnham (27) which stated flatly that in view of prior data on particle size distributions "the shape of such size distributions can be used as a means to determine by which mechanism the particles have grown and that the accumulated evidence points almost unequivocally in favor of coalescence growth as distinct from Ostwald ripening." The observation was based on a prior theory of the authors (28) picturing the coalescence growth process as a series of discrete events in which only two particles coalesce at a time and the change of volume at each step in the sequence as a random function of the particle volume after coalescence. The resulting prediction of an asymptotic particle size distribution, after many coalescence events, says that the logarithm of the particle volumes is Gaussian. The number of particles ΔN per logarithmic size interval $\Delta(\ln x)$ is given by

$$\Delta N = f_{\ln}(x) \, \Delta(\ln x) \tag{5-21}$$

where

$$f_{\ln}(x) = \frac{1}{\sqrt{2\pi} \ln \sigma} \exp\left[-\frac{1}{2} \left(\frac{\ln(x/\bar{x})}{\ln \sigma} \right)^2 \right] \tag{5-22}$$

Here x is the diameter of a spherical or half-spherical particle, $\bar{x}$ the median diameter, and σ the standard deviation. The point made by Granqvist and

Burnham was that the lognormal distribution of Eqs. (5-21) and (5-22) resulted in distributions which in general were skewed toward the large-diameter side, whereas distributions predicted from Ostwald ripening theory were skewed toward the small-diameter side. Experimental data on distributions were cited that all tended to be skewed to the large-diameter side, hence the claim.

Wanke (69) took issue with this, stating among other things that the atomic migration (Ostwald) mechanism does not inevitably lead to distributions skewed to the small-diameter side and citing some additional literature, primarily reports of bimodal distributions (51, 53) which were not lognormal. And the controversy continued (29). By now, however, there are enough data in the literature on metal crystallite size distributions to indicate that there can be considerable variations in size and shape. Whether one can make anything of these for interpretation of mechanism is probably best summed up by a comparison of distribution functions obtained from the various models in Fig. 5-12. Save for the case of coalescence control these

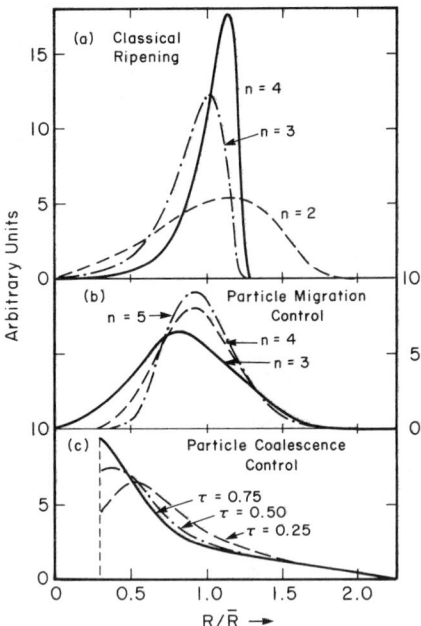

Fig. 5-12. Particle size distribution functions from various models plotted as number of particles of size $R/\bar{R}$ (arbitrary units) versus $R/\bar{R}$. (a and b) Quasi-stationary distributions. (c) Time evolution of a given distribution for growth by coalescence control, where τ is a reduced time scale. [From Wynblatt and Ahn (76). Reproduced by permission of Plenum Publishing Corp.]

II. Particle Growth Models

are all very similar, and even in that case the distribution could be similar to the other models if the sample was examined within the appropriate time scale. Most distributions obtained and reported in the literature have been via transmission electron microscopy, as we stated before, where what you see is fine but what you do not see is bothersome. We conclude that size distribution data are, in the long run, going to provide rather tricky quantitative information regarding sintering mechanisms.

At the beginning of this section we stated the basis (or one of the bases) of the difficulty in identifying sintering mechanisms, namely the similarity of particle size or dispersion versus time correlations. Indeed this is so; however, sintering-time data are not a complete dead end. Bett *et al.* (7) have pointed out that both Eqs. (5-16) and (5-19) can be written in the common form

$$1/S^n = 1/S_0^n + kT \tag{5-23}$$

where S_0 and S are the metal surface areas per unit volume initially and at time t, and k is an appropriate rate constant. The exponent n, as pointed out earlier, is generally time-variant and can assume a wide range of values. So far, the same as before. However, Bett *et al.* also show that the constant k is directly proportional to ϕ, the volume of metal per unit area of support, when the mechanism of sintering is crystallite migration and is independent of that quantity when the mechanism is Ostwald ripening. Some references to various experiments demonstrating this effect are given in Table 5-11. It is apparent that, even though wide ranges of n are involved in these different studies, there does appear to be a well-defined relationship between k and θ depending on the mechanism of sintering. The experiments subsequently conducted by Bett *et al.* to further test this hypothesis led to some unexpected results which reveal yet another complexity of this already sufficiently complex topic.

Three catalysts, containing 5, 12, and 20 wt. % Pt on carbon black (Vulcan XC-72, 80 m^2/g), were prepared and reduced in H$_2$ at 500°C. Subsequently, constant temperature-variable time experiments were carried out, primarily at 600°C, and sintering behavior was measured in both H$_2$ and N$_2$ atmospheres. (It is interesting that on the carbon support no difference in these two atmospheres was observed; however, this observation is not the point of the present story.) The basic data, in the form of Pt surface area versus time, are shown in Fig. 5-13. The main point to note is the very large and rapid initial loss of surface area for the 20% sample—not observed for the two others of lower loading. Interpretation of these data according to Eq. (5-23) led to no "best" value of n; variation between 2 and 8 yielded generally similar results, which, in fact, indicated that *neither* of the mechanisms detailed in Table 5-11 was being obeyed. The pertinent parameters are

TABLE 5-11

Equations for Sintering Rates of Supported Metal Catalysts[a]

$$1/S^n = 1/S_0^n + kt$$

Mechanism	Rate-determining process	n	k constant[b]	References[c]
Crystallite migration	Coalescence	2–3	$\propto \phi$	A, B
	Surface diffusion	3–5	$\propto \phi$	
Crystallite migration	Surface diffusion	7/2	$\propto \phi$	C
Ostwald ripening	Surface diffusion	4	Independent of ϕ	D
Ostwald ripening	Interface transfer	2	Independent of ϕ	
Ostwald ripening	Surface diffusion	3	Independent of ϕ	E[d]

[a] From Bett et al. (7). Reprinted by permission of Academic Press, Inc.

[b] ϕ = total volume of metal per unit area of support.

[c] A, E. Ruckenstein and B. Pulvermacher, *AIChE J.* **19**, 356 (1973); B, E. Ruckenstein and B. Pulvermacher, *J. Catal.* **29**, 224 (1973); C, W. B. Phillips, E. A. Desloge, and J. G. Skofronick, *J. Appl. Phys.* **39**, 3210 (1968); D, B. K. Chakraverty, *J. Phys. Chem. Solids* **28**, 2401 (1967); E, W. J. Dunning, in "Particle Growth in Suspensions" (A. L. Smith, ed.), p. 1. Academic Press, New York, 1973.

[d] Solution for three-dimensional process.

tabulated in the first three entries of Table 5-12, where it is seen that neither k nor k/ϕ is a constant, as required by one or the other of the theories. Closer examination of these results indicates that k/ϕ is reasonably constant for the 5 and 12% samples; 20% is out of line, as it was in the general sintering behavior shown in Fig. 5-13. The key to the anomalous behavior of the 20% sample seems to lie in the crystallite size distributions of the three (in spite of our skepticism above as to the usefulness of such data). The similarity of k/ϕ for 5 and 12% suggests a crystallite migration mechanism. Prior theoretical studies, particularly those of Swift and Friedlander (65) and Ruckenstein and Pulvermacher, indicated that the size

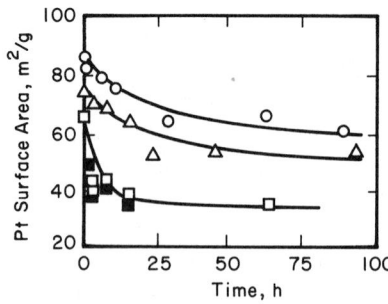

Fig. 5-13. Decrease in Pt crystallite surface area for crystallites supported on graphitized carbon black, $T = 600°C$. (○) 5% Pt in N_2; (△) 12% Pt in N_2; (□) 20% Pt in N_2; (■) 20% Pt in H_2. [From Bett et al. (7). Reproduced by permission of Academic Press, Inc.]

II. Particle Growth Models

TABLE 5-12

Rate Constants for Sintering of Platinum Crystallites on Carbon at 600°C using Eq. (5-23), $n = 4$ [a]

Wt. % Pt	ϕ (cm³ Pt/cm² surface) $\times 10^8$	S_0 (m²/g Pt)	k (m²/g)⁻⁴ hr⁻¹ $\times 10^9$	k/ϕ
5	0.3	85	0.72	0.23
12	0.8	76	1.45	0.18
20	1.4	70	72.0	5.1
20[b]	1.4	45	3.1[b]	0.21

[a] From Bett et al. (7).
[b] Based on an initial rate, assuming $S_0 = 45$ m²/g.

distributions for this mechanism assume a unique spectrum after short initial sintering times, which then remain unchanged. This "self-preservation" of the distribution requires that plots of

$$\sum_{k=1}^{\infty} \frac{N_{vk}}{N_\infty} \quad \text{versus} \quad \frac{V_k N_\infty}{\phi}$$

should be coincident, where N_∞ is the total number of particles and N_{vk} the number of particles of volume greater than V_k. Application of this self-preservation test to the 20% sample data is shown in Fig. 5-14, where it is apparent that the initial distribution and its corresponding preservation function are quite different from the stabilized distribution resulting after N_2, 600°C, 16 hr. One would then be inclined to discount the initial sintering data for the 20% sample, which indeed was done, and estimate values for k and k/ϕ on the basis of an extrapolated initial sintering rate reflecting the self-preserving distribution. The results of this analysis are shown in the fourth entry of Table 5-12, where now the k/ϕ values for 20% are quite in line with the two lower-loading samples. Hence, if we accept the extrapolation procedure, the conclusion is that sintering occurs via crystallite migration in this case.

Still one might wonder why the initial sintering of the 20% sample was so different from the other two. The conjecture was that for graphitized carbon the boundaries of basal planes are sites for carbon oxide species, which in turn are trapping sites required for Pt crystallite immobilization. For the 20% sample, the number of initial crystallites exceeded this trapping site density and hence a substantial number of very unstable crystallites were produced initially which would rapidly sinter until the crystallite population and trapping site population were in balance. At this point, the

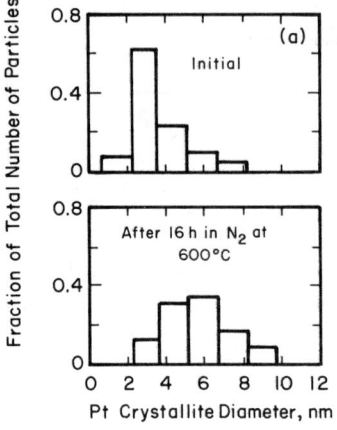
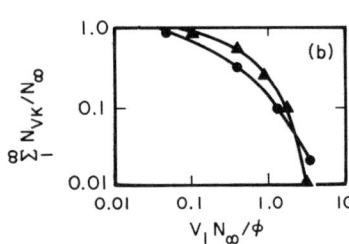

Fig. 5-14. Test for self-preservation of the Pt crystallite distribution. Sample 20 wt. % Pt on graphitized carbon, 600°C. (a) Histograms of initial and final (16 hr) distributions. (b) Circles, initial; triangles, final. [From Bett *et al.* (7). Reproduced by permission of Academic Press, Inc.]

self-preserving distribution would be established. This seems a reasonable explanation and, indeed, might be the basis for understanding similar anomalous metal sintering rate behavior on other supports, although to our knowledge the point has not been pursued very extensively in the literature.

III. CRYSTALLITE SPLITTING AND REDISPERSION

In recent years there has been an accumulation of increasingly convincing experimental evidence for the existence of the process opposite to sintering—that of redispersion. In general, we may think of this simply as a decrease in particle size accompanied by a net increase in specific surface area. Since sintering and redispersion must have a common origin, again we find similar questions concerning the mechanism of migration of entities over the surface and, in this case, questions concerning the mechanism of particle breakup (rather than coalescence). While several theories exist, in general they can again be divided into two main categories. In one, the increase in surface area and decrease in dimension are attributed to a physical splitting of the individual particles, while the other visualizes the formation and spreading of metal monolayers (molecular or atomic species) over the surface. Again, the basic division is between crystallite and atomic migration models. Most studies have dealt with the specific system Pt/Al_2O_3, although different aluminas have been employed, and hence our understanding of the phenomenon of redispersion is quite limited. From a general

III. Crystallite Splitting and Redispersion

point of view it may be stated that there appears to exist a "window" of experimental conditions encompassing certain ranges of metals loading, oxygen partial pressures (the phenomenon is not generally observed in other atmospheres), and temperatures where it is possible to redisperse Pt supported on Al_2O_3. The process is of industrial importance in catalytic reforming, and there has been a steady supply of references in the patent literature pertaining to methods for redispersion of commercial supported Pt catalysts; see Ref. (47) for a typical claim.

The following general types of redispersion procedures have been suggested in the literature:

(1) Thermal treatment in an oxygen atmosphere.
(2) Thermal treatment in oxygen followed by reduction in hydrogen.
(3) High-temperature treatment followed by rapid cooling.
(4) Chemical dissolution and redispersion of the metal.

The first detailed work appearing in the open literature appears to have been that of McHenry *et al.* (43), who postulated the formation of a $Pt/Al_2O_3/Cl$ complex to increase the amount of soluble platinum (redispersed) leading to enhanced activity of Pt/Al_2O_3 for the dehydrocyclization rate of *n*-heptane at 490°C. Much subsequent patent literature follows this general idea, chemical dissolution and redispersion, via various proposals for treatment of aged catalysts with steam and chlorine. However, direct thermal treatment in oxygen atmospheres seems to be quite effective if conducted at the proper conditions. Johnson and Keith (35) redispersed a commercial Pt/Al_2O_3 on heating in dry air between 510 and 580°C (temperatures higher than 580°C led to sintering), and Kraft and Spindler (37) reported similar findings on heating Pt/Al_2O_3 in air at 500°C. On the other hand, Weller and Montagna (73) found that cyclic treatments in oxygen and hydrogen at 550°C produced redispersion of the metal only under *reducing* conditions, while sintering was observed during the oxygen portion of the cycle.

Such studies as mentioned above have involved catalysts from a wide variety of sources and with different pretreatment conditions, so it would be unreasonable to seek a consensus on the precise events required to obtain redispersion. It does seem that the proper combination of oxidizing atmosphere and temperature is important, and in the following we will survey in a very limited manner some typical results.

A. Experimental Observations

Quite representative of many results reported concerning redispersion of Pt/Al_2O_3 are the experiments of Fiedorow and Wanke (15), who investigated the sintering-redispersion behavior of a series of five catalysts differing

TABLE 5-13

Properties of the Pt/Al$_2$O$_3$ Catalysts of Fiedorow and Wanke (15)

Catalyst	Wt. % Pt	Support	Preparation	BET (m^2/g)
1	0.50	Commerical; Engelhard Lot 18-381		250^a
2	0.53	Kaiser KA-201, −8+10 mesh	H$_2$PtCl$_6$	230–240^a
3	1.0			
4	2.0			
5	4.0			

^{a} After treatment in air at 500°C for 3 hr.

in metal loading over the range 300–700°C in oxygen atmospheres. Some properties of the catalysts investigated are given in Table 5-13. Sintering experiments were conducted at constant temperature for various times, with metal surface area or dispersion determined by hydrogen chemisorption using a pulse method.

The effect of 1 hr in an oxygen atmosphere at varying temperatures is shown in Fig. 5-15, together with the reference values of dispersion for the fresh catalyst sample. Interpretation of these results is rendered difficult by

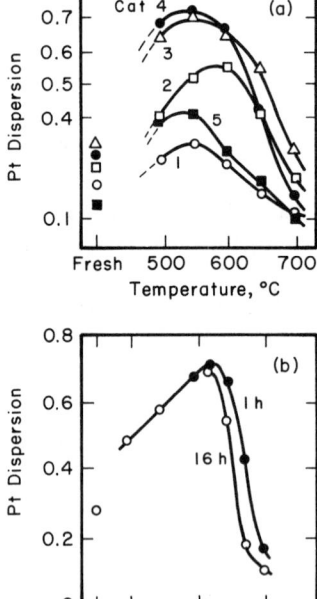

Fig. 5-15. Sintering and redispersion of Pt/Al$_2$O$_3$ in an oxygen atmosphere. (a) Effect of treatment in O$_2$ for 1 hr at various temperatures. (b) Effect of treatment time and temperature on dispersion of the 2.0% catalyst (sample 4). [From Fiedorow and Wanke (15). Reproduced by permission of Academic Press, Inc.]

III. Crystallite Splitting and Redispersion

the fact that of the laboratory-prepared catalysts (2-5), only sample 5 had a uniform loading of platinum within the individual particles. In samples 2 and 3 the metal was concentrated in the outer one-third of the support spheres, while for sample 4 the penetration was from one-third to one-half the radius. Pretreatment–preparation conditions for the various samples also differed somewhat; however, all saw H_2, 500°C and were then stored in air prior to the sintering experiments. In spite of these individual differences, it is quite clear that the dispersion for all catalysts increases after treatment in the range 500–600°C, with a maximum enhancement roughly in the vicinity of 550°C. This enhancement is quite large, too, being a factor of two to three times the initial dispersion. The temperature window is also clearly indicated by the data of Fig. 5-15a. At 700°C for all samples, redispersion has disappeared and some sintering is beginning to appear again.

The effect of treatment is shown in Fig. 5-15b for one of the samples. Basically there is no difference from 1 to 16 hr for temperatures less than about 550°C. Since that temperature was the one for maximum enhancement in dispersion, one would infer that the initial redispersion process occurs rather rapidly (at least on a relative scale). At higher temperatures, significant time effects are noted which again reflect a trend back toward the prevalance of sintering. Decreases in support area were also determined after the various thermal treatments to see if this was a significant factor in the loss of metal surface area at higher temperatures. In the most severe case a decrease in support surface of 25% was measured, while the corresponding metal dispersion decreased by 65%; hence it was concluded that loss in dispersion was indeed mainly the result of metal crystallite growth and not some other process associated with support surface area.[2]

Another type of experiment which has been popular in study of redispersion has been the use of model catalysts consisting of metal particles (again, Pt in most studies) supported on thin, dense alumina films with direct observation of changes in particle size via TEM. This has the advantage of avoiding whatever hidden pitfalls there may be in particle size determination via chemisorption methods and has the disadvantage that the alumina films employed must meet some particular requirements for successful electron microscopy and thus can only be an approximation of a typical porous alumina support. Studies of this nature have been reported by Ahn and Tien (1), Ahn et al. (2), Ruckenstein and Malhotra (55), and Stulga et al. (64). The last two, in particular, present an interesting study in contrasts which in turn might have something to say about the state of the art.

---

[2] This point has not been investigated sufficiently and must certainly play a role in some of the many investigations reported in the literature. Relatively few workers determine support surface area changes accompanying changes in metal dispersion.

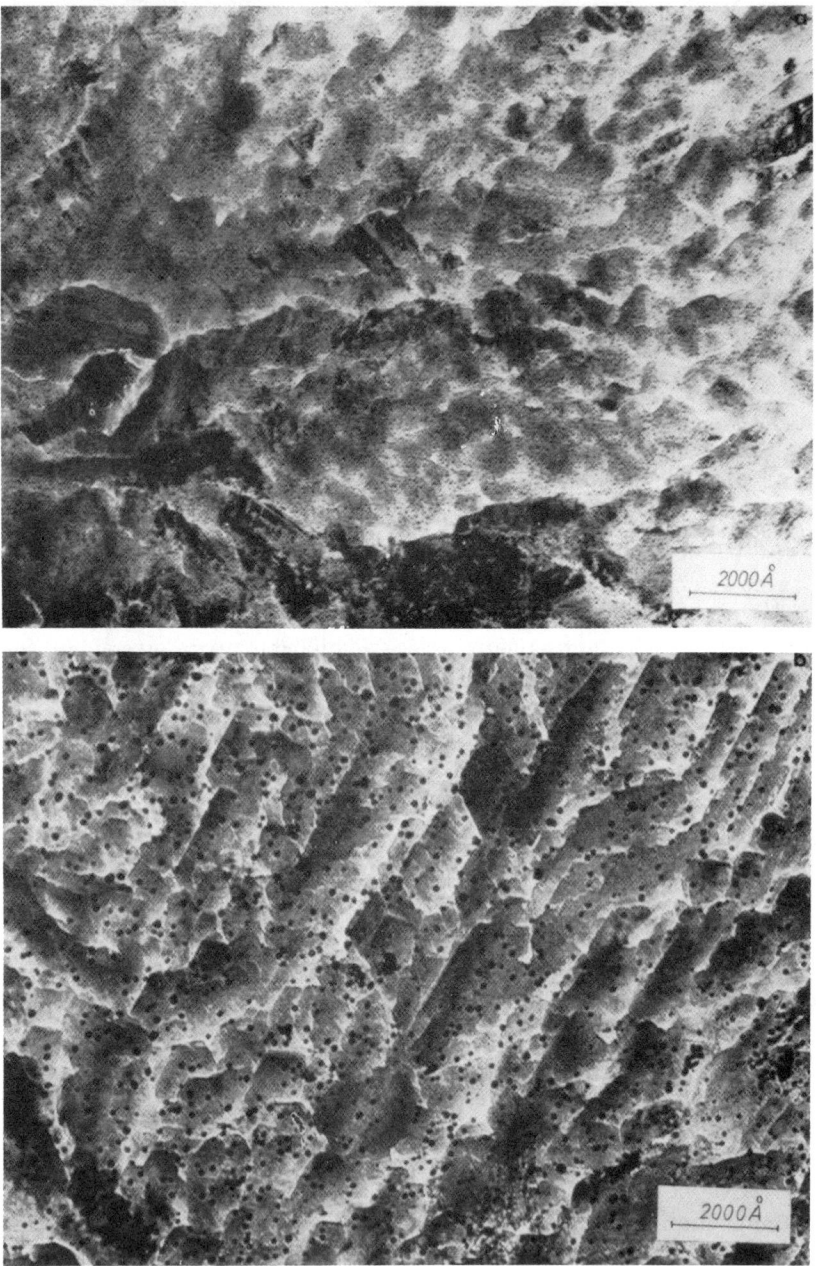

Fig. 5-16. (a) TEM of the specimen of Fig. 5-1 after further heating in air at 500°C for 24 hr and cooling to room temperature. Average crystallite size decreased from 10.7 to 4.1 nm. (b) TEM of the specimen of Fig. 5-1 after further heating in air at 400°C for 24 hr and cooling to room temperature. Average crystallite size decreased from 10.7 to 8.3 nm. [From Ruckenstein and Malhotra (55), Figs. 6 and 7. Reproduced by permission of Academic Press, Inc.]

III. Crystallite Splitting and Redispersion 219

Ruckenstein and Malhotra reported a detailed procedure for the preparation of appropriate alumina films (annealed at 600°C to form the γ phase) and the deposition of metal crystallites on them by a sputtering procedure. Figure 5-1 presents a TEM picture of the metal crystallites after treatment of the sample in air at 600°C for 24 hr. The average size of the metal particles in that figure is 10.7 nm. Figure 5-16a shows a TEM picture of the same sample after subsequent treatment at 500°C in air for 24 hr. The average size of the crystallites has decreased to about 4.0 nm and the distribution of particle sizes is much narrower. Treatment of the sample from Fig. 5-1 at 400°C instead of 500°C led to a modest decrease in particle size, from 10.7 to 8.8 nm, as shown in Fig. 5-16b. It should be emphasized that only the cycle of air heat treatment at 600°C followed by air heat treatment at 500 or 400°C was employed in these experiments, hence any explanation of the redispersion observed would have to be based on an interaction of platinum with oxygen in this temperature range. As in the case of sintering, several theories for redispersion have been proposed, which will be discussed in the next section.

The work of Stulga et al. was essentially a repeat of the experiments of Ruckenstein and Malhotra, although in addition to alumina films prepared in the same manner, a set prepared by a sputtering technique was also investigated. The heat treatment employed was air, 600°C, 18 hr followed by air, 500°C, 18 hr. Another series of very pretty TEM pictures was presen-

TABLE 5-14

Results of the Redispersion Experiments of Stulga et al. (64)

Alumina[a]	System[b]	After air, 600°C, 18 hr		After air, 500°C, 18 hr		Change (%)	
		Diameter nm[c]	Area (%)[d]	Diameter (nm)	Area (%)	Diameter	Area
Sputtered (A)	Open	14.5	9.1	15.0	8.6	+3.4	−5.5
Sputtered (A)	Closed	12.2	10.8	11.0	8.7	−9.8	−19.4
Sputtered (B)	Closed	37.3	23.8	36.7	21.2	−1.6	−10.9
Anodized (C)	Open	11.0	22.5	12.5	24.3	+13.6	+7.0
Anodized (C)	Closed	11.3	19.4	11.3	22.9	0	+15.3
Sputtered (C)	Open	16.0	18.5	14.9	17.0	−6.9	−8.1
Sputtered (C)	Sealed	15.0	15.4	17.9	16.0	+19.3	+3.8

[a] Three different experimental procedures, A, B, and C, were employed. These involved various thicknesses of the initial Pt film and the samples studied.

[b] Open—heated in an open dish exposed to laboratory air.
 Closed—heated in a sealed silica capsule, air environment, 1 atm at temperature.

[c] Arithmetic mean from 400 to 800 particle projections on electron micrograph.

[d] Calculated assuming hemispherical particles; percentage of total substrate area.

ted by Stulga et al. for "before" and "after"; however, in this case there was no indication whatsoever of any redispersion. A summary of results for all the samples investigated is given in Table 5-14. Within the probable precision of experimental measurements, the results of percentage change in diameter and percentage change in surface would probably be best interpreted as indicating that nothing happened—neither redispersion nor sintering. Since Stulga et al. were at obvious pains to reproduce the experimental procedures used previously, one can only conclude that the redispersion observed by Ruckenstein (or the lack thereof observed by Stulga et al.) is not a universal phenomenon. As of this writing, there the matter stands; clearly there are some critical factors involved in redispersion, even for these idealized model surfaces, which are unknown and uncontrolled. Perhaps even extremely low (ppb) impurity levels may be important here.

B. Some Theoretical Proposals

As mentioned in the introduction to this section, two different general concepts have been used to explain the phenomenon of redispersion. In one, a physical splitting of the individual particles is envisioned, while in the other some type of two-dimensional migration is proposed. One of the few points on which most workers are in agreement is that the temperature range 500–600°C is critical for redispersion of platinum, and the phenomenon is primarily observed only in oxidizing atmospheres. Since one generally observes sintering at temperatures greater than about 600°C in the same atmospheres, the two processes should be explained from a common origin. Now the critical temperature range for redispersion corresponds approximately to that for the decomposition of solid platinum oxides in air, hence it is reasonable to suppose that the stability of platinum oxide is important to redispersion—regardless of the detailed mechanism one may choose to describe the process.

In the particle-splitting mechanism, it is proposed that under oxidizing conditions a platinum–alumina complex is formed which, as temperature is increased, will reach a condition at which the decomposition pressure of the complex equals the partial pressure of oxygen. Ruckenstein and Malhotra report this critical temperature to be near 510°C at 0.21 atm oxygen pressure and near 580°C at 1 atm. For splitting to occur the complex existing at the support–metal (oxide) interface induces in the metal a strain energy which is relaxed by the fracture of the crystallite. The internal pressures which normally exist in sintered crystallites would also facilitate this splitting process.

There are objections to this view. Stulga et al. point out that a balance exists between relaxation of strain energy and the creation of new surface

III. Crystallite Splitting and Redispersion

area which in effect would define a lower limit of particle size for splitting to be observed. However, redispersion has been observed in catalysts with relatively high dispersions (ca. 0.4) and correspondingly small particles (3 nm). These workers also question the necessity that it be the particle which is fractured on decomposition of the complex. Viewed another way, one would envision that the crystallite is constrained in position by the support, and on oxidation the normal volume expansion is suppressed, creating tensile stress in the *support* with subsequent fracture of that phase. In any event, the properties of platinum oxide in the form of very small particles, or of any possible platinum–oxygen–alumina complexes, are not sufficiently well known to permit quantitative estimates of the magnitude of such possible effects, so the possibility of redispersion via particle splitting cannot be ruled out *a priori.*

The platinum–oxygen–support system is indeed a very complex one. Transport of Pt as either a molecular or an atomic species can occur via a number of steps, which Fiedorow and Wanke have postulated in the following:

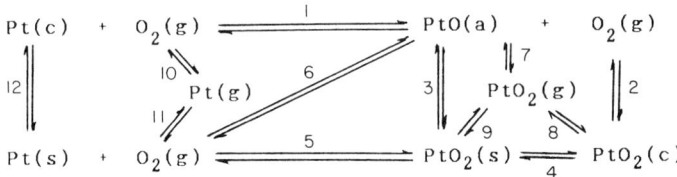

where (c) is the crystalline or solid phase, (g) the vapor phase, (a) the adsorbed phase, and (s) the species on the surface existing as individual atoms or molecules. Some proposals for these individual steps, admitted by these workers to contain "significant speculation," seem worthy of discussion. *Step 1* is adsorption of oxygen by Pt; while the stoichiometry may vary somewhat with crystallite size and conditions, this is a relatively well-understood process. *Step 2* indicates the conversion of Pt to solid oxides of uncertain stoichiometries in the temperature range 400–550°C. *Steps 3 and 4* are transfer steps for molecules of the oxide from the solid phase onto the support surface. These are thought to be fairly rapid at the temperature levels required for redispersion. *Step 5* represents decomposition of the surface oxide (molecular) species. $PtO_2(c)$ is unstable at temperatures >560°C, so it is postulated that $PtO_2(s)$ would also decompose in this temperature range. The precise temperature of decomposition is not known, but the experimental evidence on redispersion would suggest 550–600°C at 1 atm oxygen pressure. *Step 6* represents migration of Pt(s) atoms across the surface with collision and capture by PtO(a). *Steps 7, 8, and 9* are based on data for the vapor pressure of PtO_2 over Pt (58), which for

O_2 at 1 atm and 500°C is about 10^{-9} atm. Even though this is a very small number, Fiedorow and Wanke claim that transfer from PtO(a) and $PtO_2(c)$ to $PtO_2(s)$ via $PtO_2(g)$ can be appreciable for small crystallites at $T \geq 600°C$. Finally, *Steps 10, 11, and 12* can probably be neglected in comparison with the $PtO_2(g)$ equilibrium steps owing to the very low vapor pressure of Pt metal. Redispersion is attributed to the formation of $PtO_2(s)$ at temperatures <600°C, where some of the migrating $PtO_2(s)$ molecules may be trapped at high-energy sites on the support surface, resulting in the formation of new particles. Conversely, at $T > 600°C$, Pt transport occurs via Steps 3, 5, 6, 7, and 9, which involves transport of Pt(s) species which then coalesce with PtO(a), resulting in a loss of dispersion.

It is probably asking too much to suppose that the above picture is correct in detail. However, it does have the benefit of elucidating one possible common origin for both sintering and redispersion and clearly points out the central role that oxide formation can play in these phenomena.

From the foregoing it is evident that, while a number of plausible models exist for the rationalization of redispersion processes, they are hardly quantitative and it is questionable whether we should dignify them in this section with the title of theories. Dadyburjor (12) has, in fact, developed a simple theory of particle splitting which seems to include most of the essential features of that process. It is assumed that hemispherical particles of metal are oxidized in a shell-progressive manner; when the thickness of the oxide layer attains a certain critical depth, l_{crit}, the strain induced by the mismatch between oxide and metal lattices will produce a crack which rapidly spreads through the particle, leaving two quarter-spheres. (In fact, the analysis assumes this crack propagation to be infinitely fast.) Subsequently, these quarter-spheres, containing an outside layer of oxide and an inner core of metal, rearrange again to form hemispheres of smaller dimension. The external uniform oxide layer on the reformed hemispheres is now smaller than the critical thickness, so oxidation continues until the critical dimension is again attained, whereupon the splitting is repeated.

Consider the hemispherical geometry depicted in Fig. 5-17. If there is no plastic flow during oxidation, then the overall diameter d_0 must remain constant as oxidation proceeds, although l will increase. The force resulting

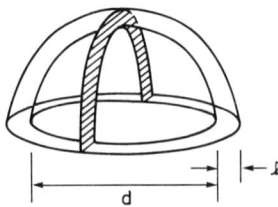

Fig. 5-17. Geometry of a hemispherical particle containing an inner metal core of diameter d and an outer oxide shell of thickness l. Overall diameter is $d_0 = d + 2l$. [From Dadyburjor (12). Reproduced by permission of Academic Press, Inc.]

III. Crystallite Splitting and Redispersion

from the difference between oxide and Pt lattices can be written (for the shaded area of the figure) as

$$F_d = E(\Delta a/a_{Pt})(\pi/8)[d_0^2 - (d_0 - 2l)^2] \quad (5\text{-}24)$$

where E is Young's modulus for PtO, $\Delta a = a_{PtO} - a_{Pt}$, and a is the lattice parameter. The surface force can be written in terms of the oxide surface tension σ as

$$F_s = (\sigma\pi)(d_0/2) \quad (5\text{-}25)$$

and splitting will occur when $F_s - F_d$ is less than some critical value F_c which defines $l = l_{crit}$. At this point, the oxide surface has attained some critical value of surface tension, σ_c, so that F_c is defined by an expression similar to Eq. (5-25) with σ replaced by σ_c. Combining the above expressions, we obtain for the critical thickness of the oxide

$$l_{crit} = \frac{d_0}{2}\left\{1 - \left[1 - \frac{4(\sigma - \sigma_c)}{d_0 E(\Delta a/a_{Pt})}\right]^{1/2}\right\} \quad (5\text{-}26)$$

A limiting case which admits of simple treatment is that of large particles, in which event $d_0 \gg [4(\sigma - \sigma_c)/E(\Delta a/a_{Pt})]$. In this case, Eq. (5-26) reduces to

$$l_{crit} \approx (\sigma - \sigma_c)/E(\Delta a/a_{Pt}) \quad (5\text{-}27)$$

and the critical thickness is, reasonably, independent of particle diameter. If we assume that the new particles formed after splitting are equal in size, then the new particle diameter is

$$d_0' = (1/2)^{1/3} d_0 \quad (5\text{-}28)$$

and the oxide thickness is

$$l' = (1/2)^{1/3} l_{crit} \quad (5\text{-}29)$$

The corresponding increase in surface area is

$$\Delta s = s' - s = 2[2\pi(d'/2 + l')^2] - 2\pi(d/2 + l_{crit})^2 \quad (5\text{-}30)$$

which is approximately

$$\Delta s = 0.26 s \quad (5\text{-}31)$$

The time required for this process to occur is essentially that required for growth of the oxide layer to the required critical thickness, since it is assumed that crack propagation is very rapid. Denoting by t_p the time for the oxide layer on the newly formed particles to increase from l to l_{crit} (still assuming the latter is independent of diameter), we may write

$$t_p = \int_{l'}^{l_{crit}} \left(\frac{dx}{k_1}\right) = \frac{l_{crit} - l}{k_1} \tag{5-32}$$

where the growth rate of the oxide layer has been taken to be linear.

We may now translate these individual particle-splitting kinetics to an assembly (distribution) of particles whose average diameter is D and surface area per unit support area is

$$S = (\pi D^2/2) N \tag{5-33}$$

with N the number per unit area of support. From Eq. (5-31) we can write

$$\Delta S = 0.26 S \tag{5-34}$$

and the rate of change of surface area is given by

$$dS/dt = \Delta S/t_p = 0.26 k_1 S/(L_{crit} - L') \tag{5-35}$$

where L_{crit} and L' are defined for the average particle diameter D. Combining with Eq. (5-27) we obtain

$$dS/dt = 1.3 k_1 E(\Delta a/a_{Pt}) S/(\sigma - \sigma_c) = kS \tag{5-36}$$

The redispersion rate which then may be expressed from this follows a familiar form. In general, one might not expect all of the restrictive assumptions in this particular model to be met, in which case an adequate generalized correlation might well be

$$dS/dt = kS^n \tag{5-37}$$

This is, of course, identical to the general correlation employed for sintering kinetics analogous to Eq. (5-9).

A final aspect of this theory considers the opposite of the assumption of large d_0 employed to obtain Eq. (5-27). Instead, let us assume $d_0 = 4(\sigma - \sigma_c)/E(\Delta a/a_{Pt})$. In this case,

$$l_{crit} = d_0/2 \tag{5-38}$$

and the entire particle must be oxidized before a crack can form. Particles of dimension smaller than d_0 would require an oxide layer greater than their dimension in order to split or, in other words, they will not split. Hence there is a minimum diameter required for splitting to occur, which is given by

$$d_{min} = 4(\sigma - \sigma_s)/E(\Delta a/a_{Pt}) \tag{5-39}$$

It may well be that this result will ultimately provide the best basis for evaluation of whether splitting is an important process in redispersion. As pointed out earlier, Stulga *et al.* criticized this view in noting that redispersion has been observed for relatively small crystallites. At present, however, there is no sound basis on which one might evaluate d_{min} in order to compare with experimental results.

IV. METAL–SUPPORT INTERACTION EFFECTS ON SINTERING AND REDISPERSION

It seems appropriate to conclude this chapter with a few rather specific observations regarding metal–support interactions and the phenomena of sintering and redispersion. Geus (25) reviewed data to 1971 on the interaction of metal atoms and crystallites with various supports from a large number of film growth studies.

Both mechanical methods and contact angle measurements have been used to determine adhesion between metal crystallites or films and ionic or oxide surfaces. From a number of studies cited by Geus, the following picture emerges: for alkali halide, oxide, and glass substrates, when ultrahigh-purity conditions and thorough reduction are employed, interaction between metal and support is due to van der Waals forces. In these cases, typical adhesion energies of 20 kcal/g-at. or less are observed. However, the presence of oxygen, either gaseous or chemically available at the interface, can change the interaction from physical adsorption to chemisorption, with much higher energies of adhesion. These effects are particularly significant on glass and oxide substrates.

Atmosphere and surface contamination also affect the adhesion of individual metal atoms to supports. In general, cleavage of salt crystals in air gives faces with higher sticking probabilities for metal atoms than crystallites cleaved *in vacuo*. Similarly, hydrocarbon deposits on the surface increase the adhesion of metal atoms to a surface.

Adhesion of metal atoms during condensation onto supports is also sensitive to the heterogeneity of the support surface. The well-known decorating effect, where nuclei concentrate around surface ledges or dislocations in film growth studies, is attributed to an increased metal atom interaction with these areas of the surface and hence a higher metal adatom population.

Thus the interaction of metal atoms and crystallites with typical supports is extremely complex. In ultrahigh vacuum on homogeneous surfaces, binding energies are low enough to be attributed to van der Waals interactions. However, in the presence of atmospheres, particularly oxygen, stronger interactions arise which are not necessarily removed by reduction or evacuation. Surface contamination and heterogeneity also increase the energy of interaction.

Although the effect of gas atmosphere and temperature on the increase in dispersion has been studied in detail, other variables are suspected to have an important role. In the subsequent sections the influence of chlorine and "chemical traps" will be discussed in detail.

A. Effect of Chloride

Much of the literature is rather vague or completely forgetful of the role of chlorine in the sintering or redispersion process, although important classes of supported metal catalysts are prepared from chlorine-containing precursors or treated with chlorine-containing compounds to maintain or enhance an acidic function. The use of H_2PtCl_6 in the preparation of Pt/Al_2O_3 catalysts is well known, and even after calcination and reduction there is appreciable retention of Cl^- by the alumina. In the sintering data of Table 5-4b it was pointed out that all catalysts had been prepared from chlorine-containing species, although there was not presumably the same chloride level among the various samples reported. We can, however, at least make some qualitative comments on the presence or absence of chloride in sintering and redispersion.

As long ago as 1970, Figueras et al. (17) reported that, in the presence of oxygen, chlorine enhances the formation of an oxidized form of platinum, indicated by the dissolution of the metallic phase at room temperature. This may be related to the redispersion process, as McHenry et al. pointed out earlier. Dautzenberg and Wolters (13) and Lieske et al. (39) both reported that redispersion of Pt/Al_2O_3 requires the presence of chlorine either in the gas phase or on the support; this is also indicated by the data for the 0.5 and 2.0% Pt catalysts in Table 5-4b. Straguzzi et al. (62) found that hydrogen-sintered Pt/Al_2O_3 (600°C) is easily redispersed by air treatment at 500°C, the extent of redispersion being dependent on sintering time and the level of residual chloride. Successive sintering-redispersion cycles performed on a given sample indicated a slow decrease in the final dispersion, particularly for heavily sintered samples. This trend was reversed by restoring the original level of chlorine. There was an apparent tendency to have a lower redispersion capacity as time increased at constant chloride content, however, for any given time there was an apparent threshold level of chlorine beyond which a redispersion capacity of 100% was attained. The higher the chlorine content, the longer the hydrogen treatment could be carried out without affecting the residual dispersion capacity.

According to Straguzzi et al., the support provides sites where Pt atoms can be localized and the concentration of those sites is related to the level of chlorine. During sintering the metal crystallites increase in size and at the same time decrease in number. Thus, we may reach a situation where the concentration of sites in the vicinity of large metal particles becomes insufficient to localize all the Pt atoms on the support. When more chlorine is added, new sites are created and the conditions for redispersion reestablished. The nature of such sites is subject to some speculation, but the following seems reasonable. When Al_2O_3 is calcined between 300 and 800°C, oxidizing sites are formed as a result of the condensation of adjacent surface hydroxyl groups. Such sites are possibly exposed aluminum atoms with a localized positive charge, in which case ionic species can be formed via electron transfer. If platinum atoms or platinum-oxygen species become trapped in these sites, the metal will be present in an electron-deficient state. Chlorine seems to enhance this oxidative property of alumina by replacing OH^- groups and attracting electrons from the Al atoms, thus increasing the strength of the sites. This is a rather detailed argument on the basis of limited data, but it is in accord with what is known concerning the dehydroxylation of alumina surfaces. A similar view has been set forth by Iida and Shirasaki (34) for the effects of chlorine on Pd/Al_2O_3 catalysts.

So far in this section, we have been concerned primarily with redispersion and not much has been said about sintering. It has been reported that a high chloride content of alumina reduces sintering rates in oxidizing atmospheres (40), and differences in chloride content in the samples investigated by various workers have undoubtedly led to some of the discrepancies in the literature. Wanke *et al.* (71) give an interesting example, among many possibilities. Lee and Kim (38) found sintering to occur in oxygen at temperatures above 600°C, while Fiedorow and Wanke (15) observed redispersion even at temperatures as high as 650°C. Both catalysts were Pt/Al_2O_3, but the chloride contents were quite different.

Perhaps the most quantitative study as of this writing on the influence of chlorine on sintering and redispersion (as usual, only for Pt/Al_2O_3) is that of Wanke *et al.* (71). In this work chlorine-free and chlorided catalysts were compared for sintering and redispersion in oxygen. The chlorided catalysts (via HCl impregnation) had a chlorine content of 1.5 wt. %. The results of hydrogen chemisorption studies with these two sets of catalysts are given in Table 5-15. It is seen that treatment of the chlorine-free catalysts (CAT-1) in oxygen at 550°C gives very little change in the H/Pt ratios. This would indicate little change in metal dispersion, and this result was verified by x-ray diffraction line broadening. The result of run 6, in which the H/Pt ratio of 0.03 after H_2 at 700°C was determined, was not in agreement with x-ray results, but this is probably the effect of strongly bound hydrogen from the high-temperature treatment. The hydrogen adsorption behavior for the chlorided catalysts (CAT-2) was significantly different. Reduction

TABLE 5-15

Influence of Chlorine Content of Alumina on Hydrogen Adsorption by 1.07 wt. % Pt/γ-Al$_2$O$_3$ Catalysts after Thermal Treatments[a]

Catalyst	Run	Treatment conditions[b] [atmosphere, temperature (°C), time (hr)]	Chlorine (wt. %)	H/Pt	Pt detected by XRD (wt. %)
CAT-1	1	Fresh, H$_2$, 500, 1	<0.05	0.30	65
	2	O$_2$, 550, 1; H$_2$, 300, 1		0.31	65
	3	H$_2$, 500, 1		0.31	65
	4	O$_2$, 550, 16; H$_2$, 300, 1		0.29	65
	5	H$_2$, 500, 1		0.26	65
	6	H$_2$, 700, 1		0.03	65
CAT-2	7	H$_2$, 300, 1[c]	1.4	0.09	65
	8	H$_2$, 500, 1	—	0.11	65
	9	O$_2$, 550, 1; H$_2$, 300, 1	—	0.38	60
	10	H$_2$, 500, 1	—	0.33	60
	11	O$_2$, 550, 16; H$_2$, 300, 1	—	0.53	50
	12	H$_2$, 500, 1	—	0.46	50
	13	H$_2$, 700, 1	0.5	0.10	50

[a] From Wanke et al. (71).
[b] Treatments done sequentially on each catalyst.
[c] Fresh reduced sample of CAT-1 treated in aqueous HCl and dried before indicated treatment.

of this material without prior oxygen pretreatment at high temperature gives very low H/Pt ratios (runs 7 and 8). There seems to be a general inhibiting effect of chlorine on hydrogen uptake since the metal dispersion in the low H/Pt runs was not affected on the basis of interpretation of x-ray results. However, oxygen treatment at 550°C both restored the hydrogen adsorption capacity and resulted in some increase in apparent Pt dispersion.

Some modest conclusions might be in order at this point. It seems that chorine, present either on the support or in the gas phase, is needed for redispersion of Pt on Al$_2$O$_3$. Be reminded that this is a very specific conclusion, since definitive data for metals other than Pt, or supports other than Al$_2$O$_3$, are not available. For example, it is well known that the response of a Pt/SiO$_2$ catalyst to chloride might well be different from that reported for aluminas, but we do not know much about this. Given this, one can say that from the data available (again, mostly for Pt/Al$_2$O$_3$) conclusions would come mostly in the order that (i) chlorine, present either on the support or in the gas phase, is required for redispersion, and oxygen alone cannot accomplish this (39), (ii) chlorine treatment in the absence of oxygen may cause redispersion (21), with chlorine in the gas phase even at temperatures around 350°C, and (iii) the appropriate chlorine treatment (or chloride content) can retard sintering but the extent of such retardation is very much dependent on conditions of pretreatment. At lower temperatures sintering may be retarded, whereas at higher temperatures the chlorine obviously leads to increased rates of redispersion.

Pt/SiO$_2$ does appear to exhibit some significant differences from Pt/Al$_2$O$_3$. Foger and Jager report that redispersion of Pt/SiO$_2$ was observed only in the temperature range 150-200°C and then only in gaseous atmospheres containing more than 25% chlorine. The redispersion was postulated to occur via formation of a [Pt(IV)Cl$_x$]/SiO$_2$ complex, although it is rather surprising that such complex formation would occur at such low temperatures and with the low dispersions of the Pt/SiO$_2$ catalysts investigated. On the other hand, Frank and Martino (22) report that redispersion of Pt/SiO$_2$ occurs using the usual oxychlorination procedures even at temperatures of 500-555°C.

B. An Example: Surface Area Stabilization of Ir/Al$_2$O$_3$ Catalysts by Group IIA Oxides

It is known that Ir will sinter at much lower temperatures than required for Pt/Al$_2$O$_3$ catalysts. Iridium in the presence of oxygen is also known to form thermally stable iridates (MIrO$_3$) with the oxides of Ca, Sr, and Ba. Because of its smaller ionic radius, Mg does not form a stable iridate. Thermo-gravimetric absorption (TGA) measurements have shown that Ca, Sr, and Ba iridates are reduced under hydrogen within the temperature range 350-450°C. The reduction products are Ir metal and the corresponding Group IIA oxide.

McVicker *et al.* (44) used these Group IIA oxides as chemical trapping agents for Ir/Al$_2$O$_3$ catalysts. Sintering and redispersion rates were determined by a combination of x-ray diffraction, chemisorption, TEM, and controlled-atmosphere electron microscopy (CAEM) measurements on both conventional and model supported Ir catalysts. These studies showed that well-dispersed Ir/Al$_2$O$_3$ catalysts are readily sintered under oxygen at temperatures above 450°C. If, however, BaO, CaO, or SrO was impregnated onto the support along with the Ir, then complete inhibition of sintering was observed up to 650°C. This effect occurred only when the concentration of Ba, Ca, or Sr oxides was in excess of the concentration of acid sites on the support. Studies of presintered Ir/Al$_2$O$_3$ catalysts to which BaO was subsequently added demonstrated that Ir could be redispersed by treating the system with oxygen at 600°C. Both the oxidative sintering suppression and oxidative redispersion of Ir are consistent with the formation of an immobile surface iridate via the reaction of a mobile molecular iridium oxide species with a well-dispersed group IIA oxide. In support of this view, Mg, which does not form an iridate, was ineffective.

The authors also calculated the crystallite size distribution from measurement of over 500 particles after the various treatments. It is clear that in calcined Ir/Al_2O_3 there was a bimodal distribution of IrO_2 crystallite sizes, with the largest fraction in all cases between 1 and 3 nm. Comparison of those crystallite size distributions shows that as the sintering temperature increases the separation between smallest and largest particles becomes wider and the number of smaller crystallites formed increases. These findings are all consistent with the notion that crystallite growth occurs by a migration mechanism whereby large crystallites increase in size at the expense of smaller ones. The effect of added BaO in suppressing the growth of Ir crystallites is probably by trapping migrating iridium oxide surface species. However, BaO, has little effect on the escape process of such species from the Ir crystallites, so the net effect is that all crystallites decrease in size and narrow in size distribution, that is, redispersion occurs.

V. SUMMARY AND EVALUATION

We began this chapter with the statement that sintering of any sort is a complicated phenomenon "where a number of mechanisms can be envisioned, all of which probably occur under one set of circumstances or another." We trust that, at this point, the reader is convinced of this.

It must be an inevitable conclusion on the basis of the examples presented here, and the discourses of many workers in the literature, that sintering is not something amenable to direct simple kinetic correlation or mechanistic interpretation. This is true regardless of whether one is speaking of the sintering of a supported metal catalyst or the loss of surface in a bulk porous solid phase. The data on sintering of supported metals, such as those provided in Tables 5-5 to 5-7, provide ample evidence of the inadequacy of simple kinetic models to correlate this mode of deactivation. In a certain sense, this is in stark contrast to the correlations of poisoning kinetics that we examined in the preceding chapter, where, given an understanding of the fundamental chemistry of the system, rather simple kinetic models seem to work fairly well.

The comments above perhaps contain the fundamental clue to why the sintering literature is at such points with itself. Basics of the mechanism(s) are postulated but not well understood. Further, in contrast to coke deposition and poisoning, one is dealing with a process primarily governed by physical rather than chemical factors. Yet, chemical factors are clearly important, or why should there be the signal influence of atmosphere on

V. Summary and Evaluation

sintering rates demonstrated in so many different experiments from so many different laboratories? In fact, one of the reasons we have not yet come to any good comprehension of the mechanism or kinetics of sintering may be that too much emphasis has been placed on understanding it as a physical—primarily thermally activated—process, at the expense of chemistry. Chemical factors may be subtle, indeed, but could also be predominant. One may ask, for example, why, after decades of research on promoted iron ammonia synthesis catalysts, the understanding of the role of various promoters is still imperfect. To be sure, one can describe quite well *what* they do (i.e., surface stabilization, phase stabilization, etc.), but *how* this is done the authors, at least, would not be able to answer in a definitive manner.

Another problem is brought forward by looking at any of the quantitative theories that have been developed for sintering on a physical basis. If we examine the case of supported metals, the principal topic of this chapter, it is apparent that data resulting from macroscopic observations cannot distinguish between the various proposals. Further, since none of the envisionments of mechanism that have found any general acceptance is unreasonable on either physical or chemical bases, one must accept the fact that any may be possible given the proper materials and set of circumstances. The problem is that we do not know what is "proper." Consider the mechanism of interparticle transport discussed in Section II,B. A very simple relationship for particle growth as a function of time, derived for diffusion control, is given by Eq. (5-19). As pointed out, however, this would in practice (i.e., in view of typical experimental error) be very difficult to distinguish from the crystallite migration model of Eq. (5-16). The situation is not unlike the familiar problem of catalytic kinetics, in which the identification of an acceptable rate equation in no general way can be taken to identify a specific reaction mechanism. Presumably one might be able to make a distinction between the characteristic constants C_1 and K_D appearing in the two equations. Yet, as also pointed out, these constants contain parameters that are not known or easily measured. It is probable, in our opinion, that much more has to be known concerning physical characteristics such as wetting angles, surface diffusion activation energies, or vapor-phase diffusion rates before more fundamental resolution of sintering kinetics, even on the basis of average particle sizes, will be possible. Progress may well be slow for such measurements are difficult and tedious to carry out and perhaps do not have the glamor associated with equally important problems in areas such as surface chemical physics. The state of current observation, thus, is adequately summarized in Table 5-11.

Using particle size distribution data as a practicable means of sorting out sintering mechanisms remains a tantalizing prospect. In our discussion, we eventually dismissed this as probably not viable in the long run. Viewed in retrospect, that may be a too harsh judgment, particularly in view of the recent advances that have been made in quantitative electron microscopy, wide-angle x-ray diffraction, and small-angle x-ray scattering (42, 46). What appears in much of the recent, more detailed work is that the shape of the metal crystallite may be as important as its size in dictating sintering behavior. The morphology of Pt crystallites on silica has been described as a distribution of equiaxed particles with minimal support interaction, so one can picture this as a collection of tiny spheres of various sizes sitting on the surface. On the other hand, Rh on titania, where the interaction between the metal and support is much stronger, has been described as consisting of nearly two-dimensional or "raftlike" structures of the metal on the surface. It is unlikely that correlations based on an average dimension, or even distribution of dimensions, would be able to reconcile these large differences in shape and support interactions (42). Ultimately, then, one might conclude that size, size distribution, and morphology are all important and the latter two are the missing factors so far in attempts to interpret distribution data. We seem to cling to the idea of little hemispherical islands sitting in the midst of a sea of support (cf. Figs. 5-11 and 5-15) when this in reality may be far from the truth.

In summary, it seems that sintering processes, admittedly complicated, are even more complex than we are at present able to evaluate. One must then modify the architecturally derived dictum "less is more" in a drastic manner to "more or less."

REFERENCES

1. T. M. Ahn and J. K. Tien, *J. Phys. Chem. Solids* **37**, 771 (1976).
2. T. M. Ahn, S. Purushothaman, and J. K. Tien, *J. Phys. Chem. Solids* **37**, 777 (1976).
3. *Prepr., Am. Chem. Soc., Div. Pet. Chem.* **30**, 132 (1985).
4. R. T. K. Baker and J. A. France, *J. Catal.* **39**, 481 (1975).
5. R. T. K. Baker, C. Thomas, and R. B. Thomas, *J. Catal.* **38**, 510 (1975).
6. J. E. Benson and M. Boudart, *J. Catal.* **4**, 704 (1965).
7. J. A. Bett, K. Kinoshita, and P. Stonehart, *J. Catal.* **35**, 307 (1974).
8. W. K. Burton, H. Cabrera, and R. C. Frank, *Philos. Trans. R. Soc. London, Ser. A* **243**, 299 (1950).
9. M. Chen and L. D. Schmidt, *J. Catal.* **55**, 348 (1978).
10. Y. F. Chu and E. Ruckenstein, *J. Catal.* **55**, 281 (1978).
11. R. D. Clay and E. E. Petersen, *J. Catal.* **16**, 32 (1970).
12. D. B. Dadyburjor, *J. Catal.* **57**, 504 (1979).
13. F. M. Dautzenberg and H. B. M. Wolters, *J. Catal.* **51**, 26 (1978).
14. G. J. den Otter and F. M. Dautzenberg, *J. Catal.* **53**, 116 (1978).
15. R. M. J. Fiedorow and S. E. Wanke, *J. Catal.* **43**, 34 (1976).
16. R. M. J. Fiedorow, B. S. Chahar, and S. E. Wanke, *J. Catal.* **51**, 193 (1978).
17. F. Figueras, B. Mercier, R. Bacand, and H. Urbain, *C. R. Acad. Sci., Ser. C*, p. 270 (1970).

18. P. C. Flynn and S. E. Wanke, *J. Catal.* **34**, 390, 400 (1974).
19. P. C. Flynn and S. E. Wanke, *J. Catal.* **33**, 233 (1974).
20. P. C. Flynn and S. E. Wanke, *J. Catal.* **37**, 432 (1975).
21. K. Foger and H. Jaeger, *J. Catal.* **92**, 64 (1985).
22. J. P. Frank and G. Martino, in "Progress in Catalyst Deactivation" (J. L. Figueiredo, ed.), p. 355. Nijhoff, The Hague, 1982.
23. J. Freel, *J. Catal.* **25**, 139 (1972).
24. P. Gallezot, in "Catalysis—Science and Technology" (J. R. Anderson and M. Boudart, eds.), Vol. 5, Chap. 4. Springer-Verlag, Berlin and New York, 1984.
25. J. W. Geus, in "Chemisorption and Reaction on Metallic Films" (J. R. Anderson, ed.), Chap. 3. Academic Press, New York, 1971.
26. A. G. Graham and S. E. Wanke, *J. Catal.* **68**, 1 (1981).
27. C. G. Granqvist and R. A. Burnham, *J. Catal.* **42**, 477 (1976).
28. C. G. Granqvist and R. A. Burnham, *Appl. Phys. Lett.* **27**, 693 (1976); *J. Appl. Phys.* **47**, 220 (1976).
29. C. G. Granqvist and R. A. Burnham, *J. Catal.* **46**, 238 (1977).
30. S. A. Hassan, G. I. Emel'yanòva, J. P. Lebeder, and N. I. Kobozev, *Zh. Fiz. Khim.* **44**, 1469 (1970).
31. S. A. Hassan, F. H. Khalil, and F. G. El-Gamal, *J. Catal.* **44**, 5 (1976).
32. R. A. Herrmann, S. F. Adler, M. S. Goldstein, and R. M. DeBaun, *J. Phys. Chem.* **65**, 2189 (1961).
33. T. R. Hughes, R. J. Houston, and R. P. Seig, *Ind. Eng. Chem. Process Des. Dev.* **1**, 96 (1962).
34. Y. Iida and T. Shirasaki, *Nippon Kagaku Kaishi* p. 1485 (1978).
35. M. F. L. Johnson and C. D. Keith, *J. Phys. Chem.* **67**, 200 (1963).
36. D. F. Klemperer and F. S. Stone, *Proc. R. Soc. London, Ser. A* **243**, 375 (1958).
37. M. Kraft and H. Spindler, *Proc. Int. Congr. Catal., 4th, 1968*, Vol. 1, p. 1252 (1971).
38. T. J. Lee and Y. G. Kim, *J. Catal.* **90**, 279 (1984).
39. H. Lieske, G. Leitz, H. Spindler, and J. Völter, *J. Catal.* **81**, 8 (1983).
40. G. Lietz, H. Lieske, H. Spindler, W. Hanke, and J. Völter, *J. Catal.* **81**, 17 (1983).
41. H. J. Maat, and L. Moscou, *Proc. Int. Congr. Catal., 3rd, 1964*, p. 1277 (1965).
42. R. J. Matyi, L. H. Schwartz, and J. B. Butt, *Catal. Rev.—Sci. Eng.* **29**, 41 (1986).
43. K. W. McHenry, R. J. Bertolacini, H. M. Brennan, J. L. Wilson, and H. S. Seelig, *Proc. Int. Congr. Catal., 2nd*, **2**, 2295 (1960).
44. G. B. McVicker, R. L. Garten, and R. T. K. Baker, *J. Catal.* **54**, 129 (1978).
45. M. Nakamura, M. Yamada, and A. Amano, *J. Catal.* **39**, 125 (1975).
46. R. K. Nandi, F. Molinaro, C. Tang, J. B. Cohen, J. B. Butt, and R. L. Burwell, Jr., *J. Catal.* **78**, 289 (1982) give applications of the latter two methods.
47. Neth. Pat. Appl. 6614074 [*C. A.* **68**, 31814b (1968)] is typical.
48. F. A. Nichols, *J. Appl. Phys.* **37**, 2805 (1966).
49. F. A. Nichols and W. W. Mullins, *Trans. AIME* **233**, 1840 (1965).
50. Z. Paál and P. G. Menon, *Catal. Rev.—Sci. Eng.* **25**, 229 (1983).
51. D. Pope, W. L. Smith, M. J. Eastlake, and R. L. Moss, *J. Catal.* **22**, 72 (1971).
52. A. E. B. Presland, G. L. Price, and D. L. Trimm, *J. Catal.* **26**, 313 (1972).
53. A. Renouprez, C. Hoang-Van, and P. A. Compagnon, *J. Catal.* **34**, 411 (1974).
54. E. Ruckenstein and D. B. Dadyburjor, *J. Catal.* **48**, 73 (1977).
55. E. Ruckenstein and M. L. Malhotra, *J. Catal.* **41**, 303 (1976).
56. E. Ruckenstein and B. Pulvermacher, *AIChE J.* **19**, 356 (1973).
57. S. R. Sashital, J. B. Cohen, R. L. Burwell, Jr., and J. B. Butt, *J. Catal.* **50**, 470 (1977).
58. H. Schäfer and A. Tebben, *Z. Anorg. Allg. Chem.* **304**, 317 (1960).
59. P. A. Sermon and G. C. Bond, *Catal. Rev.* **8**, 211 (1973).

60. J. H. Sinfelt, and G. H. Via, *J. Catal.* **56**, 1 (1979).
61. G. A. Somorjai, *in* "X-ray and Electron Methods of Analysis" (H. Van Olphen and W. Parrish, eds.), Chap. 6. Plenum, New York, 1968.
62. G. I. Straguzzi, H. R. Aduriz, and C. E. Gigola, *J. Catal.* **66**, 171 (1980).
63. *Stud. Surf. Sci. Catal.* **6** (1982).
64. J. E. Stulga, P. Wynblatt, and J. K. Tien, *J. Catal.* **62**, 59 (1980).
65. D. L. Swift and S. K. Friedlander, *J. Colloid Sci.* **19**, 62 (1964).
66. S. J. Tauster and S. C. Fung, *J. Catal.* **55**, 29 (1978).
67. S. J. Tauster, S. C. Fung, and R. L. Garten, *J. Am. Chem. Soc.* **100**, 170 (1978).
68. T. Uchijima, J. M. Herrmann, Y. Inoue, R. L. Burwell, Jr., J. B. Butt, and J. B. Cohen, *J. Catal.* **50**, 464 (1977).
69. S. E. Wanke, *J. Catal.* **46**, 234 (1977).
70. S. E. Wanke and P. C. Flynn, *Catal. Rev.—Sci. Eng.* **12**, 93 (1975).
71. S. E. Wanke, J. A. Szymura, and T. T. Yu, *Proc. Int. Symp. Catal. Deact.*, 3rd (E. E. Petersen and A. T. Bell, eds.), p. 65. Dekker, New York, 1986.
72. B. E. Warren, "X-Ray Diffraction." Addison-Wesley, Reading, Massachusetts, 1969.
73. S. W. Weller and A. A. Montagna, *J. Catal.* **20**, 394 (1971).
74. T. E. Whyte, Jr., *Catal. Rev.* **8**, 117 (1973).
75. G. R. Wilson and W. K. Hall, *J. Catal.* **17**, 190 (1970); **24**, 306 (1972).
76. P. Wynblatt and T. M. Ahn, *in* "Sintering and Catalysis" (G. C. Kuczynski, ed.), Vol. 10, p. 63. Plenum, New York, 1975.
77. P. Wynblatt, and N. A. Gjostein, *Prog. Solid State Chem.* **9**, 1 (1975).

PART II

Deactivation of Catalyst Pellets: Macroscopic Processes

Catalysts exhibit intrinsic activity that declines with age, but many factors other than the intrinsic decay can influence the overall deactivation process. The time scale of the decline can vary by many orders of magnitude depending on the system and the operating conditions, as will be seen in the following chapters. In Part I we explored the fundamental processes of decay and the various mechanisms by which it occurs. In order to be utilized in reactors, catalysts have various physical forms that can require the consideration of certain physical rate processes in conjunction with the purely chemical rate processes examined to this point. We shall now discuss the physical bases for these added complications.

Consider a catalyst whose intrinsic properties permit it to catalyze a given reaction at a rate of $\mathcal{R}$ gram-moles per second per square centimeter of surface at a specified composition and temperature of reaction mixture. The most effective utilization of this catalyst would be to develop the area within the catalytic material to as large a value as possible, since the greater surface area enhances the catalytic effect. Of course, this is done. There are two common ways to increase the specific surface area of a material: (i) reduce the particle size by grinding or pulverizing or (ii) develop a network of fine pores within the material. The latter method is used almost exclusively because large porous pellets are separated from the reaction mixture much more easily than are fine powders and because it is very difficult to grind materials fine enough to obtain large specific surface areas. Those familiar with such technology will recognize that 1-2 m^2 per gram of material is generally regarded as a low specific surface area; moderate to high surface area materials may range from 200-300 to 1000 m^2/g, respectively. In general, then, heterogeneous catalysts are porous.

The specific surface area of a porous material is roughly inversely proportional to the pore radius and, accordingly, heterogeneous catalysts also generally have small pores, often as small as 3 nm or even less. But now, to utilize all of this area, reactants must be transported into, and products out of, the interior of the catalyst through the pore structure. The more extensive the surface the more active the catalyst pellet per unit weight or volume, but also the smaller the pores and the greater the difficulty of transporting material into and out of the particle. The net result is that the interior of the pellet can be exposed to compositions and temperatures different from those at the exterior of the pellet and, as a consequence, the contribution to the overall activity of the pellet of an element of area in the interior can differ from that of the same size element on the exterior. This is, of course, a familiar problem in catalytic reaction engineering, often referred to as the "Thiele–Zeldovich" problem. Our pupose here is not to discuss this problem *per se*, as it is amply treated elsewhere,[1] but to relate the many ways in which such phenomena are affected by deactivation.[2] It is clear that if reactivity may be nonuniform throughout the pellet, deactivation may be also. The coupling between such nonuniformities greatly complicates the interpretation of deactivation data and the subsequent design of reactors employing a heterogeneous catalyst phase. We must, therefore, treat this problem rather generally to learn how rates of transport, reaction, and deactivation can interact. In so doing in Part II, we shall learn that the interaction is a mixed blessing in the sense that it is neither all good nor all bad. For example, experiments in this regime can lead to mechanistic insights difficult to identify unequivocally under gradientless conditions. We shall learn further that nonuniformity of deactivation throughout the pellet can, under certain circumstances, be preferable to uniform deactivation in operation. However, from a mathematical point of view one must deal with gradients of concentration, activity, and temperature that can evolve in shape, position, and time and the dominant feature in describing the overall pellet behavior is a considerable increase in the complexity of analysis and design.

---

[1] R. Aris, "The Mathematical Theory of Diffusion and Reaction in Permeable Catalysts." Oxford Univ. Press (Clarendon), London and New York, 1975; E. E. Petersen, "Chemical Reaction Analysis," Prentice-Hall, Englewood Cliffs, New Jersey, 1965; C. N. Satterfield, "Mass Transfer in Heterogeneous Catalysis," MIT Press, Cambridge, Massachusetts, 1970; A. Wheeler, *Adv. Catal.* **3**, 250 (1950).

[2] R. Hughes, "Deactivation of Catalysts." Academic Press, London, 1984.

CHAPTER 6

The Time Scale of Deactivation in Pelleted Catalysts

> Some problems are big, some are small, but they all seem to be important.
> *Anon.*

Practical catalysts are generally porous materials having a very large surface area to effect high catalytic activities. Such catalysts have small pores and appreciable resistance to the transport of reactants and products to and from the reaction sites. However, the usual method of preparing catalyst pellets alleviates this problem considerably. Catalysts are frequently precipitated from gels in which very fine pores—micropores—are developed on partial drying to fine powders. These powders are then pressed in dies or otherwise consolidated to make pellets of the desired size, dried, and calcined. In the interstitial spaces among the fine powders a second pore system is developed that has much larger pores. These macropores can transport materials with much less resistance than the micropores. Nevertheless, very active catalysts process large amounts of reactants in the pellets, and even in these larger pores appreciable transport resistance is possible.

I. GENERAL FORMULATION OF THE PROBLEM

In the mathematical treatment to follow, we shall consider that the process by which reactants and products are transported within catalysts is primarily by diffusion through the fluid phase. Three terms are needed in the conservation equation for species and in the conservation equation for energy: an accumulation term, diffusion of species or energy, and a generation term. The generation term contains the activity parameter, which is related to the amount of foulant on the surface. Lastly, a population balance equation completes the set of governing equations. Our purpose in this chapter is to determine the relative order of magnitude of each of the terms in the conservation equations under well-defined conditions, to investigate and review the effect of boundary conditions on the nature of the solutions, and to explore briefly methods for solving these equations.

Consider a spherical pellet immersed in an isotropic medium at a temperature T_0 and consisting of pure A at a concentration of C_{A0}. If this pellet catalyzes the reaction of $A \rightarrow B$, a reasonably general expression for the simultaneous reaction and diffusion of mass and energy to the nonisothermal deactivating pellet shown in Fig. 6-1 is given by a conservation equation for the species A:

$$-\varepsilon \frac{\partial C_A}{\partial t} \pi r^2 \, dr - \frac{\partial}{\partial r}\left(-D_{\text{eff}} \pi r^2 \frac{\partial C_A}{\partial r}\right) dr = a\pi r^2 \rho S \mathcal{R}_A \, dr \qquad (6\text{-}1)$$

where the terms from left to right represent accumulation, diffusion, and generation, respectively. A similar equation can be written for the conservation of B. For nonisothermal systems it is necessary to solve an energy equation simultaneously with the conservation of species equation. One form of the energy equation is similar which can be written for T:

$$-(1-\varepsilon)\rho C_v \frac{\partial T}{\partial t} \pi r^2 \, dr - \frac{\partial}{\partial r}\left(-k_{\text{eff}} \pi r^2 \frac{\partial T}{\partial r}\right) dr = -\Delta H_A a\pi r^2 \rho S \mathcal{R}_A \, dr \qquad (6\text{-}2)$$

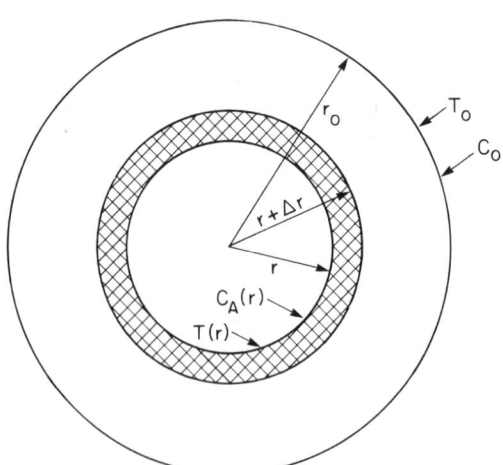

Fig. 6-1. A spherical catalyst pellet.

I. General Formulation of the Problem

A relationship between a and f is necessary, which we shall express here quite generally as

$$a = a(f) \tag{6-3}$$

A population balance is also needed to obtain f and the expression below is slightly more general than Eq. (2-81):

$$\frac{\partial f}{\partial t} = g_1(f, C_A, T) \tag{6-4}$$

The symbols in the equations above have the following functional dependence and meaning:

a	$a(r, t)$, defined by Eq. (2-4)
C_A	$C_A(r, t)$, concentration of reactant A
C_v	mean heat capacity of catalyst
D_{eff}	effective diffusivity of A in catalyst
f	$f(r, t)$, defined by Eq. (2-78)
k_{eff}	effective thermal conductivity of catalyst
ΔH_A	enthalpy of reaction of A to B, per mole of A
r	radial coordinate
S	total surface area of catalyst per gram of catalyst
$\mathscr{R}_A$	$\mathscr{R}_A(C_A, T)$, rate of reaction of A
T	$T(r, t)$, local temperature
ε	catalyst porosity
ρ	catalyst density

The equations above reveal their character more easily when rearranged to make them dimensionless. The following definitions serve to accomplish this.

$$\Psi \equiv \frac{C_A}{C_{A0}}, \quad \Theta \equiv \frac{T}{T_0}, \quad \xi \equiv \frac{r}{r_0}, \quad \tau \equiv \frac{D_{\text{eff}} t}{\varepsilon r_0^2} \tag{6-5}$$

$$\mathscr{R}_1(\Psi, \Theta) \equiv \frac{\mathscr{R}_A(\Psi, \Theta)}{\mathscr{R}_A(1, 1)}, \quad \phi \equiv r_0 \left[\frac{\rho S \mathscr{R}_A(1, 1)}{D_{\text{eff}} C_{A0}} \right]^{1/2}$$

where r_0 is the radius of the pellet and C_{A0} the concentration at $r = r_0$. Equations (6-1) and (6-2) become

$$-\frac{\partial \psi}{\partial \tau} + \frac{1}{\xi^2} \frac{\partial}{\partial \xi}\left(\xi^2 \frac{\partial \psi}{\partial \xi}\right) = a\phi^2 \mathscr{R}_1(\psi, \Theta) \tag{6-6}$$

Similarly, Eq. (6-2) becomes

$$-\delta \frac{\partial \Theta}{\partial \tau} + \frac{1}{\xi^2} \frac{\partial}{\partial \xi}\left(\xi^2 \frac{\partial \Theta}{\partial \xi}\right) = -\beta a \phi^2 \mathscr{R}_1(\psi, \Theta) \tag{6-7}$$

where

$$\delta \equiv \frac{\rho C_v D_{\text{eff}}}{k_{\text{eff}}}\left(\frac{1-\varepsilon}{\varepsilon}\right), \quad \beta \equiv \frac{(-\Delta H_A) D_{\text{eff}} C_{A0}}{k_{\text{eff}} T_1} \quad (6\text{-}8)$$

Rewriting Eq. (6-4),

$$\partial f/\partial t = g_1(f, \psi, \Theta) \quad (6\text{-}9)$$

and rearranging it, we obtain

$$\kappa(\partial f/\partial \tau) = g(f, \psi, \Theta) \quad (6\text{-}10)$$

where

$$\kappa \equiv \frac{D_{\text{eff}}}{\varepsilon r_0^2 g_1(1, 1, 1)}, \quad g(f, \psi, \Theta) \equiv \frac{g_1(f, \psi, \Theta)}{g_1(1, 1, 1)} \quad (6\text{-}11)$$

Equations (6-6) through (6-11) describe a quite general deactivating catalyst pellet. If the activity of the system does not change with time the equations are considerably simpler to solve. First of all, $g(f, \psi, \Theta)$ is zero and $f = 1$. Then the time derivatives in Eqs. (6-6) and (6-7) are zero and the resulting equations reduce to those that describe the classical system involving steady-state diffusion and reaction in porous pellets.

The complexity of the general equations for deactivation frequently can be greatly simplified by noting that the time scale for deactivation is often much longer than the time scale for filling of the catalyst pores with reactants by diffusion[1]; that is, κ is larger because $g_1(1, 1, 1)$ is small. Therefore, on a time scale for deactivation, θ, where

$$\theta \equiv \tau/\kappa \quad (6\text{-}12)$$

Eq. (6-10) can be rewritten in the form

$$\partial f/\partial \theta = g(f, \psi, \Theta) \quad (6\text{-}13)$$

and the first terms of Eqs. (6-6) and (6-7) become $(1/\kappa)(\partial \psi/\partial \theta)$ and $(\delta/\kappa)(\partial \Theta/\partial \theta)$, respectively. Therefore, on the time scale of deactivation, the transient terms in Eqs. (6-6) and (6-7) can be set equal to zero and the equations become quasi-steady-state. Thus, the behavior of the system is governed by the set of four equations below:

$$\frac{1}{\xi^2}\frac{\partial}{\partial \xi}\left(\xi^2 \frac{\partial \psi}{\partial \xi}\right) = a\phi^2 \mathcal{R}_1(\psi, \Theta) \quad (6\text{-}14)$$

---

[1] An order of magnitude calculation would be $\varepsilon r_0^2/D_{\text{eff}} \approx (0.3)(0.2)^2/0.01$ or the order of 1 sec.

$$\frac{1}{\xi^2} \frac{\partial}{\partial \xi} \left(\xi^2 \frac{\partial \Theta}{\partial \xi} \right) = \beta a \phi^2 \mathcal{R}_1(\psi, \Theta) \tag{6-15}$$

$$\frac{\partial f}{\partial \theta} = g(f, \psi, \Theta) \tag{6-16}$$

$$a = a(f) \tag{6-17}$$

Elimination of the time derivatives from Eqs. (6-14) and (6-15) greatly simplifies the general method of solution of this set of equations.

The boundary conditions on these equations need some comment. The usual conditions imposed on Eqs. (6-14) and (6-15) are

$$\psi(1, \theta) = 1, \quad \frac{\partial \psi}{\partial \xi}(0, \theta) = 0$$
$$\Theta(1, \theta) = 1, \quad \frac{\partial \Theta}{\partial \xi}(0, \theta) = 0 \tag{6-18}$$

These conditions imply that the corresponding mass and heat transfer coefficients that determine the rates of transport of mass and heat from the bulk reaction mixture phase and the external surface of the catalyst are very large, and as a consequence the concentrations of the reactants and the temperature at the surface of the catalyst are the same as their corresponding values in the bulk phase. These boundary conditions, of course, represent the simplest situation. We shall consider the case of external mass and heat transfer limitation briefly at the end of this chapter.

The initial condition imposed on Eq. (6-16) is that $f = 1$ for $\theta = 0$ for all ξ. The description of the system, then, is complete when the functional form of $a(f)$ is defined.

II. SOME ASPECTS OF THE GENERAL SOLUTION

Before proceeding with the general solution method, it is instructive to look at the limiting solution as the Thiele parameter ϕ goes to zero. This case, of course, corresponds to a frequently encountered industrial situation in which the rate of diffusion into the catalyst is high compared to the rate of chemical reaction. As ϕ goes to zero, the right-hand sides of Eqs. (6-14) and (6-15) go to zero, whereupon integration leads to constant concentrations and temperature throughout the pellet. These constant concentrations and temperature can then be substituted into Eq. (6-16) to determine the change in f with time and consequently the change in a with time utilizing Eq. (6-17).

It should be noted that this solution corresponds exactly to the cases considered in Chapter 2, where the intrinsic activity of catalysts was considered. The only difference between the result here and that of Chapter 2 is that the surface area of the pellet is included in the calculation. This limit has already been covered and needs no further discussion.

Values of ϕ greater than about one are characteristic of another class of important and interesting problems of both industrial and research importance. As shown by Eq. (6-14), the concentration will decrease with decreasing radius within the pellet. Depending on the magnitude and sign of β, the temperature within the pellet can increase or decrease with decreasing radius. It is obvious then from Eq. (6-16) that the local value of f varies with radius.

In general, the solution to this set of Eqs. (6-14) through (6-17) is not significantly more difficult than the solution to the corresponding equations for constant activity. To illustrate this, consider a pellet of uniform activity at zero time. The solutions to Eqs. (6-14) and (6-15), for a equal to unity, give the ψ and Θ profiles as functions of the radius ξ. Then, values of f at a later time increment can be calculated as a function of ξ using the corresponding values of ψ and Θ at zero time in Eq. (6-16). Equation (6-17) relates f to a to give the activity profile in ξ at the first time increment. The procedure is repeated until the desired time has elapsed. The advantage of this calculation procedure as compared to solving the original transient equations can be substantial, primarily because very much larger increments in time are possible for each step.

The quasi-steady-state equations cannot be written for all systems because κ may not always be large. A catalyst that deactivates very rapidly will have a large value of $g(1, 1, 1)$, making κ small. An example of such a system is apparently the catalytic cracking of naphthas.

In general, the solutions to the Thiele–Zeldovich problem are presented in terms of an effectiveness factor, a quantity defined as the ratio of the actual rate of reaction obtained with a catalyst pellet to the rate of reaction with the same pellet if the concentrations of all species in the reaction mixture and the temperature throughout the pellet were equal to the corresponding values at the external surface of the pellet. As used in this book, we shall add the further restriction that the above calculation be made at time = 0. Hence, we shall always consider the effectiveness as a parameter of the system at its initial and undeactivated state.[2]

In mathematical form the effectiveness factor η is

$$\eta = \frac{-4\pi r_0^2 D_{\text{eff}} \dfrac{\partial C_A}{\partial r}(r_0, 0)}{(4\pi r_0^3/3)\rho S \mathscr{R}(1, 1)} \tag{6-19}$$

---

[2] This convention is not universally adopted in the literature.

II. Some Aspects of the General Solution

which in terms of the dimensionless variables of Eq. (6-4) becomes

$$\eta = -\frac{3}{\phi^2}\frac{\partial \psi}{\partial \xi}(1, 0) \tag{6-20}$$

Evaluating these expressions for the case of an isothermal first-order reaction in a sphere, the effectiveness factor becomes

$$\eta = \frac{3}{\phi^2}(\phi \coth \phi - 1) \tag{6-21}$$

The case of finite values of the mass and heat transfer coefficients is an important consideration in the design of reactors and has been covered adequately in the literature (1). Although we shall have little direct use for these boundary conditions, the reader should be alert to use them when adapting the deactivation principles described here to a particular reactor design. To do this involves relatively simple ideas, and our goal is to show how the surface concentrations and temperature can be calculated in terms of the corresponding bulk values and the magnitudes of the mass and heat transfer coefficients, thereby reducing the problem to the simpler one described earlier.

The rate of mass transfer to the surface of a catalyst in terms of the mass transfer coefficient k_g is

$$k_g(C_g - C_0) \tag{6-22}$$

and the rate of reaction at the surface in terms of the effectiveness factor η is

$$\eta \mathcal{R}(C_0, T_0, t) \tag{6-23}$$

Equating the above and simplifying, we get

$$\frac{C_0}{C_g} = \frac{1}{1 + \eta \mathcal{R}(C_0, T_0, t)/k_g C_0} \tag{6-24}$$

where C_g and C_0 are, respectively, the concentrations of the reactant in the bulk and at the external surface of the catalyst. The system operates at the surface concentration that satisfies Eq. (6-24). External heat transfer is handled analogously in terms of a heat transfer coefficient h.

These methods and references to the literature can be found in Aris's book (1). In the chapters immediately following, we shall see that deactivation processes in pellets can appear very different from their intrinsic behavior under conditions of large values of ϕ. In addition to the analysis set forth here, the reader is also referred to the treatment given by Hughes (2).

REFERENCES

1. R. Aris, "The Mathematical Theory of Diffusion and Reaction in Permeable Catalysts," Chap. 3. Oxford Univ. Press (Clarendon), London and New York, 1975.
2. R. Hughes, "Catalyst Deactivation," Chaps. 3 and 6. Academic Press, London, 1984.

CHAPTER 7

Intraparticle Deactivation

> They cover a dunghill with a piece of tapestry
> when the procession goes by.
>
> *Cervantes*

We have now graduated up the scale from consideration of mechanisms of deactivation processes to the point where it is necessary to consider these in the context of problems in catalytic reaction engineering. The first level of scale is that of the individual catalyst particle, in which it is necessary to take the by now well-known problem of interacting transport and reaction rates and add deactivation rates. The implications are simple enough, although the analysis may not be: steady-state problems become unsteady-state or, at best, quasi-steady-state as seen in Chapter 6.

I. LIMITING TYPES OF DEACTIVATION IN PELLETS

One of the very early and now classical treatments of the poisoning of a catalyst pellet was that of Wheeler (29). It was his purpose to show that a homogeneous catalyst surface could produce a nonlinear curve on co-ordinates of rate versus fraction of surface unpoisoned even for nonselective poisons. Previously, it had been widely assumed that such curves were characteristic only of selective poisons, and chemical reasons were sought for this behavior. In Chapter 2, to be sure, we say that nonselective catalysts can give nonlinear curves when more than one active center is required for the rate-limiting step; however, this does not invalidate the subsequent Wheeler treatment. It is our first task to review this early work, not only for its historical interest but also because Wheeler's analysis leads to very simple results that serve as asymptotic limits to more sophisticated treatments that follow.

Wheeler assumed two limiting types of poison: one that adsorbs tightly with a sticking probability of essentially unity, the other with a very low sticking probability needing many collisions before adsorbing. The first type tends to collect poison at the exterior of porous catalyst pellets with a very sharp front proceeding inward as the quantity of poison adsorbed by the catalyst increases. This will be called "pore mouth poisoning." During this process, each pore is thereby divided into two zones: a catalytically inactive zone that has adsorbed its saturation amount of poison and a zone at the initial, unpoisoned activity. The second type of poison is uniformly distributed throughout the porous catalyst pellet, hence this process will be called "uniform poisoning."

Consider now the behavior of each of the above catalysts for slow reactions, that is, for Thiele parameters very much smaller than unity. If the activity versus fraction unpoisoned curve for the reaction is linear, the activity of each of the above catalysts is also linear in the amount of poison adsorbed. However, if the Thiele parameter is much larger than unity, then the two catalyst types give very different overall behavior, with the pore mouth poisoning giving the more rapid decline in activity with amount of poison adsorbed.

The physical reasons for this are straightforward. An active catalyst (one still having a Thiele parameter much greater than unity) is operating in the diffusion-influenced regime and there is a substantial concentration gradient within the pores of the catalyst. If the pore mouths are now poisoned, then all of the reactants that react in the interior on the unpoisoned zones of the pore must pass through the inactive part. The net result of this is to slow down the process much faster than would be expected on the basis of the fraction of active catalyst surface removed. In effect, the remaining active surface of the pellet has been made even more inaccessible by the zones of poisoned surface because the latter offer additional strong diffusional resistance.

The activity of a uniformly poisoned catalyst having a large Thiele parameter declines less rapidly than that of a catalyst with a small Thiele parameter. The physical reasons for this are more subtle. As the catalyst poisons, the effective rate constant decreases and the apparent Thiele parameter decreases. Thus, although the extent of the active surface in the catalyst decreases, the reactants are able to penetrate deeper into the pores to utilize more pore surface area than they could initially. The net result is that the overall catalyst pellet deactivates less rapidly than linearly with the amount of poison on the catalyst. A third limiting case [a limiting case of series poisoning as discussed by Masamune and Smith (12)] should be included with the two models of Wheeler. This case, termed "core poisoning," has the physical interpretation of a poison which deactivates the pores

I. Limiting Types of Deactivation in Pellets

from the inside out with a sharp front and, as in pore mouth poisoning, separates the pore into fully active and completely deactivated zones. The core poisoning case serves as a limit to the behavior of series poisoning, to be discussed later in this chapter.

To show the quantitative behavior of these cases, we shall use the example of first-order, irreversible reactions under isothermal conditions. The same methods can be used in general for other kinetic orders at the expense of some additional computational effort. The first-order case, however, serves well enough to display the general features of the effects of diffusional influence on the shape of the poisoning curve.

Homogeneous poisoning is the easiest to analyze. Given the pore of Fig. 7-1 of radius r and the constant boundary conditions of C_{A0} at the pore mouths, we can solve Eq. (6-14) written in Cartesian coordinates with an a of unity. Thus:

$$d^2\psi/d\xi^2 - \phi^2\psi = 0 \qquad (7\text{-}1)$$

with the boundary conditions

$$\psi(0) = 1, \qquad d\psi(1)/d\xi = 0 \qquad (7\text{-}2)$$

where

$$\psi \equiv C_A/C_{A0}, \qquad \xi \equiv x/L, \qquad \phi \equiv L\sqrt{2k/rD} \qquad (7\text{-}3)$$

Here k is the rate constant of the first-order reaction and the term $2/r$ is the surface area per unit volume of pore and corresponds to the term S in Eq. (6-4). Equation (7-1), subject to the boundary conditions of Eq. (7-2), was first solved by Thiele (26). The actual history of this problem is more

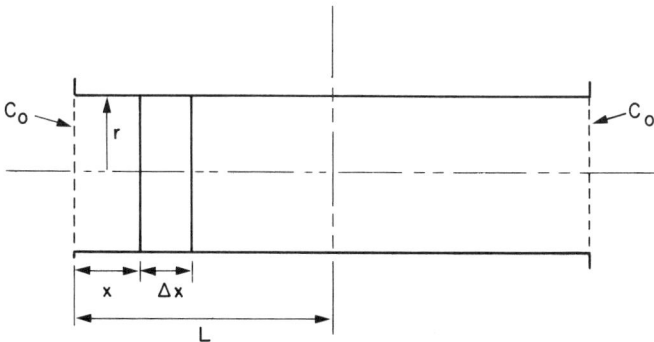

Fig. 7-1. Diffusion and reaction in a single pore.

involved than this; the reader is referred to Aris (1). The solution is

$$\psi = \cosh \phi (1-\xi)/\cosh \phi \qquad (7\text{-}4)^1$$

To use Eq. (7-4) we want to express the half-pore reaction in terms of ψ. This is

$$\mathcal{R}_p = (\text{half-pore rate, diffusion}) = -\pi r^2 D(dC/dx)_{x=0} \qquad (7\text{-}5)$$

$$\mathcal{R}_0 = (\text{half-pore rate, no diffusion}) = 2\pi r L k C_0 \qquad (7\text{-}6)$$

The effectiveness factor is by definition the ratio of these two rates, so

$$\eta = \text{effectiveness factor} = \mathcal{R}_p/\mathcal{R}_0 = -\phi^2 (d\psi/d\xi)_{\xi=0} \qquad (7\text{-}7)$$

The effectiveness factor as calculated from the derivative of Eq. (7-4) at $\xi = 0$ is

$$\eta = \tanh \phi / \phi \qquad (7\text{-}8)$$

and the reaction rate of a half-pore is given by

$$\mathcal{R}_p(\text{rate, unpoisoned}) = 2\pi r L k C_{A0} \qquad (7\text{-}9)$$

Note, however, that as the pore poisons uniformly, as shown in Fig. 7-2, we can use the same solution if we modify the rate constant to its poisoned value. Since we are discussing a case where the intrinsic activity is related to fraction unpoisoned by

$$a = f \qquad (7\text{-}10)$$

the value that should be used for the first-order rate constant of the poisoned catalyst is kf. Substituting this into Eq. (7-3) for ϕ, we obtain for the

---

[1] Briefly, Eq. (7-4) is obtained by assuming a solution to Eq. (7-1) of the form

$$\psi = A \exp(-\phi\xi) + B \exp(\phi\xi) \qquad (7\text{-}4a)$$

where A and B are arbitrary constants to be satisfied by the boundary conditions. Differentiating Eq. (7-4a) once and satisfying

$$(d\psi/d\xi)_{\xi=1} = 0$$

yields

$$A = B \exp(\phi) \qquad (7\text{-}4b)$$

Substituting Eq. (7-4b) into Eq. (7-4a) and satisfying $\psi(0) = 1$, we obtain

$$B = (1+2\phi)^{-1} \qquad (7\text{-}4c)$$

Substitution of Eqs. (7-4b) and (7-4c) back into Eq. (7-4a) and simplifying results in

$$\psi = (e^{h(1-\xi)} + e^{h(1+\xi)})/(e^{-h} + e^h) \qquad (7\text{-}4d)$$

which in turn leads directly to Eq. (7-4).

I. Limiting Types of Deactivation in Pellets

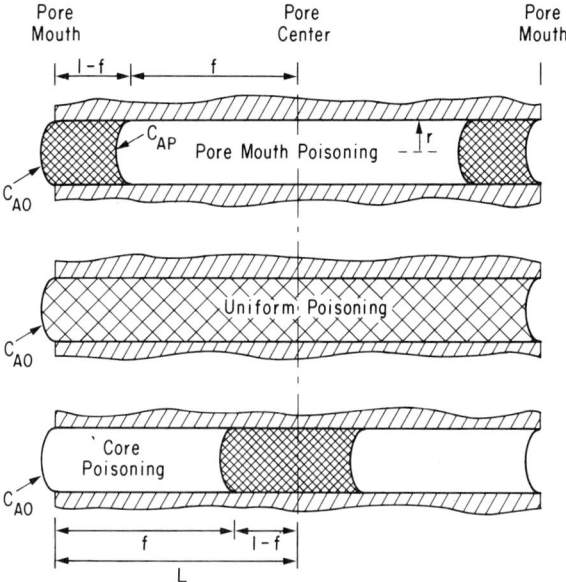

Fig. 7-2. Three limiting cases of poisoning. [From E. E. Petersen, *Exp. Methods Catal. Res.* **2**, 257 (1976). Reproduced by permission of Academic Press, Inc.]

poisoned case

$$\phi_p = \phi(f)^{1/2} \tag{7-11}$$

We can now get the variation of the rate with fraction poisoned in terms of the apparent overall activity of the pellet, $\langle a \rangle_p$ where

$$\langle a \rangle_p = \frac{\text{reaction rate of poisoned pellet}}{\text{reaction rate of unpoisoned pellet}} \tag{7-12}$$

Substituting Eq. (7-1) into Eq. (7-5) to get the effectiveness factor for the poisoned pellet, which in turn is substituted into Eq. (7-6) for the rate of reaction of the poisoned pellet, we obtain for pellet activity

$$\langle a \rangle_p = \frac{2\pi r L k f C_{A0}[\tanh(\phi\sqrt{f})]/\phi\sqrt{f}}{2\pi r L k C_{A0}(\tanh \phi)/\phi}$$

$$= \frac{\sqrt{f} \tanh(\phi\sqrt{f})}{\tanh \phi} \tag{7-13}$$

This solution is plotted as curve b in Fig. 7-3 for large values of ϕ. (It is of interest to note, though, that even for $\phi = 3$ the curve is almost identical

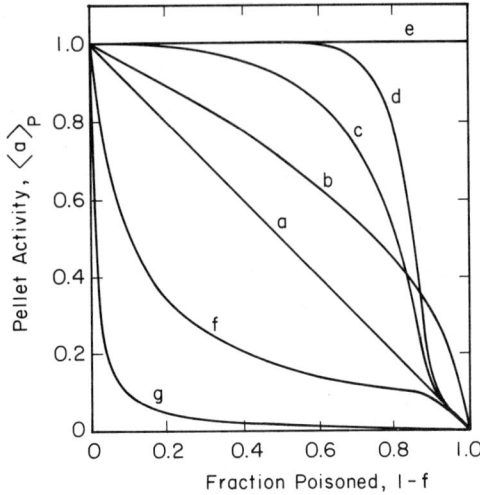

Fig. 7-3. Decrease in pellet activity with amount of poison for different types of poisoning. a, All types of poisoning, $\phi \ll 2$; b, uniform poisoning, $\phi \gg 2$; c, core poisoning, $\phi = 3$; d, core poisoning, $\phi = 5$; e, core poisoning, $\phi \gg 5$; f, pore mouth poisoning, $\phi = 10$; g, pore mouth poisoning, $\phi = 100$.

to curve b for all values of $1-f < 0.95$.) Curve a in this figure represents the pellet activity for very small values of ϕ.

To analyze the pore mouth poisoning case, we again use Eqs. (7-1) and (7-2) and their solutions, Eqs. (7-5) and (7-6). In the active zone, the rate of reaction of the poisoned catalyst is

$$[\text{reaction rate of poisoned catalyst}] = 2\pi r L f k C_{AP} \eta_p \quad (7\text{-}14)$$

where

$$\eta_p \equiv (\tanh f\phi)/f\phi \quad (7\text{-}15)$$

and for the unpoisoned catalyst

$$[\text{reaction rate of unpoisoned catalyst}] = 2\pi r L k C_{A0} \eta_u \quad (7\text{-}16)$$

where

$$\eta_u \equiv (\tanh \phi)/\phi \quad (7\text{-}17)$$

Combining Eqs. (7-14)–(7-16) we obtain

$$\langle a \rangle_p = \left(\frac{C_{AP}}{C_{A0}}\right)\left(\frac{\tanh(f\phi)}{\tanh \phi}\right) \quad (7\text{-}18)$$

I. Limiting Types of Deactivation in Pellets

The ratio C_{AP}/C_{A0} obtains from the continuity at the point $(1-f)L$ down the pore; that is, the rate of diffusion down the poisoned pore mouth is equal to the rate of reaction in the remaining unpoisoned pore. These terms are, respectively,

$$\pi r^2 D_{\text{eff}} \frac{C_{A0} - C_{AP}}{L(1-f)} = 2\pi r L f k C_{AP} \frac{\tanh(f\phi)}{f\phi} \tag{7-19}$$

Solving Eq. (7-19) for C_{AP}/C_{A0} we get

$$\frac{C_{AP}}{C_{A0}} = \frac{1}{1 + (1-f)\phi \tanh(f\phi)} \tag{7-20}$$

Combining Eqs. (7-18) and (7-20), the final result is

$$\langle a \rangle_p = \frac{1}{1 + (1-f)\phi \tanh(f\phi)} \frac{\tanh(f\phi)}{\tanh \phi} \tag{7-21}$$

Curves f and g in Fig. (7-3) are plots of Eq. (7-21) for values of $\phi = 10$ and 100, respectively.

The last case to be considered is that of core poisoning. The treatment is similar.

$$\langle a \rangle_p = \frac{2\pi r L f k C_{A0}[\tanh(f\phi)]/f\phi}{2\pi r L k C_{A0}(\tanh \phi)/\phi} \tag{7-22}$$

which reduces simply to

$$\langle a \rangle_p = [\tanh(f\phi)]/\tanh \phi \tag{7-23}$$

Obviously, according to Eq. (7-23), for large values of ϕ the activity of the pellet does not change with core poisoning. Physically this means that although the interior of the catalyst is being poisoned, the reactants cannot penetrate into this region; therefore, it is immaterial in terms of overall reaction rate whether this part of the catalyst is active or inactive. As the value of ϕ decreases, however, the activity slowly decreases with amount of poison, as shown by curves c and d in Fig. 7-3, corresponding to values of ϕ equal to 3 and 5, respectively.

The important message in this treatment is that the apparent activity of a pellet can vary greatly with the type of poisoning taking place if diffusion rates influence the overall reaction rate. In fact, we have demonstrated that the curve can be anywhere in the $\langle a \rangle_p$ versus $(1-f)$ space. Moreover, as previously stated, we anticipate that the utility of these cases is as asymptotic limits to the more realistic models to be discussed in the next section.

II. MODELING THE DEACTIVATION OF PELLETS

More realistic and sophisticated simulations of the deactivation of a catalyst pellet than that proposed by Wheeler generally result from a solution to the set of four equations given by Eqs. (6-14), (6-15), (6-16), and (6-17). The analyses to follow then will differ from those of Section I in that they use Eqs. (6-16) and (6.17). This follows because the nature of the Wheeler deactivation model made it unnecessary to know details of the deactivation kinetics. To use the set of four equations above, it is necessary to know the kinetic rate law for the main reaction and the rate at which the foulant or poison reacts with the surface active sites in addition to a relationship between activity and the amount of foulant on the surface of the catalyst. With this information, and with certain physical constants for the particular catalyst, the deactivation with time can be simulated as a straightforward calculation. This, however, is not the usual problem. In general one has a deactivation–time plot and wishes to know all of the kinetic information and the relationship between activity and the amount of foulant on the surface of the catalyst. In principle, this information can be obtained by fitting the experimental curves to sets of curves calculated for models having many different forms of the kinetic expressions and activity functions. In doing this, however, we rapidly find that the results are not very sensitive to the functional forms assumed. That is, the number of adjustable parameters to be fitted is sufficiently large that many models can be made to closely fit the data. [One is reminded here of the remark of the mathematician Cauchy to the effect that "if you give me three adjustable parameters, I'll draw you an elephant; if you give me four I'll wag his tail." For more than one wants to know on the art of elephant fitting, see Wei (27).] Nevertheless, sets of curves obtained from integrating the four equations for various forms of the kinetic and activity functions serve the additional role of showing the general behavior of deactivating systems when diffusion influences the overall reaction rate. This is the general spirit in which the work below is presented. In Chapter 8 we shall look at supplementary methods for obtaining functional forms of the kinetic and activity relations from specially designed experiments.

Masamune and Smith (13) did the pioneering work in the modeling of deactivating catalyst pellets. They considered three models, which we shall call parallel fouling, series fouling, and impurity poisoning. The first two processes occur when the reactant and product, respectively, are precursors of the foulant that ultimately deactivates the catalyst. Typical of this are the various types of coke formation. The poisoning process involves a contaminant in the reaction mixture that serves as a precursor of the catalyst poison and requires an additional conservation equation to establish the

II. Modeling the Deactivation of Pellets

poison concentration in the porous catalyst. These models correspond to the general reaction sequences outlined in Chapter 1.

The qualitative behavior of these cases is understood more easily by noting the distribution of foulants in the pores of the catalyst as shown in Fig. 7-4. Parallel fouling is shown in Fig. 7-4a, where it is noted that the foulant concentration is greater at the pore mouth than at the center. Parallel fouling occurs when the precursor is the reactant and the highest concentration of reactant is at the pore mouth. Hence, the foulant is produced at a greater rate there. Figure 7-4a also applies to independent poisoning because the poison must diffuse in from the outside. Depending on how readily it is adsorbed, the range for this type of deactivation is between Figs. 7-4a and 7-4b.

Series fouling is understood by reasoning similar to that used for parallel fouling. The foulant precursor is the product which has a higher concentration at the center, hence the foulant tends to concentrate as shown in Fig. 7-4c.

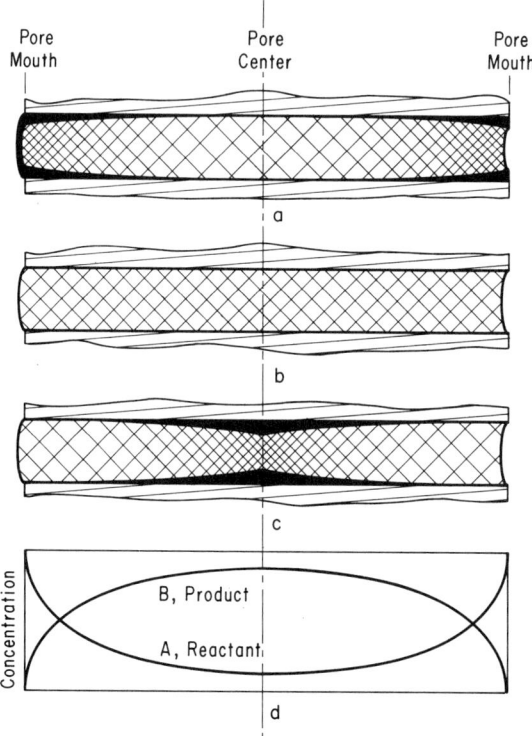

Fig. 7-4. Distribution of foulant in the catalyst pore. [From E. E. Petersen, *Exp. Methods Catal. Res.* **2**, 257 (1976). Reproduced by permission of Academic Press, Inc.]

The parallel fouling case can be represented by the stoichiometric equations

$$A \to B \text{ (main reaction, } k_A\text{)} \qquad (7\text{-}24)$$

$$A \to P \text{ (deactivation reaction, } k_p\text{)} \qquad (7\text{-}25)$$

which we will assume are also first-order in the reactant for both the fouling and the main reaction. Both reactions are also assumed to be linear in the concentration of unfouled sites $(C_{P0} - C_P)$, where C_{P0} and C_P are, respectively, the surface concentrations of foulant at saturation and at less than saturation. With these assumptions, Eq. (6-5) is

$$\mathcal{R}_A(\psi, \Theta) = \mathcal{R}_1(\psi_A) = k_A C_{A0} \psi_A$$

$$\mathcal{R}_A(1) = k_A C_{A0}$$

and

$$\phi = r_0 \sqrt{\rho S k_A C_{P0} / D_{\text{eff}}} \qquad (7\text{-}26)$$

Substituting the above into Eqs. (6-14), (6-15), and (6-17), we obtain

$$\frac{1}{\xi^2} \frac{\partial}{\partial \xi}\left(\xi^2 \frac{\partial \psi_A}{\partial \xi}\right) - a\phi^2 \psi_A = 0$$

$$\frac{\partial f}{\partial \theta_A} = -\psi_A f \qquad (7\text{-}27)^2$$

---

[2] That is,

$$\partial C_P/\partial t = k_{pA}[C_A][C_{P0} - C_P] \qquad (7\text{-}27\text{a})$$

Let

$$f = (C_{P0} - C_P)/C_{P0} \quad \text{and} \quad \psi = C_A/C_{A0}$$

Then

$$\partial f/\partial t = -k_{pA} C_{A0} \psi_A f \qquad (7\text{-}27\text{b})$$

Comparing with Eq. (6-9):

$$g_2 = k_{pA} C_{A0} \psi_A f \qquad (7\text{-}27\text{c})$$

and

$$g = g_2(f, \psi_A)/g_2(1, 1) = \psi_A f \qquad (7\text{-}27\text{d})$$

Moreover, from Eqs. (6-4) and (6-11):

$$\theta_A = \tau/\kappa = k_{pA} C_{A0} t \qquad (7\text{-}27\text{e})$$

Therefore

$$\partial f/\partial \theta_A = -\psi_A f \qquad (7\text{-}27\text{f})$$

II. Modeling the Deactivation of Pellets

The assumptions made by Masamune and Smith lead to the equality

$$a = f \tag{7-28}$$

for the form of Eq. (6-17). The boundary conditions on Eqs. (7-26)–(7-28) are

$$\left.\begin{array}{ll} \text{At } t = 0, & a = f = 1 \text{ for all } r \\ \text{At } r = r_0, & \psi_A = 1 \\ r = 0, & \dfrac{\partial \psi_A}{\partial r} = 0 \end{array}\right\} \text{ for all } t \tag{7-29}$$

If, in addition, we specify the magnitude of ϕ, a solution follows.

At this point we should be clear about what it is we seek for the solution to these equations. Of primary importance is the variation in the apparent activity $\langle a \rangle$ with time for various values of the Thiele parameter. The next question is, how would we use these results to calculate the reaction rate as a function of time for a particular system? Lastly, how do these results relate to, and compare with, the limiting models developed in Section I?

Masamune and Smith carried out the numerical calculations for this model. Their results have been cast in the form shown in Fig. 7-5. The coordinates all have their usual meanings. The pellet activity $\langle a \rangle$ is defined as

$$\langle a \rangle = \left[\frac{\text{rate of reaction at } t}{\text{rate of reaction at } t = 0} \right] = \frac{\mathcal{R}(t)}{\mathcal{R}(0)} \tag{7-30}$$

where

$$\mathcal{R}(0) = (4\pi r_0^3/3)\rho S \cdot k_A C_{A0} \eta \tag{7-31}$$

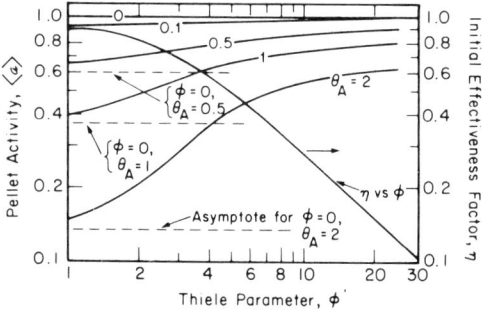

Fig. 7-5. Pellet activity for parallel fouling.

and, of course,

$$\mathcal{R}(t) = (4\pi r_0^3/3)\rho S \cdot k_A C_{A0} \eta \langle a \rangle \qquad (7\text{-}32)$$

Most of the quantities in these equations are defined in Chapter 6. Thus, to use Fig. 7-5 to calculate the reaction rate of a pellet from Eq. (7-32) is straightforward.

Figure 7-5 shows that the pellet activity changes in a regular and predictable manner with time, θ_A, which is a dimensionless time defined by Eq. (7-27e) as

$$\theta_A = k_{pA} C_{A0} t \qquad (7\text{-}33)$$

This figure reveals further that the activity falls off more rapidly with time when the Thiele parameter is small, which is understandable because, for small ϕ, the foulant is being deposited at a greater rate in the interior.

The foulant, of course, is not deposited uniformly in the pellet because, as assumed, the rate of foulant deposition is proportional to the reactant concentration. For the particular case of $\phi = 5$, the concentration profiles for ψ_A and f are shown in Fig. 7-6. The amount of foulant, $1 - f$, is greatest at r/R near unity where ψ_A is large. After very long times the activity is severely reduced even at the center of the pellet.

It is of interest to compare the results of this work with Wheeler's limiting case of pore mouth fouling discussed in Section I. We choose to make the comparison by plotting, in each case, the average activity $\langle a \rangle$ as a function of the amount of foulant in the pellet. The results should give us some insight into the sensitivity of the activity to the distribution of foulant in the pellet. To do this, it is necessary to calculate the amount of foulant on the catalyst as a function of time. This is readily done for the case of parallel fouling because the fouling kinetics are proportional to the main reaction kinetics. If we let $\langle 1 - f \rangle$ be defined as the fraction of the surface fouled on

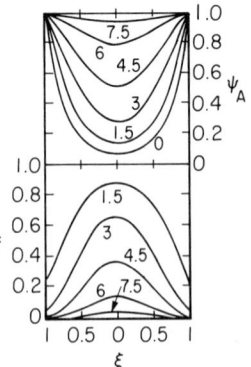

Fig. 7-6. Profiles for self-fouling, parallel mechanism, $\phi = 5$. [From Masamune and Smith (12). Reproduced by permission of the American Institute of Chemical Engineers.]

II. Modeling the Deactivation of Pellets

the catalyst at a particular time, that is, an average fraction fouled, then at any time the rate of fouling is proportional to the main reaction rate, or

$$\partial \langle 1-f \rangle / \partial \theta_A = \eta \langle a \rangle \qquad (7\text{-}34)^3$$

where the activity $\langle a \rangle$ is expressed as a function of θ_A. Integrating once, we get our final result

$$\langle 1-f \rangle = \eta \int_0^{\theta_A} \langle a \rangle \, d\theta_A \qquad (7\text{-}35)$$

---

[3] The details of this derivation are as follows: Eq. (7-27a)

$$\frac{\partial C_P}{\partial t} = k_{pA} C_A (C_{P0} - C_P) \qquad (7\text{-}27a)$$

represents the rate of fouling at a particular time and position in the spherical catalyst pellet where the concentrations of reactants and foulant are C_A and C_{P0}, respectively. To obtain the rate of fouling in the catalyst at a particular time we integrate with respect to position after normalizing by the surface area. This is represented below.

[rate of production of foulant in pellet at time t]

$$= \int_0^{r_0} 4\pi r^2 \rho S \frac{\partial C_P}{\partial t} \, dr \qquad (7\text{-}34a)$$

$$= \int_0^{r_0} 4\pi r^2 \rho S k_{pA} C_A (C_{P0} - C_P) \, dr \qquad (7\text{-}34b)$$

$$= \frac{4}{3} \pi r_0 \rho S C_{P0} \frac{\partial \langle 1-f \rangle}{\partial t} \qquad (7\text{-}34c)$$

Equating Eqs. (7-34b) and (7-34c) and simplifying, we get

$$\frac{1}{3} \frac{\partial \langle 1-f \rangle}{\partial \theta_A} = \int_0^1 \psi_A f \left(\frac{r}{r_0}\right)^2 d\left(\frac{r}{r_0}\right) \qquad (7\text{-}34d)$$

Similarly, the rate of the main reaction at a particular time and position is

$$\text{rate of main reaction} = k_A C_A (C_{P0} - C_P) \qquad (7\text{-}34e)$$

Integrating with respect to position to get the main reaction rate of the pellet,

$$[\text{rate reaction of pellet at time } t] = \int_0^{r_0} 4\pi r^2 \rho S k_A C_A (C_{P0} - C_P) \, dr \qquad (7\text{-}34f)$$

$$= \tfrac{4}{3} \pi r_0^3 \rho S k_A C_{A0} C_{P0} \eta \langle a \rangle \qquad (7\text{-}34g)$$

Equating Eqs. (7-34f) and (7-34g) and simplifying results in

$$\frac{\eta \langle a \rangle}{3} = \int_0^1 \psi_A f \left(\frac{r}{r_0}\right)^2 d\left(\frac{r}{r_0}\right) \qquad (7\text{-}34h)$$

Equating Eqs. (7-34d) and (7-34h) leads us to Eq. (7-34) above.

The quantity $\langle 1-f \rangle$ can be calculated from Fig. 7-5 using (7-35) and the corresponding values of $\langle a \rangle$ can be read from the figure. Some results for $\phi = 5$ and $\phi = 10$ are plotted in Fig. 7-7. A comparison to the Wheeler model for a spherical pellet is made using Eq. (7-36) below:

$$\langle a \rangle = \frac{\bar{\xi}(\bar{\xi}\phi \coth \bar{\xi}\phi - 1)}{[1+(1-\bar{\xi})(\bar{\xi}\phi \coth \bar{\xi}\phi - 1)](\phi \coth \phi - 1)} \qquad (7\text{-}36)^4$$

where $\bar{\xi} = r_i/r_0$.

The fraction of foulant on the surface as a function of r_i is

$$1 - f = \frac{(4\pi/3)(r_0^3 - r_i^3)}{(4\pi/3)r_0^3} = 1 - \left(\frac{r_i}{r_0}\right)^3 = 1 - \bar{\xi}^3 \qquad (7\text{-}37)$$

[4] Equation (7-36) is derived as follows for a first-order irreversible and isothermal reaction:

$\mathcal{R}_i = $ [reaction rate for a pellet fouled to radius r_i from r_0]

$$= \tfrac{4}{3}\pi r_i^3 \rho S k C_{Ai} \eta_i \qquad (7\text{-}36\text{a})$$

where

$$\eta_i = (3/\phi_i^2)(\phi_i \coth \phi_i), \qquad \eta = (3/\phi^2)(\phi \coth \phi - 1)$$
$$\phi_i = r_i(\rho Sk/D_{\text{eff}})^{1/2}, \qquad \phi = r_0(\rho Sk/D_{\text{eff}})^{1/2} \qquad (7\text{-}36\text{b})$$

and C_{Ai} is the concentration of reactant at $r = r_i$. However, $\mathcal{R}_i$ is also given by

$$\mathcal{R}_i = -4\pi r_i^2 D_{\text{eff}} \left(\frac{\partial C_A}{\partial r}\right)_{r=r_i} \qquad (7\text{-}36\text{c})$$

where $(\partial C_A/\partial r)_{r=r_i}$ is the derivative obtained from the solution to the equation

$$\frac{1}{r^2}\frac{\partial}{\partial r}\left(r^2 D_{\text{eff}} \frac{\partial C_A}{\partial r}\right) = 0 \qquad (7\text{-}36\text{d})$$

with the boundary conditions

$$r = r_0, \qquad C_A = C_{A0}$$
$$r = r_i, \qquad C_A = C_{Ai}$$

Solving:

$$\left(\frac{\partial C_A}{\partial r}\right)_{r=r_i} = \left(\frac{C_{A0} - C_{Ai}}{1/r_i - 1/r_0}\right)\left(\frac{1}{r_i^2}\right) \qquad (7\text{-}36\text{e})$$

Substitution of Eq. (7-36e) in (7-36c) and equating to Eq. (7-36a) leads to an expression for C_{Ai} of the form

$$C_{Ai}/C_{A0} = [1 + (1 - r_i/r_0)(\phi_i^{2/3})\eta_i]^{-1} \qquad (7\text{-}36\text{f})$$

From Eq. (7-36a), the ratio of $\mathcal{R}_i$ and $\mathcal{R}_0$ leads to $\langle a \rangle$, that is,

$$\langle a \rangle = \frac{(4/3)r_i^3 \rho S k C_{Ai} \eta_i}{(4/3)r_0^3 \rho S k C_{A0} \eta} \qquad (7\text{-}36\text{g})$$

which, on substitution of Eqs. (7-36b) and (7-36f) and simplification, leads directly to Eq. (7-36).

II. Modeling the Deactivation of Pellets

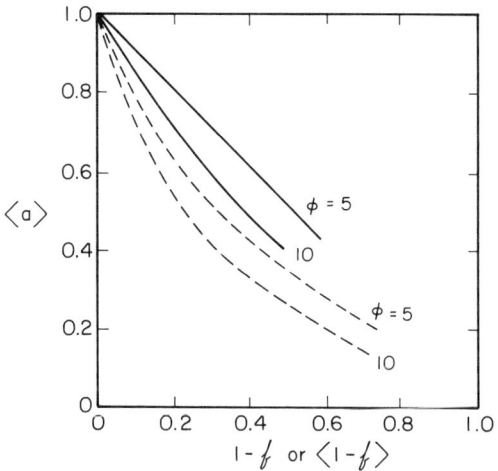

Fig. 7-7. Activity as affected by foulant distribution within the pellet. (——) Results using Fig. 7-5 and Eq. (7-35) [Masamune and Smith (13)]; (- - - -) results using Eqs. (7-36) and (7-37) [Wheeler (29)].

The curves for $\langle a \rangle$ versus $\langle 1-f \rangle$ as calculated from Eqs. (7-36) and (7-37) are also shown in Fig. 7-7. From the results of these two models, we learn that the activity is sensitive to the distribution of foulant and that distributions represented by Figs. 7-1 and 7-4 do not give the same activity for equivalent amounts of fouling. The Wheeler model for fouling reduces the activity much faster. [A note in passing is that the activity calculated from Eq. (7-36) is virtually identical to that from Eq. (7-35) for $r_i/r_0 > 0.75$.]

The case of series fouling is simulated using Eqs. (7-26) and (7-28). The fouling equation differs from the parallel case according to a series fouling mechanism. The rate of fouling is again assumed to be proportional to $(C_{P0} - C_P)$ but also proportional to the concentration of B, the product. Thus:

$$\partial C_P/\partial t = k_{pB}(C_B)(C_{P0} - C_P) \tag{7-38}$$

By a series of transformations analogous to those presented in Eqs. (7-27a)–(7-27f), Eq. (7-38) reduces to

$$\partial f/\partial \theta_B = -\psi_B f \tag{7-39}$$

where

$$\theta_B = k_{pB} C_{A0} t \quad \text{and} \quad \psi_B \equiv C_B/C_{A0} \tag{7-40}$$

Masamune and Smith integrated Eqs. (7-26), (7-28), and (7-39) to simulate series fouling and reported results for the case of $C_{B0}/C_{A0} = 1$ and for equal diffusivities of A and B. These results are shown in Figs. 7-8 and 7-9.

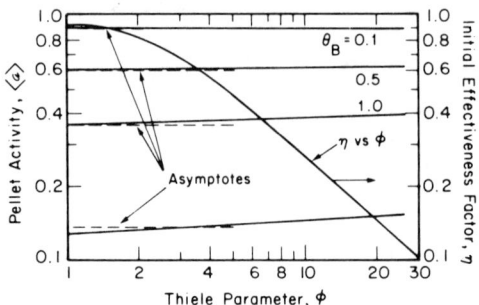

Fig. 7-8. Pellet activity for series fouling, $C_{B0}/C_{A0} = 1$.

Values of $\langle a \rangle$ read off Fig. 7-8 as functions of θ_B for a given value of the Thiele parameter ϕ can be used directly in Eq. (7-32) to calculate the reaction rates as a function of time, θ_B. The reaction rates can then be related to real time using Eq. (7-40).

Figure 7-8 shows the expected qualitative behavior that the activity $\langle a \rangle$ is not a strong function of ϕ, particularly in the region of small to moderate values. This is expected because C_B is relatively constant throughout the pellet and the rate of deactivation can be approximated by the rate of deactivation at $\phi = 0$. When $\phi = 0$, Eq. (7-39)

$$df/d\theta_B = -\psi_B f$$

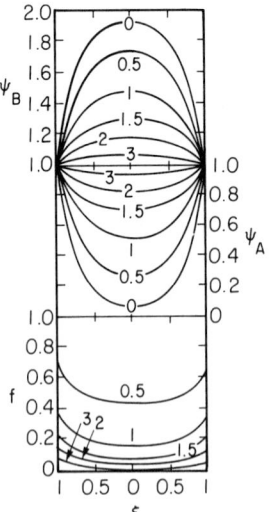

Fig. 7-9. Profiles for self-fouling, series mechanism, $\phi = 5$. Parameter is θ_B. [From Masamune and Smith (12). Reproduced by permission of the American Institute of Chemical Engineers.]

II. Modeling the Deactivation of Pellets

is readily solved directly because ψ_B is constant. Integration leads to

$$f = \exp(-\psi_B \theta_B) \qquad (7\text{-}41)$$

This solution almost exactly corresponds to the values for $\phi = 1$ in Fig. 7-8. Although it is difficult to read the values presented in Masamune and Smith's paper with precision, it appears that $\langle a \rangle$ is essentially constant out to $\phi \approx 10$. It would seem, therefore, that an excellent approximation of the reaction rate with time could be calculated from Eqs. (7-41), (7-36b), and (7-32). One can, of course, use the same technique for various values of C_{B0}.

When $C_{B0} = 0$, the activity remains unchanged with time for $\phi = 0$. For small to intermediate values of ϕ, only very small reductions in $\langle a \rangle$ would be predicted with time. For larger values of ϕ, C_B will build up in the interior and foul the inner surface, but the activity will remain essentially unchanged for a long period of time.

Comparing these results to the limiting core fouling case, we would expect excellent agreement only when C_{B0} is very small. Then both models will predict essentially no reduction in activity for long periods of time. When C_{B0} is larger, the discrepancy between the predictions of the two models grows and the prediction of the limiting core model is not useful for quantitative work, as was shown by Masamune and Smith. Fortunately, Eq. (7-41) will give good results in this region.

Impurity poisoning is the last case to be considered. We again use Eqs. (7-26) and (7-28) with boundary conditions of Eq. (7-29), but Eq. (7-27) must again be modified, this time to account for deactivation by another species in the reaction mixture which is neither a reactant nor a product. This equation takes the analogous form:

$$\partial C_P / \partial t = k_{pI} C_I (C_{P0} - C_P) \qquad (7\text{-}42)$$

which becomes

$$\partial f / \partial \theta_1 = -\psi_1 f \qquad (7\text{-}43)$$

where

$$\psi_1 \equiv C_I / C_{I0}, \qquad \theta_1 \equiv k_{pI} C_{I0} t \qquad (7\text{-}44)$$

and where C_{I0} is the concentration of poison in the ambient reaction mixture. However, we cannot use Eq. (7-43) unless we know the concentration of poison at various radii in the catalyst pellet. This requires another material balance equation to Eq. (7-26). In real quantities this material balance equation in C_I is

$$\frac{1}{r^2} \frac{\partial}{\partial r}\left(r^2 D_I \frac{\partial C_I}{\partial r}\right) - \rho S k_{pI} C_I (C_{P0} - C_P) = 0 \qquad (7\text{-}45)[5]$$

---

[5] In writing Eq. (7-45), it is again assumed that the time derivative can be set to zero because the diffusion relaxation time is much shorter than the deactivation relaxation time.

and it reduces to the form

$$\frac{1}{\xi^2}\frac{\partial}{\partial \xi}\left(\xi^2\frac{\partial \psi_1}{\partial \xi}\right) - a\phi_1^2\psi_1 = 0 \tag{7-46}$$

using the definitions

$$\left.\begin{array}{c} \psi_1 \equiv C_1/C_{10}, \quad \xi \equiv r/r_0, \quad f \equiv (C_{P0}-C_P)/C_{P0}, \quad a = f \\ \phi_1 \equiv r_0\sqrt{\rho Sk_{pI}C_{P0}/D_I} \end{array}\right\} \tag{7-47}$$

The boundary conditions on Eq. (7-46) are generally

$$\left.\begin{array}{cc} \text{at } r = r_0, & \psi_1 = 1 \\ r = 0, & \partial \psi_1/\partial r = 0 \end{array}\right\} \text{ for all } t \tag{7-48}$$

The impurity poisoning system of equations requires specification of both ϕ and ϕ_1 to obtain a solution. Masamune and Smith integrated the equations for various values of ϕ, holding ϕ_1 constant at 10. These results are shown in Figs. 7-10 and 7-11. Evidently for $\phi_1 = 10$ the apparent pellet activity $\langle a \rangle$ is relatively independent of ϕ except for large values of ϕ_1. However, this case is more complex than those of series and parallel fouling and the characteristics of its solution cannot be represented by the solution for a single value of ϕ_1. It is of value to look at the changes in the solution as the value of ϕ_1 is varied.

We can compare the exact solutions to some limiting solutions to understand better how ϕ_1 affects the behavior of this system. Two limiting cases are of interest, those for $\phi_1 \to 0$ and $\phi_1 \to 100$. The former corresponds to uniform poisoning and the latter to pore mouth poisoning.

Consider first uniform poisoning. Equation (7-46) predicts $\psi_1 = 1$ as $\phi_1 \to 0$, whereupon Eq. (7-43) readily integrates to

$$f = e^{-\theta_1} \tag{7-49}$$

It follows then that the activity is constant, $a = f$, for all positions within the catalyst and known as a function of time. This permits us to obtain the solution to Eq. (7-26) analytically as a function of ϕ_a, a modified Thiele parameter of the form

$$\phi_a = r_0^2\sqrt{\rho Ska/D_I} = \sqrt{a\phi^2} = e^{-\theta_1/2}\phi \tag{7-50}$$

In terms of the effectiveness factor, the rate of reaction of the pellet, $\mathcal{R}_p$, is

$$\mathcal{R}_p = 4\pi r_0^3 \rho Ska C_{A0} \eta_a \tag{7-51}$$

where η_a is a modified effectiveness factor based on ϕ_a for spherical coordinates.

II. Modeling the Deactivation of Pellets

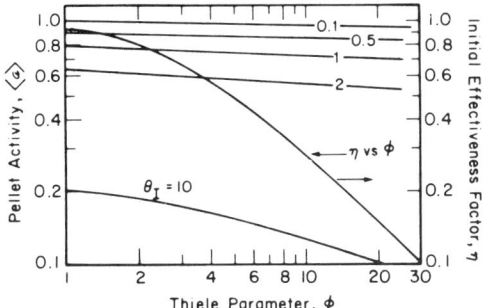

Fig. 7-10. Pellet activity for impurity poisoning.

Defining the activity in the usual way,

$$\langle a \rangle \equiv \mathcal{R}_p/\mathcal{R}_0 = a\eta_a/\eta = e^{-\theta_1}\eta_a/\eta \tag{7-52}$$

The results of Eq. (7-52) are plotted in Fig. 7-12 as the solid lines.

We can readily get another set of curves for the case where $\phi_1 = \phi$. These are obtained from the solution to the parallel case when we realize that the solutions to Eqs. (7-26) and (7-46) are identical when $\phi_1 = \phi$, hence the curves from Fig. 7-5 represent solutions and are also plotted in Fig. 7-12.

The limiting solution of $\phi_1 \to \infty$ is somewhat more involved to obtain, but is important in understanding the impurity poisoning case. As $\phi_1 \to \infty$,

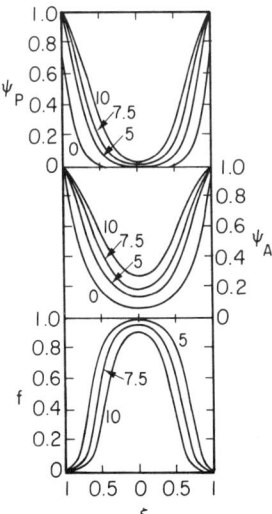

Fig. 7-11. Profiles for independent poisoning, $\phi = 5$, $\phi_I = 10$. Parameter is θ_I. [From Masamune and Smith (12). Reproduced by permission of the American Institute of Chemical Engineers.]

the poison reacts instantaneously with the surface so that the outside of the pellet is completely poisoned and the inside is completely unpoisoned. Therefore, not only must the reactants diffuse through a completely dead zone but also the poison must diffuse through this dead zone to the border between the zones, where it adsorbs to move the border deeper into the catalyst. The method of computation is to calculate as a function of time the value of the dimensionless radius $\bar{\xi}$ at which the border between zones is located for $\phi_1 \to \infty$. Then Eq. (7-36) is used to calculate the corresponding values of $\langle a \rangle$ as functions of $\bar{\xi}$. The variation of $\bar{\xi}$ with time is given by

$$\theta_1/\phi_1^2 = (1-\bar{\xi}^2)/2 - (1-\bar{\xi}^3)/3 \qquad (7\text{-}53)^6$$

where, as before,

$$\theta_1 \equiv k_{p1}C_{10}t, \qquad \phi_1 \equiv r_0\sqrt{\rho S k_{p1} C_{P0}/D_1}, \qquad \bar{\xi} \equiv \bar{r}/r_0 \qquad (7\text{-}54)$$

Figure 7-13 shows curves of $\langle a \rangle$ versus θ_1/ϕ_1^2 for $\phi = 0.5$ and 10 as obtained from Eqs. (7-36) and (7-53). The simplicity of the forms of this asymptotic solution suggests their possible utility for use in calculations of integral reactor behavior. Accordingly, it is of interest to investigate how large ϕ_1 must be before the asymptotic solution gives acceptable accuracy.

Figure 7-14 shows the asymptotic solution and the numerical solutions of Masamune and Smith for $\phi_1 = 10$, $\phi = 5$, 10. Obviously, the agreement is not acceptable even for very crude calculations for $\phi_1 = 10$. It appears that ϕ_1 would have to be of the order of 100 before reasonable accuracy is

---

[6] Equation (7-53) is derived as follows. In the completely deactivated zone there is no loss of poison; therefore,

$$(1/\bar{\xi}^2)\,\partial(\xi^2\,\partial\psi_1/\partial\xi)/\partial\xi = 0 \qquad (7\text{-}53\text{a})$$

subject to the boundary conditions

$$\psi_1(1) = 1, \qquad \psi_1(\bar{\xi}) = 0 \qquad (7\text{-}53\text{b})$$

The solution is

$$\psi_1 = [\bar{\xi}/(1-\bar{\xi})](1/\bar{\xi} - 1/\xi) \qquad (7\text{-}53\text{c})$$

and the derivative of ψ_1 at $\bar{\xi}$ is

$$(\partial\psi_1/\partial\xi)_{\xi=\bar{\xi}} = [(1-\bar{\xi})(\bar{\xi})]^{-1} \qquad (7\text{-}53\text{d})$$

A material balance at $\xi = \bar{\xi}(r = \bar{r})$ is

$$D_1(\partial C_1/\partial r)_{r=\bar{r}} = \rho S C_{P0}\,d\bar{r} \qquad (7\text{-}53\text{e})$$

which, on substitution of Eq. (7-53d) and nondimensionalization, yields

$$(\bar{\xi} - \bar{\xi}^2)\,d\bar{\xi} = d\theta_1/\phi_1^2 \qquad (7\text{-}53\text{f})$$

Integration of Eq. (7-53f) subject to the boundary condition that $\theta_1 = 0$, $\bar{\xi} = 1$ leads to Eq. (7-53).

II. Modeling the Deactivation of Pellets 265

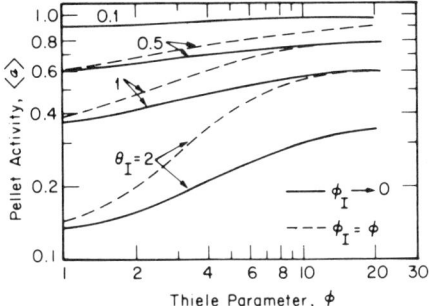

Fig. 7-12. Asymptotic solutions to independent poisoning model.

obtained. This is unfortunate because use of the asymptotic solution would greatly simplify the analysis of integral reactor behavior.

In this section we have presented representative cases of parallel fouling, series fouling, and impurity poisoning. Although the results shown have been based on the assumptions of a first-order main reaction and fouling reactions first-order in the foulant precursor and activity, it will be shown in Chapter 8 that these results will also represent the qualitative behavior of systems in which other orders have been chosen.

The similarity of parallel fouling and impurity poisoning has been demonstrated. It has been further shown that there are substantial differences between the pore mouth model of Wheeler and the more sophisticated models. Using the Wheeler model would of course be conservative, but

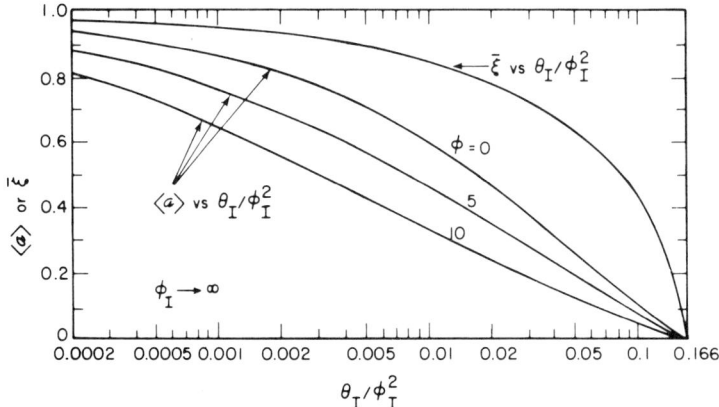

Fig. 7-13. Asymptotic solution for impurity poisoning: case of pore mouth poisoning.

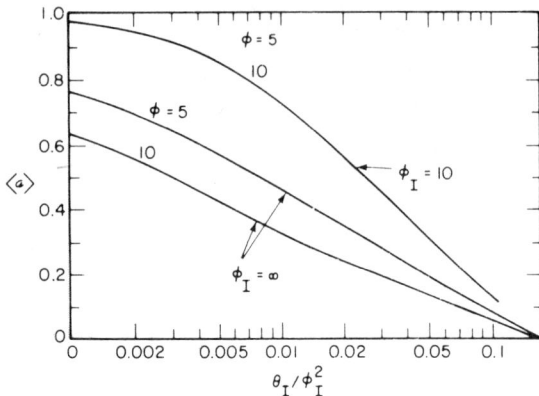

Fig. 7-14. Impurity poisoning case: comparison of asymptotic solutions for $\phi_1 \to \infty$ and numerical solution for $\phi_1 = 10$.

highly inaccurate. The series fouling case appears to be rather closely approximated by the uniform fouling model, but not by the core fouling model.

There is an additional model that can be explored for the deactivation of a pellet based on the supposition that catalysts deactivate as a function of time alone and not the concentrations of the various species locally in the pellet. There is strong evidence that this assumption is not a good generalization; however, models based on it often serve to predict the behavior of integral reactors, as we shall show in later chapters. As before, let us suppose that the deactivation process is slow compared to the rate at which diffusion can occur in the pellet. Under these conditions the pseudo-steady-state assumption provides solutions that closely approximate the transient solutions. Accordingly, we write Eq. (7-55) to represent conservation of species within the pellet:

$$\frac{1}{\xi^2}\frac{\partial}{\partial \xi}\left(\xi^2 \frac{\partial \psi_A}{\partial \xi}\right) - a\phi^2 \psi_A = 0 \qquad (7\text{-}55)$$

This equation is the same as Eq. (7-26) except that now the local activity a is a function of time only, which for purposes of illustration we shall assume is an exponential function of the form

$$a = k/k_0 = e^{-k_d t} \qquad (7\text{-}56)$$

$$a = e^{-\theta_d} \qquad (7\text{-}57)$$

We could have assumed the hyperbolic form discussed in Chapter 3. The

II. Modeling the Deactivation of Pellets

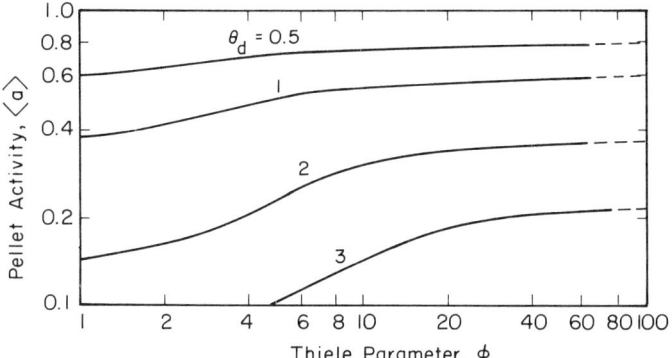

Fig. 7-15. Pellet activity for time-dependent deactivation.

value of $\langle a \rangle$ is given in terms of η as

$$\langle a \rangle = \frac{\frac{4}{3}\pi R_0^3 k_0 e^{-\theta_d} \rho S C_{A0} \eta(\theta_d)}{\frac{4}{3}\pi R_0^3 k_0 \rho C_{A0} \eta(0)} \tag{7-58}$$

which on simplification becomes

$$\langle a \rangle = \frac{\phi \exp(-\theta_d/2) \coth[\phi \exp(-\theta_d/2) - 1]}{\phi \coth \phi - 1} \tag{7-59}$$

As

$$\phi \to \infty, \quad \langle a \rangle = \exp(-\theta_d/2)$$

$$\phi \to 0, \quad \langle a \rangle = \exp(-\theta_d)$$

The solution to these equations is shown in Fig. 7-15. The significant feature of this figure is its similarity to Fig. 7-5 for parallel fouling, particularly if the values of the Thiele parameter are not too large. Of course, as expected, it also closely resembles independent poisoning models when the magnitude of the Thiele parameter for the poisoning species is small. This rather fortunate happenstance perhaps accounts in part for the success of the time-dependent deactivation assumption in the modeling of deactivating reactors, because many deactivating reactor systems involve either parallel fouling or independent poisoning.

A somewhat similar analysis, admitting the inclusion of more complicated kinetic expressions, has been developed by Lee and Butt (9), again using the concept of a pellet effectiveness factor representing the combined effect of diffusion and deactivation. The results are expressed in a manner such that this effectiveness factor can then be directly included in reactor conservation equations. Again we define the activity $\langle a \rangle$ as expressed in

Eq. (7-30) and a pellet effectiveness ε_p as

$$\varepsilon_p = \left[\frac{\text{observed rate}}{\text{intrinsic rate at pellet surface conditions}} \right]$$

so

$$\varepsilon_p = \eta_f \langle a \rangle$$

where η_f is the effectiveness factor of the fresh catalyst in terms of surface conditions. The value of ε_p so defined can be used directly in the normal formulation of reactor conservation equations. This approach simplifies the analysis, although numerical quadrature of integrals involving complex kinetic expressions is required in general. As in the other analyses presented in this chapter, one conveniently dissects the results into cases such as uniform parallel deactivation and pore mouth parallel deactivation. The approach, however, does not account for deactivation intermediate to uniform or pore mouth mechanisms.

We have not, to this point, attempted analytical description of the deactivation of multifunctional catalysts, including the effects of diffusional retardation of main reaction rates. We do not feel that such description is required in full mathematical detail here, since the basic conservation equations follow generally the formalism of Masamune and Smith; additional reaction steps require additional conservation equations for other reaction paths as well as additional relationships for the deactivation of individual functions, such as the separate deactivations of the individual functions shown for 1-butane hydrogenation and isomerization in Chapter 4.

Analysis of analogous situations, with intraparticle diffusion but *no* deactivation, was described in early work by Gunn and Thomas (7) for reaction networks such as

$$A \underset{k_2}{\overset{k_1}{\rightleftharpoons}} B \overset{k_3}{\longrightarrow} C \quad (U, V) \qquad \qquad \textbf{I}$$

with the first step occurring on the U function of the catalyst and the second on the V function (nomenclature similar to that in Chapter 10). Another reaction network considered by Gunn and Thomas was

$$A \underset{k_2}{\overset{k_1}{\rightleftharpoons}} B \overset{k_3}{\underset{k_4}{\diagdown}} \begin{array}{l} C \quad (U) \\ E \quad (V) \end{array} \qquad \qquad \textbf{II}$$

Such networks would have pertinence, in the first case, to the "non-trivial" scheme of Weisz (28), where the reversible first step is involved in the sequential transformation of reactant to an intermediate that then decomposes as in **I**. A reaction scheme such as **II** is illustrated, for example, by

II. Modeling the Deactivation of Pellets

the interception of the course of cyclohexene dehydrogenation to benzene to yield methylcyclopentane. The work of Gunn and Thomas, however, concentrates on the efficient yield of the product, C or E in the above, and concludes that discrete rather than composite (i.e., coimpregnated or coprecipitated) catalysts give better performance if the final reaction product is the desired one. Catalyst composition in terms of the relative amounts of the functions U and V will, of course, have an important influence on the product distribution and optimization of catalyst performance. This in turn depends on the relative values of the rate constants, and the relationship between these factors is discussed in some detail by Gunn and Thomas.

Snyder and Matthews (21) have considered the scheme of II with deactivation and diffusion, in which the U function (reaction to undesired product, in this case foulant) was considered the variable. Activity was considered a linear function of the concentrations of foulant, and the resulting conservation equations were solved numerically for isothermal conditions. Although the original analysis of Snyder and Matthews pertains to both intraparticle and intrareactor problems, we concentrate on the former here. In this case it is important to note that the ratio of rate constants, k_3/k_4, is central to the deactivation behavior as well as the catalyst formulation. The additional diffusional resistance of the discrete formulation results in lower conversions than the composite case; however, the latter deactivates more slowly because the intermediate has a higher probability of escape before formation of foulant. This point was specifically investigated for the discrete formulation by variation of individual particle sizes; the U function particle size was fixed and the V size allowed to vary. Decreasing the size of the V particles produced the expected increase in conversion. Nonetheless, the intermediate must still diffuse out of the U particle to react to desired product, however small the V particle; thus the discrete case does not attain the limiting efficiency of a composite catalyst. An interesting compensation was reported between the absolute activity of a function and its relative amount required for maximum conversion. As k_4 is increased, the relative rates of deactivation do not change very much, whereas one might expect a decrease in deactivation because of a decrease in the ratio k_3/k_4. However, in reactor operation there is a corresponding increase in the optimal fraction of U sites that offsets this tendency. On the other hand, as k_1 is increased the deactivation increases significantly, presumably because of the higher concentration of precursor B.

This analysis was subsequently enlarged on by Lee and Butt (10) for bifunctional reactions of type I (with $k_2 = 0$), in which deactivation could occur either by parallel impurity poisoning or by coke formation in which either reactant and intermediate or intermediate and product were coke precursors. Both intraparticle and interphase diffusional resistances were included for consideration in an isothermal reaction, and results were compared with the diffusion-free case. Perhaps the most important observation was that for all three mechanisms of deactivation the inclusion of some degree of diffusional limitation benefited both catalyst efficiency and life, as shown in some of the results of Masamune and Smith for monofunctional formulations. Further, Lee and Butt show how relatively reliable estimates of the magnitude of such benefits can be made from experimental results with fresh catalysts. A second important result was the very different response of poisoning and fouling mechanisms to catalyst composition and to the selectivity for deactivation. Composition is an effective variable for improving yield (in this case of the intermediate B) when poisoning occurs if one can identify the limiting function, while fouling mechanisms are relatively insensitive to composition variation. In the former case, if one can identify the limiting function, increasing its amount can lead to improvement in yield simply because one has increased the catalyst tolerance (i.e., capacity) for a particular poison. However, in fouling, the same strategy simply leads to increases in reactivity for the main reaction and corresponding increases in fouling rates without much change in yield.

It is clear from such studies that interactions of diffusion, deactivation, and reaction in bi- or polyfunctional catalysts are complex and that one's intuition is not always well served. Parametric calculations such as those of Snyder and Matthews, and Lee and Butt serve to define general trends but are, indeed, so highly parametrized that complete generalization is not possible. It certainly appears, though, from results of the analysis of diffusion, deactivation, and reaction in both mono- and bifunctional catalysts that the intrusion of deactivation on traditional diffusion–reaction theory leads to strategies of catalyst design, formulation, or operating procedures that in many cases are substantially different from conventional wisdom.

III. NONUNIFORM DISTRIBUTION OF CATALYTIC MATERIALS IN PELLETS

There are two principal reasons for distributing the catalytically active material nonuniformly within a catalyst pellet: to improve selectivity and to improve deactivation characteristics. Papers in the literature that deal exclusively with the former topic include the experimental findings of Mars and Gorgels (11) on the selective hydrogenation of acetylene in a large excess of ethylene and the work of Fredrickson (5) on the selective oxidation of o-xylene to phthalic anhydride. Both of these reactions are of the series

III. Nonuniform Distribution of Catalytic Materials in Pellets

type $A \to B \to C$, where B is the desired product, and selectivity is enhanced in such situations when the active centers are easily available to the reaction mixture. This is facilitated by deposition of the active catalyst in a thin layer in the exterior region of the pellet. Another study of nonuniform distribution of catalytic material was reported by Rutkin and Petersen (18). This theoretical paper supposes that the functions of a dual-function catalyst are separated and the selectivities are compared when the U function is in the interior and the V function outside, and the inverse. These selectivity problems are important and interesting in themselves, but the deactivation properties of nonuniformly distributed catalysts, in keeping with the central theme of this book, will be the primary concern of this section.

Shadman-Yazdi and Petersen (19) looked at self-fouling by a series mechanism in catalysts in which the distribution of catalytic material was highest on the exterior and fell to zero at the center according to the equation

$$a(\xi) = a_0 \xi^\delta \tag{7-60}$$

where a_0 is the activity at the exterior of a spherical catalyst, ξ is the dimensionless radius r/r_0, and δ is a distribution parameter.

The following equations recast the results of Shadman-Yazdi and Petersen in a slightly different form; to compare catalysts of the same average activity, therefore, the value of a_0 is computed such that the function is normalized, that is,

$$\int_0^1 a(\xi) \, d\xi = 1 \tag{7-61}$$

Substituting Eq. (7-60) into (7-61) and carrying out the integration, the form of Eq. (7-60) becomes

$$a(\xi) = (\delta + 1)\xi^\delta \tag{7-62}$$

$$k = \bar{k} a(\xi) \tag{7-63}$$

where $\bar{k}$ is the average activity.

The rate of reaction at any time, θ_B, is given by

$$\mathcal{R}(\theta_B) = V\rho S \bar{k} C_{A0} \eta \langle a \rangle \tag{7-64}$$

which reduces to

$$\mathcal{R}(0) = V\rho S \bar{k} C_{A0} \eta \tag{7-65}$$

where V is the pellet volume, ρS the surface area of catalyst per unit volume, η the effectiveness factor of the unfouled catalyst, and $\langle a \rangle$ the apparent activity of the catalyst pellet.

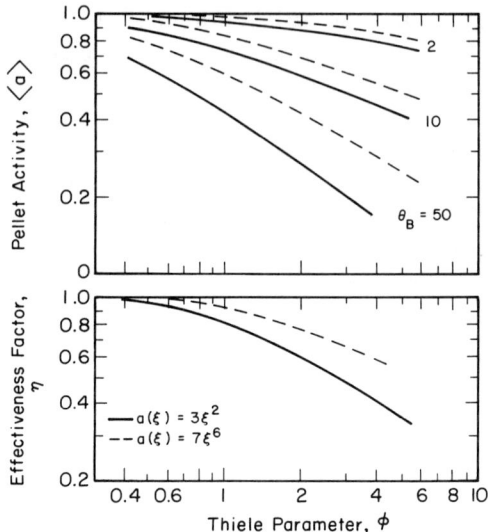

Fig. 7-16. Deactivation characteristics of nonuniformly impregnated catalysts for series fouling. [From Shadman-Yazdi and Petersen (19). Reproduced by permission of Pergamon Press, Ltd.]

Values of $\langle a \rangle$ and η have been computed as functions of ϕ, and the variation of the former with respect to θ_B is presented in Fig. 7-16 for Cartesian coordinates. The abscissa and parameter are defined accordingly as

$$\phi = L\sqrt{\rho S k/D} \qquad (7\text{-}66)$$

$$\theta_B = \bar{k}_B C_{A0} t \qquad (7\text{-}67)$$

Figure 7-16 shows that distributing the catalyst toward the exterior improves the catalyst performance with respect to poisoning if a series mechanism operates.

In similar work, Corbett and Luss (3) investigated the performance of nonuniform distributions of catalytic activity in single spherical pellets. They chose to compare catalysts of equal average activity, which leads to

$$k = \bar{k} a(\xi) \qquad (7\text{-}68)$$

where the normalizing function for a spherical particle is

$$\frac{\int_0^1 \xi^2 a(\xi)\, d\xi}{\int_0^1 \xi^2\, d\xi} = 1 \qquad (7\text{-}69)$$

III. Nonuniform Distribution of Catalytic Materials in Pellets

Using Eq. (7-69), four distribution functions were chosen:

$$a(\xi) = 4\xi^9$$
$$a(\xi) = 4\xi/5$$
$$a(\xi) = 1$$
$$a(\xi) = 2.5 - 2\xi$$
(7-70)

The rate of reaction is again defined analogously to Eqs. (7-64) and (7-65):

$$\mathcal{R}(\theta_A) = 4\pi r_0^3 \rho S \bar{k}_A \eta \langle a \rangle \quad (7\text{-}71)$$

$$\mathcal{R}(0) = 4\pi r_0 \rho S \bar{k}_A \eta \quad (7\text{-}72)$$

and

$$\phi = r_0 \sqrt{\bar{k}_A/D_A}, \qquad \theta_1 = kC_{A0}t \quad (7\text{-}73)$$

The results of calculations for ϕ and ϕ_1 equal to 10 and 1, respectively, are shown in Fig. 7-17. From this figure, it is apparent that the activity of the catalyst is protected better by impregnating the active material predominantly in the inner parts of the pellet. When an impurity deactivation mechanism operates, the effects would be even more pronounced if ϕ_1 were larger. To calculate the reaction rates as a function of time from Fig. 7-17, it is necessary to know the effectiveness factor for the unpoisoned catalyst, η. For the conditions of Fig. 7-17, values of η are 0.5, 0.31, 0.27, and 0.21, respectively, in the order shown in Fig. 7-16, the largest value of η being, of course, associated with the distribution closest to the surface.

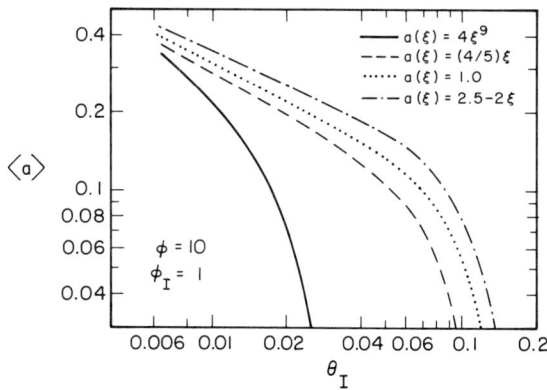

Fig. 7-17. Deactivation characteristics of uniformly impregnated catalysts for impurity poisoning. [From Corbett and Luss (3). Reproduced by permission of Pergamon Press, Ltd.]

From these two examples, it is clear that the lifetime of a catalyst can be strongly influenced by the distribution of active material within the catalyst pellet. The most effective distribution in turn depends on the manner in which the fouling or poisoning process occurs. In the cases presented, it was shown that if series fouling prevails, the active catalytic material should be concentrated on the outside of the pellet, whereas if impurity poisoning prevails, there are advantages to placing the active catalytic material toward the interior of the pellet. In the latter case, however, the strategy is not as clear-cut, because we must consider the values of two Thiele parameters, one for the impurity, ϕ_p, and one for the main reaction, ϕ. In simplest terms, if ϕ is small and ϕ_p large, the active catalytic material should be placed in the interior, the "egg yolk" model shown in Fig. 7-18a. The reason is that in this configuration the active catalytic material is readily available to the main reactants because ϕ is small, but the impurity is not easily accessible to the active catalytic material because ϕ_p is large. Accordingly, the active catalytic material is protected from deactivation by the impurity until the sharp wave front of poison reaches it, at which time deactivation commences.

The strategy is not as clear for other combinations of ϕ and ϕ_p (except for both ϕ and ϕ_p small) and an analysis of these cases was one of the

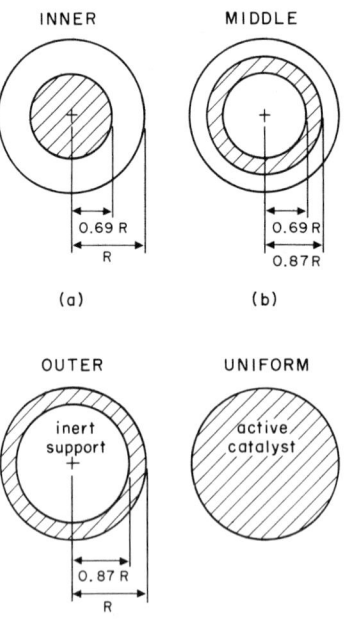

Fig. 7-18. Nonuniform distributions of catalytic materials in a sphere. [From Becker and Wei (2). Reproduced by permission of Academic Press, Inc.]

III. Nonuniform Distribution of Catalytic Materials in Pellets

main purposes of the work of Becker and Wei (2). They studied the behavior of four nonuniform catalyst distributions as shown in Fig. 7-18 for an impurity poison exhibiting linear poison deposition kinetics and an exponential relationship between amount of poison and activity. A linear main reaction was also assumed. To compare catalysts, the values of ϕ and ϕ_p were held constant for each of the catalysts and the time required to reach $\eta\langle a\rangle = 0.4$ (an arbitrary but reasonable value) was determined. The results are shown qualitatively on the ϕ and ϕ_p plane of Fig. 7-19.

Each catalyst shown in Fig. 7-18 has its particular "best" region. In keeping with the discussion above, when ϕ is small and ϕ_p is large, the inner or egg yolk type of distribution is best. For ϕ small and ϕ_p small any distributions behave in about the same way; therefore a uniform distribution might be used. When ϕ is large and ϕ_p small, the catalytic ingredient is best placed close to the outer surface of the pellet, where it is available to

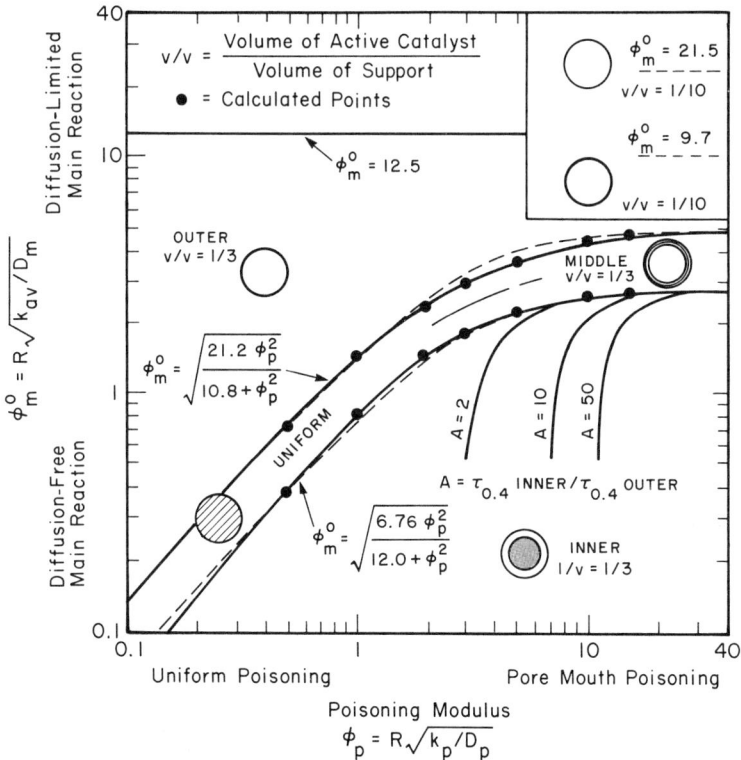

Fig. 7-19. Catalyst selection chart based on maximum operating time to specified final effectiveness. [From Becker and Wei (2). Reproduced by permission of Academic Press, Inc.]

the main reactants. It will be available to the impurity wherever it is located within the pellet because ϕ_p is small. The performance of the various pellets for intermediate values of ϕ and ϕ_p suggests that an intermediate distribution of active catalytic material gives the best result. This is, of course, in qualitative agreement with the physical requirements for a system that places the catalytically active material in the interior to protect it from the impurity yet at the same time near the surface to be available to the reaction mixture. The location of the active material is a compromise and the best location depends on the specific values of ϕ and ϕ_p. Note that in Fig. 7-19 both uniform and middle distributions are confined to a rather narrow range of values of ϕ and ϕ_p.

A further example of the effects of distribution of active components of catalysts on their resistance to poisons was presented by Summers and Hegedus (24) using a platinum and palladium/alumina automobile exhaust catalyst. They prepared five catalysts with the following configurations: Pt(exterior)/Pd(interior), Pd(exterior)/Pt(interior), coimpregnated Pt and Pd (both exterior), Pt(exterior), and Pd(exterior). The details are shown in Table 7-1.

In accelerated endurance tests these catalysts performed rather differently. First, the catalysts had different initial hydrocarbon conversion activities; then, during exposure to exhaust gases from engines using fuel

TABLE 7-1

Properties of Pt–Pd Distributions in the Catalysts of Summers and Hegedus (24)

Property	Pt/Pd	Pd/Pt	Pt–Pd	Pt	Pd
Pt band[a]					
Begins at (μm)	0	77 ± 18	0	0	—
Width (μm)	82 ± 36	To center	>100	82 ± 36	—
Pd band[a]					
Begins at (μm)	107 ± 16	0	0	—	0
Width (μm)	37 ± 7	100 ± 18	~100	—	100 ± 18
Pt (wt %)	0.036	0.038	0.040	0.036	—
Pd (wt %)	0.021	0.018	0.016	—	0.018
Metal dispersion (%)[b]					
Fresh catalysts	61	62	55	68	53
Sintered catalysts	10	15	8	3	39

[a] Values given are the mean of 10 pellets ± standard deviation, determined by the $SnCl_2$ technique.

[b] From CO chemisorption assuming 1:1 stoichiometry.

III. Nonuniform Distribution of Catalytic Materials in Pellets

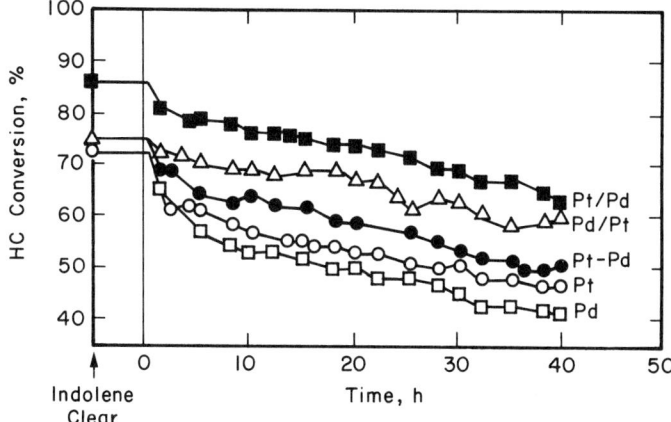

Fig. 7-20. Hydrocarbon conversion during accelerated poisoning experiments. [From Summers and Hegedus (24). Reproduced by permission of Academic Press, Inc.]

containing phosphorus and lead, the differences in their relative activities became more pronounced. This is shown in Fig. 7-20. The catalysts were aged until the exhibited integral averaged poison levels resembled those of an operating catalyst after use for 32,000 km.

Further tests of these catalysts in both their fresh and aged conditions included "light-off" experiments. The aged catalyst was taken from the front of the bed of the previous test; this material corresponded to 80,000 km "old." In these tests the temperature at which there is 50% CO conversion over the catalysts under comparable conditions is measured. The results in Fig. 7-21 show that the Pt- and Pd-containing catalysts gave the smallest changes in the light-off temperatures. Similar tests comparing fresh catalysts and catalysts sintered for 7 hr at 870°C in air are shown in Fig. 7-22. These experiments demonstrate the profound effect of catalyst distribution on performance.

In a theoretical paper, Wolf (30) shows that a marked improvement in resistance to poisoning of an automobile catalyst is obtained by using a composite catalyst in which the outside layer contains no active catalytic material, the active material being contained in the inner layer. The outer layer acts as a scavenger for the impurity poisons and thus protects the inner layer. In such a design it is possible to produce the outer layer with larger pores to minimize transport limitations in the catalyst and facilitate access of the poison to the scavenger section of the composite. A further development of this concept in application to automotive catalysis has been presented by Hegedus and Summers (8).

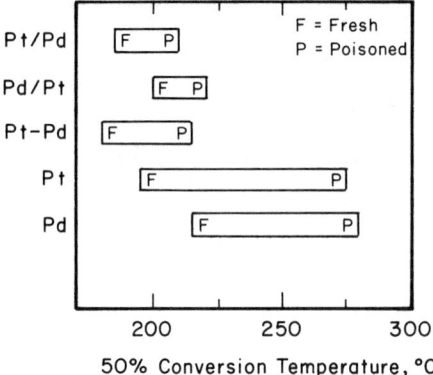

Fig. 7-21. Fifty percent conversion temperatures for CO, fresh and poisoned catalysts. [From Summers and Hegedus (24). Reproduced by permission of Academic Press, Inc.]

All of these examples establish the important effect of the distribution of active catalytic material within catalyst particles on performance in the face of deactivation. Knowledge of the deactivation mechanism can in many instances lead one to design catalysts with improved performance by the simple expedient of properly distributing the active ingredients through the pellet. The term "simple" in the above may not be so simple. For example, it is not intuitively obvious how the middle distribution of Fig. 7-18 would be attained reproducibly in large-scale manufacturing operations.

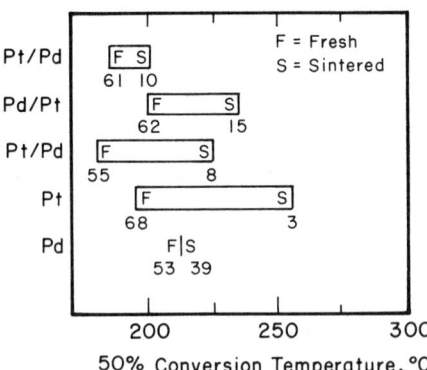

Fig. 7-22. Fifty percent conversion temperatures for CO, fresh and sintered catalysts. The numbers indicate percentage of exposed metal. [From Summers and Hegedus (24). Reproduced by permission of Academic Press, Inc.]

IV. FOULING OF HYDRODESULFURIZATION CATALYSTS

Hydrodesulfurization (HDS) as a process is important in removing organic sulfur-containing compounds prior to further processing. Details of the desulfurization process are available elsewhere (6). The literature on the topic is enormous, but we shall focus here on a particular aspect that is especially pertinent to the content of this chapter. Deactivation of these HDS catalysts furnishes an excellent example of complex and interactive mechanisms via coke formation and impurity poisoning with various degrees of diffusional interaction. These unique characteristics are particularly pronounced when the feedstocks are heavy residuum stocks or coal-derived liquids.

Such "heavy feeds" contain large trace concentrations of organometallic compounds that deposit on the catalyst under reaction conditions. The organometallic molecules are large and approach the same order of magnitude as the pore dimensions; as a consequence, they diffuse very slowly into the interior of the catalyst particles. The net result is that such compounds react (decompose) close to the pore mouths of the catalyst and tend to restrict the passage of the sulfur-containing organic compounds into the interior. This can be viewed as a combination of pore mouth poisoning and pore mouth blockage, affecting both the activity and the transport properties of the catalyst. Coking occurs simultaneously with the demetallation process and contributes to aging; in most evaluations of HDS deactivation it appears that the coke deposition is not diffusion-limited, but this could be a subject of debate depending on specific catalysts and feedstocks. However, the coke can be removed by burning in oxygen (Chapter 12), but the metals are not removed by such a procedure.

Overall, then, the hydrodesulfurization process presents an interesting and complex deactivation case in which the main reaction is moderately influenced by diffusion, coking occurs in parallel with the main reaction, and metal compounds plate out metal sulfides at the pore mouths and lead to pore plugging after many regenerations. This problem is more complex to analyze than the coking case described in Chapter 3; however, the methods used are similar. Our purpose here is not to make a detailed analysis of the pellet problem such as that presented by Rajagopalan and Luss (16), but rather to look at the physical processes occurring during the aging of these catalysts, in particular demetallation and the interaction between coking and demetallation.

Tamm *et al.* (25) present an excellent introduction to the demetallation problem. The experiments typically used in such studies are reactions run at constant conversion in which the temperature is increased to maintain the conversion. Details of this strategy will be presented in Chapter 10; at present, it is sufficient to say that the temperature versus time-on-stream curve so obtained is some representative measure of catalyst deactivation. Figure 7-23 shows a typical deactivation curve for a pilot plant integral reactor used for hydrodesulfurization. The arrows along the curve indicate the points where catalyst samples were removed for analysis during the experiment. The radial distribution of metals found in the aging catalyst is shown in Fig. 7-24 for samples taken from the reactor entrance and exit. Although these two samples were not taken at the same elapsed time, they do serve to show the general shapes of the intraparticle distribution curves for iron, nickel, and vanadium. The main details are that iron deposits primarily on the exterior, vanadium penetrates somewhat deeper and passes through a maximum at a radius quite close to the exterior, and nickel penetrates most deeply (almost uniformly) but does exhibit a slight maximum close to the exterior of the catalyst, like vanadium.

The metal distributions found in the samples taken from the pilot plant reactor are shown in Fig. 7-25. However, the results here are presented in a different way. If $M(r)$ represents the metal distribution function, such as those shown on the previous figure, then the average amount of metal on a cylindrical pellet, M, can be determined from

$$M = \frac{2\pi \int_0^R M(r) r \, dr}{2\pi \int_0^R r \, dr} \qquad (7\text{-}74)$$

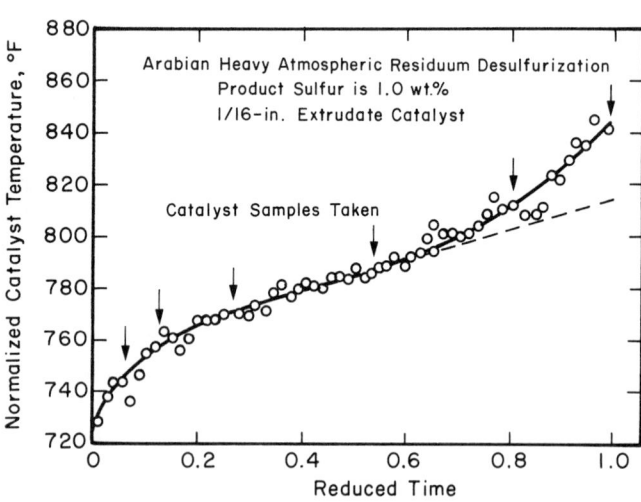

Fig. 7-23. Typical deactivation curve for residuum hydroprocessing catalyst. [From Tamm et al. (25). Reproduced by permission of the American Chemical Society.]

IV. Fouling of Hydrodesulfurization Catalysts

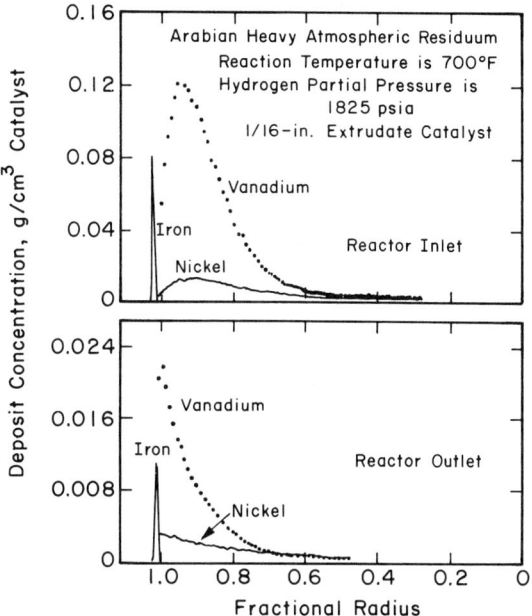

Fig. 7-24. Typical deposition patterns for nickel, vanadium and iron in residuum hydroprocessing catalyst. [From Tamm *et al.* (25). Reproduced by permission of the American Chemical Society.]

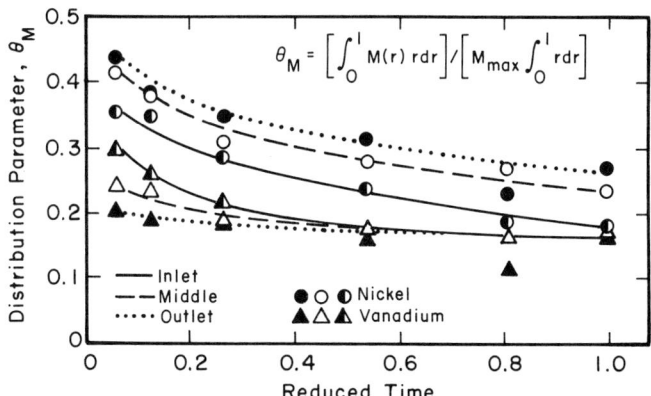

Fig. 7-25. Nickel and vanadium distribution parameters as a function of reactor position and time. [From Tamm *et al.* (25). Reproduced by permission of the American Chemical Society.]

We then define $\theta_M = \bar{M}/M_{max}$, where M is the maximum on the metal distribution curve. This is used to get the final form for θ_M:

$$\theta_M = \frac{2\pi \int_0^R M(r)r\,dr}{M_{max} 2\pi \int_0^R r\,dr} \tag{7-75}$$

In Eq. (7-75), θ_M has the same interpretation as the effectiveness factor if the maximum in the metal distribution curve is at the outer radius. With this interpretation, Fig. 7-25 shows that the organometallic compounds have less access to the interior of the pellet as aging progresses. Also, as observed, nickel has more access to the interior of the catalyst than vanadium. This might well be interpreted as indicating that the nickel-containing organometallic molecules are smaller than those containing vanadium.

The carbon deposits show a quite different pattern, as revealed in Fig. 7-26. The distribution as well as the absolute amount of coke changes as time progresses. However, with the exception of a small layer near the surface, coke deposition is relatively uniform, indicative of relatively small diffusion limits in this (these) reaction(s). On comparison of the coke and metals profiles, as the metals build up on the catalyst at the entrance to the reactor, the coke level there decreases. Since coke levels at the reactor exit are always greater than at the entrance, this may be indicative of some participation of metals deposition in coke removal.

According to the results of Tamm *et al.*, higher temperatures and higher hydrogen partial pressures increase the deposition of vanadium. This conclusion is tempered by the fact that, in real life, the nature and amount of metals in the crude can have substantial effects on the distribution of metal in the catalyst pellets. Smaller-pore catalysts seem to have a higher maximum vanadium concentration than large-pore catalysts. Also, smaller pellets

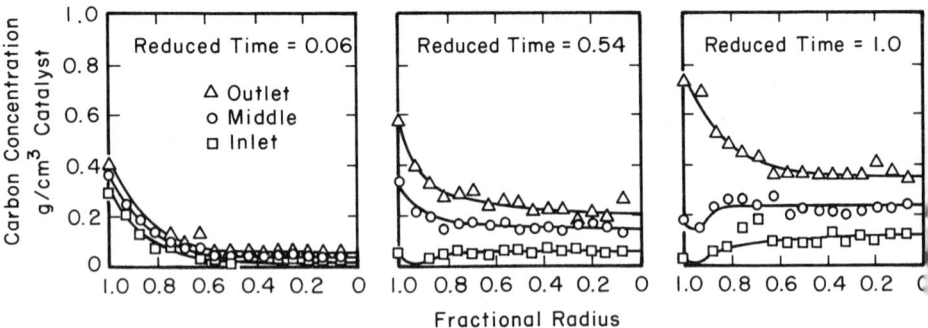

Fig. 7-26. Carbon distribution in the catalyst as a function of reactor position and time. [From Tamm *et al.* (25). Reproduced by permission of the American Chemical Society.]

IV. Fouling of Hydrodesulfurization Catalysts

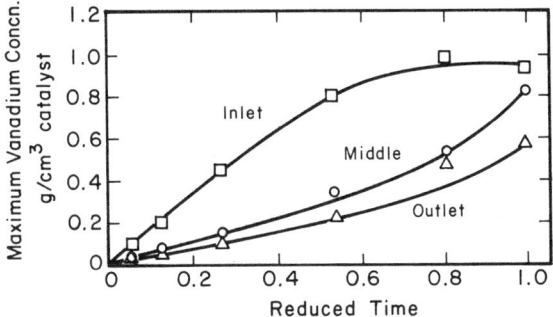

Fig. 7-27. Maximum vanadium deposition as a function of reactor position and time. [From Tamm et al. (25). Reproduced by permission of the American Chemical Society.]

exhibit the same metal distributions as larger pellets when plotted in terms of concentration as a function of absolute distance from the outer surface.

The maximum vanadium concentrations at the inlet, the middle, and the outlet of the reactor as the catalyst ages are shown in Fig. 7-27, and the corresponding final activity distribution is shown in Fig. 7-28. It can be seen that the metals deposition, and concomitant poisoning, proceeds down the bed as a kind of wave. Such behavior will be discussed in detail in Chapter 10. Pore plugging owing to the deposition of metals is a serious, irreversible deactivation process. It is a matter of debate whether coke or

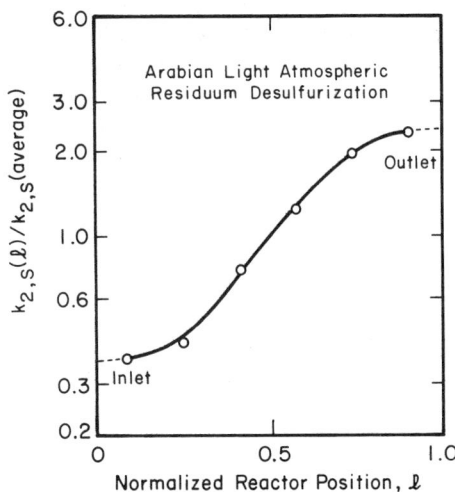

Fig. 7-28. Catalyst activity at end of run as a function of reactor position. [From Tamm et al. (25). Reproduced by permission of the American Chemical Society.]

metals deposition is the limiting factor in decay of HDS catalysts; however, as stated before, the coke can be removed by burning, while demetallation is not so easily accomplished. [Some progress on the latter is being made, though. Mohan *et al.* (14) have reported the use of aqueous solutions of heteropoly acids to remove selectively most of the nickel and vanadium from metallized HDS catalysts. Their process also resulted in complete recovery of the original surface area and pore volume.]

It is the interaction of several different deactivation processes that makes the analysis of intraparticle decay of HDS catalysts such a challenging problem. To reiterate, we have coke deposition, normally only mildly diffusion-limited; metals deposition, highly diffusion-limited, leading to pore mouth plugging; rehydrogenation of coke by the metals deposited (after all, nickel is a very good hydrogenation catalyst); and complex kinetics of the main reaction, some steps of which may be diffusion-limited and others not. Further, the molecular dimensions of the metal-containing porphyrinic molecules are such that they approach the pore size of the catalyst support, and the diffusion process, sometimes termed "configurational diffusion," differs from normal conceptions of diffusion within porous media. Recall also that the effective pore diameter is changing with time. While we cannot present anything approaching a survey of the large amount of work done on HDS catalyst deactivation, some specific studies of the modeling and experimental results on this interwoven intraparticle problem can be addressed.

Of note is the work of Nitta *et al.* (15). Their model assumes a diffusion-limited second-order reaction for the main HDS kinetics as well as for vanadium removal, both reactions also first-order with respect to hydrogen concentration in the liquid phase. Nickel deposition was not considered separately from vanadium; however, the coke deposition reaction was taken to be reversible with the reverse step first-order in hydrogen and coke content on catalyst. The configurational diffusion effect was included using the approach of Spry and Sawyer (22), and pore blocking (change in diffusivity) was taken proportional to the volume of vanadium and coke deposited per unit pore volume. This intraparticle model was then incorporated into the design equations for a plug flow reactor for the simulation of typical temperature-increase requirement (TIR; see Chapter 10) experiments. Some typical results of this effort are shown in Fig. 7-29. It is noteworthy that most of the kinetic parameters of the model were determined in separate experimentation, so while the model might not be considered completely extrapolatable, it is more than a simple interpolative parametric curve fit. No feedstock reactivity correlation is included in the model, however, so short-term HDS tests are required for individual cases to determine the initial catalyst activity. Overall, one could conclude that the model of Nitta

IV. Fouling of Hydrodesulfurization Catalysts

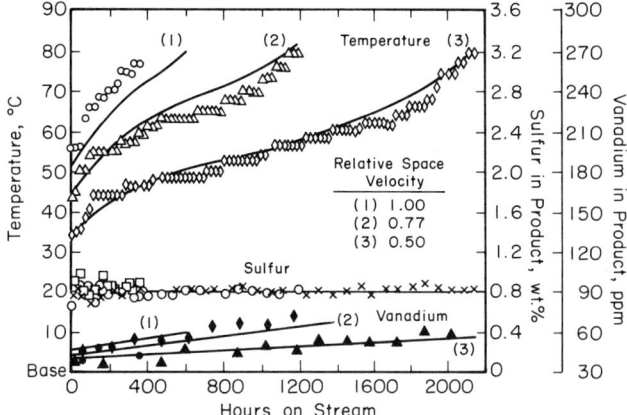

Fig. 7-29. Simulation of hydrodesulfurization of Iranian heavy vacuum resid in a bench-scale fixed-bed reactor. [From Nitta *et al.* (15). Reproduced by permission of the American Institute of Chemical Engineers.]

et al. is rather successful in view of the limitations of its assumptions. Another model in apparent accord with experimental data is that of Dautzenberg *et al.* (4), based on a pore plugging mechanism; see also Sie (20).

An extensive series of experiments concerned with the interaction between coke and metals deposition and the activity of HDS catalysts has been reported by Riley and Silbernagel (17). In these studies HDS was examined for three feedstocks: (i) a whole Venezuelan crude (Jobo), (ii) its deasphalted oil (DAO), and (iii) the whole crude after (partial) demetallation by HF treatment. Properties of the resulting materials are shown in Table 7-2. All three have comparable sulfur contents, while the distribution

TABLE 7-2

Feedstock Properties[a]

Property	Whole crude	HF-treated	Deasphalted oil
Sulfur, wt. %	4.00	3.75	3.81
Conradson carbon, wt. %	15.5	14.1	9.6
Asphaltenes, wt. % C_5 insoluble	18.4	19.5	4.1
Nickel, ppm	123	44	51
Vanadium, ppm	423	70	196
(V/Ni) atomic	4.0	1.8	4.4

[a] From Riley and Silbernagel (17).

of the heavier molecular species, reflected in the asphaltenes and Conradson carbon levels, appears to be similar for the whole crude and the HF-treated material. By contrast, the DAO has an appreciably lower asphaltene and Conradson carbon content. As seen, the total metals content in the three varies considerably, as does the V/Ni ratio. Separate electron spin resonance (ESR) studies indicated that the porphyrin metals were removed by the HF treatment, but carbon radicals (thought to be an indicator of coke precursors) were not affected, being concentrated primarily in the asphaltene fraction.

Reaction experiments (TIR mode) were conducted under similar conditions of temperature, pressure, and space velocity for all three feeds over a small pore (diameter ~ 10 nm) Co-Mo/Al_2O_3 catalyst; typical results with the whole crude are shown in Fig. 7-30a. In comparison, the relative behavior of the DAO and HF feeds is shown in Fig. 7-30b. It appears that the initial deactivation (roughly 20 days) is similar for all feeds, although the desulfurization activity is higher for the DAO. The HF feed is similar to the whole crude in both activity and deactivation rate, although the metals remaining in the HF feed are much more refractory in deactivation than those in the whole crude, based on a metal loading on catalyst, indicating a potential important role of coke as well as metals in determining long-term deactivation. Comparison of the net reactivity of the different feeds under these similar conditions is also informative. These data are given in Table 7-3. The net desulfurization rates (or final levels, since these are integral reactor experiments) are about the same but there are large differences in the removal of metals. One would infer immediately from such results that there is much more chemistry involved, associated with rather complex reaction networks, than is expressed by simple second-order lumping models

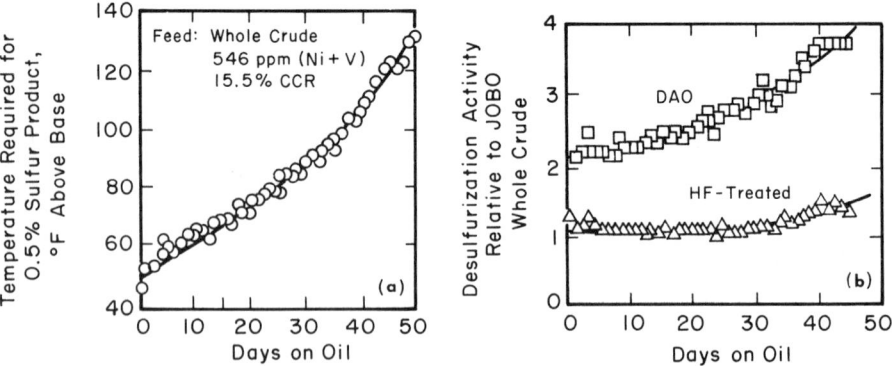

Fig. 7-30. (a) Deactivation behavior for whole crude. (b) Comparison of activity and deactivation rates for DAO and HF feeds with whole crude.

IV. Fouling of Hydrodesulfurization Catalysts

TABLE 7-3

Net Reactivity of the Three Feedstocks

	Whole crude	HF-treated	Deasphalted oil
Desulfurization, %	85	88	96
Nickel removal, %	59	35	90
Vanadium removal, %	60	20	90
Relative metals deposition rate[a]	100	8	58

[a] Metals in feed times demetallation.

such as that of Nitta *et al.* The idea of nth-order lumping persists, however, and really seems to work very well even for realistic feedstocks (23), but in the absence of long-term deactivation.

More detailed characterization of the nature of the carbonaceous and metallic deposits yields additional insight into how complex the problem really is. It has been shown, in fact, that the vanadium exists in several chemical states, changing with both time of operation and position in the bed. Results from the study of Riley and Silbernagel are shown in Fig. 7-31. At low loadings the predominant species is VO^{2+}, but at the higher loadings near the bed entrance one sees primarily diamagnetic V and V_2S_3. Again, an overall wavelike motion of metals through the bed is observed, similar to that reported by Tamm *et al.*, but now the wave has components. What this has to do with the overall deactivation behavior is not clear, but in detailed modeling one probably cannot afford to ignore it. A similar situation pertains for coke formation, as shown in Fig. 7-32. Considering the carbon radical density as some measure of the nature of the coke deposited (increasing graphitization leading to more refractory coke and lower radical density), it appears from the figure that there is a clear interaction between metals

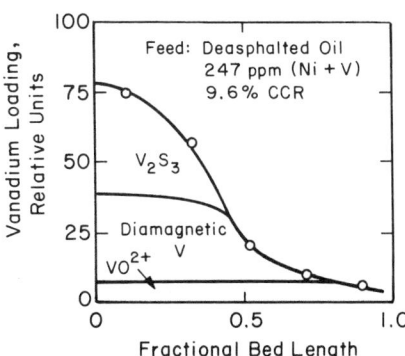

Fig. 7-31. Chemical characterization of vanadium deposits in the HDS reaction with DAO feed.

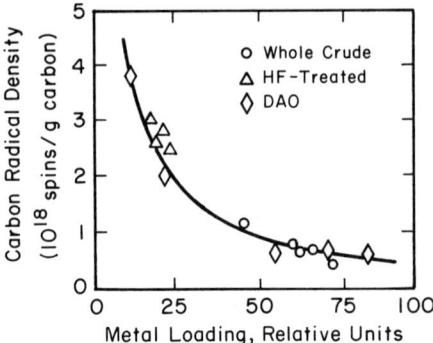

Fig. 7-32. Interactions between coke and metals deposition in HDS. Aging of coke not influenced by nature of the feed.

loading and the nature of the coke. Thus, in actual operation the level of foulants and their chemical nature both change, and the chemistry is clearly interactive. Surely this problem of metals deposition, coke deposition, pore plugging, and evolutionary chemistry involved in HDS catalyst deactivation represents one of the most challenging problems (intellectual and technological) in chemical reaction engineering. It is probably safe to say that 10 years from the time this is read (whatever the date) this will remain so.

V. SUMMARY AND EVALUATION

It is apparent that the analysis of limiting types of deactivation, either shell mechanisms or uniform deposition of deactivant, is very informative in establishing the working range of the effects, overall, on porous catalysts. Indeed, the studies of Wheeler, Masamune and Smith, and Weisz and Goodwin (Chapter 9) probably tell us most of what we need to know about the shell mechanisms. However, the studies of Masamune and Smith also tell us much about the gray area between the limits of shell and uniform deactivation. Note also that the kinetic network involved in deactivation is of signal importance in determination of the course of fouling. In cases where the reactant acts as the foulant precursor, the tendency (with the exception of uniform deactivation) is to deactivate the individual particle from the outside to the inside, leading eventually to the shell mechanism, while in cases where the product is the coke precursor it is possible to have fouling from the inside to the outside. Analogous processes exist in the deactivation of fixed beds, where it becomes apparent that these pathways correspond to various types of wave motion, either of catalyst activity or

V. Summary and Evaluation

foulant/poison concentration, in the bed. In all cases considered in this chapter, it is apparent that the distinction between parallel and series fouling is important in the analysis of overall behavior; this will also be examined in specific detail in Chapter 8.

For diffusion-limited reactions in individual particles the distinction between the limit on the main reaction and the fouling or poisoning reaction is important. Various cases of this have been discussed in Section II. In the limit it would be constructive to discover whether one might be able to control the extent of deactivation by a clever design of pore structure. This may be possible in cases where the coke precursors tend to be large molecules while the desired reactant/products are smaller. Then one can resort to shape-selective catalysts, such as certain types of zeolites, to exclude undesirable components from the pore structure. Important examples of this exist, but in general one is a prisoner of the molecular structure of the main and deactivation reactions and the pore structure of suitable catalysts. Such strategies are mainly confined to deactivation by coking rather than by poisoning, since in the latter case there is often not much differentiation in molecular geometry between the reactant and poison molecules.

There appears to be some possibility to mitigate intraparticle deactivation effects by controlling the distribution of active ingredient within the particle. In certain instances selectivity may also be controlled in this manner. The study of Becker and Wei is a good example of the control of this interplay between diffusionally limited main and poisoning reactions by the proper positioning of active ingredient within the particle. However, one sees that when both main and poisoning reactions are severely diffusion-limited the analysis calls for an annular deposition of active catalyst that might be difficult to obtain in large-scale manufacture. Nonetheless, in more straightforward cases of deactivation, such as severe shell poisoning, such strategies are feasible. Here one concentrates the active ingredient in the outer shell, providing in effect extra sites for the poison as a guard bed for the reaction. "Sacrificial lambs" would be a good description.

Variation of pore structure also seems to be a possibility for controlling deactivation, as discussed by Wolf. The idea is derivative from shape-selective sieve catalysts, but more sophisticated in the sense that one produces a graded variation of porosity with position in the particle. This combines with the above in the concept that it might be possible to obtain a porous catalyst composite with both spatially variant active ingredient and pore structure in a way that would allow one to control both reaction and diffusion in a favorable manner. Again, whether this could be done on a practical manufacturing scale must be considered. Further, while a number of investigations have appeared in the literature, they are theoretical, and not much experimentation along these lines has appeared.

Our primary example in this chapter, the deactivation of hydrodesulfurization catalysts, may well be the most complicated in the book. Here we deal with interactive chemical mechanisms of deactivation, demetallation, and coke deposition, with both chemical and physical factors playing a role in overall deactivation. In addition, one of the metals deposited participates in the eventual removal of the coke. Examination of this should have been fun, overall a plot that Verdi would have been proud of.

REFERENCES

1. R. Aris, "The Mathematical Theory of Diffusion and Reaction in Permeable Catalysts," p. 37. Oxford Univ. Press (Clarendon), London and New York, 1975.
2. E. R. Becker and J. Wei, *J. Catal.* **46**, 372 (1977).
3. W. E. Corbett, Jr. and D. Luss, *Chem. Eng. Sci.* **29**, 1473 (1974).
4. F. M. Dautzenberg, J. van Klinken, K. M. H. Pronk, S. T. Sie, and J. B. Wyffels, *Chem. React. Eng.—Houston, Int. Symp., 5th, ACS Symp. Ser.* No. 65, p. 254 (1978).
5. W. Frederickson, *Chem.-Ing.-Tech.* **41**, 967 (1969).
6. B. C. Gates, J. R. Katzer, and G. C. A. Schuit, "Chemistry of Catalytic Processes," McGraw-Hill, New York, 1979.
7. D. J. Gunn and W. J. Thomas, *Chem. Eng. Sci.* **20**, 89 (1965).
8. L. L. Hegedus and J. C. Summers, *J. Catal.* **48**, 345 (1977).
9. H. H. Lee and J. B. Butt, *AIChE J.* **28**, 405 (1982).
10. J. W. Lee and J. B. Butt, *Chem. Eng. J.* **6**, 111 (1973).
11. P. Mars and M. J. Gorgels, *Proc. Eur. Symp. Chem. React. Eng. 3rd*, p. 55 (1964).
12. S. Masamune and J. M. Smith, *AIChE J.* **12**, 384 (1966).
13. S. Masamune and J. M. Smith, *AIChE J.* **12**, 389 (1966).
14. R. R. Mohan, B. G. Silbernagel, and G. H. Singhal, personal communication, 1982.
15. H. Nitta, T. Takatsuka, S. Kodama, and T. Yokoyama, *AIChE Natl. Meet., Houston, 1979.*
16. K. Rajagopalan and D. Luss, *Ind. Eng. Chem. IEC Process Des. Dev.* **18**, 459 (1979).
17. K. L. Riley and B. Silbernagel, *Stud. Surf. Sci. Catal.* **6**, 313 (1980).
18. D. R. Rutkin and E. E. Petersen, *Chem. Eng. Sci.* **34**, 109 (1979).
19. F. Shadman-Yazdi and E. E. Petersen, *Chem. Eng. Sci.* **27**, 227 (1972).
20. S. T. Sie, *Stud. Surf. Sci. Catal.* **6**, 545 (1980).
21. A. C. Snyder and J. C. Matthews, *Chem. Eng. Sci.* **28**, 291 (1973).
22. J. C. Spry and W. H. Sawyer, *AIChE Annu. Meet., Los Angeles, 1975.*
23. R. Stephan, G. Emig, and H. Hofmann, *Chem. Eng. Process* **19**, 303 (1985).
24. J. C. Summers and L. L. Hegedus, *J. Catal.* **51**, 185 (1978).
25. P. W. Tamm, H. F. Harnsberger, and A. G. Bridge, *Ind. Eng. Chem. Process Des. Dev.* **20**, 262 (1981).
26. E. W. Thiele, *Ind. Eng. Chem.* **31**, 916 (1939).
27. J. Wei, CHEMTECH, p. 128 Feb. (1975).
28. P. B. Weisz, *Proc. Int. Congr. Catal., 2nd*, p. 937 (1961).
29. A. Wheeler, *Adv. Catal.* **3**, 307 (1950).
30. E. E. Wolf, *J. Catal.* **47**, 85 (1977).

CHAPTER 8

Direct Measurement of Reaction Nonuniformity in Catalyst Pellets

> Don't tell me facts; I never believe facts.
> *Sidney Smith*

The poisoning and fouling models of the previous chapter should be verified experimentally before they can be applied with confidence. Particularly convincing is the evidence from a reactor that can give specific information concerning gradients that evolve during reaction. Direct verifications are given by single-pellet reactors designed specifically to produce gradients that can be measured. As implied in the name, the diffusion reactor is devised to operate under conditions where appreciable concentration and/or temperature gradients are manifest within the pores of the catalyst. This class of reactor differs inherently from the class of "gradientless reactors" in which the designer endeavors to minimize the influence of transport properties on reaction rates. With these diffusion-controlled reactors the observer obtains information on the concentrations and temperatures in the interior of the pellet, in addition to the usual global rate data. These additional data have been of considerable utility in diagnosing deactivation mechanisms. In this chapter we shall be concerned in particular with the analysis of results obtained from two types of single-pellet gradient reactors designed for the analysis of concentration or temperature gradients produced during the deactivation of individual catalyst particles. The particle dimensions are exaggeratedly large compared to commercial practice, but appropriate scaling renders the experimental results relevant. Both concentration and temperature measurements are informative, as will be seen.

One may well question the opportunity in designing such specified gradient reactors, given the difficulty of measuring such gradients and the well-known efficiency of gradients in camouflaging the true values of kinetic parameters. However, it should become clear that it is worth the effort as long as one follows the philosophy of designing the experiment to fit the model rather than searching for a model that will fit the experiment.

I. THE SINGLE-PELLET DIFFUSION REACTOR: CONCENTRATION MEASUREMENTS

If the relaxation time for chemical reaction is of the same order of magnitude or less than the relaxation time for diffusion, a concentration gradient will develop within a catalyst pellet. This has been shown in many examples in previous chapters and needs no further rationalization here. As shown in Chapter 6, the Thiele parameter ϕ is a measure of the ratio of these relaxation times. The isothermal diffusion reactor provides a means of measuring the composition of the reaction mixture at the center of the pellet and the bulk composition near the outer face. The technique for measuring the composition at the center of the pellet is shown in Fig. 8-1. Figure 8-1a illustrates the Thiele–Zeldovich model. The boundary conditions for the second-order differential equation governing the behavior of the model are

$$x = 0, \quad C_A = C_{A0}$$
$$x = L, \quad dC_A/dx = 0$$

where the latter derives from a symmetry argument. In Fig. 8-1b the boundary conditions are the same, but the derivative condition is obtained from recognition of the fact that no reaction takes place in the centerplane chamber because it contains no catalyst. The composition of the centerplane chamber therefore must be the same as at $x = L$ in the pellet of Fig. 8-1a. The single pellet of Fig. 8-1b is therefore a realistic experimental representation of the Thiele–Zeldovich model.

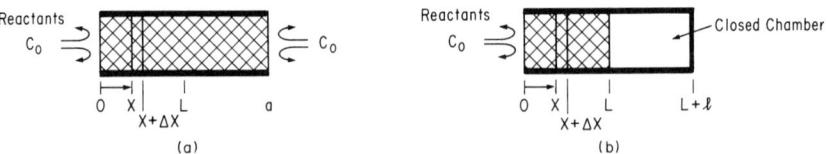

Fig. 8-1. Comparison of (a) the Thiele–Zeldovich model with (b) the single-pellet diffusion reactor. [From Petersen (21). Reproduced by permission of Academic Press, Inc.]

I. The Single-Pellet Diffusion Reactor: Concentration

The centerplane chamber composition in reactant can vary in magnitude from the bulk value to zero for an irreversible reaction. Neither limit gives useful information; however, for values of the Thiele parameter in general of about unity, the reactant concentration will be between those limits and will supply additional information as interpreted with the Thiele–Zeldovich theory. The utility of the isothermal single-pellet diffusion reactor and its design and operation in the study of reaction systems of constant activity have been presented elsewhere and will not be discussed further here (10, 21).

When deactivating catalyst pellets are studied the centerplane composition changes with time, as represented in Fig. 8-2. As deactivation progresses, the concentration of reactant in the centerplane increases because the effective value of the Thiele parameter is decreased by the decay process. Mechanistic information is derived from the shape of a plot of pellet activity $\langle a \rangle_p$ versus a dimensionless centerplane concentration defined by

$$\Omega \equiv \frac{\psi(\tau, 1) - \psi(0, 1)}{1 - \psi(0, 1)} \tag{8-1}$$

where $\psi(\tau, 1)$ and $\psi(0, 1)$ are the centerplane concentrations at times τ and zero, respectively. This variable may be thought of as the fractional movement of $\psi(\tau, 1)$ toward the zero-activity limit. This variable has no simple physical interpretation and must not be confused with $(1-f)$ of the Wheeler analysis; while Ω and $(1-f)$ are qualitatively related, the relationship is

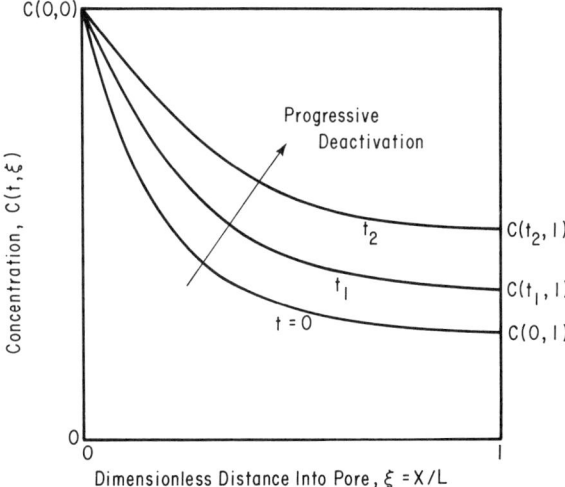

Fig. 8-2. Changes in centerplane concentration with progressive deactivation.

neither linear nor simple. Nevertheless, Ω is useful because it can be measured directly and because it permits the observer to follow the course of deactivation without explicit reference to time. Of course, in Section I of Chapter 7 we saw the utility of the variable $(1-f)$ in the $\langle a \rangle$ versus $(1-f)$ plots as used by Wheeler. The problem with this approach is that there is no direct way experimentally to measure $(1-f)$. For reasons that will be clearer shortly, we seek a variable to replace time because the latter is influenced by both mechanistic and kinetic effects. The variable Ω appears to be affected mainly by mechanistic factors.

The first work to illustrate this was a report by Balder and Petersen (1) on the deactivation of a Pt/η-Al$_2$O$_3$ catalyst during the hydrogenolysis of cyclopropane. They developed the single-pellet diffusion reactor to measure the reaction rates and the centerplane concentrations simultaneously as functions of time as the catalyst deactivated. This permitted characterization of the deactivation process on a plot of $\langle a \rangle = \mathcal{R}(t)/\mathcal{R}(0)$ versus Ω as shown by the experimental points in Fig. 8-3.

The reference lines in Fig. 8-3 derive from the pore models discussed in Section I of Chapter 7. For uniform fouling $\langle a \rangle$ is given as a function of f by Eq. (7-13), whereas

$$\Omega = \frac{1/\cosh(\phi\sqrt{f}) - 1/\cosh \phi}{1 - 1/\cosh \phi} \qquad (8\text{-}2)$$

Thus, $\langle a \rangle$ can be represented as a function of Ω. Similarly, for pore mouth fouling $\langle a \rangle$ is given by Eq. (7-21) and Ω is calculated from its definition and $\psi(f, 1)$ by

$$\psi(f, 1) = \frac{C(\bar{x})}{C(0)} \frac{C(1)}{C(\bar{x})} = \frac{C(\bar{x})}{C(0)} \frac{1}{\cosh \phi(f\phi)}$$

$$= \left[\frac{1}{\cosh(f\phi)}\right]\left[\frac{1}{1+(1-f)\phi \tanh(f\phi)}\right] \qquad (8\text{-}3)$$

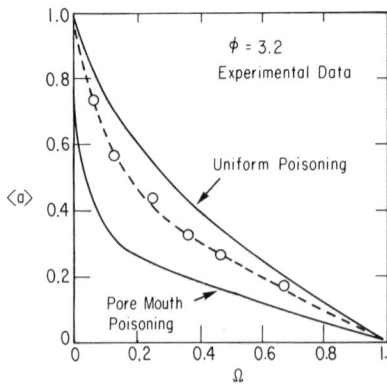

Fig. 8-3. Comparison of experimental deactivation behavior with uniform and pore-mouth poisoning. [From Balder and Petersen (1). Reproduced by permission of Pergamon Press, Ltd.]

I. The Single-Pellet Diffusion Reactor: Concentration

Curves of these limiting cases are presented in Fig. 8-3, and the experimental data lie in between. This would indicate that the fouling process is neither uniform or pore mouth; that is, there is a nonuniform foulant distribution with a greater amount of foulant near the pore mouth.

It was possible to verify this by an experiment whose results required no modeling for their interpretation. Hahn and Petersen (9) modified the diffusion reactor to permit either face of the catalyst to be exposed to the reaction mixture. In this way a nonuniformly fouled catalyst would exhibit different activity to the same reaction mixture when passed consecutively over each of the faces. The results of such a test are shown in Fig. 8-4. The ordinate is the appearance of propane during the hydrogenolysis of cyclopropane over a $Pt/\eta\text{-}Al_2O_3$ catalyst. The slope of the line in the figure is the reaction rate. During period I the initial rate of the catalyst is measured with reaction mixture past the bulk face; then the catalyst is deactivated on stream for 555 min. After deactivation, the activity of the catalyst with the reaction mixture past the bulk face is again measured during period II. The reaction mixture is now directed to pass the centerplane chamber face and the rate of reaction again measured in period III. The fact that the rate measured in period II is greatly reduced from the initial value and that the

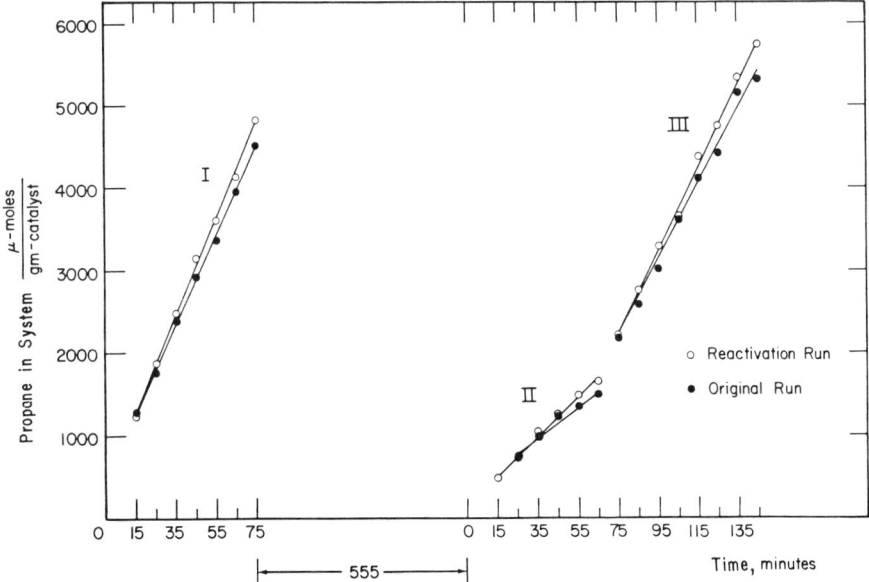

Fig. 8-4. Activity plot showing non-uniform distribution of foulant in the catalyst pellet. I, Initial activity from bulk face; II, activity of bulk face after deactivation; III, activity of centerplane face after deactivation. [From Hahn and Petersen (9). Reproduced by permission of the Chemical Institute of Canada.]

rate measured in period III is almost as high as the initial rate demonstrates directly and conclusively that the catalyst was operating with a high Thiele parameter and that the fouling was nonuniform.

Dougharty (7) interpreted the results of Balder and Petersen using various parallel fouling models similar to those used by Masamune and Smith (19). Dougharty varied the orders with respect to f and ψ_A to attempt a specific fit to the data but concluded that although all parallel models gave qualitative agreement, none fit exactly, but one model represented by

$$\partial f / \partial \theta_A = -\psi_A f^2 \tag{8-4}$$

gave the "best" fit. These modeling results also suggest that the $\langle a \rangle$ versus Ω curve is not extremely sensitive to kinetic details. Hegedus and Petersen (11) conducted a systematic study of model behaviour on the $\langle a \rangle$ versus Ω plot. They integrated the Cartesian coordinate forms of Eqs. (7-26) (namely $\partial^2 \psi_A / \partial \xi^2 - a\phi^2 \psi_A = 0$), (7-28), and (7-46). Their results can be represented typically by Fig. 8-5, which summarizes all of the model calculations based on general deactivation models of the form

$$\partial f / \partial \theta_A = -\psi_A^\delta f^\alpha \tag{8-5}$$

for parallel fouling,

$$\partial f / \partial \theta_B = -\psi_B^\delta f^\alpha \tag{8-6}$$

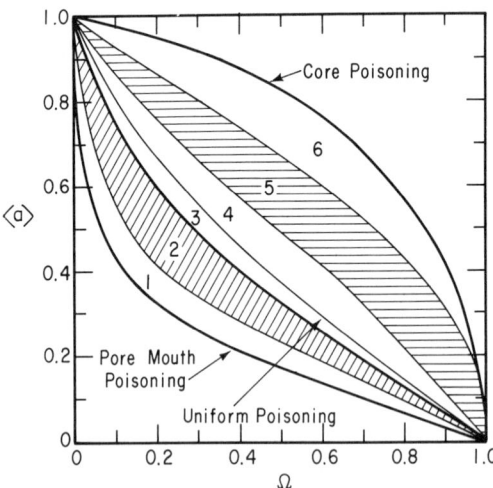

Fig. 8-5. Plot to discriminate among poisoning models, $\phi = 2.5$. [From Hegedus and Petersen (11). Reproduced by permission of Pergamon Press, Ltd.]

I. The Single-Pellet Diffusion Reactor: Concentration

for series fouling, and

$$\partial f / \partial \theta_1 = -\psi_1^\delta f^\alpha \qquad (8\text{-}7)$$

for independent poisoning. For values of $\phi = 2.5$ and 4.5 and values for α and δ of 2.5, 1, and 2, the curves fall in well-defined zones. The results are illustrated in Fig. 8-5 for the case $\phi = 2.5$ as falling in zones 1 through 6. Parallel fouling cases are confined to zone 2; series fouling cases are confined to zone 5. Zones 2 and 5 are intermediate between limiting cases of pore mouth, uniform, and core fouling. Independent poisoning can be anywhere in zone 1, 2, or 3. To complicate matters somewhat more, a combination of parallel and series poisoning, which we shall refer to as triangular fouling, can lead to a curve anywhere in zone 2, 3, 4, or 5. It should be noted that the curves for $\phi = 4.5$ are similar, but the limiting curves tend closer toward the corners of the plot.

Clearly, then, the position of a curve on this kind of plot does not uniquely determine the fouling or poisoning precursor except for zones 1 and 4, which belong to independent poisoning and triangular fouling, respectively. However, a maximum of three experiments can, in principle, distinguish uniquely among the models. Briefly, fit the first curve, say in region 2, to a given set of δ and α for parallel fouling. Increase the temperature; if the same δ and α correlate the data, then it is parallel fouling. If not, one is left with either impurity or independent fouling. To distinguish between these, add product to the reaction mixture at the original conditions and carry out the reaction. If the curve remains unchanged, it is impurity poisoning; if the curve moves toward zone 3 it is triangular. Obviously, further tests and different techniques can be used to distinguish deactivation models. The main feature of this technique is that curves such as those shown in Fig. 8-5 have the potential of identifying the deactivation mechanism in a very few experiments. As we shall see later, the power of the method derives from being able to eliminate the time variable, which is much more complex and more difficult to scale than the centerplane concentration variable.

Let us now show how this technique can be utilized by an example. Hegedus and Petersen (12) studied the deactivation of a reforming-type 0.25 wt. % Pt/$\eta - Al_2O_3$ catalyst used to accelerate the hydrogenolysis of cyclopropane to propane. Their experimental results are represented by the points in Fig. 8-6. In the same figure are the curves calculated for various δ and α in Eq. (8-5) for the parallel fouling model with $w = \psi_A + \psi_B = 1$. In a second experiment at a higher temperature, $\phi = 4.47$, the experimental points fall outside the range of parallel poisoning, as shown in Fig. 8-7, so this mechanism does not fit. In a third experiment at $\phi = 3.32$, with product in the reaction mixture, the curves move to almost superimpose on the

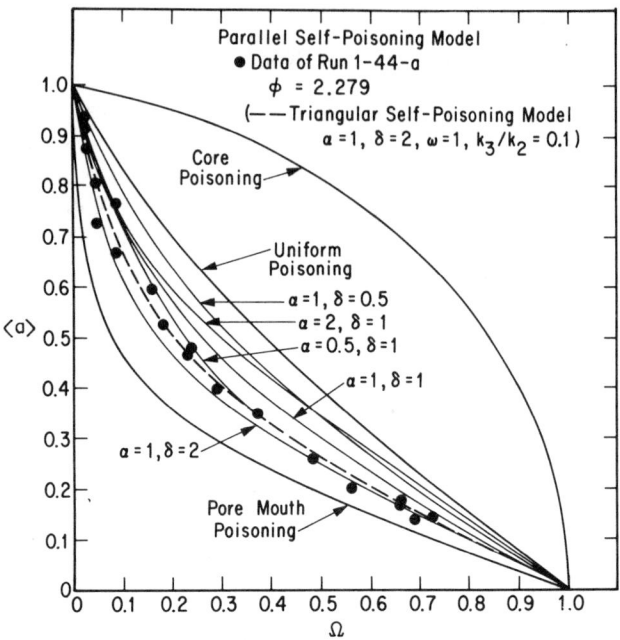

Fig. 8-6. Relative overall rate versus centerplane concentration for cyclopropane hydrogenolysis, $T = 50°C$. [From Hegedus and Petersen (12). Reproduced by permission of Pergamon Press, Ltd.]

uniform fouling line, Fig. 8-8, thereby eliminating impurity poisoning and leaving us with a triangular fouling mechanism. Thus in only a few experiments one has learned how the catalyst deactivates by a self-fouling mechanism, that both reactant and product appear to be precursors of the foulant, and that the activation energy for product fouling is greater than that for reactant fouling. It is important to emphasize again that this kind of mechanistic information is obtained readily from this kind of experiment, but the method appears to be relatively insensitive to kinetic details such as the order of the fouling reactions with respect to precursors. These can be more accurately determined under gradientless conditions.

It was stated earlier that the elimination of time as a reaction variable enabled mechanistic details of the deactivation process to be investigated independently of their time scales. This is illustrated by further experiments of Hegedus and Petersen (13) on cyclopropane hydrogenolysis on Pt/Al_2O_3. Two runs on the same catalyst pellet under identical conditions were carried out. The only difference between runs was in the regeneration of the catalysts—one was reduced in hydrogen for 10 hr and the other for 13 hr.

I. The Single-Pellet Diffusion Reactor: Concentration

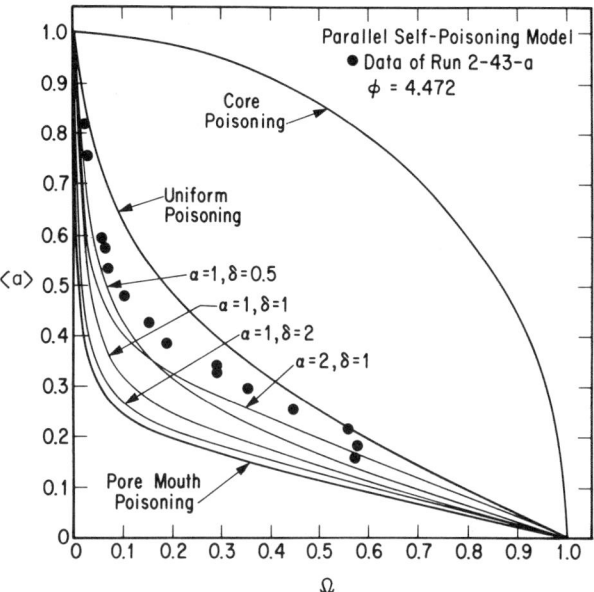

Fig. 8-7. Relative rate versus normalized centerplane concentration for cyclopropane hydrogenolysis, $T = 75°C$. [From Hegedus and Petersen (12). Reproduced by permission of Pergamon Press, Ltd.]

Figure 8-9 shows these two runs in centerplane concentration versus time coordinates. Note that both catalysts have the same initial activity, that is, the same initial centerplane concentration, but differ in the rate at which they deactivate. The catalyst that was reduced longer exhibited a deactivation curve delayed in time compared to the catalyst having the shorter reduction period. It is difficult to decide from this curve whether the mechanism of deactivation has changed, or indeed to make any other statement than that the larger reduction period results in a more stable material. However, the same data shown as $\langle a \rangle$ versus Ω in Fig. 8-10 suggest strongly that, although the second catalyst is more stable, the same deactivation mechanisms are operating.

These examples have shown that certain kinds of information are advantageously obtained from the diffusion reactor. However, to obtain the maximum information, experimental data on $\langle a \rangle$ versus Ω plots must be compared to the limiting cases of pore mouth, uniform, and core poisoning models. In all cases considered so far only a first-order main reaction has been treated. The effect of a main reaction not of first order has been investigated by Wolf and Petersen (25). Using numerical solutions to the

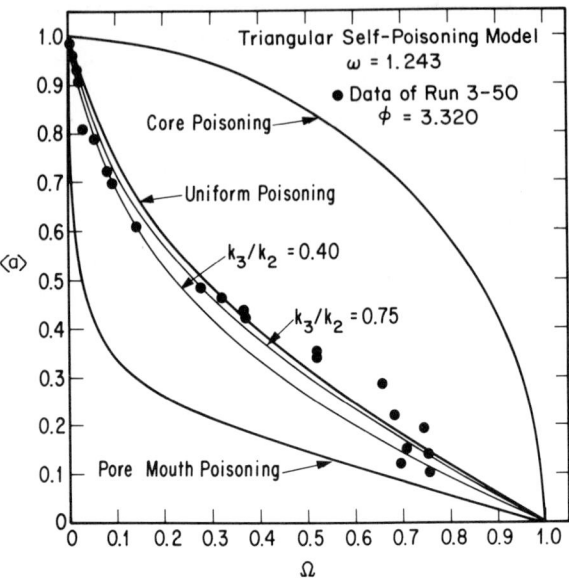

Fig. 8-8. Relative overall rate versus normalized centerplane concentration with 0.86×10^{-6} mol/cm^3 initial propane concentration. [From Hegedus and Petersen (12). Reproduced by Permission of Pergamon Press, Ltd.]

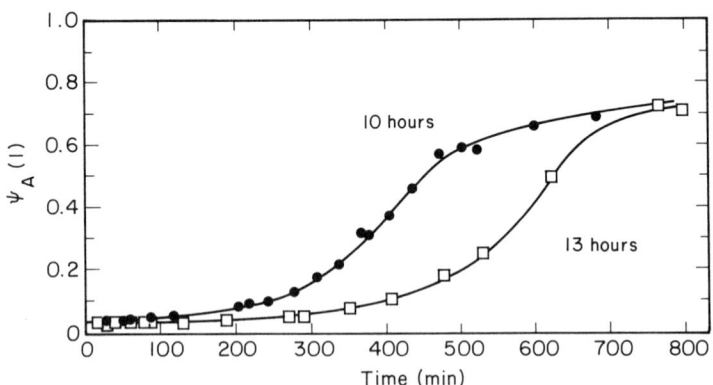

Fig. 8-9. Effect of H$_2$ pretreatment on the time scale of the deactivation process. [From Hegedus and Petersen (13). Reproduced by permission of Academic Press, Inc.]

II. The Single-Pellet Diffusion Reactor: Temperature

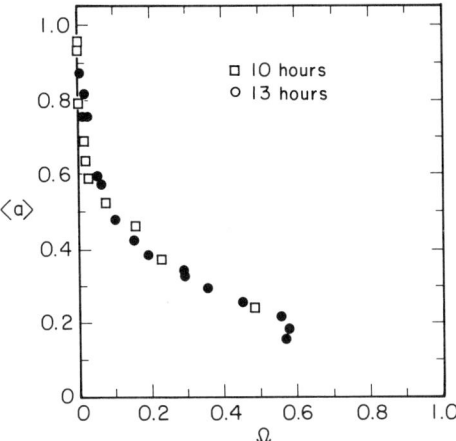

Fig. 8-10. Effect of H_2 pretreatment on the mechanism and kinetics of the deactivation process. [From Hegedus and Petersen (13). Reproduced by permission of Academic Press, Inc.]

conservation equations, they investigated main reaction orders of $\frac{1}{2}$, 1, and 2. It was found that although order gives rise to differences in the $\langle a \rangle$ versus Ω plots and their limiting forms, the differences are orderly and not dominant, so the qualitative characteristics are approximately the same. However, the limiting models for zero-order main reactions do differ considerably, as would be expected (26).

In summary, the single-pellet isothermal diffusion reactor appears to be a tool for the general study of deactivation of catalysts. Its particular configuration permits the construction of the time-implicit $\langle a \rangle$ versus Ω plots from which diagnostic information on the mechanism and deactivation precursors can be obtained.

II. THE SINGLE-PELLET DIFFUSION REACTOR: TEMPERATURE MEASUREMENTS

In the last section we showed how concentrations measured at the center of a pellet can be used within the framework of the Thiele–Zeldovich model to learn mechanistic information about the deactivation of heterogeneous catalysts. In that analysis an implicit assumption was made that the catalyst was isothermal throughout. In many cases this assumption is valid, but in others it is not. Intuitively one would expect that a high heat of reaction would differentiate the two cases—a large enthalpy of reaction suggesting a need for a nonisothermal treatment. In principle, a concentration decrease in the pellet is always accompanied by a temperature change unless the reaction is absolutely thermal neutral. The magnitude of the temperature change is given by an expression first derived by Damköhler (6) and shown below:

$$\Delta T = (-\Delta H) D_{\text{eff}} (C_0 - C) / K_{\text{eff}} \tag{8-8}$$

where C is the point intraparticle concentration of reactant corresponding to the value of ΔT, and K_{eff} is an effective thermal conductivity. In practice, the maximum intraparticle ΔT will occur when $C = 0$, so an estimate of the maximum influence of nonisothermality can be obtained from

$$(\Delta T)_{\text{max}} = (-\Delta H)(D_{\text{eff}}) C_0 / K_{\text{eff}} \tag{8-9}$$

These expressions are quite general and true for all geometries and kinetics (22). We note that, in addition to the enthalpy of the reaction and the change in concentration, the ratio of effective diffusivity and effective thermal conductivity is also important. Since there is normally some uncertainty involved in the measurement of these parameters, there was for considerable time speculation in the literature as to whether such predicted temperature effects would indeed be found experimentally. This has largely been resolved, as will be seen in this section. Subsequent to the work of Damköhler and Prater, a number of other criteria were set forth for the possible existence of intraparticle temperature gradients. These have been summarized elsewhere (2). These criteria have not, however, had explicit reference to deactivated catalyst pellets.

In Chapter 6 the relative magnitudes of intra- and interphase resistances to heat transfer were discussed, and it was shown that in general interparticle resistance was controlling in the overall heat transfer process. Notable exceptions exist, though, and our purpose in this section is to explore how deactivation phenomena affect temperature profiles both inside and outside catalyst pellets.

It is convenient to begin this consideration by seeing whether the apparent thermal effects indicated by Eqs. (8-8) and (8-9) are in fact observed experimentally. Early work was reported by Cunningham et al. (5) for ethylene hydrogenation on a mixed oxide catalyst, Miller (20) for H_2-O_2 on a Pt/Al_2O_3 catalyst, and Würzbacher (27) for the Pt and Ag-catalyzed H_2-O_2 reaction. All these workers observed temperature differences between surface and particle center on the order of 20–40°C. Subsequently, a single-pellet reactor designed to allow measurement of radial temperature profiles was developed (14) and used in the investigation of benzene hydrogenation on Ni/kieselguhr. In agreement with the prior work, they observed gradients on the order of 20°C in a number of cases, but in analysis of data were

II. The Single-Pellet Diffusion Reactor: Temperature

unable to obtain a good correlation between pellet density and ΔT_{max}, presumably because of the uncertainties in values of D_{eff} and K_{eff}.

Kehoe and Butt (15) developed a much-improved version of the Irving (14) single-pellet temperature gradient reactor design, insofar as possible, to mimic experimentally the model of a cylindrically symmetric, infinitely long catalyst pellet with uniform temperature and concentration boundary layers. The philosophy here was very similar to that of the isothermal single-pellet design produced to simulate the Thiele–Zeldovich model. Typical experimental temperature profiles, again for the unpoisoned benzene hydrogenation on Ni/kieselguhr, are shown in Fig. 8-11. These are steady-state measurements reflecting variations as a function of feed composition and flow rate through the reactor. The temperature difference from surface (radial position = 1.0) to bulk represents thermal resistance in the external boundary layer; note that in these experiments the external gradients are somewhat less than the internal gradients. We shall return to similar experiments for deactivated pellets presently.

Turning to intraparticle heat and mass transfer effects under conditions of catalyst decay, it appears that the first substantial modeling results were

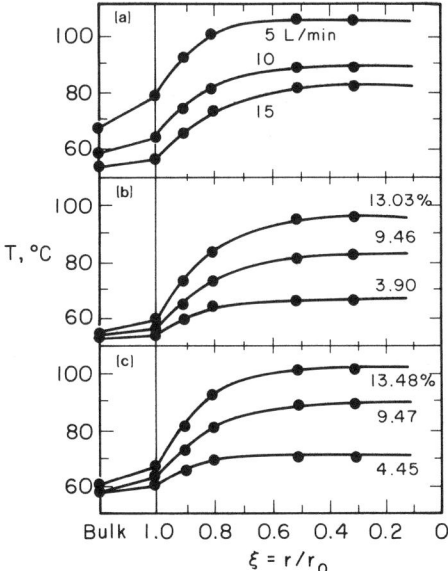

Fig. 8-11. Measured internal and external profiles for pellet 1 with feed temperature of 52°C. (a) Profiles as a function of flow rate for feed composition of ~10% C_6H_6. (b) Profiles as a function of feed composition at high flow, 15 liters/min. (c) Profiles as a function of feed composition at intermediate flow, 10 liters/min. [From Kehoe and Butt (15). Reproduced by permission of the American Institute of Chemical Engineers.]

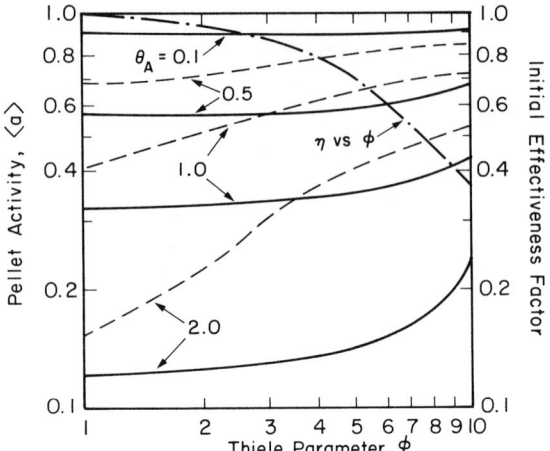

Fig. 8-12. Pellet activity for parallel fouling. $\gamma = 20$, $\gamma_f = 30$, $\beta = 0.1$. Dashed lines are for isothermal conditions. [Adapted from Sagara *et al.* (23). Reproduced by permission of the American Institute of Chemical Engineers.]

reported by Sagara *et al.* (23). They considered the two cases of parallel and series fouling, with the results summarized in Figs. 8-12 and 8-13. These two plots differ from the corresponding curves in Sagara *et al.* in that the results are recast in terms of an initial effectiveness factor and an activity as described by Eqs. (7-30)–(7-32). The corresponding isothermal curves

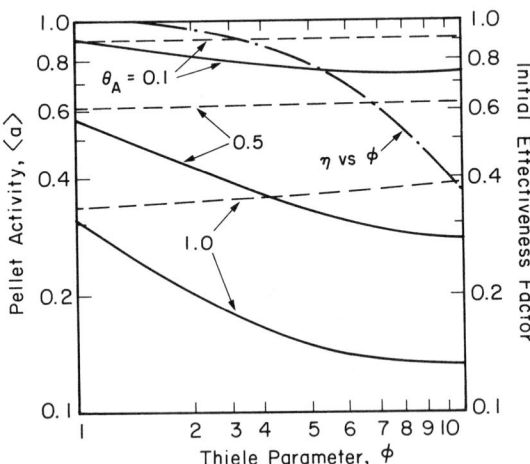

Fig. 8-13. Pellet activity for series fouling. $\gamma = 20$, $\gamma_f = 30$, $\beta = 0.1$, $C_{B0}/C_{A0} = 1$. Dashed lines are for isothermal conditions. [Adapted from Sagara *et al.* (23). Reproduced by permission of the American Institute of Chemical Engineers.]

II. The Single-Pellet Diffusion Reactor: Temperature

from Figs. 7-5 and 7-8 are included on these figures for comparison. The parameters on these figures are defined as follows:

$$\gamma = E/RT, \qquad \gamma_f = E_f/RT, \qquad \beta = (-\Delta H)D_{eff}C_0/K_{eff}T_0 \qquad (8\text{-}10)$$

It will be seen that β corresponds to the parametrization of Eq. (8-9), where the temperature rise is written as $(\Delta T)_{max}/T_0$, γ_f is the normalized activation energy for the fouling reaction, and θ_A is given by Eq. (7-33). The general overall nonisothermal behavior is such that the activity is always smaller than in the corresponding isothermal case. This is quite different from results obtained for nonisothermal effectiveness in nondeactivating systems, but is a result of the increased fouling at higher temperatures within the pellet, as shown in Fig. 8-14. (Symbols in the figure have the same meaning as those presented in Chapter 7; T/T_0 is the reduced temperature.) The interaction of heat transfer and mass transfer resistances is evident. Aside from the usual increase in temperature and decrease in concentration, the amount of foulant on the surface goes through a maximum, as shown in the upper part of this figure. The explanation is straightforward. Close to the outer surface of the pellet the temperature is higher than at the outer radius and the concentration of reactant is high. Hence, the rate of foulant production increases with positions progressively deeper into the pellet

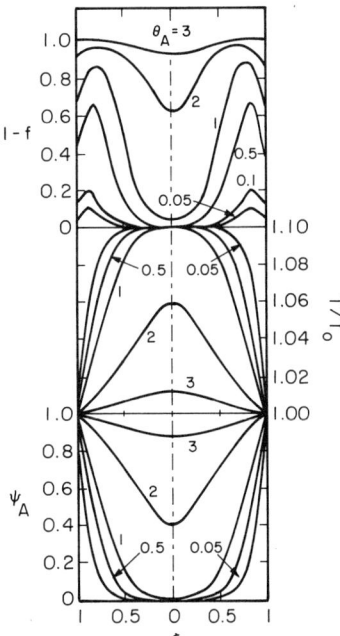

Fig. 8-14. Radial profiles for parallel fouling, nonisothermal intraparticle conditions. $\gamma = 20$, $\gamma_f = 30$, $\beta = 0.1$, $\phi = 5$; $\theta_A =$ scaled time $= tk_A C_{A0}/q_0$. [From Sagara *et al.* (23). Reproduced by permission of the American Institute of Chemical Engineers.]

owing primarily to the temperature rise in the region close to the pellet surface. However, the rate of foulant production also depends on the concentration of reactant. Eventually, this decrease in reactant concentration offsets the temperature effect and the amount of foulant production must go through a maximum. The effect of heat of reaction on this behavior, in terms of effectiveness, is shown in Fig. 8-15.

The behavior of the series fouling system of Fig. 8-13 also shows activities smaller than for the corresponding isothermal system. The shapes of the activity profiles deviate markedly as time increases, largely because the rate of production of foulant in the interior of the pellet is much faster owing to the high temperatures and high concentrations there, as shown in Fig. 8-16. Therefore, for a given time, the activities are much lower than they would be for comparable isothermal conditions.

Now, the models of deactivating pellets under nonisothermal conditions make definite predictions about the temperature profiles as well as quantitative predictions about the rate of overall activity decline as a function of time. A number of experiments presented in the literature have been performed to test the validity of these models. An early paper presenting intraparticle temperature profiles under deactivating conditions was that of Koh and Hughes (16). They studied the hydrogenation of ethylene over a catalyst pellet containing 5% Ni supported on silica–alumina. Seven 0.0025-cm Pt/Pt–Rh thermocouples were embedded within a single pellet 1.26 cm in diameter in experimental configuration similar to those of Irving and Kehoe. Poisoning of the system occurred if the ethylene stream was not purified of traces of oxygen. The temperature profiles from an unpoisoned pellet are shown in Fig. 8-17. The temperature level in each of the experiments is quite different, but the temperature difference between the center

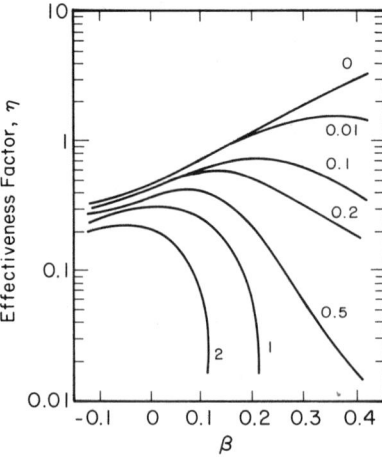

Fig. 8-15. Effect of heat of reaction for parallel fouling. $\gamma = 20$, $\gamma_f = 30$, $\phi = 5$; parameter = θ_A. [From Sagara et al. (23). Reproduced by permission of the American Institute of Chemical Engineers.]

II. The Single-Pellet Diffusion Reactor: Temperature

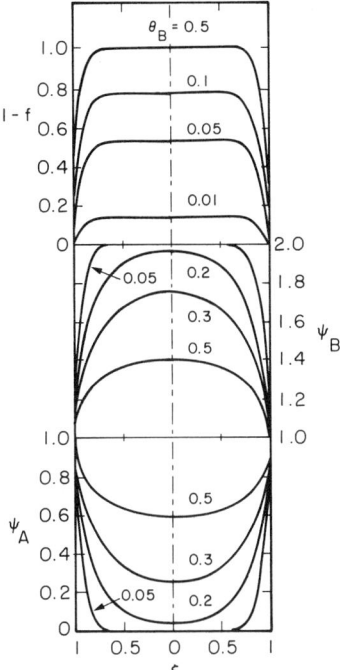

Fig. 8-16. Radial profiles for series fouling, nonisothermal intraparticle conditions. $\gamma = 20$, $\gamma_f = 30$, $\beta = 0.1$, $\phi = 10$, $\theta_A = 1.0$, $\theta_B = tk_A C_{B0}/C_{p\infty}$. [From Sagara et al. (23). Reproduced by permission of the American Institute of Chemical Engineers.]

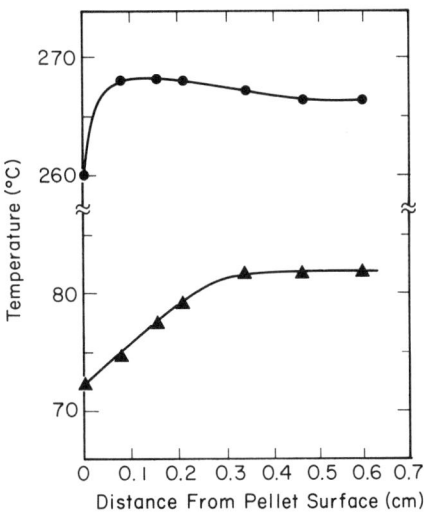

Fig. 8-17. Comparison of intraparticle temperature profiles for active and poisoned pellet. (●) Active pellet, 24% C_2H_4, T(initial) = 20°C; (▲) poisoned pellet, 20.7% C_2H_4, T(initial) = 54°C. [From Koh and Hughes (16). Reproduced by permission of the American Institute of Chemical Engineers.]

of the pellet and the surface is the same in both cases. However, the temperature gradient for the high-temperature active catalyst is very much higher than for the deactivated catalyst. The former observation would be predicted from Eq. (8-8) and the latter from the general theory presented in Chapter 6. Further verification of Eq. (8-8) follows from Figs. 8-18 and 8-19, which show that in general the temperature rise in the pellet is independent of the state of deactivation of the pellet and depends only on the concentration of ethylene in the reaction mixture. The linearity of the rising portion of the temperature profiles in Fig. 8-18 suggests nonuniform poisoning, probably close to the pore mouth type.

Further information and interpretation of intraparticle temperature profiles under deactivating conditions have been reported by Butt *et al.* (3, 8, 18). This work employed benzene hydrogenation on Ni/kieselguhr under experimental conditions essentially identical to those of Kehoe, using the same reactor, but with the addition of thiophene in the reaction mixture as a poison for the Ni catalyst. The properties of the catalyst are given in Table 8-1. The pellet was 1.27 cm in diameter and about 6 cm long, and six iron-constantan thermocouples, 0.005 cm in diameter, were embedded in it. An additional thermocouple monitored temperature in the well-mixed

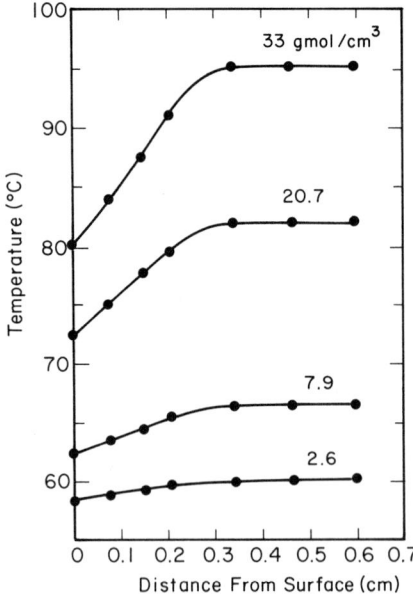

Fig. 8-18. Variation of intraparticle temperature profiles with ethylene concentration, poisoned pellet. T(initial) = 54°C. Parameter is ethylene concentration. [From Koh and Hughes (16). Reproduced by permission of the American Institute of Chemical Engineers.]

II. The Single-Pellet Diffusion Reactor: Temperature

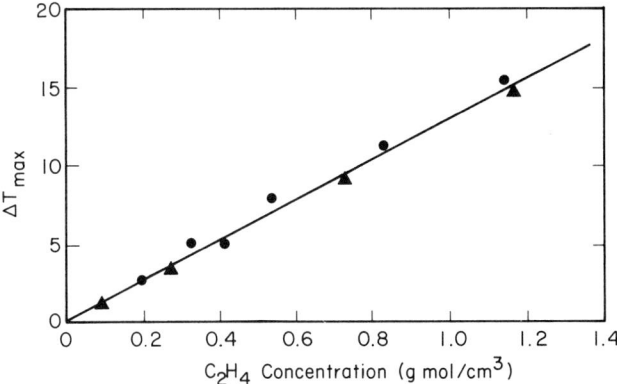

Fig. 8-19. Comparison of surface temperature and maximum intraparticle temperature rise. (●) Active pellet; (▲) poisoned pellet. [From Koh and Hughes (16). Reproduced by permission of the American Institute of Chemical Engineers.]

gas phase surrounding the pellet. The catalysts were prepoisoned to the desired activity level by exposure to thiophene (1%) in the benzene–hydrogen reaction mixture. Thiophene is adsorbed on nickel strongly and nearly irreversibly at the temperatures employed; thus one should expect behavior characteristic of pore mouth poisoning.

TABLE 8-1

Catalyst Properties[a]

Properties	Values
Physical form	1.27 cm diameter by 6–6.5 cm cylinders, prepared from Harshaw Ni-0104 P Ni/kieselguhr, 58% (wt.) Ni, 24 m² Ni/g (reduced)
Density	$1.71 \pm 0.03 \times 10^{-3}$ kg/m³
Heat capacity	0.604×10^3 J/kg-°C
Effective diffusivity	$3\text{–}4 \times 10^{-6}$ m²/sec
Effective thermal conductivity	$17.6 \pm 0.8 \times 10^{-2}$ J/m-sec-°C
Micropore characteristics	
Porosity	0.22
Average pore radius	6 nm
Macropore characteristics	
Porosity	0.34
Average pore radius	315 nm

[a] From Lee *et al.* (18). Reprinted by permission of the American Institute of Chemical Engineers.

One of the main objectives of the work was to establish the effect of poisoning on the transient behavior of the catalyst pellet with the latter stages of the transient, of course, giving the steady-state behavior. Catalysts were typically prepoisoned to overall activity levels of 0.40 and 0.16 of initial activity, then maintained in a hydrogen atmosphere at some desired initial temperature level. A hydrogen–benzene feed (thiophene-free) was then introduced into the single-pellet reactor and the development of thermal profiles followed as a function of time. Typical data for three such experiments are given in Fig. 8-20. The time on stream required to approach steady state appears to be 15 min, as shown in the figure, perhaps decreasing somewhat for the very highly deactivated catalyst. One can readily justify this value in terms of the time constant for the transient heat transfer equation. The dimensionless time will take the form

$$K_{\text{eff}} t / \rho C_0 r_0^2 = 1 \tag{8-11}$$

where t is the characteristic time for the system. Using the values given by Lee *et al.*, we obtain $t = 200$ sec. The value of 15 min represents approximately five time constants and should be very close to steady state. The fact that the extent of deactivation has little effect on the time required to attain steady state may seem a bit surprising at first; however, on reflection there is really no reason for this not to be so unless the deposit of poison is sufficient to alter appreciably the thermal characteristics of the pellet.

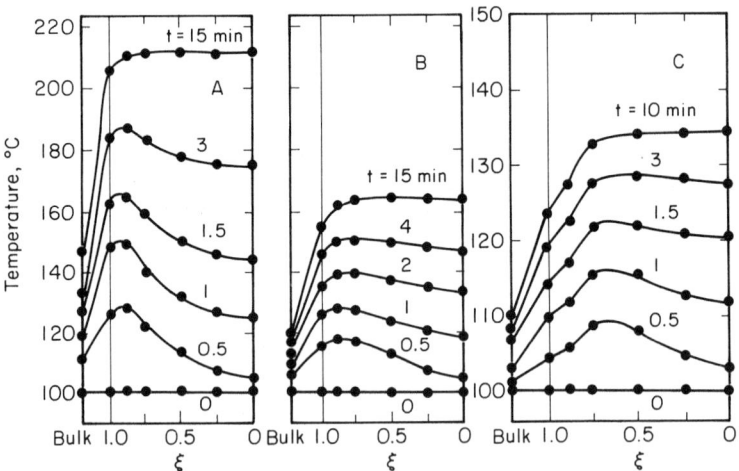

Fig. 8-20. Effect of deactivation on start-up transients. Initial temperature, 100°C; benzene mole fraction, 0.16; catalyst prepoisoned with 1% thiophene, 65°C. (A) Fresh; (B) activity = 0.40; (C) activity = 0.16. [From Lee *et al.* (18). Reproduced by permission of the American Institute of Chemical Engineers.]

II. The Single-Pellet Diffusion Reactor: Temperature

As expected, the magnitude of the exotherm is severely decreased as the extent of poisoning increases, but in addition Fig. 8-20 reveals a number of important characteristics of a system in which heat transfer from the pellet to the ambient reaction mixture is important. First, the left side of the figure for the fresh catalyst shows that at steady state the temperature difference between the pellet surface and bulk phase is considerably greater than the temperature difference within the pellet. This is, of course, in accordance with the rule of thumb mentioned in Chapter 6. However, as the catalyst deactivates, as shown in the center and at the right side of the figure, the difference between internal and external gradients decreases. This is also the expected result because as the catalyst deactivates the amount of reactants transferred to the catalyst surface decreases and the amount of heat generated decreases correspondingly. Note, though, that the difference in temperature between the pellet surface and its center remains constant, in keeping with Eq. (8-8). The point at which the temperature reaches its maximum value is pushed progressively farther into the pellet as it deactivates, and eventually the point at which the maximum temperature is attained will be at the center of the pellet. Further deactivation beyond this will result in a decreased temperature rise. The characteristic shape of the temperature profiles in the highly deactivated catalyst is reflective of shell-progressive poisoning.

Given the trend in the ratio of internal to external temperature gradients described above, however, one becomes a little uneasy as to whether the major thermal resistance will *always* be in the external boundary layer. To explore this further, one additional pellet was deactivated to an activity of 0.10 that of the fresh pellet and the transient experiment repeated. The result is shown in Fig. 8-21, where even after 3 min the ratio of intra- to interphase gradients exceeds two.

The relative importance of intra- and interphase gradients has been shown to reside in the ratio of mass to thermal Biot numbers. The upper bound in overall temperature rise for an exothermic reaction in terms of observable quantities was derived by Lee and Luss (17), although Carberry (4) subsequently pointed out that individual values of mass and thermal Biot numbers were required and proposed an alternative treatment in terms of the observable quantity $\bar{\eta}\mathrm{Da}$, a rate parameter, and the ratio of mass to thermal Biot numbers. This leads to the following relationship between internal and overall gradients:

$$\Delta T_x/\Delta T_0 = r(\bar{\eta}\mathrm{Da})/[1 + \bar{\eta}\mathrm{Da}(r-1)] \qquad (8\text{-}12)$$

where

$$\bar{\eta}\mathrm{Da} = R_0/k_g a C_0, \qquad r = \mathrm{Bi(mass)}/\mathrm{Bi(heat)}$$

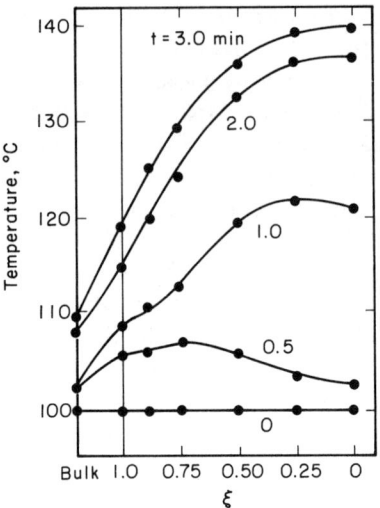

Fig. 8-21. An extreme example of alteration of inter- and intraphase gradients by deactivation. Conditions as for Fig. 8-20 except benzene mole fraction = 0.18. [From Lee et al. (18). Reproduced by permission of the American Institute of Chemical Engineers.]

In the above R_0 is the observed rate of reaction, $k_g a$ the mass transfer coefficient, and C_0 the reactant concentration. The value of r can be calculated from

$$r = (K_{\text{eff}}/D_{\text{eff}})(1/\rho C_p) \tag{8-13}$$

where ρ and C_p are the density and heat capacity of the gaseous reaction mixture. This was tested by Butt et al. using data for both fresh and deactivated catalysts reported by Kehoe and Butt and Lee et al., with the results shown in the parity plot of Fig. 8-22. The agreement with the theoretical expression is excellent, even exceptional, and would verify the applicability of Eq. (8-12) over a very wide range of experimental results. Note that there are a number of very small values of $(\Delta T_x / \Delta T_0)$ both predicted and observed near the origin of the coordinates in the figure, where presumably the internal gradients are larger than those in the boundary layer.

A final point of interest associated with these temperature gradients and temperature transients is the apparent migration of "hot spots" as the pellet evolves to its final steady-state value. One sees in Fig. 8-20 the development of an apparent maximum temperature near the surface of the pellet that remains in that position until steady state is attained in the fresh catalyst. Careful attention to the partially deactivated catalyst of relative activity 0.40, however, reveals that the maximum temperature is initially displaced somwhat to the interior of the pellet and then migrates toward the surface. A wide variety of classes of these wandering temperature maxima have been observed, as shown in Fig. 8-23. These include maxima that remain near the surface until steady state; maxima that migrate toward the center,

II. The Single-Pellet Diffusion Reactor: Temperature

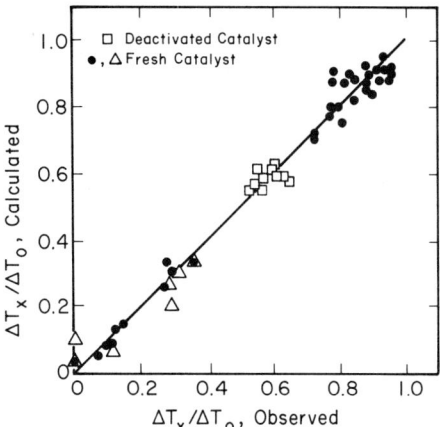

Fig. 8-22. Comparison of theory and experiment for $(\Delta T_x/\Delta T_0)$ in fresh and deactivated catalysts. (●) Fresh catalyst (Lee *et al.*); (□) deactivated catalyst (Lee *et al.*); (△) fresh catalyst (Kehoe and Butt). [From Butt *et al.* (3). Reproduced by permission of the American Chemical Society.]

then reverse direction toward the surface, and finally return to the center; and maxima that proceed uniformly toward the center. The behavior seems roughly to be correlated with the Damköhler number of the reacting system, $\phi^2 \text{Bi}$ (mass), but simulations as discussed below do not seem to track this behavior very well. In some sense, these wandering profiles are reminiscent

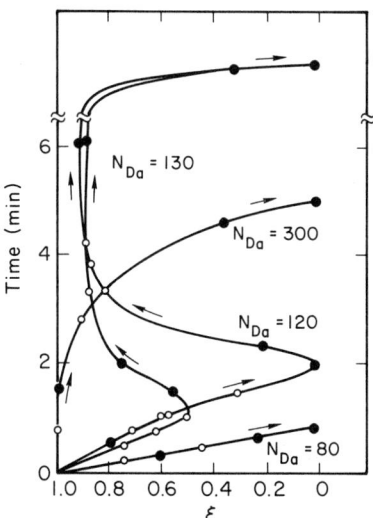

Fig. 8-23. Variations in hot spot migration patterns in start-up experiments. [From Lee *et al.* (18). Reproduced by permission of the American Institute of Chemical Engineers.]

of the "wrong way" behavior of fixed beds, but any simple explanation would appear elusive at best.

From an engineering point of view, it is of considerable interest to attempt simulation of these transient results reflecting the influence of deactivation. This is of interest not only because of the desire to predict the magnitude and duration of the transients but also because such experiments provide a very sensitive test of the model used in the simulation. As stated before, the experiment is established to conform to the model of a symmetric, infinitely long cylinder with uniform external composition of reaction mixture. Even at this, the mathematical representation is rather complicated. We establish the following conservation relations (hydrogen is omitted since in the experiments it was present in great excess).

Benzene:

$$\partial C_B/\partial t = D_{\text{eff}}\nabla^2 C_B - ar_B(T, C_B) \tag{8-14}$$

Cyclohexane:

$$\partial C_c/\partial t = D_{\text{eff}}\nabla^2 C_c + ar_B(T, C_B) \tag{8-15}$$

Thiophene:

$$\partial C_T/\partial t = D_{\text{eff}}\nabla^2 C_T + \rho r_T(T, C_T) \tag{8-16}$$

Energy:

$$\rho C_p \, \partial T/\partial t = -K_{\text{eff}}\nabla^2 T + (-\Delta H)ar_B(T, C_B) \tag{8-17}$$

In this case we use the separable form of deactivation kinetics, with the net rate of reaction given at any time by $\langle a \rangle r_B(T, C_B)$. A similar system has been used for the investigation of fixed-bed reactor transients, and pertinent parameters are given in the example of Chapter 10. Briefly, we have that

$$r_d = da/dt = k_d^0 \, e^{-E_d/RT} P_T a \tag{8-18}$$

and the kinetics of the benzene hydrogenation reaction used in the experiments are given by

$$r_B(T, P_B) = \frac{k_1^0 K_0 \exp[(Q-E)/RT] P_B P_{H_2}}{1 + K_0 \exp(-Q/RT) P_B} \tag{8-19}$$

In the above C_i is the concentration of benzene, cyclohexane, or thiophene, K_{eff} and D_{eff} are the effective transport properties, a is the point activity, r_B is the rate of reaction of benzene as defined in Eq. (8-19), ρ and C_p are the effective catalyst densities and heat capacities, and r_d is the net rate of deactivation defined by Eq. (8-18). There is an important distinction between

II. The Single-Pellet Diffusion Reactor: Temperature

r_d and r_T, the rate of thiophene uptake as used in Eq. (8-16). In essence

$$r_T = M_T r_d \tag{8-20}$$

where M_T is the poison adsorption capacity (i.e., amount of poison per unit of catalyst required to completely deactivate the catalyst).

A number of initial and boundary conditions are required to complete formulation of the simulation. For the initial (fresh) catalyst:

$$T = T_0(r), \quad a = 1, \quad C_i = C_{i_0}(r) \tag{8-21}$$

and for the deactivated catalyst one must make the modification

$$a = a(r)$$

Boundary conditions, fresh or deactivated, require that

$$(\partial C_i/\partial r) = (\partial T/\partial r) = (\partial a/\partial r) = 0 \tag{8-22}$$

at the center of the pellet, and

$$D_{\text{eff}}(\partial C_i/\partial r)_{R_B} = k_g[C_{i_b}(t) - C_i]$$
$$K_{\text{eff}}(\partial T/\partial r)_{R_p} = h_T[T_b(t) - T] \tag{8-23}$$

at the boundary. In Eqs. (8-23), k_g and h_T are transfer coefficients normalized to the external area of the pellet.

The overall set of equations is reminiscent of those in Chapter 10 used to model a similar problem in fixed beds, and more quantitative details will be reserved until that point. However, concerning simulation, several factors should be pointed out here. First is the large parametrization of the problem; if one counts up all the parameters required, including those in the kinetic model for the main reaction, the total comes out to about 13, which is generally considered to be an unlucky number. This may not be a trivial generalization, when one considers that even from independent experimentation such parameters are rarely known individually to better than ±10%. More than this, the parametric sensitivity is very unevenly distributed, as discussed below. Nontheless, the simulation of experiments such as shown in Fig. 8-20 was carried out using a Crank-Nicolson method with a shifted boundary technique (24). The results of this, using the parameters reported by Lee *et al.*, are shown in Fig. 8-24 for the transient start-up experiments of Fig. 8-20 and for tracking the course of deactivation in steady-state profiles obtained in the prepoisoning results (Fig. 8-25). The simulation results of Fig. 8-24 indicate very good agreement with the shapes of the temperature profiles measured experimentally, but cross-reference with the data of Fig. 8-20 will reveal that the time scales are woefully wrong. From the order of magnitude calculation according to Eq. (8-11) one must

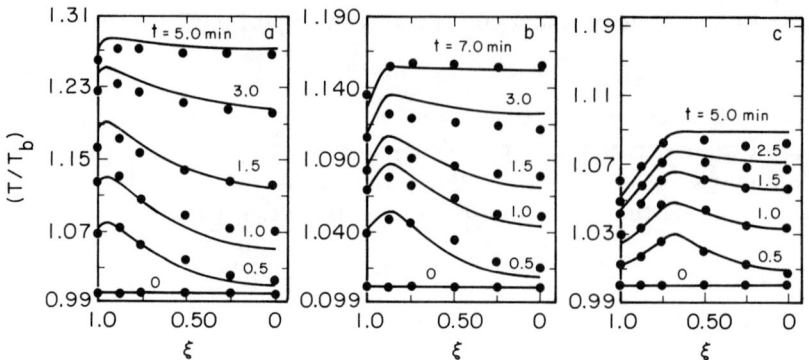

Fig. 8-24. Simulation results for the experiments of Fig. 8-20. [From Lee *et al.* (18). Reproduced by permission of the American Institute of Chemical Engineers.]

conclude that the simulation, not the experiment, is wrong. It will be amplified that the culprit is the value of M_T, poison adsorption capacity, used in the calculation. This single number is instrumental as a "time driver" and is central in defining the time scale of the response. It is thus an important and very sensitive parameter in determining the time of approach to steady state. Nonetheless, the simulation, which fails rather badly in keeping up with individual transients, follows the overall course of deactivation fairly well, as shown in Fig. 8-25. The reason is probably in the differences in averaging between a, the point activity involved in individual

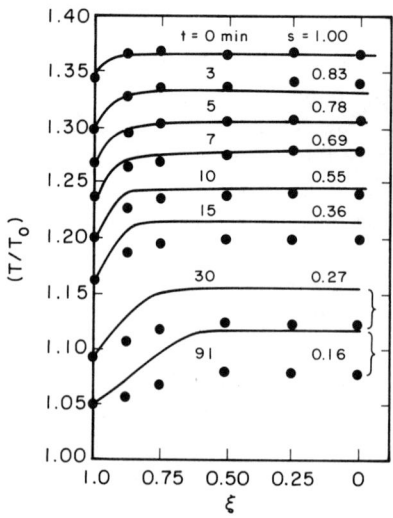

Fig. 8-25. Simulation of the history of intraphase gradients upon progressive deactivation. Deactivation of benzene hydrogenation on Ni/kieselguhr by 1% thiophene, $T = 65°C$. [From Lee *et al.* (18). Reproduced by permission of the American Institute of Chemical Engineers.]

II. The Single-Pellet Diffusion Reactor: Temperature

response simulation, and the integrated average $\langle a \rangle$ that is ultimately involved in following the time evolution of profiles. This is also reflected in the prediction of overall activity and effectiveness, as shown in Fig. 8-26.

The question of parametric sensitivity in such simulations is troublesome overall, however; we see in the above that M_T is essential in defining the time scale of response. The question then arises of whether there are other sensitive parameters with respect to modeling the *magnitude* of the exotherms. As expected, the answer is yes. The study of Downing *et al.* addressed this point directly in the Orwellian sense that "all parameters are equal, but some parameters are more equal than others." Indeed, aside from M_T in the simulation of Eqs. (8-14)–(8-23), it was found that the effective diffusivity D_{eff} and external heat transfer coefficient h_T were of particular sensitivity. Figure 8-27 shows this result. The square point represents the values of the pair D_{eff}–h_T used in the original simulation, as determined in independent experiments. However, in subsequent computational exploration it was found that a trajectory of these pairs would yield similar simulations of the final steady-state temperature level (judged acceptable if within ±5°C of the experimentally observed value). This is shown by the line in Fig. 8-27 with the specific pairs investigated indicated by the points. The particular pair of D_{eff}–h_T has little effect on the time scale of the response, since M_T is by far the dominant parameter in this regard. The extreme parametric sensitivity is shown by the points off the line, where the figures in parentheses refer to the deviations in degrees Celsius of the computed profile at steady state compared to the experimental value.

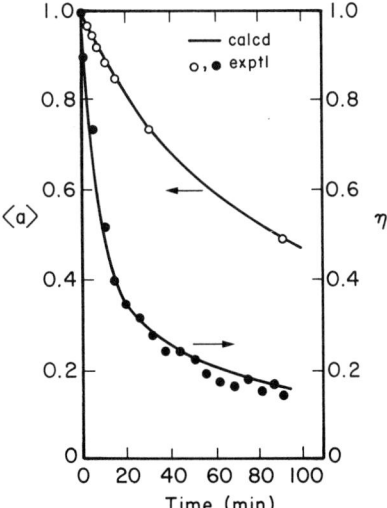

Fig. 8-26. Prediction of the history of overall activity and effectiveness on progressive deactivation. [From Lee *et al.* (18). Reproduced by permission of the American Institute of Chemical Engineers.]

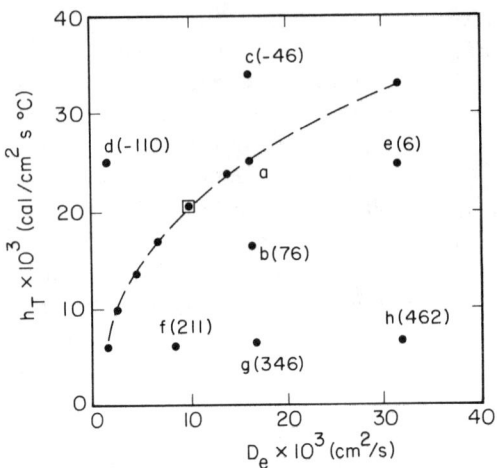

Fig. 8-27. Steady-state parameter analysis—sensitivity to D_{eff} and h_T. [From Downing et al. (8). Reproduced with permission of the American Institute of Chemical Engineers.]

Unfortunately, there is no consistent pattern to these results, other than to recognize that one is dealing with coupled, highly nonlinear equations for which intuition is not always a very good guide.

In summary, such efforts at simulation appear to teach that global averaged quantities such as steady-state activity and overall rates of deactivation can be determined using independently evaluated parameter sets, but detailed simulation of transients involving both profile shapes and time scales requires parameters to be known to a degree of accuracy probably beyond current capabilities in measurement or estimation.

III. SUMMARY AND EVALUATION

It is clear that these single-pellet diffusion reactors offer considerable information concerning mechanisms of deactivation by taking advantage of the gradients engendered by simultaneous transport processes. As stated, they stand at the opposite end of the scale from so-called gradientless reactors. The isothermal single-pellet reactor appears almost unique in its ability to screen the various deactivation possibilities, such as parallel and series, with considerable economy in experimentation. The nonisothermal single-pellet reactor is perhaps less useful for such diagnostics; however, it is seen to provide detailed information on profile shapes as well as overall gradients. Along the way, it also provides information concerning the relative

III. Summary and Evaluation

sensitivity of the many parameters involved in the diffusion-reaction-deactivation process.

We reiterate that such approaches derive from the philosophy of designing the experiment to fit a model rather than the other way around, although it is seen that the model being mimicked can be rather complex. The success in doing this in the isothermal Thiele-Zeldovich case allows one to eliminate time as a reaction variable, permitting the deactivation to be studied independently of its time scale. This implies that the method has access to decay processes of widely different time scales, the importance of which we have discussed in Chapter 6.

It is interesting that from the nonisothermal work, over rather wide ranges of conditions and for different catalysts and reactions, strong support is provided for the Damköhler relationship and Prater's subsequent analysis. On somewhat more shaky ground is the generalization of the relative magnitude of intraparticle and boundary layer thermal gradients. While the latter may be generally dominant in fresh catalysts, it appears that deactivation can alter this. One might also expect that particular types of deactivation can affect the time scale of catalyst transients, although this was not seen in the work discussed here. Such cases would be found when significant alterations of the thermal or transport properties of the catalyst are affected by decay, for example, via pore plugging through heavy coke or metals deposition.

What are the limitations or drawbacks to the use of such single-pellet reactors? For the isothermal single-pellet diffusion reactor it may be isothermality itself. The pellet is, after all, of macroscopic dimension and for reactions with appreciable heats of reaction it may be difficult to avoid various kinds of thermal gradients. Also, the temperature level itself may present some problems, since the design discussed by Hegedus and Petersen relies on a good seal between pellet and containing vessel to fit the Thiele-Zeldovich model. For deactivations with a short time scale, it is necessary to have a rapid, specific, and unobtrusive method for centerplane concentration measurements. For the nonisothermal single-pellet reactor the most obvious point is first; it is not easy to embed very fine wire thermocouples in a large pellet and know precisely where the thermojunction is. Here temperature level can also be a problem, since the cylindrically symmetric model experimentally calls for a good seal between the pellet and the top and bottom of the containing vessel. Nonetheless, all these problems, for both isothermal and nonisothermal designs, can be and have been met in the work described here. We conclude that the single-pellet reactor provides a useful, if somewhat unorthodox, technique for the investigation of various aspects of deactivation.

REFERENCES

1. J. R. Balder and E. E. Petersen, *Chem. Eng. Sci.* **23**, 1287 (1968).
2. J. B. Butt and V. W. Weekman, Jr., *AIChE Symp. Ser.* No. 143, 27 (1974).
3. J. B. Butt, D. M. Downing, and J. W. Lee, *Ind. Eng. Chem. Fundam.* **16**, 270 (1977).
4. J. J. Carberry, *Ind. Eng. Chem. Fundam.* **14**, 129 (1975).
5. R. A. Cunningham, J. J. Carberry, and J. M. Smith, *AIChE J.* **11**, 636 (1965).
6. G. Damköhler, *Z. Phys. Chem.* (*Leipzig*) **193**, 16 (1943).
7. N. R. Dougharty, *Chem. Eng. Sci.* **25**, 489 (1970).
8. D. M. Downing, J. W. Lee, and J. B. Butt, *AIChE J.* **25**, 461 (1979).
9. J. L. Hahn and E. E. Petersen, *Can. J. Chem. Eng.* **48**, 147 (1970).
10. L. L. Hegedus and E. E. Petersen, *Catal. Rev.—Sci. Eng.* **9**, 245 (1974).
11. L. L. Hegedus and E. E. Petersen, *Chem. Eng. Sci.* **28**, 69 (1973).
12. L. L. Hegedus and E. E. Petersen, *Chem. Eng. Sci.* **28**, 345 (1973).
13. L. L. Hegedus and E. E. Petersen, *J. Catal.* **28**, 150 (1973).
14. J. P. Irving and J. B. Butt, *Chem. Eng. Sci.* **22**, 1859 (1967).
15. J. P. G. Kehoe and J. B. Butt, *AIChE J.* **18**, 347 (1972).
16. H. Koh and R. Hughes, *AIChE J.* **20**, 395 (1974).
17. J. C. M. Lee and D. Luss, *Ind. Eng. Chem. Fundam.* **8**, 596 (1969).
18. J. W. Lee, J. B. Butt, and D. M. Downing, *AIChE J.* **24**, 212 (1978).
19. S. Masamune and J. M. Smith, *AIChE J.* **12**, 384 (1966).
20. F. W. Miller, Ph.D. Thesis, Rice Univ., Houston, 1964.
21. E. E. Petersen, *Exp. Methods Catal. Res.* **2**, 257 (1976).
22. C. D. Prater, *Chem. Eng. Sci.* **8**, 284 (1958).
23. M. Sagara, S. Masamune, and J. M. Smith, *AIChE J.* **13**, 1226 (1967).
24. D. U. von Rosenberg, "Methods for the Numerical Solution of Partial Differential Equations," Elsevier, New York, 1969.
25. E. E. Wolf and E. E. Petersen, *Chem. Eng. Sci.* **29**, 1500 (1973).
26. E. E. Wolf and E. E. Petersen, *Chem. Eng. Sci.* **32**, 493 (1973).
27. G. Würzbacher, *J. Catal.* **5**, 476 (1966).

CHAPTER 9

Regeneration of Coked Particles

> Where there's smoke, there must be fire...
> *Anon.*

The process of coke removal or regeneration is intertwined with that of coke formation in many instances, so it is appropriate at this point to review (at the expense of some redundancy with the material of Chapter 3) some of the governing kinetics. Indeed, let us begin the discussion here by considering the suppression or inhibition of coke formation and then train our attention on the subsequent removal of coke from a catalytic surface.

I. REACTIONS OF COKE WITH HYDROGEN AND OXYGEN

A. Hydrogen Inhibition

In the simplest instance, if we represent the coke "molecule" via the empirical formula CH, then the inhibition process can be represented as

$$y\text{CH} + x\text{H}_2 \rightleftarrows \text{C}_y\text{H}_{2x+y} \qquad \text{I}$$

where $\text{C}_y\text{H}_{2x+y}$ represents a hydrogenated form of coke that has been removed from the surface. Written in the opposite direction, this species could be considered the coke precursor molecule. As shown, this reaction is a reversible one; if we assume that some effective equilibrium level of coke can be established by the hydrogenation, then

$$K_1 = [\text{C}_y\text{H}_{2x+y}]/[\text{CH}]^y[\text{H}_2]^x \qquad (9\text{-}1)$$

or

$$[\text{CH}]^y = [\text{C}_y\text{H}_{2x+y}]/K_1[\text{H}_2]^x \qquad (9\text{-}2)$$

Now let us further assume the same typical relationship between the effective level of coke on the catalyst surface and the rate constant for the main reaction, k, say

$$ka = k(1 - C_c/C_c^0) \qquad (9\text{-}3)$$

where C_c is coke on catalyst and C_c^0 a scaling parameter in the normal fashion. Then on combination of Eqs. (9-2) and (9-3) we may say that

$$a \propto (1 - [CH]^y) \qquad (9\text{-}4)$$

or

$$ka \propto \{1 - k[C_y H_{2x+y}]/K_1[H_2]^x\} \qquad (9\text{-}5)$$

From Eq. (9-5) it is clear that the kinetic implications of hydrogen inhibition will appear as a term in the denominator of the rate equation of the main reaction. Whether this is in power law form, as illustrated here, or as an adsorption inhibition term in a Langmuir-Hinshelwood-Hougen-Watson (LHHW) rate equation is not of special concern in the present discussion; both have been employed in the literature, as will be illustrated in some of our subsequent analysis. However, if the inhibition process is viewed as a chemical reaction the approach given here is probably closer to reality than that of competitive adsorption. High hydrogen partial pressures, thus, are an effective means for suppressing coke formation; even in the event that K_I is small some control may be obtained, since x is often greater than one. Of course, as we have pointed out a number of times, it is quite difficult to discuss "the mechanism" of coke formation in any but rather general terms, so K_I remains a kinetic parameter which must be identified for each catalyst-reaction system of interest.

Commercial realizations of the implications of Eq. (9-5) are found in many places, but especially in the very large scale hydrotreating and reforming processes in the petroleum and synthetic fuels industry. As an example, for the hydrodesulfurization of heavy residual oils, where one typically wishes to go from about 4 to 0.5 wt. % S (or less), hydrogen partial pressures on the order of 150 atm are routinely employed. The kinetics of the desulfurization reaction are ordinarily far less demanding than these conditions might imply, but the effective catalyst life is roughly proportional to the square of hydrogen partial pressure; hence, in spite of hydrogen costs, process equipment (high pressure) costs, and operating costs (compression), there remains a substantial economic incentive for operation at high hydrogen partial pressures.

Care should be taken, however, to avoid the too-literal interpretation of coke suppression via reactions proposed to be as simple as the example of I above. The discussion in Chapter 3 indicated clearly the possibility that

I. Reactions of Coke with Hydrogen and Oxygen

very complex reaction networks can be involved in coke formation—via reactant, product, or side reaction precursors, and so on. Unfortunately, we have a tendency to draw all the arrows in the same downhill direction towards the ultimate disaster termed "coke." Yet, if many different pathways (i.e., different reactions) are involved, it is certainly unreasonable to expect the result to be the same chemical entity, or even related ones. In fact, the chemical characterizations we discussed previously attest to this—via such diverse means as aromatics content, basicity, and hydrogen content. Another distinction that is important to make for the presentation here is reactivity. Wide variations in the reactivity of coke, particularly with respect to hydrogen, have been observed in many cases. Often such reactivity variations are split into two lumps, known in diverse jargons as "fast coke" and "slow coke," or "refractory" and "nonrefractory," or "equilibrium" and "nonequilibrium." The situation is qualitatively illustrated in Fig. 9-1, where we consider a series of experiments run at constant space velocity, temperature, and reactant concentrations, but at varying hydrogen partial pressure. Characteristically, the relative activity will decline with time on stream as shown, but the details of this decline are quite different. In experiment 1, conducted at low hydrogen partial pressure, the equilibrium in reaction I is far to the left, and we are essentially in a kinetically controlled region of coke deposition throughout the duration of the run. The effective activity of the catalyst is destroyed before the equilibrium can play any role in inhibition. Now as P_{H_2} is increased in the successive experiments 2, 3, and 4, a different pattern begins to emerge. Experiment 2 is also characterized by a large, relatively rapid decline in activity, but with the suggestion of a tail on the activity curve (which in practice would probably be masked by experimental error). Going higher in hydrogen partial pressure in experiments 3 and 4 reveals the affair in more detail. One sees here an initial

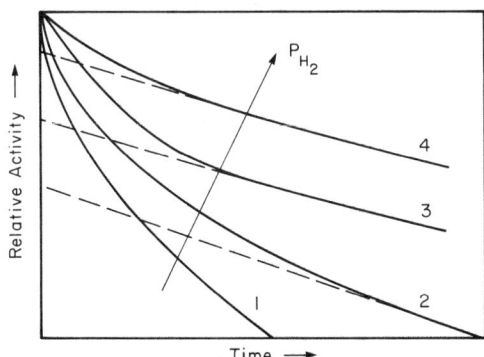

Fig. 9-1. Inhibition of coke formation via hydrogen—two lumps.

relatively rapid deactivation followed by a long-term slower and near-linear deactivation. The initial deactivation region is strongly dependent on hydrogen partial pressure, while long-term deactivation is independent of it. A simple conceptual model for such behavior, thus, is to break the coke formation process into two parts, one associated with the rapid formation of a deposit and corresponding deactivation which can be controlled in extent by hydrogen partial pressure, and one with a slow deactivation which cannot. Correspondingly, we identify the two coke species on the basis of reactivity to hydrogen as "nonrefractory" and "refractory." As indicated in Fig. 9-1, an extrapolation to zero time from the long-term deactivation region could provide, at least in principle, a means for identifying the separate contributions of refractory and nonrefractory material and for estimating separately the kinetic/equilibrium constants associated with each region.

Apparent small changes in the nature of curves such as those of Fig. 9-1 may also be important in the identification of reaction schemes for coke formation. For example, in the paragraph above we considered the case in which long-term deactivation was independent of hydrogen partial pressure. From experimental observation, this would be suggested by the fact that the slopes of the long-term activity–time curves were the same and that some degree of internal consistency would be obtained in the two-lump analysis via the extrapolation procedure. As a result, we can write the following reaction model for coke formation:

$$(\text{Precursor})_\text{I} \rightleftarrows H_2 + (\text{Coke})_\text{I} \quad\quad (a)$$
$$(\text{Precursor})_\text{II} \rightarrow (\text{Coke})_\text{II} \quad\quad (b)$$

II

where the first step in **II** is the same as **I** (written in the opposite direction) and the second step represents the long-term deactivation process. Now suppose that the experimental observation was that long-term deactivation rates were also dependent on hydrogen partial pressure, but less so than the initial rates. This would obviously imply the simultaneous participation of hydrogen in both reaction steps and perhaps even reversibly to some extent in the long-term deactivation as well. Hence:

$$(\text{Precursor})_\text{I} \rightleftarrows H_2 + (\text{Coke})_\text{I} \quad\quad (a)$$
$$(\text{Precursor})_\text{II} \rightleftarrows H_2 + (\text{Coke})_\text{II} \quad\quad (b)$$

III

and the net effects of hydrogen inhibition will be rather more complex than those envisioned in Eq. (9-5).

The discussion above concentrated on an interpretation of the varying response of coke to hydrogen inhibition in terms of a two-lump scheme postulating two different types of coke. An important variation on this,

I. Reactions of Coke with Hydrogen and Oxygen

particularly for long-term deactivation in processes such as catalytic hydrotreating, is a change in the reactivity of a single type of coke species as the coke is aged. This has been shown in the work of Riley *et al.*[14] via the use of electron spin resonance (ESR) measurements. They have shown that ESR determination of the carbon radical density, essentially unpaired aromatic π electrons embedded in the coke, indicates a substantial decrease with time on stream in heavy crude hydrodesulfurization on Co-Mo catalysts. The change is indicative of a change from "fresh" coke, relatively high in free-radical density and correspondingly reactive, to "aged" coke with much lower free-radical density and lower reactivity. Indeed, the latter as characterized by the ESR technique is similar to graphite. Thus, a series sequence might be written to represent this process in place of III, in which $(Coke)_I$ is transformed into $(Coke)_{II}$. However, the chemistry of this has not been elucidated, as discussed in Chapter 7.

Numerous examples of the hydrogen inhibition of coke formation have been documented in the literature. A typical study involving rather complex interactions of coke and coke precursor has been reported by Wolf and Petersen (21) for the dehydrogenation of methylcyclohexane on a Pt/γ-Al$_2$O$_3$ catalyst. In the temperature range 350–400°C, with $15 < P_{MCH} < 60$ torr, $0 < P_{H_2} < 800$ torr, the reaction is approximately first-order with respect to both methylcyclohexane and hydrogen; however, very complex deactivation behavior was encountered, dependent sensitively on the partial pressure of hydrogen. In general, the rate of deactivation was decreased on increasing hydrogen partial pressure, as would be expected; however, deactivated catalysts could be only partially reactivated by treatment in hydrogen. Furthermore, when the reaction was conducted under conditions of $P_{H_2} = 0$, the deactivation was very rapid, while at $P_{H_2} = 100$ torr it was very slow. (In a sense, these last two observations are exploring the difference between curves 1 and 4 of Fig. 9-1.) The sum of observations concerning main reaction and deactivation behavior could be explained via the following kinetic scheme:

$$\text{MCH} + \text{S} \underset{k_{-1}}{\overset{k_1}{\rightleftharpoons}} [\text{MCH} \cdot \text{S}] \overset{k_2}{\longrightarrow} [\text{TOL} \cdot \text{S}] + 3\text{H}_2 \underset{k_{-3}}{\overset{k_3}{\rightleftharpoons}} \text{TOL} + \text{S}$$

$$k_4 \updownarrow k_{-4}$$

$$[\text{C} \cdot \text{S}] + \text{H}_2 \qquad\qquad \text{IV}$$

$$k_5 \updownarrow k_{-5}$$

$$\text{C} + \text{H}_2 \overset{k_6}{\longrightarrow} \text{W}$$

in which the rate-determining step (RDS) of the surface reaction is dehydrogenation of intermediate [MCH · S] to [TOL · S]. The adsorbed intermediate [MCH · S] plays a pivotal role here in both dehydrogenation and deactivation, since it forms the coke precursor [C · S] as well. In the deactivation scheme, we see a sequential process involving the formation of "reversible" (nonrefractory) coke C as well as the ultimate irreversible (refractory) coke W.

Analysis of the kinetics of a scheme such as **IV** is more complicated than the simple equilibrium arguments with which we started off this discussion, since several stages, both reversible and irreversible, are involved in the deactivation. We must, in setting a model, look carefully at the relative time scales that are involved in the different sequences. The steps involved in the main reaction can be considered rapid with respect to any other time scale, hence a steady-state concentration of [MCH · S] is established, with subsequent deactivation via formation of the precursor [C · S]. Hence if we write a surface site balance for an "initial" state of the catalyst:

$$[S]+[MCH \cdot S]+[TOL \cdot S]=[S]_0-[C \cdot S]_0=[S]_{t=0} \quad (9\text{-}6)$$

where [S] represents vacant sites, $[S]_0$ total initial site concentration per unit volume, $[C \cdot S]_0$ initial poison precursor concentration, and $[S]_{t=0}$ the net effective site concentration for the initial catalyst surface. Now in the second and longer time scale the reversible coke has attained equilibrium and deactivation is dominated by the formation of the irreversible product W. Under these conditions a surface balance yields

$$[S]+[MCH \cdot S]+[TOL \cdot S]+[C \cdot S]=[S]_{t=0}-([C]+[W]) \quad (9\text{-}7)$$

The [C]+[W] terms are omitted from the initial balance owing to the differing time scales of the reversible and irreversible deactivation; hence, Eq. (9-6) corresponds to the initial portion of curve 4, say, of Fig. 9-1 and Eq. (9-7) to the long-term portion. Now, according to the surface dehydrogenation of [MCH · S] as the RDS, we have the following ordering among the rate constants:

$$k_2 < k_1, k_{-1}, k_3, k_{-3} \quad (9\text{-}8)$$

but

$$k_4, k_{-4} < k_2 \quad (9\text{-}9)$$

so

$$[MCH \cdot S] = K_1[MCH][S] \quad (9\text{-}10)$$

$$[TOL \cdot S] = K_3[TOL][S] \quad (9\text{-}11)$$

and

$$[C \cdot S] = K_4[MCH \cdot S]/[H_2] \quad (9\text{-}12)$$

I. Reactions of Coke with Hydrogen and Oxygen

where

$$K_1 = k_1/k_{-1}, \quad K_3 = k_3/k_{-3}, \quad K_4 = k_4/k_{-4} \tag{9-13}$$

Combining Eqs. (9-7), (9-10), (9-11), and (9-12), we obtain for [S]:

$$[S] = \frac{[S]_{t=0} - ([C] + [W])}{1 + K_1[\text{MCH}] + K_1 K_4[\text{MCH}]/[H_2] + K_3[\text{TOL}]} \tag{9-14}$$

The rate of the dehydrogenation reaction follows as

$$(-r)_{\text{MCH}} = k_2[\text{MCH} \cdot S] = \frac{k_2 K_1 \{[S]_{t=0} - ([C] + [W])\}[\text{MCH}]}{1 + K_1[\text{MCH}] + K_1 K_4[\text{MCH}]/[H_2] + K_3[\text{TOL}]} \tag{9-15}$$

Now in order to be in accord with the experimental observation of first-order kinetics with respect to [MCH], for $P_{H_2} > 0$ the three adsorption terms in the denominator must be small with respect to unity, so Eq. (9-15) reduces to

$$(-r)_{\text{MCH}} = k_2 K_1 [S]_{t=0} \left(1 - \frac{[C] + [W]}{[S]_{t=0}}\right)[\text{MCH}] \tag{9-16}$$

The rate of formation of reversible coke is given by

$$d[C]/dt = (-r)_c = k_5[C \cdot S] - k_{-5}[C][H_2] - k_6[C] \tag{9-17}$$

Substituting for $[C \cdot S]$ from Eq. (9-12):

$$(-r)_c = \frac{k_5 K_1 K_4}{[H_2]} \left(1 - \frac{[C] + [W]}{[S]_{t=0}}\right)[S]_{t=0}[\text{MCH}] - k_5[C][H_2] - k_6[C] \tag{9-18}$$

This relationship is reminiscent of a parallel deactivation (type II—reactant) mechanism save for the last two terms. However, again considering time scales, these will be small contributors to $(-r)_c$ over much of the effective life of the catalyst.

In summary, we can see from this example that the mechanisms of coke formation and its inhibition by hydrogen are closely tied together. The general concept of considering various ranking of reactivity of "coke" with respect to hydrogen is valid, and simple two-lump models might form a reasonable starting point of the analysis of deactivation–inhibition in very complicated reaction systems.

A number of other studies for well-defined reaction–catalyst systems similar to the above in which hydrogen inhibition effects have been elucidated include the work of Inami et al. (7) on 1-butene isomerization over Ag–Pd films, Sinfelt and Rohrer (16) for methylcyclohexane decomposition on supported Pt, Pitkethly and Goble (13) for hydrocarbons on Pt, and Myers *et al.* (12) for various hydrocarbons on Pt reforming catalysts.

B. Kinetics of Coke Oxidation

While we are able to describe in an overall way the nature of hydrogen inhibition effects on coke formation in terms of some combination of equilibrium and nonequilibrium steps in a kinetic sequence, the nature of these steps in terms of chemistry and associated rate constants will be highly individual to each catalyst and feedstock combination. Hence, there can be no generality associated with such analyses and the determination of rate constants must be conducted on an individual basis.

This particular state of affairs, however, does not seem to extend universally to the other reaction of coke in which we are primarily interested—that with oxygen. The oxidation of coke is a rapid reaction, for essentially we are talking about a combustion process, and we shall see later that in many practical applications having to do with the regeneration of coked catalysts it is completely diffusion-limited. Intrinsic kinetics are of interest since they establish the lower bound for diffusional limits, on the one hand, and establish attainable rates in certain cases of lower-temperature regeneration of temperature-sensitive materials (some zeolites) on the other. In addition, there is a large literature dealing with the kinetics of combustion of various forms of pure carbon, hence comparison of coke burning with carbon burning may lead to some further insight into the nature of the material itself.

The intrinsic kinetics of carbon burning were reported by Bondi *et al* (3) to be first-order in carbon and first-order in oxygen according to

$$(-r)_c = kPy_{O_2}C_c \tag{9-19}$$

where P is the total pressure, y_{O_2} the mole fraction oxygen, and C_c the weight of coke on catalyst. No detailed data were given to support this correlation; however, it obviously had been distilled from prior experience. However, there are some difficulties with Eq. (9-19) that must be pointed out from the beginning. First, C_c is a gross measure of carbon plus hydrogen that conceivably can vary from one application to the next owing to variation in stoichiometry. More important, though, is the basic assumption of first-order kinetics with respect to carbon. The validity of this is certainly dependent on the amount of coke present initially. Consider the situation depicted in the two parts of Fig. 9-2. In Fig. 9-2a we illustrate multilayer deposition of coke within an individual pore. Oxidation of this material would initially remove an outer layer only to uncover more of the same below. Since the reaction environment remains unchanged insofar as an

I. Reactions of Coke with Hydrogen and Oxygen

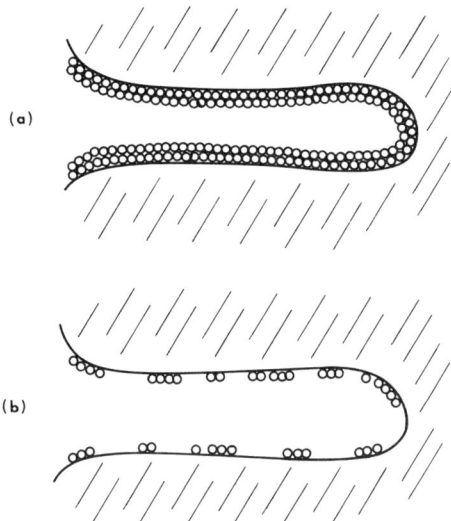

Fig. 9-2. Two limits for coke deposition within a pore. (a) Multilayer coke deposition; (b) monolayer or less coke deposition.

incoming molecule of oxygen is concerned, the oxidation reaction in this regime of C_c must be zero-order in coke and Eq. (9-19) is wrong. On the other hand, when coke concentrations attain the level of monolayer or less, we have the situation shown in Fig. 9-2b; here obviously the reaction environment changes as coke is depleted and a first-order dependence on C_c is reasonable.

Much of the work concerning coke burning kinetics which we shall discuss has been done by those concerned with the regeneration of catalysts used in catalytic cracking. Here typical amounts of coke on deactivated catalysts may be on the order of 5 wt. %, and monolayer or less coverage seems a reasonable assumption. Some analytical investigation of this point was reported early by Haldeman and Botty (6), who measured the dispersion of coke deposits with an electron microscope. They found these deposits to be highly dispersed, which suggested that a large weight fraction of coke could exist on a catalyst in a state such that the entire amount is accessible to the oxygen reactant. For example, the carbon atom in graphite occupies an area of about 4 Å^2, and if this is taken to be the area of a coke unit, then at one monolayer coverage on a catalyst of $250 \text{ m}^2/\text{g}$ area there is 12.5 wt. % carbon. Experimentally, Weisz and Goodwin (19) confirmed the uniform accessibility of coke on a $250 \text{ m}^2/\text{g}$ catalyst up to about 6 wt. % carbon; thus for a given temperature and oxygen partial pressure within this uniformly accessible region:

$$dC_c/d\theta = -kC_c \quad (9\text{-}20)$$

An illustration of typical first-order results for carbon burning on an SiO_2/Al_2O_3 cracking catalyst is given in Fig. 9-3. The initial flatter portion of curve B in the figure is the result of the total surface carbon not being accessible to the oxygen and kinetically represents the transition from zero-order to first-order in C_c. Other results of Weisz and Goodwin concerning intrinsic kinetics are reported in terms of the quantity θ_{85}, the time required for reaction of 85% coke burnoff (weight basis). From Eq. (9-20):

$$\ln(C_c/C_{c_0}) = -k\theta \quad (9\text{-}21)$$

$$\theta_{85} = -(1/k)\ln 0.15 \quad \text{or} \quad k = 1.89/\theta_{85} \quad (9\text{-}22)$$

Hence, intrinsic burning rates in terms of θ_{85} should be independent of initial coke level—the experimental verification is shown in Fig. 9-4a.

Combustion behavior was found to be independent of the structural properties or history of the SiO_2/Al_2O_3 catalysts investigated and independent as well of the source of the coke, including laboratory cracking of LETGO, commercial cracking in a moving-bed unit, and so forth, as shown in Figs. 9-4b and 9-4c. Further, it was concluded that the combustion reaction as observed on silica–alumina was uncatalyzed, by comparison of the intrinsic rate constant on SiO_2/Al_2O_3 with those measured on a variety of other oxides, as shown in Fig. 9-4d. The activation energy corresponding to the lines drawn in Fig. 9-4 is 37.6 ± 1.6 kcal/mol, and the intrinsic burning rate constant in all these cases is

$$k = (4 \times 10^7)\exp[-(37{,}600 \pm 1{,}600)/RT]\,\text{sec}^{-1} \quad (9\text{-}23)$$

The reported activation energy is in good agreement with the value of 34.5 kcal/mol previously reported by Haldeman and Botty. Further comparison of the value of the specific rate constant determined by Weisz and

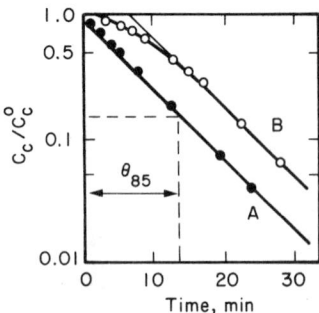

Fig. 9-3. Typical examples of rate plots of carbon remaining versus burning time; SiO_2/Al_2O_3 in air, 538°C. (●) Normal sample, 3% carbon (A); (○) initial flattening due to partial inaccessibility, mix 7–20% carbon (B). [From Weisz and Goodwin (19). Reproduced by permission of Academic Press, Inc.]

I. Reactions of Coke with Hydrogen and Oxygen

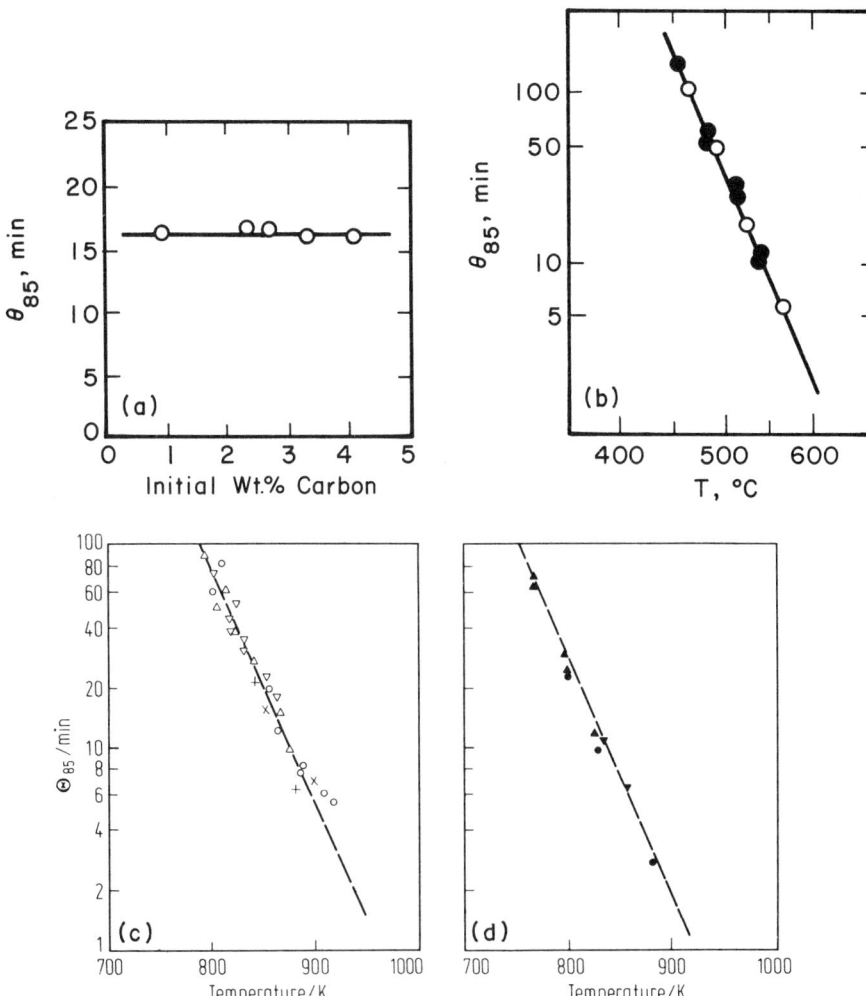

Fig. 9.4. (a) Independence of the intrinsic burning rate from initial coke level for $SiO_2/10\%$ Al_2O_3 at 527°C, 100–500 m²/g. (b) Burning rate constant on different SiO_2/Al_2O_3 catalysts. (○) 1951 SiO_2/Al_2O_3; (●) 1958 Durabead. (c) Burning rate independence of origin of the coke. Symbols represent laboratory cracking of LETGO commercial TCC adiabatic reactor, cumene cracking at 420°C, C_6-C_{10} naphtha cracking at 535°C, propylene at 420°C. (d) Comparison of intrinsic combustion rate constants on noncatalyzing oxide bases. (▲) Filtrol 110; (●) Fuller's earth. [From Weisz and Goodwin (19). Reproduced by permission of Academic Press, Inc.]

Goodwin with that for graphite oxidation determined by Gulbransen and Andrew (5) (for conditions of $Py_{O_2} = 0.21$ atm and 4.1×10^{-16} cm² per graphite atom) gives

$$k \text{ (Weisz and Goodwin)} = 4 \times 10^7 \exp(-37{,}600 \pm 1{,}600/RT) \quad \text{sec}^{-1} \quad (9\text{-}24)$$

$$k \text{ (Gulbransen and Andrew)} = 3 \times 10^7 \exp(-36{,}700/RT) \quad \text{sec}^{-1} \quad (9\text{-}25)$$

This comparison provides additional support for the contention that the coke oxidation is independent of carbon structure and is not catalyzed by the silica–alumina.

A very striking example of the catalysis of coke oxidation, however, was provided by incorporating a transition metal oxide such as chromia into the SiO_2/Al_2O_3. Similar activation energies were observed, but the absolute rate was approximately a factor of four greater under identical conditions on a catalyst containing only 0.15 wt. % Cr_2O_3. When one deals with other types of transition oxide-containing catalysts, such as chromia–alumina, the enhancement of the burning rate is even more pronounced; differences by orders of magnitude from those on conventional SiO_2/Al_2O_3 are to be expected.

The general first-order dependence of burning rate on carbon tested so thoroughly by Weisz and Goodwin appears well established and has been investigated in addition by Massoth (9) and Walker et al (17). Kinetics with respect to oxygen have also been well established as first-order by Walker et al. and corroborated in several subsequent studies, notably that of Massoth.

These results on coke burning kinetics are remarkable for their simplicity; however, a detailed mechanistic explanation of such observations is complicated in particular by the extreme sensitivity of rate to very small amounts of transition metal incorporated into the SiO_2/Al_2O_3 matrix and the similarity of activation energies between the catalyzed and noncatalyzed oxidation. Weisz and Goodwin concluded that the observed temperature coefficient did not reflect the activation energy of a simple kinetic step, but was the result of either a mechanism proceeding via a mobile transition state with an unfavorable free energy of formation or a mechanism producing a product that strongly inhibits the oxidation by removing carbon sites from the reaction. We conclude only by restating that the overall kinetics of coke combustion are comparatively simple, but the mechanism is apparently not.

Another aspect of carbon burning kinetics in regeneration is that a number of workers, including Bondi et al., Massoth, and Weisz and Goodwin, report initial exotherms and rates of reaction which are considerably in excess of those to be expected on the basis of Eq (9-23). Massoth and Menon (10)

I. Reactions of Coke with Hydrogen and Oxygen

in particular investigated the initial regime of coke combustion and associated such transients with the oxidation of strongly adsorbed hydrocarbon species containing a higher ratio of hydrogen to carbon than normal coke (whatever that is). The detailed study by Massoth (9) yielded evidence indicating that the carbon and hydrogen components of coke are oxidized at different rates. Under conditions corresponding to chemical control of the coke oxidation, below 475°C with 2-mm-diameter beads, the data shown in Fig. 9-5a were reported by Massoth. Clearly, the carbon and hydrogen are being oxidized at different rates, and one is therefore involved with two separate, parallel gas–solid reactions which may or may not be coupled. A

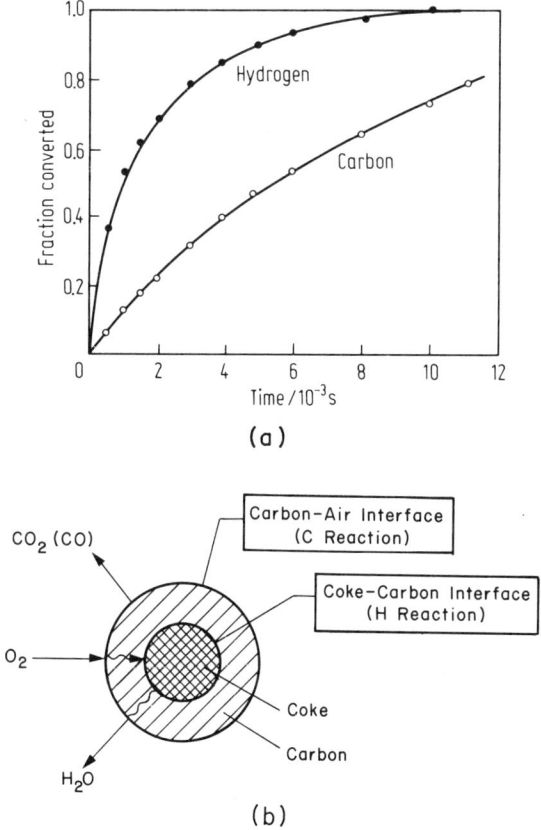

Fig. 9-5. (a) Carbon and hydrogen conversion with time for a sample of 9.9 wt. % coke, 480°C, 0.062 atm O_2, 1.05 atm total. (b) Model for simultaneous carbon and hydrogen reaction. [From Massoth (9). Reproduced by permission of the American Chemical Society.]

conventional chemical rate correlation for the coke oxidation data was obtained; however, the hydrogen did not fit a surface reaction model. The final analysis given proposed a double-core model for the overall oxidation (Fig. 9-5b) in which two interfaces develop as the reaction proceeds, the outer that of the carbon–oxygen and the inner that of hydrogen–oxygen. The hydrogen burning rates are taken to be controlled by oxygen diffusion through the unreacted carbon layer to the inner core, and the carbon burning rates are chemically controlled. Quantitative development of this double-core model resulted in a satisfactory correlation of data on both burning rates and product distribution variations with time. However, the problem has now become one of the analysis of diffusion-coupled reaction processes. In any event, the combustion of carbon is the slow reaction, and for purposes of regeneration analysis or design this complication may be ignored provided due account is taken of the enhanced exotherms involved in the initial stages of oxidation.

II. THE INTRAPARTICLE PROBLEM

A. Coupling of Diffusion and Reaction Rates

We concluded the prior section by stating that a full correlation of intraparticle regeneration should perhaps be approached as a diffusion-coupled process involving both rates of hydrogen and carbon burning. However, most quantitative approaches to the problem have considered the latter only and, indeed, have been successful probably because carbon burning is in most cases rate-determining.

It was shown in early work (15) that intraparticle coke burning was influenced more strongly by the distinction between reaction of carbon on the external surface versus that in the interior than by the distinction between individual chemical species. Thus, one could identify an apparent dual rate regime in which initial burning occurred at a high rate that subsequently decreased to a lower, pseudosteady rate that persisted until almost all the carbon (coke) had been removed. This was interpreted to be the result of initial burning on the external surface of the particles, with subsequent reaction limited by the diffusion of oxygen to the combustion zone located within the particles. A mathematical analysis of this has been given by Ausman and Watson (1) who envisioned external burning of coke (called the "constant rate" period) continuing until all surface material was removed, then a second period ("falling rate" period) commencing in which all reaction was completely within the particle. Two separate analytical formulations of the problem are thus required. For a spherical particle

II. The Intraparticle Problem

during the constant rate period, we have for the oxygen balance:

$$\frac{1}{\xi^2}\frac{d}{d\xi}\left(\xi^2 \frac{d\psi}{d\xi}\right) = N_r^2 \psi \qquad (9\text{-}26)$$

where

$$\xi = r/R_p, \qquad \psi = y_{O_2}/y_{O_2,g}, \qquad N_r^2 = kPR_p^2/D_{O_2}CN$$

in which y_{O_2} and $y_{O_2,g}$ are oxygen mole fractions in the pellet and the bulk gas, respectively, D_{O_2} is the oxygen intraparticle diffusivity, k the carbon burning rate constant, C the total (gas) molal concentration, P the total pressure, R_p the pellet radius, and N the molal ratio of carbon to oxygen in the reaction. The carbon burning kinetics are taken to be dependent only on the oxygen partial pressure, corresponding to more than monolayer coverage of carbon (a reasonable assumption in this regime); thus, for oxygen

$$(-r)_{O_2} = kPy_{O_2}N \qquad (9\text{-}27)$$

Corresponding boundary conditions for this situation are

$$\xi = 1, \qquad d\psi/d\xi = (1/\text{Sh})(1-\psi) \qquad (9\text{-}28)$$

$$\xi = 0, \qquad \psi = \text{finite}$$

Where the Sherwood number in Eq. (9-28) defines the ratio of external to internal mass transport rate coefficients as

$$\text{Sh} = D_{O_2}/RT_g k_g R_p \qquad (9\text{-}29)$$

where R is the gas constant, T_g the bulk gas temperature, and k_g the external mass transfer coefficient. The solution to this problem, in terms of the intraparticle oxygen profiles, is

$$\psi = \frac{1}{\beta_m \xi}\left[\frac{\sinh(N_r \xi)}{\sinh(N_r)}\right] \qquad (9\text{-}30)$$

where

$$\beta_m = 1 + \text{Sh}\left(\frac{N_r}{\tanh N_r} - 1\right)$$

During the constant rate period, the O_2 profile of Eq. (9-30) is time-invariant, so the carbon burning rate at any particular point will be constant (though, of course, it will vary throughout the structure). After θ time of regeneration, the carbon concentration at any point is

$$C_c = C_{c_0} - (kPy_{O_2})(M_c/\rho_s)\theta \qquad (9\text{-}31)$$

in which M_c is the molecular weight of carbon, ρ_s is the catalyst particle density, and C_c and C_{c_0} in the original analysis of Ausman and Watson are in units of pounds of carbon per pound of catalyst. In dimensionless form, one obtains

$$\frac{C_c}{C_{c_0}} = X = 1 - \frac{N_r^2 \tau}{\beta_m \xi}\left[\frac{\sinh(N_r \xi)}{\sinh(N_r)}\right] \qquad (9\text{-}32)$$

where

$$\tau = ND_{O_2}Cy_{O_{2,g}}M_c\theta/\rho_s C_{c_0}R_P^2$$

The fraction of initial carbon remaining in the pellet at τ is

$$X_F = \frac{\int_0^1 N\xi^2\, d\xi}{\int_0^1 \xi^2\, d\xi} = 1 - \left(\frac{3}{\beta_m}\right)\left(\frac{N_r}{\tanh N_r} - 1\right)\tau \qquad (9\text{-}33)$$

and the length of the constant rate period ($X = 0$ at $\xi = 1$ when the constant rate period ends) is from Eq. (9-32)

$$\tau_{CRP} = \beta_m/N_r^2 \qquad (9\text{-}34)$$

One may well ask at this point what these results indicate as far as overall carbon burning rates are concerned; that is, what would the *observed* rate of carbon depletion be during the constant rate period? Assume that such a rate could be correlated by

$$(-r')_c(\text{overall}) = k'_c f(y_{O_2})f(X_F) \qquad (9\text{-}35)$$

where $f(X_F)$ depicts some dependence of rate on average carbon concentration. Now $(dX_F/d\tau)$ can be obtained from Eq. (9-33) as

$$\frac{dX_F}{d\tau} = 1 - \left(\frac{3}{\beta_m}\right)\left(\frac{N_r}{\tanh N_r} - 1\right) \qquad (9\text{-}36)$$

and terms of Eq. (9-35) can be shown by comparison to be

$$f(y_{O_2}) = y_{O_{2,g}}$$
$$f(X_F) = 1$$

$$k'_c = \frac{3ND_{O_2}C}{R}\left\{\frac{[(N_r/\tanh N_r) - 1]}{1 + \text{Sh}[(N_r/\tanh N_r) - 1]}\right\} \qquad (9\text{-}37)$$

N_r in this analysis is a type of Thiele parameter, and the result above shows that the apparent rate constant is strongly dependent on its value. Indeed, Eq. (9-37) is quite suggestive of results reported in later years by McGreavy and Cresswell (11) for the combined effects of inter- and intraphase gradients on reaction effectiveness factors.

II. The Intraparticle Problem

In the falling rate period a more complex situation develops, since as this period progresses, a carbon-free zone develops within the particle where the only process occurring is the diffusion of oxygen to that portion of the particle still containing coke. This is illustrated generally in Fig. 9-6, where for values of $\tau > \tau_{CRP}$ one sees that the carbon profiles fall to zero at radial positions within the external surface of the particle.

In the diffusion zone, $\xi > \xi'$ in the figure, we have

$$\frac{1}{\xi^2}\frac{d}{d\xi}\left(\xi^2 \frac{d\psi}{d\xi}\right) = 0 \tag{9-38}$$

$$\xi = 1, \quad d\psi/d\xi = (1-\psi)/\text{Sh}$$

$$\xi = \xi', \quad \psi = \psi'$$

whence

$$\psi = \psi'\left\{\frac{\text{Sh}+[(1/\xi)-1]}{\text{Sh}+[(1/\xi')-1]}\right\} + \frac{(1/\xi')-(1/\xi)}{\text{Sh}+[(1/\xi')-1]} \tag{9-39}$$

In the diffusion–reaction zone, Eq. (9-26) applies with the first boundary condition of Eq. (9-28) changed to $\psi = \psi'$ at $\xi = \xi'$. This result is

$$\psi = \psi'\left[\frac{\xi' \sinh(N_r\xi)}{\xi \sinh(N_r\xi')}\right]$$

The interfacial oxygen concentration ψ' can be obtained by equating the gradients of ψ and ψ' at ξ':

$$\psi' = \frac{1}{1+[(1-\xi')+\text{Sh }\xi']\{[N_r\xi'/\tanh(N_r\xi')]-1\}}$$

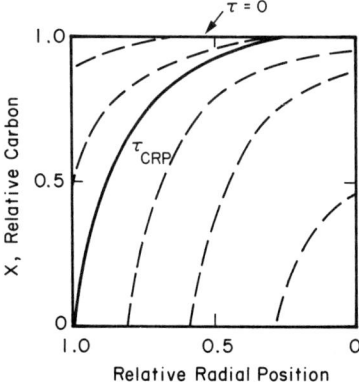

Fig. 9-6. Progression of carbon profiles during diffusion-limited intraparticle regeneration, $N_r = 6.80$. [From Ausman and Watson (1). Reproduced by permission of Pergamon Press, Ltd.]

The carbon concentration at any point in the reaction zone is

$$X = X_{CRP} - N_r^2 \int_{\tau_{CRP}}^{\tau} \psi \, d\tau \qquad (9\text{-}40)$$

with

$$X_{CRP} = 1 - \left(\frac{1}{\xi}\right) \frac{\sinh(N_r \xi)}{\sinh(N_r)}$$

which allows one to compute profiles as shown in Fig. 9-6. If one tries to cast the overall kinetics of the falling rate period into the form of Eq. (9-35), the result is a very messy expression in which $f(y_{O_2})$ is again given by $y_{O_{2,g}}$, but the factor $k_c' f(X_F)$ is a very complicated implicit function of N_r, ξ', Sh, D_{O_2}, N, and C. Perhaps the greatest value of the result is the clear demonstration of just how complex overall burning rates are—in the more general cases not amenable to simple kinetic interpretation.

A similar type of analysis, concerned with differences in burning rates, has been given by Bondi et al (3), who calculated transients in temperature as a function of burning time. Very large exotherms were indicated, and these were used as a basis for interpreting the observed rapid sintering of fresh cracking catalyst on addition to a fluid catalytic cracking (FCC) reactor/regenerator unit.

B. Shell-Progressive Burning

A fortuitous combination of engineering practice and the rates of reaction involved in coke combustion allows in many cases a considerable simplification of the analysis detailed above. Under typical conditions encountered in commercial operation (at least for catalytic cracking) the kinetics of carbon burning are very rapid. The benefit of this is to decouple the mass transfer rate from the reaction rate to the extent that the rate of reaction now becomes a boundary condition on the intraparticle transport of oxygen. The fundamentals of this have already been illustrated in Fig. (3-43) for the intraparticle oxygen concentration profiles in such a case. One sees here the diffusion of oxygen through a previously regenerated external shell to a core containing the carbonaceous deposit, with reaction occurring exclusively at the interface between the core and the regenerated zone. It has been shown (18) that such will indeed be the case when the following criterion is met:

$$(R^2/D_{O_2})(-r)_{O_2}/C_{O_{2,g}} \gg 1 \qquad (9\text{-}41)$$

where $(-r)_{O_2}$ is the observed rate of consumption of oxygen per unit volume of catalyst. If we use typical values of the parameters for amorphous cracking

II. The Intraparticle Problem

catalysts (but not fluid-bed catalysts), then $R \simeq 0.2$ cm, $D_{O_2} \simeq 5 \times 10^{-3}$ cm^2/sec, $C_{O_{2,g}} \simeq 3 \times 10^{-6}$ mol/cm^3, and one has

$$(-r)_{O_2} \gg 4 \times 10^{-7} \text{ mol } O_2/\text{cm}^3\text{-sec} \tag{9-42}$$

In practical applications, the $\gg$ in the above should perhaps be thought of as requiring a number approximately two orders of magnitude greater to ensure the existence of shell-progressive burning. The distinctions in observed $(-r)_{O_2}$ as a function of particle size, contrasting shell-progressive with gradientless burning, are shown in Fig. 9-7.

One recognizes the situation as being entirely analogous to prior discussions of shell-progressive poisoning or coke formation, except now the regenerated zone rather than the poisoned zone is contained in the growing shell. The rate of reaction is just equal to the rate of diffusion of oxygen at the particle surface, so the following simple balance may be written:

$$(-r)_{O_2} = -D_{O_2} \left(\frac{dC_{O_{2,g}}}{dr}\right)_{R_p} 4\pi R^2 = \frac{4\pi R D_{O_2} P}{R y_{O_2}} \left(\frac{d\psi}{d\xi}\right)_{\xi=1} \tag{9-43}$$

and the rate of carbon removal is N times this. The fraction of carbon remaining, X_F, is given by the ratio of the volume of the core containing coke to the volume of the total particle, corresponding to ξ' and ξ in Fig. 3-43; thus:

$$X_F = (\xi')^3 \tag{9-44}$$

The gradient at the surface, $(d\psi/d\xi)_{\xi=1}$, can be computed from the solution of the diffusion without reaction problem treated by Ausman and Watson,

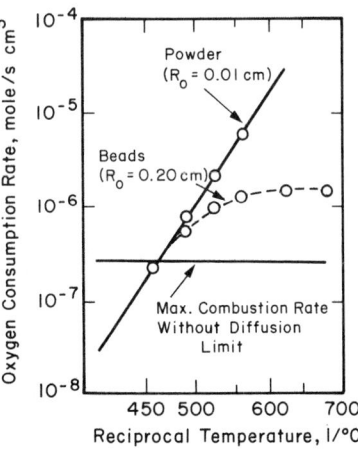

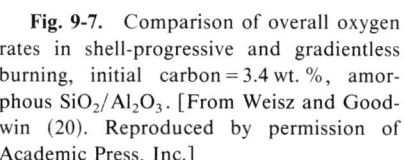

Fig. 9-7. Comparison of overall oxygen rates in shell-progressive and gradientless burning, initial carbon = 3.4 wt. %, amorphous SiO_2/Al_2O_3. [From Weisz and Goodwin (20). Reproduced by permission of Academic Press, Inc.]

and the final result is the implicit expression

$$\tfrac{1}{2}(1 - X_F^{2/3}) - \tfrac{1}{3}(1 - X_F) = \frac{ND_{O_2}C_{O_{2,g}}}{R_p^2 C_{c_0}} \theta \tag{9-45}$$

where θ is chronological time[1]. The above analysis, given by Weisz and Goodwin (20), has been compared by the same authors to a body of experimental data for coke combustion at 975 K on 350 m^2/g SiO$_2$/Al$_2$O$_3$ particles differing in initial coke content and radius under conditions where shell-progressive burning would be expected. Since it is not practical in commercial operation to remove all the coke deposited in the regeneration step, they modified Eq. (9-45) to express θ_{85}, the time required for 85% coke removal, and made comparison with experimental data on that basis [cf. Eq. (9-22)]. Substituting $X_F = 0.15$ in Eq. (9-45) gives

$$\theta_{85} = 0.076 R_p^2 C_{c_0} / ND_{O_2} C_{O_{2,g}} \tag{9-46}$$

where θ_{85} is in minutes. An example of the excellent correlation between this simple theory via the predicted dependences of θ_{85} on R_p^2 and C_{c_0} and experimental results are given in Fig. 9-8. The experimental data shown represent burn-off measurements with commercial cracking catalyst (10 wt. % Al$_2$O$_3$, 0.15 wt. % Cr$_2$O$_3$). For the C_{c_0} functionality, uniform beads 0.4 cm in diameter were coked to initial levels between 1 and 5 wt. % carbon. The dependence on particle size was tested with the same catalyst material coked to a uniform initial level of 3.0 ± 0.2 wt. %. Other functionalities as indicated by the theory were also verified.

The very appealing simplicity of the shell-progressive model invites comparison with the results of more detailed calculations. An extensive discussion of this sort for the results presented by Masamune and Smith (8) has been given in Chapter 7. The conclusions here are similar; parallel mechanism regeneration processes with Thiele moduli on the order of several tens lie within the region where the shell-progressive approximation is satisfactory. As always, however, any such approximation must be used with caution.

Somewhat more conservative applications of the shell-progressive approach have been advocated by Carberry and Gorring (4). Their analysis presumes no single step to be rate-controlling from among boundary layer transport, intraparticle diffusion (through either regenerated or deactivated regions), and reaction in the central core (however, the case of diffusion limitation in the reaction zone is not considered). Pure shell-progressive poisoning or regeneration is claimed to be expected in parallel mechanisms only when the Thiele modulus is greater than about 200; this indeed ensures near-discontinuous profiles of the coke or of the poison; however, the

---

[1] A detailed derivation of the shell-progressive model has been given in Chapter 3.

II. The Intraparticle Problem

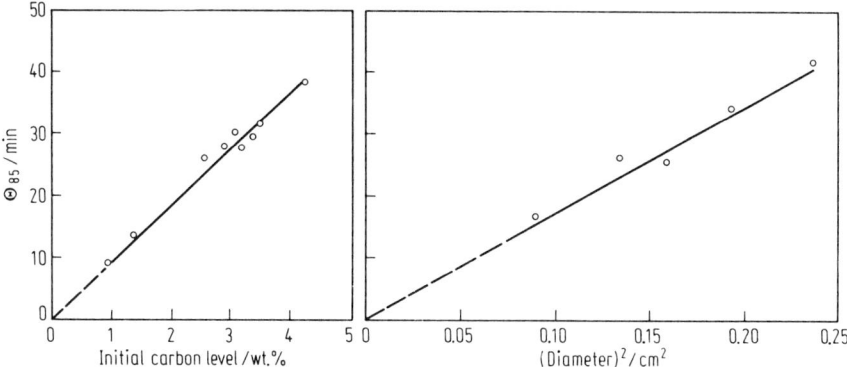

Fig. 9-8. Two tests of the shell-progressive proportionalities suggested by Eq. (9-46). [From Weisz and Goodwin (20). Reproduced by permission of Academic Press, Inc.]

analysis of the results of Masamune and Smith indicates that the existence of a small interface between active and inactive zones of the catalyst does not seriously affect the usefulness of the shell-progressive approximation.

If no single step is rate-controlling, it is most fruitful to approach the problem by defining some characteristic time norms associated with the rates of different processes. For poisoning or coke formation in a catalyst particle of slab geometry let us consider three possibilities: (i) external mass transfer, (ii) shell-progressive diffusion, or (iii) chemical reaction. If any one of the three is rate-controlling, the following characteristic times can be derived (in the order listed above):

$$t_m = \rho L(1 - X_F)/Nk_g C_{O_2} \tag{9-47}$$

$$t_d = \rho L^2(1 - X_F)^2/2NDC_{O_2} \tag{9-48}$$

$$t_c = \rho L(1 - X_F)/NkC_{O_2} \tag{9-49}$$

in which L is the half-thickness, ρ is the saturation concentration of poison or coke, k_g and k are mass transfer and reaction rate constants, and D is an effective diffusivity. In applications to regeneration, X_F would again be the fraction of carbon remaining, N a stoichiometric coefficient, and C_{O_2} the external (gas-phase) oxygen concentration. The overall time-dependent conversion is expressed simply as the sum of the three characteristic times:

$$\theta = t_m + t_d + t_c \tag{9-50}$$

The results above are readily generalized to spherical coordinates and written in nondimensional form, whence

$$\theta = \frac{\rho R_p^2}{NDC_{O_2}} \left[\frac{(1 - X_F)}{3} \left(\frac{1}{Nu} - 1 \right) + \frac{(1 - X_F^{2/3})}{2} + \frac{(1 - X_F^{1/3})}{Da} \right] \tag{9-51}$$

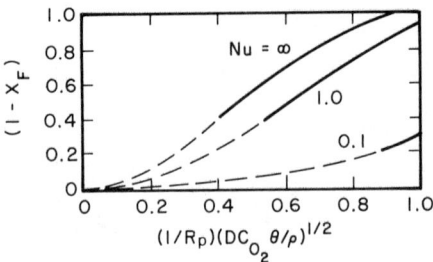

Fig. 9-9. Fraction of spherical particle poisoned or regenerated versus dimensionless time, $Da = \infty$. [From Carberry and Gorring (4). Reproduced by permission of Academic Press, Inc.]

where the Nusselt number is $Nu = (k_g R_p / D)$ and the Damköhler number is $Da = (k R_p / D)$. A graphical representation of some typical results from Eq. (9-51) is given in Fig. (9-9) for the case in which the mechanism is shell-progressive ($Da = \infty$). Note that the relationship between the fraction of particle regenerated, $1 - X_F$ and chronological time θ is linear with respect to $\theta^{1/2}$ over a large portion of the parameter region; hence a time dependence somewhat akin to the Voorhies correlation is a possible approximation even for shell-progressive burning.

Finally, it should be pointed out that the pseudo-steady-state assumption is invoked in all these shell-progressive analyses, where the interface is taken to be stationary and the steady-state diffusion–reaction problem is solved. The validity of this assumption has been examined in detail by Bischoff (2), who showed that the accuracy of the approximation depends on the ratio of reactant concentration in the bulk gas phase to the bulk density of the solid, C_O / ρ_s, being a small number. Since this is the case in most gas–solid systems, the pseudo-steady-state assumption would not appear to constitute a serious limitation to the shell-progressive analysis.

III. SUMMARY AND EVALUATION

It is clear from the discussion of this chapter that when one deals with the scale involved in the regeneration of coked particles the analysis involves a general class of diffusion–reaction problems that are a sort of "deactivation in reverse." Even the dual rate regime identified by Rudershausen and Watson has its analogy in intraparticle deactivation, although the shell-progressive mechanism would appear to be prevalent in most cases of regeneration of practical import. We must remember, however, that dual kinetic regimes for coke burning are not necessarily completely associated with reaction–diffusion interactions. One recalls that coke is a nominal term used to refer to a variety of materials with an empirical formula of about (CH), and evidence has been given that the carbon and hydrogen components may not be oxidized at the same rate.

The unexpected kindness of the analysis of shell-progressive regeneration is, of course, due to the fact that the kinetics of coke (carbon) oxidation are rapid compared to transport rates. Thus, the mass transfer and reaction processes are neatly decoupled. Catalyst ingredients, designed for other means, may also be effective in promoting the coke oxidation reaction; thus generically similar catalysts differing only in the degree of incorporation of some promoter may exhibit vastly different behavior on regeneration. Coke oxidation is obviously a very exothermic process, and the development of thermal waves in the course of oxidative regeneration in fixed beds is a cause of concern, as will be seen in the next chapter. Here the absolute rate of oxidation is of concern, whether or not transport limits are of interest, since the magnitude of the exotherms encountered is dependent on the reaction kinetics (i.e., rate of heat generation).

REFERENCES

1. J. M. Ausman and C. C. Watson, *Chem. Eng. Sci.* **17**, 323 (1962).
2. K. B. Bischoff, *Chem. Eng. Sci.* **18**, 711 (1963).
3. A. Bondi, R. S. Miller, and W. G. Schlaffer, *Ind. Eng. Chem. Process Des. Dev.* **1**, 196 (1962).
4. J. J. Carberry and R. L. Gorring, *J. Catal.* **5**, 529 (1966).
5. E. A. Gulbransen and E. A. Andrew, *Ind. Eng. Chem.* **44**, 1034, 1039 (1952).
6. R. G. Haldeman and M. C. Botty, *J. Phys. Chem.* **63**, 489 (1959).
7. H. Inami, B. J. Wood, and H. Wise, *J. Catal.* **13**, 397 (1969).
8. S. Masamune and J. M. Smith, *AIChE J.* **12**, 384 (1966).
9. F. E. Massoth, *Ind. Eng. Chem. Process Des. Dev.* **6**, 200 (1967).
10. F. E. Massoth and P. G. Menon, *Ind. Eng. Chem. Process Des. Dev.* **8**, 383 (1969).
11. C. McGreavy and D. L. Cresswell, *Chem. Eng. Sci.* **24**, 608 (1969); *Can. J. Chem. Eng.* **47**, 583 (1969).
12. C. G. Myers, W. H. Lang, and P. B. Weisz, *Ind. Eng. Chem.* **53**, 299 (1961).
13. R. C. Pitkethly and G. C. Goble, *Proc. 2nd Int. Congr. Catal.*, Vol. 2, p. 1851 (1960).
14. K. L. Riley, B. G. Silbernagel, and J. B. Butt, *North Am. Meet., 7th, Catal. Soc.*, Boston 1981.
15. C. G. Rudershausen and C. C. Watson, *Chem. Eng. Sci.* **3**, 110 (1954).
16. J. H. Sinfelt and J. C. Rohrer, *J. Phys. Chem.* **66**, 1193 (1962).
17. P. L. Walker, Jr., F. Rusinko, Jr., and L. G. Austin, *Adv. Catal.* **11**, 134 (1959).
18. P. B. Weisz, *Z. Phys. Chem.* **11**, 1 (1957).
19. P. B. Weisz and R. B. Goodwin *J. Catal.* **6**, 227 (1966).
20. P. B. Weisz and R. B. Goodwin, *J. Catal.* **2**, 397 (1963).
21. E. E. Wolf and E. E. Petersen, *J. Catal.* **46**, 190 (1977).

PART III

Deactivation in Chemical Reactors: Global Processes

We have now progressed far enough along the increasing scale of magnitude in our adventures with catalyst deactivation to consider the final level, that of effects on the catalytic reactor in its performance and operation. In Part I our concerns were largely chemical in nature, while in Part II there was a fair interplay between chemical problems and reaction engineering analysis. In Part III the analysis will be largely based in catalytic reaction engineering, with perhaps more mathematics overall, but also with more reliance on numerical simulation than analytical solution. The reader with some grounding in reaction engineering will recognize many familiar formulations such as plug flow reactors, axial dispersion models, and the like. However, as in the intraparticle problems of Part II, the comfort of steady-state conservation equations is displaced by the transients induced by deactivation and the entire analysis becomes one of unsteady-state behavior. The time scale of deactivation is also of signal importance. Here one is concerned primarily with the relative rates associated with residence time within the reactor in comparison to kinetics of the main and deactivation reactions. As one example, it will be seen for fixed beds that in cases where the time scale of deactivation is not long compared to reactor residence times (i.e., not many orders of magnitude greater) there develop sharp gradients in activity, concentration, and temperature that are displaced in both magnitude and position, and the analysis becomes something of a mathematical as well as a reaction engineering challenge. Fluid beds have their problems as well, but they come in a somewhat disguised form because of the differences in mixing behavior.

One particular feature of attempting to treat deactivation/reactor problems is the large number of parameters generally involved. Thus even the simplest analysis of, say, a plug flow fixed bed requires the set of reactor parameters such as wall heat transfer coefficient and bed porosity (perhaps bed effective thermal conductivity if significant thermal effects are present) as well as the complete set of main and deactivation reaction kinetic and selectivity parameters, activation energies, and so forth. Further detail in modeling such as consideration of axial dispersion or intrareactor radial heat transfer appends additional parameters in a hurry. The end result, unless one resorts to relatively simple empirical parametrization, is a formidable system of coupled nonlinear differential equations with perhaps a dozen parameters (always imperfectly known) and with generally different parametric sensitivity to each. It is enough to make one rather rapidly a philosopher of simulation in the sense of "what exactly do I want to know and how much effort is required (or am I willing to expend) to get it?" We have already had a taste of this in the analysis of nonisothermal intraparticle deactivation considered in Chapter 8, and more is forthcoming by way of comparison between Chapters 10 and 12. Most of the material in what follows is devoted to deactivation by poisoning or fouling; we do not have anything against sintering, but the literature has not gone very much in that direction.

CHAPTER 10

Deactivation in Fixed Beds

> Whatsoever Aristotle may have thought on this, I do not care.
>
> *Occam*

Until the past few years, the consequences of catalyst deactivation for fixed-bed reactor performance was largely an academic, theoretical exercise—and even at that, one of relatively recent origin. However, recent years have seen a pronounced increase in experimentation pertaining to fixed-bed performance in the face of deactivation. As we saw earlier, the existence of catalyst deactivation changes an entire class of chemical/catalytic reaction engineering problems from steady-state to unsteady-state ones; nowhere is this more striking than in the analysis of fixed-bed processes. In this chapter we shall follow, more or less in historical chronology, the development of theory and confrontation with experiment, starting with the concept of moving reaction zones and their occurrence in both coking and poisoning processes. We shall then examine some aspects of the influence of deactivation on the dynamics of fixed-bed reactors and the consequences of some particular operational strategies such as constant-conversion operation when decay is occurring. First, however, let us consider the qualitative anatomy of a fixed bed undergoing deactivation.

I. ACTIVITY DISTRIBUTIONS IN FIXED BEDS—A SNAPSHOT

In 1961 Anderson and Whitehouse (1) reported a study of the influence of various internal activity distributions on overall observed performance of fixed beds. They were not concerned with the development of such distributions, the topic we address in the next section, but rather in exploring models in which concentrations and activity profiles were arbitrarily set and performance determined in terms of averages over the bed. Hence we may "look in" on what is going on at various times by considering the averaged behavior resulting from different profiles.

The mechanism of deactivation is considered to be poisoning and two balances are written, one for the distribution of poison within the bed and a second for the relationship between poison concentration and catalyst activity. Various combinations of linear, exponential, and hyperbolic relationships were considered for each of the two. Thus we have for poison distribution:

$$f(C_p/C_p^0) = (1 - b'w), \qquad b'w \le 1$$
$$f(C_p/C_p^0) = \exp(-b'w) \qquad (10\text{-}1)$$
$$f(C_p/C_p^0) = (1 + b'w)^{-1}$$

where w is the length variable, b' is some constant, and $f(C_p/C_p^0)$ is the poison distribution function. Similarly, the relative activity functions considered were

$$s/s_0 = 1 - (\alpha_1/s_0)C_p$$
$$s/s_0 = (1 + s_0\alpha_3 C_p)^{-1} \qquad (10\text{-}2)$$
$$s/s_0 = \exp(-\alpha_2 C_p)$$

in which α_1, α_2, α_3 are appropriate scaling constants and s is the activity function. Characteristics of overall performance can be obtained by suitable averaging:

$$\bar{F} = \frac{1}{W} \int_0^W (s/s_0) \, dw$$

where $\bar{F}$ is the average relative activity of the reactor with respect to initial conditions, and

$$\overline{(C_p/C_p^0)} = \frac{1}{W} \int_0^W f(C_p/C_p^0) \, dw \qquad (10\text{-}3)$$

where $\overline{(C_p/C_p^0)}$ is the average poison concentration in the reactor. For example, if Eq. (10-3) is integrated assuming the exponential relationship for $f(C_p/C_p^0)$ the following result is obtained:

$$\bar{C}_p = [C_p^0(1 - C_p^W/C_p^0)]/\ln(C_p^0/C_p^W) \qquad (10\text{-}4)$$

where C_p^W is the poison concentration at the reactor exit. This in turn will establish an average activity level within the bed and hence dictate the conversion obtained. A typical result is shown in Fig. 10-1a for the combina-

I. Activity Distributions in Fixed Beds—a Snapshot

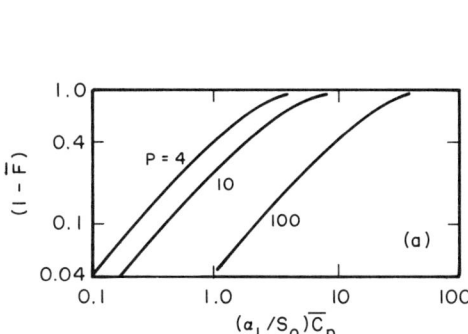
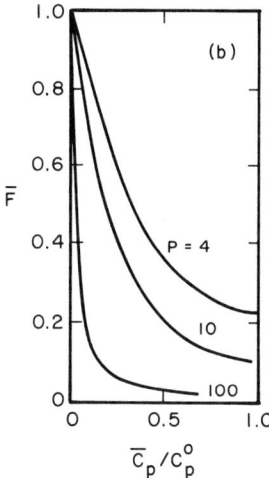

Fig. 10-1. (a) Average activity decay in terms of bed-averaged poison concentration. $s/s_0 = (1 + s_0\alpha_3 C_p)^{-1}$; $f(C_p/C_p^0) = \exp(-b'w)$; $P = (\alpha_1/s_0)C_p^0$. (b) Dependence of activity on poisoning parameters. [From Anderson and Whitehouse (1). Reproduced by permission of the American Chemical Society.]

tion of hyperbolic activity distribution and exponential poison distribution. The parameter P represents different inlet levels of poison concentration, and the abscissa written in terms of $\bar{C}_p$ in effect represents a measure of time on stream. Further detail is provided by Fig. 10-1b, essentially a cross-plot displaying the sensitivity of $\bar{F}$ to this inlet concentration of poison or to the parameters of the particular system as expressed by α_1.

Perhaps the most interesting result of this analysis is not shown in Fig. 10-1, namely that results in terms of average activities or average poison concentration, or the corresponding averaged conversion, tend to look alike for all the combinations investigated. We note from Fig. 10-1a that $(1-\bar{F})$ and $(\alpha_1/s_0)\bar{C}_p$ are linear on log–log coordinates over a very wide range of these variables. This is characteristic of all systems; in addition, the sense of the parametric dependence on P is also characteristic of all systems. Thus all mechanisms of catalyst poisoning, at least in their correlation via Eqs. (10-1) and (10-2), look the same on the basis of overall reactor performance. The snapshots so provided, then, while they are not informative as to the dynamics resulting from deactivation, clearly indicate that the fixed bed is not going to be an optimum device for inferring kinetic data or building kinetic models for deactivation phenomena.

II. MOVING REACTION ZONES—THEORETICAL DEVELOPMENT FOR ISOTHERMAL REACTORS

A. A Plug Flow Model

The analysis of Froment and Bischoff (20) stands as the first to elucidate quantitatively the nature of reactor transients and activity profiles based on solution of reactor continuity equations. The reactor model employed was that of one-dimensional plug flow, isothermal with constant density and molal concentration and with irreversible first-order reaction. Both type II parallel- and series-type deactivation kinetic models were investigated, corresponding to deactivation via coke formation. Allowing for the possibility that time scales of deactivation might be of the same order as the main reaction, we may write for reactant A, mole fraction x:

$$\partial x/\partial \tau + \partial x/\partial z = -(\Omega \rho_B d_p/F) r_A \tag{10-5}$$

in which the nondimensional variables τ and z are defined as

$$\tau = (F/\varepsilon \rho_A \Omega d_p)\theta, \quad z = w/d_p, \quad Z = W/d_p$$

where F is the feed rate in mass/time, ε the void fraction, d_p the catalyst particle diameter, Ω the reactor (total) cross section, ρ_A the mass density of reactant A, ρ_B the bulk density of catalyst, W the reactor length, θ the time, and r_A the rate of disappearance of A. For the rate of deactivation, which will be measured by the rate of accumulation of coke on the catalyst, we have

$$\partial C_c/\partial \tau = (\varepsilon \rho_A \Omega d_p/F) r_c \tag{10-6}$$

where C_c is the carbon content of the catalyst in weight/total weight. By changing variables, Eqs. (10-5) and (10-6) are simplified to

$$\partial x/\partial z = -(\Omega \rho_B d_p/F) r_A, \quad \partial C_c/\partial \eta' = (\Omega \rho_A \varepsilon d_p/F) r_c$$

in which $\eta' = \tau - z$, where η' is the variable along a characteristic of Eq. (10-6).

Now for the mechanism of coke formation via a reactant precursor we may write the following expressions for r_A and r_c:

$$r_A = k'_A Px + k'_{A,d} Px \tag{10-7}$$

$$r_c = k'_{A,d} Px \tag{10-8}$$

and for a product precursor

$$r_A = k'_A Px \tag{10-9}$$

$$r_c = k'_{A,d} P(1-x) \tag{10-10}$$

II. Moving Reaction Zones—Theoretical Developments for Isothermal Reactors

where P is total pressure. We complete building the model for reaction and deactivation kinetics by expressing rate constants for reaction and deactivation indirectly in terms of coke concentration through the activity variable s via

$$k'_A = k'_{A_0} s \quad \text{or} \quad k'_{A,d} = k'_{A,d_0} s \tag{10-11}$$

$$s = \exp(-\alpha_2 C_c), \quad s = 1/(1 + K_c C_c) \tag{10-12}[1]$$

Of the four possible combinations of deactivation mechanism (reactant or product precursor) and activity relationship [Eq. (10-12)], we will illustrate the development for reactant precursor and exponential activity function. Further, $k'_{A,d}$ will be taken to be independent of s. Under these conditions, the conservation equations become

$$\partial x/\partial z = -a[\exp(-\alpha_2 C_c) + \nu]x \tag{10-13}$$

$$\partial C_c/\partial \eta' = bx \tag{10-14}$$

in which

$$a = \frac{\Omega \rho_B d_p P}{F} k'_{A_0}, \quad b = \frac{\Omega \rho_A \varepsilon d_p P}{F} k'_{A,d_0}, \quad \nu = \frac{k'_{A,d_0}}{k'_{A_0}}$$

with the boundary conditions

$$x(0, \eta') = 1 \tag{10-15}$$

$$C_c(z, 0) = 0 \tag{10-16}$$

Equation (10-16) expresses the initial condition of zero coke deposition within the bed. When $\nu \ll 1$, the time scale of deactivation is slow with respect to the main reaction and an analytical solution may be obtained as

$$x = \{1 + \exp(-\alpha_2 b \eta')[\exp(az) - 1]\}^{-1} \tag{10-17}[2]$$

$$\exp(-\alpha_2 C_c) = \{1 + \exp(-az)[\exp(\alpha_2 b \eta') - 1]\}^{-1} \tag{10-18}$$

Figure 10-2 shows the reactant composition and carbon deposition profiles computed from Eqs. (10-17) and (10-18). Note that in general these are nonlinear and time-variant in shape as well as magnitude. Hence sim-

---

[1] It is important to note that this approach implies an explicit relationship between coke level and activity. We spent a considerable amount of effort in Chapter 3 to show that such a relationship is generally not correct. However, if one simply changes coke to foulant in the fixed-bed process the contradiction is avoided and the mathematical treatment is analytically correct. The same point appears in various disguises in a number of other formulations for coke formation in fixed-bed reactors.

[2] Note that $\nu \ll 1$ amounts to a quasi-steady state in which the time derivative, $(\partial x/\partial \tau)$, of Eq. (10-5) may be discarded.

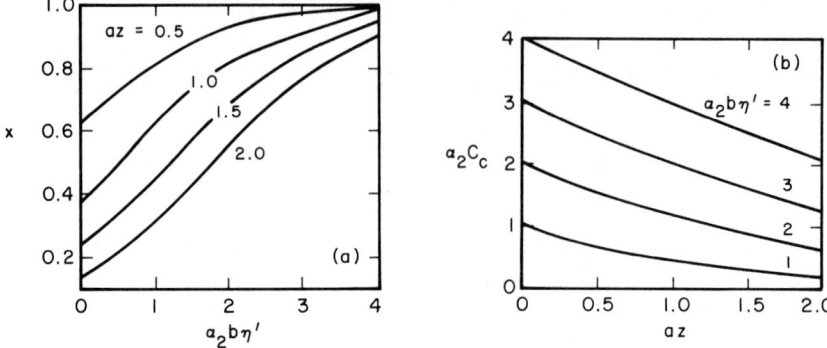

Fig. 10-2. (a) Reactant mole fraction versus time group for parallel reaction mechanism with exponential activity function. (b) Corresponding coke profiles. [From Froment and Bischoff (20). Reproduced by permission of Pergamon Press, Ltd.]

plifications based on assumed profile shapes representing average behavior can mislead. Note also that the coke deposition profiles are decreasing from bed inlet to exit; this is directly the result of the fact that the reactant is the precursor for coke formation so that, all other factors being the same, the maximum rate of coke deposition will occur where the reactant (precursor) is present in greatest concentration, that is, at the bed inlet.

For the hyperbolic activity function with reactant precursor, the conservation equations are

$$\partial x/\partial z = -ax/(1+K_c C_c) \quad (10\text{-}19)$$

$$\partial C_c/\partial \eta' = bx/(1+K_c C_c) \quad (10\text{-}20)$$

where a and b are as defined before except that k'_{A_0} is written as $(k'_{A_0} + k'_{A,d_0})$ in a. These equations may also be solved for the same boundary conditions to give

$$x = \exp[-az + (1+2K_c b\eta')^{1/2} - 1 - K_c C_c] \quad (10\text{-}21)$$

$$K_c C_c \exp[K_c C_c] = [(1+2K_c b\eta')^{1/2} - 1]$$
$$\times \exp[-az + (1+2K_c b\eta')^{1/2} - 1] \quad (10\text{-}22)$$

The reactant composition and coke deposition profiles computed from these relationships are qualitatively very similar in appearance to those of Fig. 10-2 for the exponential case.

Similar solutions can be obtained for the various activity relationships using the product precursor model. Some results for this case are shown in Fig. 10-3, where the parameters are as defined previously and again $k'_{A,d}$ is

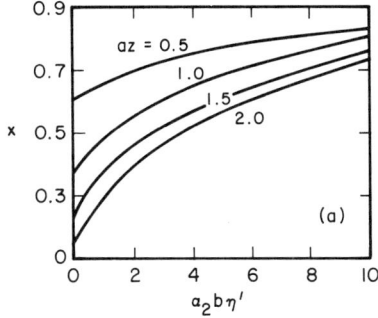 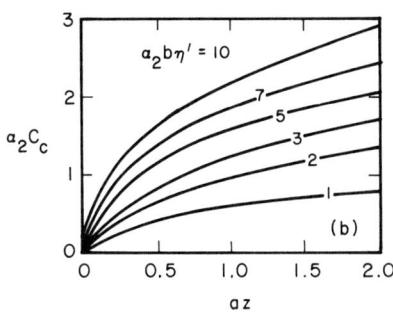

Fig. 10-3. (a) Reactant mole fraction versus time parameter for series mechanism, exponential activity. Curves at constant az. (b) Corresponding carbon profiles. Curves at constant $\alpha_2 b\eta'$. [From Froment and Bischoff (20). Reproduced by permission of Pergamon Press, Ltd.]

taken independent of s. Now the situation with respect to coke deposition is just the opposite, with coke concentration increasing with reactor length. However, we may apply the same reasoning in explanation of this; since no product is in the feed [Eq. (10-15)], no coke can be formed at the bed inlet and greater amounts are formed only as the reaction product concentration increases.

Now the topic of this section is moving reaction zones, and it is not yet apparent from the discussion so far in what way this analysis involves such zones. Let us reexamine the results obtained from Eqs. (10-17) and (10-18) depicted in Fig. 10-2, but now in terms of the rate of the main reaction per unit volume. For the reactant precursor mechanism, where coke concentration and activity decrease with length, it is implied that the locus of maximum reactivity may change as time progresses since the reactor inlet, which is normally the most reactive portion of the bed because of the high concentration of reactant, is preferentially deactivated. The rate of reaction relative to that on the fresh catalyst is

$$r_A/r_A^0 = k_A' Px/k_{A_0}' P = x \exp(-\alpha_2 C_c) \tag{10-23}$$

This can be evaluated as a function of time and distance on substituting for x and C_c from Eqs. (10-17) and (10-18). The rate surface is shown in Fig. 10-4. The development of a maximum in the reaction rate is clearly demonstrated here; the maximum passes along the length of the bed and eventually out the end. One thus has an activity wave within the reactor, and if the operation is not isothermal (as has been assumed here) this activity wave will appear as a thermal one in the form of continuously varying temperature profiles within the bed. This point is discussed in full

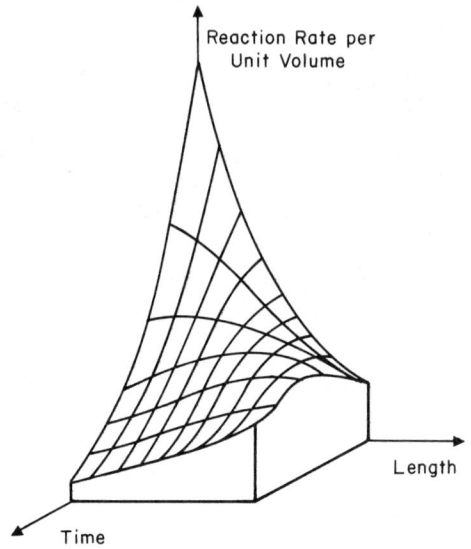

Fig. 10-4. Rate surface for parallel mechanism with exponential activity function. [From Froment and Bischoff (20). Reproduced by permission of Pergamon Press, Ltd.]

detail in the next section. The activity wave shown in Fig. 10-4 is a result of the effects of the opposing trends of reactant concentration profile and carbon concentration profile within the reactor on the rate of deactivation and ultimately on the net rate of the main reaction per unit volume. This behavior would be observed, then, only for the "descending" carbon profile with reactor length.

If we apply the same reasoning to the carbon profiles depicted in Fig. 10-3, we would anticipate a progressive narrowing of the zone of activity progressing from the exit, which is preferentially deactivated in this case, to the inlet. Here the zone of reaction is literally squeezed out of the front end of the bed at sufficiently long times of reaction.

In a simple way we may summarize these effects of moving reaction zones by saying that for the reactant precursor mechanism the bed is deactivated from front to back and for product precursor from back to front. The similarity of such behavior to corresponding situations in individual catalyst particles is striking. It will be remembered that in intraparticle deactivation, reactant precursors led to deactivation from outside to inside, product precursors from inside to outside.

B. A Poison Wave Model

The analysis described above is based on the quasi-steady-state assumption with respect to relative rates of deactivation and the main reaction. A

II. Moving Reaction Zones—Theoretical Developments for Isothermal Reactors

partial consequence of this was that coke profiles and hence activity profiles extend throughout the length of the reactor during most of the process. Now, particularly in some cases of chemical poisoning, as pointed out in Chapter 7, the uptake of poison on the catalyst surface is rapid and essentially irreversible. Here an activity profile develops such that the inlet portion of the bed will be completely deactivated, the exit portion will retain initial activity, and a central portion will display a sharp change from no activity to full activity (s from 0 to 1 in terms of our activity factor). The situation is entirely analogous to breakthrough waves encountered in fixed-bed adsorption or ion exchange, and a theoretical analysis of fixed-bed poisoning can be made on this basis. Wheeler and Robell (39) carried out such an analysis for type I poisoning of an isothermal, irreversible first-order reaction in a plug flow reactor, in which the poison is distributed in the bed according to the fixed-bed adsorption theory of Bohart and Adams (5). Their adsorption model postulates a kinetic model of the following form:

$$\partial C_p / \partial \theta = r_{ads} = k_{ads} C_{p,g}(1 - C_p / C_{p\infty}) \qquad (10\text{-}24)$$

and the conservation equation, presuming isothermality and plug flow with no dispersion, is

$$\partial C_{p,g} / \partial w = r_{ads} \qquad (10\text{-}25)$$

where k_{ads} is the adsorption rate constant and $C_{p,g}$ the concentration of poison in the gas phase. The adequacy of kinetics such as these in practical applications can be viewed as of a semiempirical nature that is capable of reproducing a number of physically reasonable situations via variation of the parameters k_{ads} and $C_{p\infty}$. The solution for a bed with no poison adsorption initially is

$$\frac{C_p}{C_{p\infty}} = \frac{1 + \exp(-N_t \theta / \theta_\infty)}{1 + \exp(-N_t \theta / \theta_\infty)[\exp(N_t w / W) - 1]} \qquad (10\text{-}26)$$

where $N_t \equiv k_{ads} W / v$ is the number of adsorption transfer units in the reactor and θ_∞ is the ratio of the total capacity of the catalyst for poison adsorption to the rate at which poison is introduced into the reactor:

$$\theta_\infty = \rho_B C_{p\infty} W / M v C^0_{p,g} \qquad (10\text{-}27)$$

Here ρ_B is the bulk density of catalyst, M the molecular weight of poison, and $C^0_{p,g}$ the inlet concentration of poison.

Turning to the chemical reaction part of the problem, for the irreversible first-order model we have

$$\ln \frac{C_w}{C_0} = \frac{1}{v} \int_0^W k \, dw \qquad (10\text{-}28)$$

where k is the first-order rate constant, C_0 and C_w are the inlet and outlet concentrations of reactant, v is the superficial velocity, k_0 is the initial rate constant, and W is the reactor length. The rate constant is a function of poison concentration, which in turn is a function of position and time. Again we have an option as to how the value of the rate constant (i.e., activity) is related to poison concentration. In the simplest case we may assume a linear function (nonselective poisoning):

$$k/k_0 = s = 1 - C_p/C_{p\infty} \qquad (10\text{-}29)$$

A generalization of this is the selective poisoning model of Wheeler accounting for diffusional disguise, which we discussed previously. In this case,

$$\frac{k}{k_0} = s = \frac{1}{1 + h_0 C_p/C_{p\infty}} - \frac{C_p/C_{p\infty}}{1 + h_0} \qquad (10\text{-}30)$$

where h_0 is the Thiele modulus for the unpoisoned catalyst and, in the context of Eq. (10-30), is the parameter which determines the degree of selectivity of the poisoning, C_p is the poison concentration, and $C_{p\infty}$ is the saturation poison concentration ($s = 0$ when $C_p = C_{p\infty}$). It is seen that Eq. (10-29) is simply the limiting value of Eq. (10-30) as $h_0 \to 0$.

Equations (10-26) and (10-30) can be combined and the integral of Eq. (10-28) evaluated to give the following general solution for exit conversion as a function of time:

$$\ln\left(\frac{C_w}{C_0}\right) = -\frac{k_0/k_{ads}}{1+h_0}\left\{\ln\left[1 + \exp\left(-N_t\frac{\theta}{\theta_\infty}\right)(\exp N_t - 1)\right]\right.$$
$$\left. + h_0 \ln\left[1 + \frac{\exp(-N_t\theta/\theta_\infty)(\exp N_t - 1)}{1 + h_0[1 - \exp(-N_t\theta/\theta_\infty)]}\right]\right\} \qquad (10\text{-}31)$$

For the nonselective poisoning model of Eq. (10-29):

$$\ln\left(\frac{C_w}{C_0}\right) = -\left(\frac{k_0}{k_{ads}}\right)\ln\left\{1 - \exp\left(-N_t\frac{\theta}{\theta_\infty}\right) + \exp\left[N_t\left(1 - \frac{\theta}{\theta_\infty}\right)\right]\right\} \qquad (10\text{-}32)$$

The solution provided by Eqs. (10-31) or (10-32) can be used to evaluate the parameters θ_∞, N_t, (k_0/k_{ads}), and h_0 from various limiting cases. Convenient limits are at zero time, conversion before breakthrough of poison, conversion at long time ($\theta/\theta_\infty \gg 1$), and conversion at $\theta = \theta_\infty$. Some typical profiles obtained from the Bohart–Adams equation are shown in Fig. 10-5a; it is clear that after an initial period of transient operation a wave or front of poison concentration is established, invariant in shape, which passes through the bed at essentially constant velocity. This wave can be rather diffuse or very sharp, depending on the parametric values, of course. In particular, sharp poisoning fronts are associated with small values of (k_0/k_{ads}), and in such cases one would observe a catastrophic decrease in

II. Moving Reaction Zones—Theoretical Developments for Isothermal Reactors

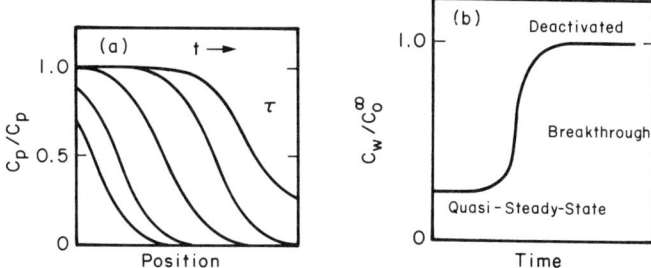

Fig. 10-5. (a) Bohart-Adams generalized poisoning wave profiles. The parameter $\tau = (N_t \theta / \theta_\infty)$. (b) Conversion-time behavior for a very sharp poisoning wave.

conversion over a short period of time associated with breakthrough of poison from the bed. This is shown qualitatively in Fig. 10-5b.

Now the Wheeler-Robell analysis envisions the main reaction to be diffusion-controlled, but not the poisoning reaction. Whether or not this is so is, of course, a question of relative molecular dimensions, but dual diffusion control would seem more typical. The analysis has been extended to diffusion-controlled poisoning reactions by Haynes (23) who used a shell model as an approximation for the poisoning process in a type I system.[3] The Thiele modulus for the poisoning reaction is h_L, and $h_L \gg h_0$ for the shell model assumption to hold. For a spherical catalyst particle the fraction of original activity, s, is related to the radius of the poison-free zone by

$$s = 1 - \xi^3 \tag{10-33}$$

where ξ is the ratio r_1/R of the poison-free zone to the total radius. The effectiveness factor as a function of the degree of poisoning is obtained by mass balances on the poison and poison-free zones as

$$\eta_L = \frac{3}{h_L^2 \xi^2} \left[\frac{h_L \xi \coth h_L \xi - 1}{h_L(1-\xi) \coth h_L \xi + 1} \right] \tag{10-34}$$

The equation for poisoning kinetics and the conservation equation giving the poison distribution in the bed are

$$\frac{\partial (C_{p,g}/C_{p,g}^0)}{\partial z} = \frac{3 N_t (C_{p,g}/C_{p,g}^0)}{h_L^2} \left[\frac{h_L \xi \coth h_L \xi - 1}{h_L(1-\xi) \coth h_L \xi + 1} \right] \tag{10-35}$$

$$\frac{\partial \xi}{\partial \tau} = \frac{(C_{p,g}/C_{p,g}^0)}{\xi^2 h_L^2} \left[\frac{h_L \xi \coth h_L \xi - 1}{h_L(1-\xi) \coth h_L \xi + 1} \right] \tag{10-36}$$

[3] Note the distinction here between "shell model" and "shell-progressive" model. The former admits both diffusion and reaction in the poison-free zone of the catalysts, while the latter implies uniform accessibility of reactant to the poison-free zone.

where N_t is the number of adsorption transfer units as defined by Wheeler and Robell, z the dimensionless reactor length, and τ a dimensionless time defined as $\tau \equiv k_{ads} C_{p,g}^0 t / C_{p\infty}$. Analytical solution in terms of the diffusion-free forms is possible as $h_L \to 0$, but in general numerical solution will be required. If the poison distribution in the bed is determined as a function of time and position from Eqs. (10-35) and (10-36), then the corresponding bed activity is

$$(s)_{bed} = \left[\frac{h_0 \xi \coth h_0 \xi - 1}{h_0(1-\xi) \coth h_0 \xi + 1} \right] \left(\frac{1}{h_0 \coth h_0 - 1} \right) \quad (10\text{-}37)$$

The corresponding conversion relationship for first-order irreversible kinetics is

$$\ln\left(\frac{C_W}{C_0}\right) = -k'\left(\frac{W}{v}\right) \int_0^1 (s)_{bed} \, dz \quad (10\text{-}38)$$

where $k' = k[(3/h_0^2)(h_0 \coth h_0 - 1)]$ is the effective rate constant of the unpoisoned catalyst. Generalized graphical plots of the solutions to Eqs. (10-35)–(10-38) are given in the original reference. The properties of the system that must be known are k and k_{ads}, h_0, and h_L, and the limiting poison capacity, $C_{p\infty}$.

C. Alternative Arithmetic

Some alternative methods for treating the mathematics of moving zones in isothermal fixed-bed deactivation have been reported by Bischoff (3) and by Ozawa (33). [See also Chow *et al.* (12).] Following Bischoff, we recognize that the conservation equations representing the isothermal fixed bed with deactivation are of the general form

$$\partial u / \partial z = -g(v) u \quad (10\text{-}39)$$

$$\partial v / \partial \tau = g(v) u \quad (10\text{-}40)$$

where u and v are dimensionless reactant and poison concentration variables, z is a dimensionless position variable, and τ is a dimensionless process time variable. The function $g(v)$ describes the effect of poison on the reaction rate constants (taken to be the same for both), and the normal boundary conditions are

$$u(0, \tau) = u_0(\tau) \quad (10\text{-}41)$$

$$v(z, 0) = v_0(z) \quad (10\text{-}42)$$

We assume a solution of the form

$$u(z, \tau) = \frac{v(z, \tau) - v_0(z)}{v(0, \tau) - v_0(z)} u_0(\tau) \quad (10\text{-}43)$$

II. Moving Reaction Zones—Theoretical Developments for Isothermal Reactors

substituting Eq. (10-40) in (10-43), we determine that

$$\frac{\partial v(z,\tau)}{\partial \tau} = g(v)\frac{v(z,\tau)-v_0(z)}{v(0,\tau)-v(z)}u_0(\tau) \tag{10-44}$$

For $z=0$ we have from this

$$\partial v(0,\tau)/\partial \tau = g[v(0,\tau)]u_0(\tau) \tag{10-45}$$

Integrating Eq. (10-45) with respect to τ:

$$\int_0^\tau u_0(\tau')\,d\tau' = \int_{v(0,0)}^{v(0,\tau)}\frac{dv'}{g(v')} = \int_{v_0}^{v(0,\tau)}\frac{dv'}{g(v')} \tag{10-46}$$

Now Eq. (10-46) defines a relationship between $v(0,\tau)$ and τ which appears in the limits of both sides of the equation. If we repeat the process with Eq. (10-39) and (10-43) the result obtained is

$$\frac{\partial u(z,\tau)}{\partial z} = -g[v(z,\tau)]\frac{v(z,\tau)-v_0(z)}{v(0,\tau)-v_0(z)}u_0(y) \tag{10-47}$$

which can be integrated, for $v_0(z)=v_0$, to

$$-z = \int_{v(0,\tau)}^{v(z,\tau)}\frac{dv'}{(v'-v_0)g(v')} \tag{10-48}$$

Equation (10-48) gives the relationship between $v(z,\tau)$, z, and $v(0,y)$, so that we may obtain an overall solution to the fixed-bed deactivation via Eqs. (10-43), (10-46), and (10-48). Application of this method is in principle not difficult, although the ease (or difficulty) of integration will certainly be influenced by the nature of the function $g(v)$.

In the procedure suggested by Ozawa we start with the same set of Eqs. (10-39) and (10-40) but use a Legendre transformation to convert this pair into ordinary differential equations. Let the right side of both be represented in general form by the function $f(u,v)$, and define a new function ϕ such that

$$u = \partial \phi/\partial \tau = \phi_\tau \tag{10-49}$$

$$v = \partial \phi/\partial z = \phi_z \tag{10-50}$$

so the original equations become

$$\phi_{z\tau} = f(\phi_z, \phi_\tau) \tag{10-51}$$

If we assume that the deactivation kinetics in $f(u,v)$ are separable and at least one portion of the kinetic expression is linear in the dependent variable, then

$$f(u,v) = g(u)h(v) \tag{10-52}$$

where $g(u)$ is a linear function of u. If we make the transformation

$$\Phi = u\tau + vz - \phi \tag{10-53}$$

so

$$\Phi_v = z, \qquad \Phi_u = \tau \tag{10-54}$$

then we can eventually reduce the original pair of equations to

$$\frac{du}{u} = \frac{dv}{(v - v_0) + [(\partial v_0/\partial z)/h(v_0)]} \tag{10-55}$$

$$dz = \frac{dv}{h(v)[(v - v_0) + (\partial v_0/\partial z)/h(v_0)]} \tag{10-56}$$

and

$$\tau = \Phi_u = \frac{\partial}{\partial u}\left[\int z\, dv + \text{const}\right] \tag{10-57}$$

Equations (10-55) to (10-57) provide an implicit solution to the problem. This method turns out to be particularly convenient to use for type II deactivations and in the analysis of associated selectivity problems. An example of particular interest is that of first-order reaction and first-order deactivation with a catalyst of dual functionality, U and V:

$$\begin{array}{ll}(U) & A + S_1 \rightarrow B + S_1 \\ (V) & A + S_2 \rightarrow C + S_2\end{array} \quad \text{(main reactions)}$$

$$\begin{array}{ll}(U) & A + S_1 \rightarrow A \cdot S_1 \\ (V) & A + S_2 \rightarrow A \cdot S_2\end{array} \quad \text{(fouling reactions)}$$

I

For an isothermal, fixed-bed reactor with plug flow, the conservation equations for this reaction system assume the following nondimensional forms:

$$\partial a/\partial z = -K_{A,1} a s_1 - K_{A,2} a s_2 \tag{10-58}$$

$$\partial s_1/\partial \theta = -K_{d,1} a s_1 \tag{10-59}$$

$$\partial s_2/\partial \theta = -K_{d,2} a s_2 \tag{10-60}$$

where s_1 and s_2 are the activity variables for the U and V functions, respectively, a is the dimensionless reactant concentration $[A]/[A]_0$, $K_{A,1}$ and $K_{A,2}$ are rate constants for the two main reactions, and $K_{d,1}$ and $K_{d,2}$ are rate constants for the deactivations. These rate constants are defined as follows:

$$K_{A,1} = k_{A,1} \tau (1 - \varepsilon), \qquad K_{A,2} = k_{A,2} \tau (1 - \varepsilon) \tag{10-61}$$

$$K_{d,1} = k_{d,1} \tau \rho, \qquad K_{d,2} = K_{d,2} \tau \rho \tag{10-62}$$

II. Moving Reaction Zones—Theoretical Developments for Isothermal Reactors

in which τ is the total residence time in the reactor based on empty cross section, ε the bed void fraction, and ρ the molal volume of the gas feed (assumed not to change with position). As usual, z is the normalized length, w/W, and θ is the time of operation in terms of the number of residence times. Note that the quasi-stationary-state assumption is invoked here via the absence of the time derivative from Eq. (10-58). By combining Eqs. (10-59) and (10-60) we have

$$\frac{ds_1}{ds_2} = \frac{K_{d,1}}{K_{d,2}}\left(\frac{s_1}{s_2}\right) = n\left(\frac{s_1}{s_2}\right) \qquad (10\text{-}63)$$

so

$$s_1 = s_2^n \qquad (10\text{-}64)$$

The three original equations may be reduced to two using this result:

$$\partial a/\partial z = -K_{A,1} a s_2^n - K_{a,2} a s_2 \qquad (10\text{-}65)$$

$$\partial s_2/\partial \theta = -K_{d,2} a s_2 \qquad (10\text{-}66)$$

If we define $r \equiv (K_{A,1}/K_{A,2})$, we may write the pair as

$$\partial a/\partial z = -K_A a s_2 \qquad (10\text{-}67)$$

$$\partial s_2/\partial \theta = -K_{d,2} a s_2 \qquad (10\text{-}68)$$

where $K_A = K_{A,2}(rs_2^{n-1}+1)$. Analytical solution to the problem requires some form of linearization for K_A in Eq. (10-67), in accordance with the requirements expressed in Eq. (10-52). The easiest thing to do is assign a value to s_2 which represents some appropriate average over the time interval considered in solution. Then the integration can be carried out over stepwise time intervals with new values for s_2 and K_A assigned for each interval. Now with K_A constant, Eqs. (10-67) and (10-68) conform precisely to the conditions required for the Legendre transformation. The boundary and initial conditions are

$$\begin{array}{lll} a = 1, & z = 0, & \theta > 0 \\ s_2 = 1, & \theta = 0, & 0 \leq z \leq 1 \end{array} \qquad (10\text{-}69)$$

and the final solutions are

$$s_2 = -\frac{\exp(z)}{1 - \exp(z)[1 + \exp(K_{d,2}\theta/K_A - z)]} \qquad (10\text{-}70)$$

$$s_1 = s_2^n \qquad (10\text{-}71)$$

$$a = s_2 \exp(K_{d,2}\theta/K_A - z) \qquad (10\text{-}72)$$

$$c = \frac{[C]}{[C]_0} = K_{A,2} + \frac{K_{A,2}\exp(K_{d,2}\theta/K_A)}{1-\exp(K_{d,2}\theta/K_A)-\exp(z)} \quad (10\text{-}73)$$

$$\text{Selectivity} = c/a \quad (10\text{-}74)$$

The result for c shown in Eq. (10-73) is actually obtained from the solutions for a and s_2 via

$$dc/dz = K_{A,2} a s_2 \quad (10\text{-}75)$$

$$c = K_{A,2}\exp\left(\frac{K_{d,2}\theta}{K_A}\right)\int_0^z \frac{\exp(z)\,dz}{\{[(1-\exp(K_{d,2}/K_A)]-\exp(z)\}^2} \quad (10\text{-}76)$$

A number of related problems can be treated in this fashion for various combinations of main reaction and deactivation kinetics using the type II kinetic model. Of particular interest is the case in which the deactivation kinetics are integer order greater than unity and independent of conversion; we saw earlier that this is the form of correlation most often attempted with respect to sintering (albeit rather shoddy at times). A second case of interest is that in which the kinetics of the main reaction are zero order, as encountered in the hydrogenation or hydrogenolysis of hydrocarbons in some instances. In these cases analytical solutions can be obtained with no approximations, and a summary of some solutions for type II systems under isothermal conditions is given in Table 10-1.

D. Intraphase and Interphase Resistances

We have seen in the work of Haynes one approach to fixed-bed deactivation in which intraparticle effectiveness is accounted for, but the analysis is restricted to the assumptions of Bohart–Adams adsorption theory. Olson (32) reported a parametric investigation in which both intra- and interphase mass transfer resistance were included in the poisoning kinetic model for an isothermal, plug flow reactor. Shell-progressive deactivation is assumed, and the parameters of importance in determining the overall reactor behavior are mass Biot number, $\text{Bi}_m = (k_g R/D_K)$, the Damköhler number for the poisoning reaction, $\text{Da} = (k'_{A,d} R/D_K)$, and a solid diffusion transfer unit parameter, $N_s = (3 D_K W/R^2 v')$, where k_g is a fluid phase mass transfer coefficient, R the catalyst pellet (spherical) radius, D_K the Knudsen diffusivity of poisoning reactant within the catalyst, W the reactor length, and v' the interstitial linear velocity in the bed. For the reaction model, the effectiveness corresponding to a particle in which the unpoisoned core radius R_p is

$$\eta = 3\psi^{-2}\left[\text{Bi}'_m + \frac{1-R_p}{R_p} + \frac{1}{R_p(R_p\psi\coth(R_p\psi)-1)}\right]^{-1} \quad (10\text{-}77)$$

II. Moving Reaction Zones—Theoretical Developments for Isothermal Reactors

TABLE 10-1

Some Analytical Solutions for Activity and Selectivity, Type II Systems in Isothermal Fixed-Bed Deactivation[a,b]

I. Main reaction zero order, deactivation zero order

$$\partial a/\partial z = -K_{A,1}s_1 - K_{A,2}s_2$$

$$\partial s_1/\partial \theta = -K_{d,1}, \qquad \partial s_2/\partial \theta = -K_{d,2}$$

$$s_1 = 1 - K_{d,1}\theta, \qquad s_2 = 1 - K_{d,2}\theta$$

$$a = 1 - K_{A,1}z(1 - K_{d,1}\theta) - K_{A,2}z(1 - K_{d,2}\theta)$$

$$c = K_{A,2}z(1 - K_{d,2}\theta)$$

II. Main reaction zero order, deactivation first order

$$\partial a/\partial z = -K_{A,1}s_1 - K_{A,2}s_2$$

$$\partial s_1/\partial \theta = -K_{d,1}s_1, \qquad \partial s_2/\partial \theta = -K_{d,2}s_2$$

$$s_1 = \exp(-K_{d,1}\theta), \qquad s_2 = \exp(-K_{d,2}\theta)$$

$$a = 1 - K_{A,1}z \exp(-K_{d,1}\theta) - K_{A,2}z \exp(-K_{d,2}\theta)$$

$$c = K_{A,2}z \exp(-K_{d,2}\theta)$$

III. Main reaction zero order, deactivation second order (sintering)

$$\partial a/\partial z = -K_{A,1}s_1 - K_{A,2}s_2$$

$$\partial s_1/\partial \theta = -K_{d,1}(s_1)^2, \qquad \partial s_2/\partial \theta = -K_{d,2}(s_2)^2$$

$$s_1 = 1/(1 + K_{d,1}\theta), \qquad s_2 = 1/(1 + K_{d,2}\theta)$$

$$a = 1 - K_{A,1}z/(1 + K_{d,1}\theta) - K_{A,2}z/(1 + K_{d,2}\theta)$$

$$c = K_{A,2}z/(1 + K_{d,2}\theta)$$

IV. Main reaction first order, deactivation zero order

$$\partial a/\partial z = -K_{A,1}s_1 a - K_{A,2}s_2 a$$

$$\partial s_1/\partial \theta = -K_{d,1}, \qquad \partial s_2/\partial \theta = -K_{d,2}$$

$$s_1 = 1 - K_{d,1}\theta, \qquad s_2 = 1 - K_{d,2}\theta$$

$$a = \exp(m_1 z)$$

$$c = K_{A,2}(1 - K_{d,2}\theta)[\exp(m_1 z)]/m_1$$

where $m_1 = -K_{A,1}(1 - K_{d,1}\theta) - K_{A,2}(1 - K_{d,2}\theta)$

(continues)

Table 10-1 (*continued*)

V. Main reaction first order, deactivation first order

$$\partial a/\partial z = -K_{A,1}s_1 a - K_{A,2}s_2 a$$
$$\partial s_1/\partial \theta = -K_{d,1}s_1, \qquad \partial s_2/\partial \theta = -K_{d,2}s_2$$
$$s_1 = \exp(-K_{d,1}\theta), \qquad s_2 = \exp(-K_{d,2}\theta)$$
$$a = \exp(m_2 z)$$
$$c = K_{A,2} \exp(-K_{d,2}\theta)[\exp(m_2 z)]/m_2$$
$$m_2 = -K_{A,1}\exp(-K_{d,1}\theta) - K_{A,2}\exp(-K_{d,2}\theta)$$

VI. Main reaction first order, deactivation first order and conversion-dependent

$$\partial a/\partial z = -K_{A,1}s_1 a - K_{A,2}s_2 a$$
$$\partial s_1/\partial \theta = -K_{d,1}s_1 a, \qquad \partial s_2/\partial \theta = -K_{d,2}s_2 a$$
$$s_1 = s_2'', \qquad s_2 = -\frac{\exp(z)}{1-\exp(z)[1+\exp(K_{d,2}\theta/K_A - z)]}$$
$$a = s_2 \exp(K_{d,2}\theta/K_A - z)$$
$$c = K_{A,2} + \frac{K_{A,2}\exp(K_{d,2}\theta/K_A)}{1-\exp(K_{d,2}\theta/K_A)-\exp(z)}$$
$$K_A = K_{A,2}[r(s_2)_{\text{avg}}^{n-1}+1]$$
$$r = K_{A,1}/K_{A,2}, \qquad n = K_{d,1}/K_{d,2}$$

VII. Main reaction first order, deactivation second order (sintering)

$$\partial a/\partial z = -K_{A,1}s_1 a - K_{A,2}s_2 a$$
$$\partial s_1/\partial \theta = -K_{d,1}(s_1)^2, \qquad \partial s_2/\partial \theta = -K_{d,2}(s_2)^2$$
$$s_1 = 1/(1+K_{d,1}\theta), \qquad s_2 = 1/(1+K_{d,2}\theta)$$
$$a = \exp\left(-\frac{K_{A,1}z}{1+K_{d,1}\theta} - \frac{K_{A,2}z}{1+K_{d,2}\theta}\right)$$
$$c = \frac{-K_{A,2}}{1+K_{d,2}\theta}\left(\frac{1}{m_3} - \frac{\exp(m_3 z)}{m_3}\right)$$
$$m_3 = -\left(\frac{K_{A,1}}{1+K_{d,1}\theta} - \frac{K_{A,2}}{1+K_{d,2}\theta}\right)$$

[a] From Butt (6). Reprinted by permission of the American Chemical Society.
[b] Boundary and initial conditions for all cases are given by Eq. (10-69).

where the prime prefers to the Biot number for the catalytic system as distinguished from the poisoning reaction and ψ is a Thiele modulus, $(k'_A R/D_K)^{1/2}$, for the main reaction. The effectiveness factor of Eq. (10-77) can be averaged over the bed length and written in terms of an activity ratio:

$$A(T) = \bar{\eta}/\eta_0 \qquad (10\text{-}78)$$

where η_0 is the effectiveness factor for the unpoisoned bed. Some results of this analysis obtained numerically for typical ranges of parametric values are shown in Fig. 10-6a for fractional times required to saturate the bed completely. The larger values of N_s, reflecting a larger D_K and more efficient penetration of poison into the catalyst, produce rapid bed deactivation at fractional times approaching unity. There is a crossover for the curves, at high ψ, between low and high N_s. Low N_s, meaning slow adsorption of poison, generally improves reactor performance, but if the Thiele modulus is large, the nearly uniformly poisoned reactor is less active than one in which a sharp poison front is present (owing to the nonlinear nature of the relationship between ψ and effectiveness). Thus an *a priori* estimate of

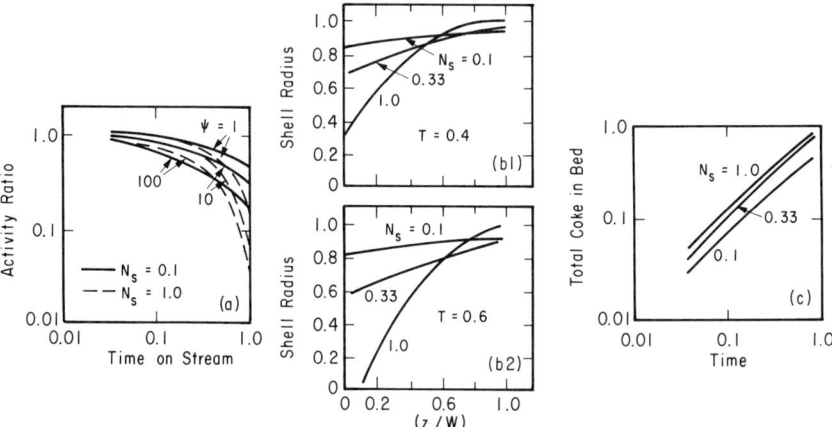

Fig. 10-6. (a) Activity of the bed as a function of time on stream. The value of N_s pertains to the transport rate of poison into the pellet while the Thiele parameter refers to the main reaction. (b1) Radial position of the poisoned shell as a function of axial position for $T = 0.4$. Catalyst at bed entrance is not totally poisoned for any value of N_s, so the constant-bandwidth assumption is not valid. (b2) Similar results for $T = 0.6$. Constant-bandwidth assumption is valid here only for $N_s = 1.0$. (c) Coke buildup. The total amount deposited is well correlated by a Voorhies approach. As N_s increases, the bed tends toward piston-flow deposition. Notes: $\text{Bi}_m = \text{Da} = 20$ in (a) and (c). $T = (v'\theta - w)/W\{\varepsilon a C_A^0/[(1-\varepsilon)q_\infty]\}$, where C_A^0 is the inlet concentration, a the stoichiometric coefficient, moles (solid)/moles (fluid), and q_∞ the equilibrium concentration on solid, moles per cubic centimeter. [From Olson (32). Reproduced by permission of the American Chemical Society.]

relative activity for different cases can be risky if the main reaction is severely diffusion-limited.

In Fig. 10-6b are illustrated poison profiles in the reactor for various values of N_s and times of operation; again note the sensitivity of the results to N_s, here manifested in the sharpness of the poisoning front. The constant pattern of bandwidth analysis, built into the approach of Wheeler and Robell, is obviously not applicable within the time period illustrated, since the activity profile extends throughout the bed.

Finally, Fig. 10-6c shows some averaged coke concentrations over the bed, determined computationally and displayed in the form of a Voorhies correlation. Again, the sensitivity of the results to the diffusion transfer unit parameter N_s is demonstrated. It is apparent from this analysis that there may be instances, as we saw before in the discussion of intraparticle deactivation, where there are advantages to *increasing* the diffusion limitation in a given reaction system, here via a decrease in D_K and hence N_s. One sacrifices some net reactivity; however, by excluding the coke precursor from the catalyst pore structure the net rate of deactivation is decreased and lifetime increased.

E. Some Experimental Implications

The averaging of coke profiles and display in the form of a Voorhies correlation has just been illustrated in Fig. 10-6c. Note that the correlation works with averaged values even in the case where the profiles are very nonlinear and sharp gradients exist within the bed (i.e., larger values of N_s).

This point has been explored in detail for the plug flow model, not only with respect to log-log coke-time correlations but also with respect to rate constant averaging, by Froment and Bischoff. For example, from the coke profiles of Figs. 10-2 and 10-3, we may average according to

$$\overline{\alpha_2 C_c} = \frac{1}{W} \int_0^W \alpha_2 C_c \, dw \qquad (10\text{-}79)$$

for various values of the parameter $\alpha_2 b\eta'$. Such average values, plotted versus time of operation on log coordinates, are

$$\overline{\alpha_2 C_c} = A'(\eta')^n \approx A\theta^n \qquad (10\text{-}80)$$

and are shown in Fig. 10-7. While the slopes of these lines vary according to the combinations of kinetic model and coke-activity relationship, they are all essentially linear over the entire range of parameters considered. Hence, owing to the qualitative similarity of the various combinations, we must conclude that they are experimentally indistinguishable. In other

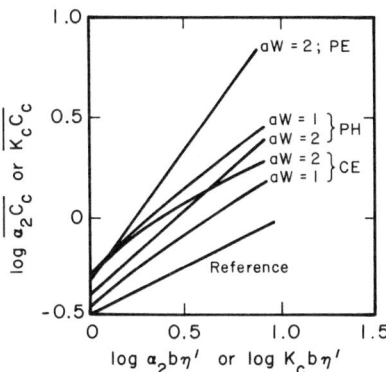

Fig. 10-7. Distance-averaged carbon content versus time groups. PE, Parallel reaction, exponential activity; PH, parallel reaction, hyperbolic activity; CE, consecutive reaction, exponential activity; reference, $\frac{1}{2}\log \eta' + $ constant. [From Froment and Bischoff (20). Reproduced by permission of Pergamon Press, Ltd.]

words, one would not be able to deduce either kinetic models or activity relationships on the basis of Voorhies-type correlations of bed-average coke concentrations. One possible mode of discrimination could be the response to space velocity (aW on the figure). It is seen that increasing space velocity suppresses coke formation somewhat in the case of parallel (reactant precursor) deactivation and enhances it for series (product precursor) deactivation. Whether this might be utilized in practice would, of course, depend on the latitude available for manipulation of this variable.

Averaging effects also present formidable problems when there are activity profiles within a reactor. For integral reactors where rate constants must be derived from conversion data we would have, for the example of a first-order irreversible reaction,

$$\frac{W}{F} = -\frac{1}{k'_A P} \int_{x_0}^{x} \frac{dx}{x} \tag{10-81}$$

Now, if activity is varying with time, it would be the first thought to obtain k'_A at different times and extrapolate these values back to zero time (that is, if activity is not varying too rapidly with time). However, Eq. (10-81) presumes k'_A to be a constant, which it cannot possibly be if there are activity profiles within the reactor, so we actually are measuring an average value:

$$\overline{k'_A} = \frac{1}{W} \int_0^W k'_A(w, \theta) \, dw \tag{10-82}$$

If coking is the means of deactivation, then the averaged rate constant would be related to carbon concentration, also an average value, as

$$\overline{C_c} = \frac{1}{W} \int_0^W C_c(w, \theta) \, dW \tag{10-83}$$

However, the implied relationship, that

$$\overline{k'_A} = k'_A(\overline{C_c}) \tag{10-84}$$

is not true when gradients are present. Let us reexamine the situation where there is reactant precursor coking and an exponential activity function (cf. Fig. 10-2). Now:

$$k'_A = (k'_A)_0 \exp(-\alpha_2 C_c) \tag{10-85}$$

where $(k'_A)_0$ is the bed entrance rate constant. The ratio plotted as a function of position in the reactor is as shown in Fig. 10-8a, and if we average this ratio as per Eq. (10-82) we obtain

$$\frac{\overline{k'_A}}{(k'_A)_0} = \frac{1}{aW}\left[\alpha_2 C_c^W - \alpha_2 C_c^0 + \ln\frac{\exp(\alpha_2 C_c^0) - 1}{\exp(\alpha_2 C_c^W) - 1}\right] \tag{10-86}$$

which is shown in Fig. 10-8b. Finally, the rate constant at the average carbon

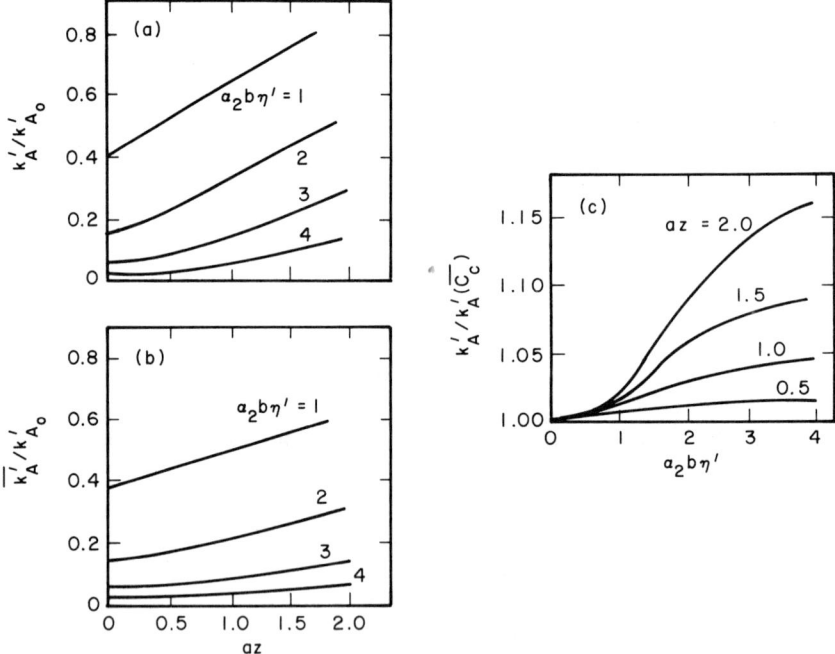

Fig. 10-8. (a) Rate coefficient versus dimensionless position in reactor; parameter = $\alpha_2 b\eta'$. (b) Average rate coefficient versus position, parallel reaction with exponential activity; parameter = $\alpha_2 b\eta'$. (c) Comparison of average rate coefficient with rate coefficient evaluated at average carbon content, parallel reaction with exponential activity; parameter = az. [From Froment and Bischoff (20). Reproduced by permission of Pergamon Press, Ltd.]

III. Moving Reaction Zones—Nonisothermal Reactors

concentration, $k'_A(\overline{C_c})$, can be obtained via combination of Eqs. (10-79) and (10-85) as

$$\frac{\overline{k'_A}}{k'_A(\overline{C_c})} = \frac{\exp(\overline{\alpha_2 C_c})}{aW}\left[\alpha_2 C_c^W - \alpha_2 C_c^0 + \ln\frac{\exp(\alpha_2 C_c^0) - 1}{\exp(\alpha_2 C_c^W) - 1}\right] \quad (10\text{-}87)$$

which is shown in Fig. 10-8c. Any deviation from unity in this plot is a direct measure of the error due to the use of Eq. (10-84). Hence, to obtain reliable kinetics in such cases, we must know the carbon profiles within the reactor or must be able to devise some reliable means for extrapolation of measurements to zero time.

The necessity of always having to work around or to account for activity profiles in fixed beds when conducting kinetic studies obviously implies that such devices are not the best means for investigating deactivation in the laboratory. Indeed, the selection of appropriate laboratory reactors is sufficient to stand as a topic in itself and has been discussed in detail by Weekman (37).

III. MOVING REACTION ZONES—THEORETICAL DEVELOPMENT FOR NONISOTHERMAL REACTORS

A. A Model for Nonisothermal–Nonadiabatic Operation

We may develop models for the nonisothermal reactor undergoing deactivation in much the same fashion as for the isothermal case, appending to the normal conservation equations the expressions for activity and activity change with time. Since we used deactivation by coke formation as the example in most of the previous section, here we will use parallel poisoning as the example.

Experience tells us that in most cases a plug flow model is adequate for mass conservation equations; axial dispersion or mixing cell models may be required if a selectivity function is involved or for short beds. However, in dealing with the energy conservation equation the bed thermal conductivities are normally sufficiently large that a dispersion term is needed. This is particularly so in deactivation problems where steep gradients, moving with time of operation, are encountered. Let us then consider a reaction of the general form A → products with deactivation by poison L. This corresponds formally to the type I scheme discussed in Chapter 4. If we permit the possibility of similar time scales for the main reaction and for deactivation, then a mass balance on the reactant is

$$\varepsilon \frac{\partial C_A}{\partial t} = \varepsilon D_A \frac{\partial^2 C_A}{\partial z^2} - \varepsilon \frac{\partial v' C_A}{\partial z} + s\rho_c(-r_A) \qquad (10\text{-}88)$$

where ε is the bed porosity, D_A an appropriate dispersion coefficient, v' the interstitial velocity, z the length coordinate, s the current activity level ($0 \le s \le 1$), ρ_c the catalyst density, $(-r_A)$ the rate of reaction per weight of catalyst, and C_A the reactant concentration. For the poison we will have a corresponding balance:

$$\varepsilon \frac{\partial C_L}{\partial t} = \varepsilon D_L \frac{\partial^2 C_L}{\partial z^2} - \varepsilon \frac{\partial v' C_L}{\partial z} + \rho_c(-r_L) \qquad (10\text{-}89)$$

in which $(-r_L)$ would denote the rate of chemisorption of poison on the catalyst per unit weight. Now $(-r_L)$ and the rate of deactivation are related via

$$ds/dt = (-r_s) \qquad (10\text{-}90)$$

and

$$(-r_L) = M_T(-r_s) \qquad (10\text{-}91)$$

where M_T is a capacity parameter reflecting the amount of poison required (say weight/weight) to completely deactivate the catalyst ($s = 0$). For the moment we need assume no particular functional form for either $(-r_A)$ or $(-r_s)$.

The energy balance, which for nonisothermal-nonadiabatic operation includes a heat transfer term at the wall, is

$$\frac{\partial T}{\partial t} = \frac{\lambda_{\text{eff}}}{\overline{\rho C_p}} \frac{\partial^2 T}{\partial z^2} - \varepsilon v' \frac{\rho_g C_{pg}}{\overline{\rho C_p}} \frac{\partial T}{\partial z} + \frac{2\alpha}{R_r \overline{\rho C_p}}(T_w - T) + \frac{s(-\Delta H_R)}{\overline{\rho C_p}} \rho_c(-r_A) = 0 \qquad (10\text{-}92)$$

where λ_{eff} is the bed thermal conductivity, α the wall heat transfer coefficient, T_w the wall temperature (assumed constant in this representation of the model), $\overline{\rho C_p}$ an average volumetric heat capacity, R_r the reactor radius, and $(-\Delta H_R)$ the heat of reaction.

In various applications, some considerable simplifications to this set of equations may be possible. Notably, if $(-r_s) \ll (-r_A)$, then all time derivatives may be discarded and ordinarily, as noted above, the dispersion terms in the mass balances for both reactant and poison are not required. Even at that, one has to be impressed at the number of parameters involved in the description of this simple type I system just as far as the reactor model is concerned. We have not yet even included the kinetic parameters involved in the reaction rate model and the deactivation model and still have accumu-

III. Moving Reaction Zones—Nonisothermal Reactors

lated some 13 parameters in the descriptions of Eqs. (10-88) to (10-92); further, note that intraparticle-intraphase gradients are not included.

Those familiar with the literature concerned with modeling of fixed-bed reactors will also realize that there is a rich menu of possible boundary conditions to go along with these equations. The most general of these for our purposes can be written:

$$z = 0, \quad (dY_i/dz)_{z=0} = (v'/D_i)(Y_{i_{z=0}} - Y_{i_0}) \tag{10-93}$$

where Y_{i_0} are feed conditions of C_A, C_L, and T, and

$$z = L, \quad (dY_i/dz)_{z=L} = 0 \tag{10-94}$$

Obviously, the solution methods required for this set of equations and boundary conditions are numerical, and the precise numerical strategy will be dependent on the magnitudes of the various parameters. Normally, it is the thermal parameters which have the most profound influence on the coupling between mass and energy balances—including with these the activation energies for both the main and deactivating reactions.

B. Adiabatic Modification

The expression of the pseudohomogeneous dispersion model for reaction-deactivation under adiabatic conditions involves a trivial modification of the equations of the previous section. Here we set the wall heat transfer coefficient α to zero, Eq. (10-92) assumes the form

$$\frac{\partial T}{\partial t} = \frac{\lambda_{\text{eff}}}{\rho C_p} \frac{\partial^2 T}{\partial z^2} - \varepsilon v' \frac{\rho_g C_{pg}}{\rho C_p} \frac{\partial T}{\partial z} + \frac{s(-\Delta H_B)}{\rho C_p} \rho_c(-r_A) = 0 \tag{10-95}$$

and this expression, together with the mass balances of Eqs. (10-88) and (10-89), kinetic correlations, and boundary conditions, constitutes the adiabatic model. In fact, the absence of the heat transfer term leads to a simpler problem to solve for the adiabatic case as compared to nonisothermal-nonadiabatic operation.

C. Numerical Considerations

Our general concern in this chapter is the concept of moving reaction zones in fixed beds as they are influenced by, or induced by, catalyst deactivation. Nowhere are these more graphically illustrated than by the motion of a hot spot through the bed, or the motion of an adiabatic temperature wave through the bed. In general, these waves will appear as illustrated in Fig. 10-9a if the poisoning is rapid and irreversible and the main reaction is exothermic. Details of the shapes of the waves will vary, of course, dependent on parametric values; however, a general characteristic of this type of poisoning model appears to be a relatively constant shape (independent of time of deactivation) and relatively constant rate of passage through the bed.

As one might expect, the coupled heat and mass balances and menu of boundary conditions, coupled with poisoning and coking kinetics, admit of no simple solution. The essence of these interactions was reported via parametric numerical calculations by Blaum (4) for deactivation in nonisothermal fixed beds based on a two-phase reactor model, using parameters for carbon monoxide on NiO with either parallel or series coke formation. The results presented by Blaum are very detailed, dealing both with rates of deactivation that are high or low compared to the main reaction, and a number of different parametric variations. At this point in our narrative, however, let us just consider one variant on the idealized profiles of Fig. 10-9a. This will refer to the results obtained by Blaum for the case in which there is slow deactivation (rate of decay insufficient to affect the time scale required to establish the steady state in the reactor) and we examine in particular the cases in which the activation energy for deactivation is less than, equal to, or greater than that for the main reaction. Modification of the idealized thermal waves shown in Fig. 10-9a for nonadiabatic operation is demonstrated in Fig. 10-9b in a generalization of the specific results presented by Blaum. The three panels of this figure show progressively the types of thermal waves obtained when the activation energy for deactivation is (i) much smaller than that for the main reaction, (ii) near that for the main reaction, and (iii) somewhat greater than that for the main reaction. When the activation energy for deactivation is zero (or much less than that for the main reaction), obviously the magnitude of the hot spot has no effect on the deactivation process and the thermal wave assumes a fixed shape and passes through the bed at essentially constant velocity. This theoretical result we will see confirmed by experimental data to be presented later. When the activation energy for deactivation is about that for the main reaction (but still less) the hot spot becomes more diffuse and decreases in magnitude with increasing time, as shown in the second part of Fig. 10-9b. The magnitude of this decrease will grow as the activation energies approach each other. Finally, as one can see in the last part of the figure, when the activation energy of the main reaction dominates that of the deactivation there is a severe decrease in the magnitude of the temperature profile with time. In the last two cases, the development of the temperature profile has been shown by Blaum to be associated with the development of minima in the activity profiles, located near (but not at) the entrance to the bed. However, in the first case (deactivation energy ≪ main reaction energy) there appears to be a uniform variation of activity throughout the bed. This is discussed in more detail subsequently.

III. Moving Reaction Zones—Nonisothermal Reactors

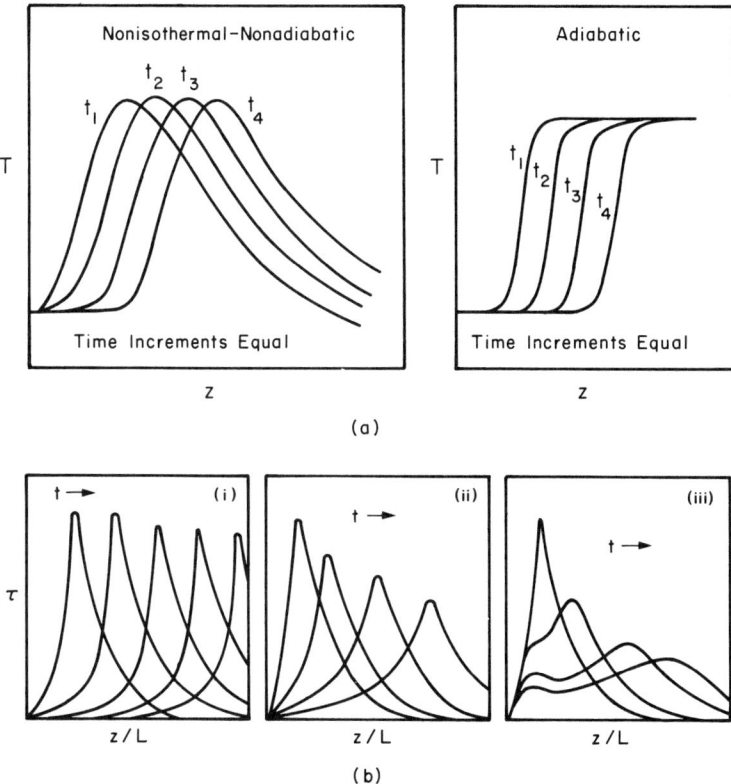

Fig. 10-9. (a) General shapes of thermal waves in fixed beds undergoing rapid poisoning—exothermic reaction. (b) Thermal waves in a nonadiabatic fixed-bed reactor: activation energy for deactivation (i) ≈ 0, (ii) slightly smaller than that for main reaction, and (iii) greater than that for main reaction. [From Blaum (4). Reproduced by permission of Pergamon Press, Ltd.]

From a computational point of view, such steep, moving profiles present a particular challenge if one is to keep computation time and storage requirements within reason. Among the large number of methods which have been proposed, two stand out as having been used extensively and with good success: the classical Crank–Nicolson combined with quasilinearization (30) and the method of orthogonal collocation (19). However, in reaction and reactor modeling for highly exothermic reactions, in which very steep concentration and temperature gradients are commonly encountered, both of these methods have significant drawbacks. The main disadvantage of the Crank–Nicolson technique is the requirement for a large number of equidistant space steps, resulting in a large amount of computer time. Drawbacks of the method of orthogonal collocation result from the fact that the spatial grid points are determined by the zeros of the orthogonal polynomial, which, of course, have nothing to do with the solution profile. This is particularly detrimental in case of unsteady or creeping profiles, where the grid points should be concentrated in the (moving!) region of maximum reaction rates. An increase in the number of grid points (i.e., in the degree of the orthogonal polynomial) again results in a better approximation but at the expense of a considerably increased amount of computer time, since a set of fully occupied matrices must be inverted instead of a set of tridiagonal matrices as in the case of the Crank–Nicolson method.

We would like to be able to place a maximum density of grid points at the place where they are most needed, along the steep portion of the profiles and at the maximum, and move them along as the profile moves along. One way to derive an approximation for the spatial difference quotient is to assume that the profiles of the dependent variables can be piecewise approximated by second-order parabolas. This is shown in Fig. 10-10a, where it is assumed that the differential equation is exactly fulfilled in the circled points. If the solution $y(z)$ is represented piecewise by the depicted parabolas, certain errors are introduced. An approximation for the maximum error in Fig. 10-10a between z_{L-1} and z_L is the difference E between the parabolas at $z' = (z_{L-1} + z_L)/2$. The knowledge of this error provides an easy means for an automatic space step control: If the error is above a specified *minimum* accuracy level, an additional grid point must be inserted at z'. If two subsequent errors are below another specified *maximum* accuracy level, the grid point in between can be omitted in subsequent calculations.

Thus the spatial accuracy will always be kept within a prespecified range and the grid points are positioned in an optimal way (contrary to methods with equidistant grid points, where one is never sure whether the accuracy in zones of steep gradients or maxima is sufficient). The general form of the conservation equations is given by

$$\frac{\partial y}{\partial t} = D_{\text{eff}} \frac{\partial^2 y}{\partial z^2} + v' \frac{\partial y}{\partial z} + S(y, z) \qquad (10\text{-}96)$$

with boundary conditions

$$\left.\frac{\partial y}{\partial z}\right|_{z=0} = \frac{v'}{D_{\text{eff}}} (y_{z=0} - y_0), \qquad \left.\frac{\partial y}{\partial z}\right|_{z=1} = 0 \qquad (10\text{-}97)$$

During the calculation y will be evaluated at certain grid points z_L, $L = 1, 2, \ldots, LM$, as shown in Fig. 10-10b. It turns out that it is useful for the derivation of the spatial difference approximation in a nonequidistant grid to distinguish between the grid points and the elements to which they belong. In Fig. 10-10b it can be seen that the grid point z_L lies almost at the boundary of its (shaded) element, which is defined to extend from $(z_{L-1} + z_L)/2$ to $(z_{L+1} + z_L)/2$. A finite-difference approximation which is

III. Moving Reaction Zones—Nonisothermal Reactors

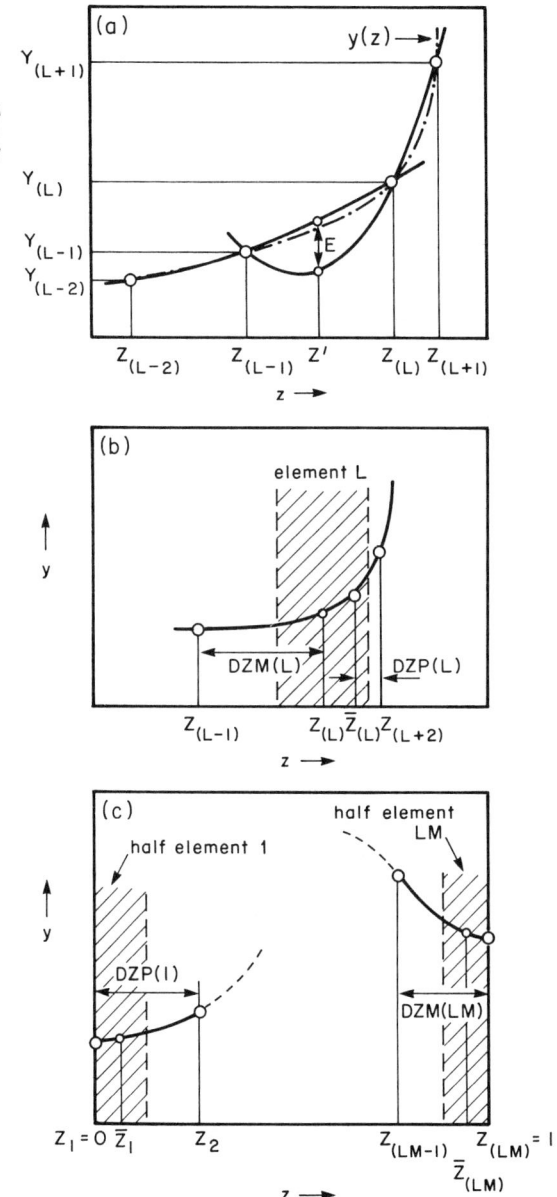

Fig. 10-10. (a) Approximation of a solution profile $y(z)$ (dashed line) by second-order parabolas (solid lines). (b) Grid points and elements in a nonequidistant grid. (c) Half elements at the beginning and end of the space domain. [From Eigenberger and Butt (18). Reproduced by permission of Pergamon Press, Ltd.]

evaluated at $z = z_L$ [as well as the source term $S(y_L, z_L)$] therefore only imperfectly represents the mean value in the element, and it seems more reasonable to evaluate the difference approximation for the middle of the element at

$$z = \bar{z}_L = (z_{L-1} + 2z_L + z_{L+1})/4 \tag{10-98}$$

instead.

The spatial differential quotient for each element can be approximated by placing a second-order parabola through the grid points $(L-1)$, L, $(L+1)$ and evaluating its derivatives at $z = \bar{z}_L$. Thus

$$D_{\text{eff}} \frac{d^2y}{dz^2} - v' \frac{dy}{dz}\bigg|_{z=\bar{z}_L} = C(L,1)y(L-1) + C(L,2)y(L) + C(L,3)y(L+1) \tag{10-99}$$

where

$$C(L,1) = \frac{2D_{\text{eff}} + 0.5v'(DZM(L) + DZP(L))}{DZM(L)(DZM(L) - DZP(L))}$$

$$C(L,2) = -\frac{4D_{\text{eff}} + v'(DZP(L) - DZM(L))}{2DZM(L)DZP(L)} \tag{10-100}$$

$$C(L,3) = -(C(L,1) + C(L,2)), \qquad \text{for } L = 2 \text{ to } LM - 1$$

The approximation is the same for all elements except for the first and the last one. As can be seen in Fig. 10-10c, only half-elements are left at the beginning and the end of the reactor and the difference approximation has to be evaluated in their middle, that is, at $\bar{z}_1 = DZP(1)/4$ and $\bar{z}_{LM} = z_{LM}$. Two values of y at the grid points and the respective boundary condition will be used to calculate the parabolas and to determine the differential quotient at $\bar{z}_1$ and $\bar{z}_{LM}$. Thus

$$D_{\text{eff}} \frac{d^2y}{dz^2} - v' \frac{dy}{dz}\bigg|_{z=z_1} = DE(1) + C(1,2)y(1) + C(1,3)y(2) \tag{10-101}$$

where

$$C(1,2) = -\frac{2D_{\text{eff}}}{DZP(1)^2} - \frac{3}{2}\frac{v'}{DZP(1)} - \frac{(v')^2}{2D_{\text{eff}}}$$

$$C(1,3) = \frac{2D_{\text{eff}}}{DZP(1)^2} - \frac{v'}{2DZP(1)}$$

$$DE(1) = -y_0(C(1,2) + C(1,3))$$

III. Moving Reaction Zones—Nonisothermal Reactors

and

$$D_{\text{eff}} \frac{d^2 y}{dz^2} - v' \frac{dy}{dz}\bigg|_{z=\bar{z}_{LM}} = C(LM, 1)y(LM-1) - C(LM, 2)y(LM) \quad (10\text{-}102)$$

where

$$C(LM, 1) = -C(LM, 2) = \frac{2D_{\text{eff}}}{DZM(LM)^2} + \frac{v'}{2DZM(LM)} \quad (10\text{-}103)$$

The difference approximation at the entrance based on the parabolic representation [Eqs. (10-101) and (10-102)] is limited to diffusivities $D_{\text{eff}} > 0$. For small values of D_{eff} a satisfactory approximation can be obtained using the first method mentioned above, that is, a balance equation for the first element (reactor cross section A):

$$A \frac{DZP(1)}{2} \frac{dy(1)}{dt}$$

$$= Av' \underbrace{\left\{ y_0 - \frac{y(1)+y(2)}{2} \right\}}_{\text{convection}} + \underbrace{D_{\text{eff}} A \frac{y(2)-y(1)}{DZP(1)}}_{\text{diffusion}} + \underbrace{A \frac{DZP(1)}{2} \bar{S}(1)}_{\text{source term}}$$

$$\frac{dy(1)}{dt} = \underbrace{\frac{2v'}{DZP(1)}}_{DE(1)} y_0 - \underbrace{\left(\frac{2D_{\text{eff}}}{DZP(1)^2} + \frac{v'}{DZP(1)} \right)}_{C(1,2)} y(1)$$

$$+ \underbrace{\left(\frac{2D_{\text{eff}}}{DZP(1)^2} - \frac{v'}{DZP(1)} \right)}_{C(1,3)} y(2) + \bar{S}(1) \quad (10\text{-}104)$$

The original diffusion equation and boundary conditions are now reduced to a set of LM ordinary differential equations in time that have the general form

$$dy(L)/dt = C(L, 1)y(L-1) + C(L, 2)y(L)$$
$$+ C(L, 3)y(L+1) + DE(L) + \bar{S}(L) \quad (10\text{-}105)$$

where

$$DE(L) = 0 \quad \text{for } L \neq 1$$

Details of the solution of Eq. (10-105) have been given by Eigenberger and Butt (18) following the Crank-Nicolson technique using the trapezoidal rule with quasi-linearization. Also, some general rules for determining whether grid points should be added, maintained, or deleted are discussed. The general approach is basically the same as the Crank-Nicolson technique with equidistant space steps except that the constants in the coefficient matrix change with position and must be recalculated if the spatial grid changes. The accuracy of the spatial grid must be determined after each time step iteration.

An alternative method (11) has been proposed based on a method of orthogonal collocation on finite elements in which the space domain was split into a number of (nonequidistant) elements, each of which was described by a low-order collocation polynomial. The amount of the residuals could be used as guideline to position the elements optimally. Compared to usual orthogonal collocation, the main advantages are optimal position of elements (grid points) and a resulting block-diagonal matrix instead of a fully occupied matrix. The orthogonal collocation method has also been discussed by Kam and Hughes (25) for the simulation of both series and parallel coking mechanisms (first-order kinetics) in an adiabatic bed with a main reaction of the Langmuir-Hinshelwood form.

In summary, we may say here that the mathematical modeling or simulation of nonisothermal deactivation problems is often quite difficult owing to the presence of steep and spatially variable profiles. While we have attempted to introduce the reader to some aspects of the problem here, we remind ourselves that this book deals with catalyst deactivation and not numerical techniques for the solution of partial differential equations. Hence those who will be concerned with details of the formulation of such simulation models will find it necessary to consult the original papers cited, the books of Lee and Finlayson, and perhaps that of Lapidus (29) for further details concerning the triadiagonalization procedures employed in the development of Crank-Nicolson methods for solution of equations of the form of (10-90). Finite element methods are also useful in these types of problems. Note that entirely analogous methods can be used for intraparticle deactivation problems; the procedures, in fact, are somewhat simpler.

IV. SOME EXPERIMENTAL RESULTS—ISOTHERMAL REACTORS

One of the more prominent void areas in the literature up to about 1970 was the lack of any detailed experimental information concerning the behavior of deactivating fixed beds and hence any good ideas about the value of the theoretical approaches which we have set forth above. In recent years this situation has changed somewhat, and Table 10-2 gives a summary (not comprehensive) of experimental studies involving catalyst deactivation in integral fixed-bed reactors—both isothermal and nonisothermal.

IV. Some Experimental Results—Isothermal Reactors

One early exception to the lack of experimental data on fixed-bed deactivation was the work of Van Zoonen (36) on coke formation in the hydroisomerization of olefins. Experiments were conducted using a silica-alumina nickel sulfide catalyst in the temperature range 300–400°C and hydrogen pressures of about 40 atm. The reaction proceeds by a complex mechanism on the silica–alumina function to yield isoparaffins and diolefins; the latter are hydrogenated by the NiS function back to monoolefins. Coke is formed from the olefins via the diolefin intermediate; hence the coking rate is proportional to the olefin partial pressure. Both the main reaction and the coke-forming reaction were taken to be first order in olefin and inversely proportional to coke concentration. Hence, $(1/C_c)$ is the corresponding form of the activity–coke relationship employed in the Froment-Bischoff analysis. Thus:

$$(-r_A) = k'_A Px/C_c \qquad (10\text{-}106)$$

and

$$(r_c) = k'_{A,d} Px/C_c \qquad (10\text{-}107)$$

where Px represents the olefin partial pressure, k'_A and $k'_{A,d}$ are rate constants for main and coking reactions, respectively, and C_c is the weight of coke on catalyst (weight per weight). Since this is basically a reactant precursor mechanism, one would expect to see coke profiles of the general nature shown in Fig. 10-2b. However, one must be a bit cautious concerning the kinetic model employed here, since $(-r_A)$ is infinite where $C_c = 0$; that is, one would detect no olefin in the gas phase when the catalyst is fresh. A consequence of this is that an observer stationed at the bed exit would observe no olefin in the effluent until the coke front reached the bed exit; hence there is an apparent breakthrough of olefin at the bed exit after an extended period of operation. While the mathematical singularities of the kinetic model are probably not true in the limit, the general characteristics of the behavior outlined above are observed. Figures 10-11a and 10-11b illustrate the olefin breakthrough and coke profiles, respectively, observed in experiments on the hydroisomerization of 1-hexene over a range of pressure, space velocity (S_v), and hydrogen/hydrocarbon ratio. The solid lines shown are best fits to the experimental data involving two parameters:

$$a = k'_{A,d} P_0 / k'_A, \qquad b = (k'_A)^2 / 2 k'_{A,d} P_0^2 \qquad (10\text{-}108)$$

where P_0 is the inlet total pressure. Quite satisfactory agreement is indicated; it is interesting that in this case the breakthrough of olefin at exit would be the quantity one would wish the model to simulate most closely (assuming one did not wish to lose reactant in the product stream), rather than any absolute measure of activity or coke concentration profile.

TABLE 10-2

Catalyst Deactivation—Integral Reactor Studies[a]

Reaction	Catalyst	Conditions	Reaction scheme
i-Pentane dehydrogenation	Cr_2O_3/Al_2O_3 (8/10, 16/32, and 40/60 mesh)	488–560°C, 1 atm, 0.6–3.6 sec R.T., $x_F = 0.14$	(A) $iC_5 \xrightarrow{1}$ (B) $iC_5^= \xrightarrow{3}$ (C) isoprene; $2\downarrow$ other products; $4\searrow$ coke $\nearrow 5$ (E)
n-Butane dehydrogenation	20% $Cr_2O_3/Al_2O_3(\gamma)$, 1% $Cr_2O_3/Al_2O_3(\gamma)$, 20% $Cr_2O_3/Al_2O_3(\alpha)$	550°C, 1 atm, $P_{nC_4} = 0.11$–0.56 atm	$nC_4 \xrightarrow{(k_1)} nC_4^= \xrightarrow{(k_2)}$ coke
n-Butane dehydrogenation	Cr_2O_3/Al_2O_3, 200 m²/g	497–592°C, 1 atm, $P_{nC_4} = 0.32$–0.71 atm, $P_{nC_4} = 0.10$–0.52 atm	$nC_4 \to nC_4^= \xrightarrow{(k_c)}$ coke
Theoretical investigation	—	—	(a) $A \to R$ (b) $A \to R \to S$
Chlorination of tetrachloroethane	SiO_2, 30–40 mesh	200°C	tetra $\xrightarrow{(k_1)}$ penta $\xrightarrow{(k_2)}$ hexa
Dehydration of 2-methyl-3-butene-2-ol to isoprene	Al_2O_3	260–300°C	$A \to R$; $A \to$ coke
1-Butene dehydrogenation	20% Cr_2O_3/Al_2O_3, 57 m²/g	490–600°C[c], $P_{C_4^=} = 0.02$–0.27 atm, $P_{H_2} = 0$–0.10 atm, $P_{diene} = 0$–0.10 atm	$C_4^= \to$ diene; $k_3 \searrow \swarrow k_2$ coke
Catalytic cracking	SiO_2/Al_2O_3, amorphous and zeolite	~900°F	Feed $\xrightarrow{k_1}$ gasoline; $k_3 \searrow \swarrow k_2$ dry gas and coke
Cumene cracking	La-Y zeolite, 20/25–100/140 mesh	360–500°C	
Benzene hydrogenation	Ni/kieselguhr, 12/20 mesh	50–200°C, $x_B = 0.10$–0.25, $x_T = 300$–3000 ppm	$C_6H_6 + 3H_2 \to C_6H_{12}$; $T + S \to T \cdot S$

[a] From Butt and Billimoria (8). Reprinted by permission of the American Chemical Society.
[b] A, H. Noda, S. Tone, and T. Otake, *J. Chem. Eng. Jpn.* **7**, 110 (1974); B, R. Toei, K. Nakanishi, K. Yamada, and M. Okazaki, *J. Chem. Eng. Jpn.* **8**, 131 (1975); C, S. Uchida, S. Osuda, and M. Shindo, *Can. J. Chem. Eng.* **53**, 666 (1975); T. Otake, E. Kunugita, and K. Kugo, *Kogyo Kagaku Zasshi* **68**, 58 (1965); D, A. Sadana and L. K. Doraiswamy, *J. Catal.* **23**, 147 (1971); E, K. B. S. Prasad and L. K. Doraiswamy, *J. Catal.* **32**, 384 (1974); F, G. Greco, Jr., F. Alfani, and F. Gioia, *J. Catal.* **30**, 155 (1973); G, F. J. Dumez and G. F. Froment, *Ind. Eng. Chem. Process Des. Dev.* **15**, 291 (1976); L. H. Hosten and G. F. Froment, *Ind. Eng. Chem. Process Des. Dev.* **10**, 280 (1971); H, V. W. Weekman, Jr., *Ind. Eng. Chem. Process Des. Dev.* **7**, 90 (1968); V. W. Weekman, Jr., *Ind. Eng. Chem. Process Des. Dev.* **8**, 388 (1969); V. W. Weekman, Jr. and D. M. Nace, *AIChE J.* **16**, 397 (1970); D. M. Nace, S. E. Voltz, and V. W. Weekman, Jr., *Ind. Eng. Chem. Process Des. Dev.* **10**, 530 (1971); S. E. Voltz, D. M. Nace, and V. W. Weekman, Jr., *Ind. Eng. Chem. Process Des. Dev.* **10**, 538 (1971); B. Gross, D. M. Nace, and S. E. Voltz, *Ind. Eng. Chem. Process Des. Dev.* **13**, 199 (1974); S. E. Voltz, D. M. Nace, S. M. Jacob, and

IV. Some Experimental Results—Isothermal Reactors

m of deactivation kinetics	Form of main reaction kinetics	Activity versus coke content	Activation energy for coke formation	Intraparticle diffusion effects	References[b]
$(k_4 C_B^p + k_5 C_c^q)$ $= 1$	$r_A = s(k_1 + k_2)C_A$ $r_B = sk_1 C_A - s(k_3 + k_4)C_B^p$	$s = 1 - \alpha C_c$	$4 = 12.6$ kcal/mol $5 = 10.6$ kcal/mol	No, for $<16/32$ mesh	A
$\dfrac{sk_2 P_{nC_4^=}}{1 + K_{C_4} P_{c_4}}$	$r_{C_4} = \dfrac{-sk_1 P_{C_4}}{(1 + K_{C_4} P_{c_4})}$	$s = 1 - \alpha C_c$	—	No	B
$k_c P_{C_4^=}$	$r_{C_4} = (sk_L + k_B) P_{C_4}$ $k_L =$ Lewis acidity $k_3 =$ Brønsted acidity	$s = 1 - \alpha C_c$ (better than exponential)	8.9 kcal/mol	Yes	C
act time ear, exponential, ebraic)	$r_A = sk_A C_A^n$	—	—	No	D
act time r, exponential)	$r_T = -k_1 C_T s$ $r_P = (k_1 c_T - k_2 c_P)s$	—	—	No	E
$_c P_{MBE}$ $_c$	$r_m = -sk_m$	Unity Unity	—	Yes (345°C) No (271°C)	F
$\dfrac{(k_{CB}^0 P_B^{n_1} + k_{CD}^0 P_D^{n_2})s}{(1 + K_{CH} \sqrt{P_H})^2}$	$r_H = \dfrac{-sk_H^0 K_B (P_B - P_H P_D / K)}{(1 + K_B P_B + K_H P_H + K_D P_D)^2}$	$s = e^{-\alpha C_c}$ $(\alpha_H = \alpha_C)$	(a) 32.8 kcal/mol ($C_4^=$) (b) 21.0 kcal/mol (diene)	No (0.4-0.7 mm) G Yes (4 mm, industrial)	G
act time onential)	$r_F = -(k_1 + k_2) s y_F^2$ $r_G = k_1 s y_F^2 - k_2 s y_G$	$s = e^{-\alpha t_c}$ $t_c =$ time	Data given as $f(T)$	Yes	H[d]
on stream rbolic)	$r_c = \dfrac{-k_2 K_1 [s][C]}{K_1[C] + K_3[Y]} \cdots$ $\dfrac{+k_{-2} K_3 [Y][S][Z]}{+K_4[Z]}$	$s = (1 + G t_c)^{-m}$ $(m = 1)$	$360° - G = 0.0806$ $430° - G = 0.0259$ $500° - G = 0.0144$	Yes and no	I[e]
$k_d C_T s$	$-r_B = \dfrac{a \exp(-E_a / RT) C_B a}{1 + b \exp(Q / RT) C_B}$	—	$E_d = 1.1$ kcal/mol	No	J

V. W. Weekman, Jr., *Ind. Eng. Chem. Process Des. Dev.* **11**, 261 (1972); S. M. Jacob, B. Gross, S. E. Voltz, and V. W. Weekman, Jr., *AIChE J.* **22**, 701 (1976); I, B. W. Wojciechowski, *Can. J. Chem. Eng.* **46**, 48 (1968); D. R. Campbell and B. W. Wojciechowski, *J. Catal.* **20**, 217 (1971); D. A. Best and B. W. Wojciechowski, *J. Catal.* **31**, 74 (1973); **47**, 11 (1977); **47**, 343 (1977); B. W. Wojciechowski, *Catal. Rev.—Sci. Eng.* **9**, 79 (1974); T. M. John and B. W. Wojciechowski, *J. Catal.* **37**, 240 (1975); **37**, 348 (1975); R. A. Pachovsky and B. W. Wojciechowski, *J. Catal.* **37**, 120 (1975); R. A. Pachovsky and B. W. Wojciechowski, *Can. J. Chem. Eng.* **53**, 308 (1975); **53**, 659 (1975); J, H.-S. Weng, G. Eigenberger, and J. B. Butt, *Chem. Eng. Sci.*, **30**, 1341 (1975); T. H. Price and J. B. Butt, *Chem. Eng. Sci.* **32**, 393 (1977).

[c] Conditions for hydrogenation kinetics, slightly different conditions employed for coking kinetics.
[d] Results formulated for fixed, moving, and fluid bed reactors; see Chapter 12.
[e] Notation in reaction scheme: C = cumene; Y, Z = products; S = site; CS, etc. = adsorbed species; k_2: CS → YS; k_{-2}: YS → CS; K_1: S⇌CS; K_3: S⇌YS; K_4: S⇌ZS.

Hence, it appears from these data that the simple reactor model and quasi-steady-state assumption provide a reasonable representation of reactor performance for deactivation via coke formation under reasonable process conditions. Subsequent experimental work on coke formation in fixed beds has been carried out in considerable detail and elegance by Froment and co-workers, in particular the studies listed in Table 10-2 for n-butene dehydrogenation and some earlier work by DePauw and Froment (13) concerned with coke formation in isomerization of n-pentane on Pt/Al_2O_3. Since the work of DePauw and Froment has been reviewed in some detail (8), here we will consider in detail the work of Dumez and Froment (16) on the dehydrogenation of 1-butene to butadiene on a commercial Cr_2O_3/Al_2O_3 catalyst, 20 wt.% Cr_2O_3 with a surface area of 57 m^2/g. This reaction proceeds in the range 490–600°C, and a suitable kinetic scheme would indicate both the olefin and the diene to be coke precursors:

$$\begin{array}{c} H\ H\ H\ H \\ |\ \ |\ \ |\ \ | \\ C=C-C-C-H \\ |\ \ \ \ \ |\ \ | \\ H\ \ \ \ H\ H \\ \downarrow \\ \text{coke} \end{array} \rightarrow H_2 + \begin{array}{c} H\ H\ H\ H \\ |\ \ |\ \ |\ \ | \\ C=C-C=C \\ |\ \ \ \ \ \ \ \ \ \ | \\ H\ \ \ \ \ \ \ \ H \\ \downarrow \\ \text{coke} \end{array}$$

The investigation involves separate studies of the kinetics of the main reaction, the kinetics of coke formation, and the incorporation of these results into a reactor model—hence we will have a kind of history in miniature of the development of a reaction–deactivation process model here.

To start with the kinetics of the main reaction, one could envision a rich variety of possibilities for reaction schemes involved in butene dehydrogenation proceeding either by atomic or molecular steps, as follows:

(a) Atomic dehydrogenation; surface recombination of hydrogen
    (1) $B + L \rightleftharpoons BL$ (a1)
    (2) $BL + L \rightleftharpoons ML + HL$ (a2)
    (3) $ML + L \rightleftharpoons DL + HL$ (a3)
    (4) $DL \rightleftharpoons D + L$ (a4)
    (5) $2HL \rightleftharpoons H_2L + L$
    (6) $H_2L \rightleftharpoons H_2 + L$

(where $B = n$-butene, $D =$ butadiene, $H_2 =$ hydrogen, $M =$ an intermediate complex)

(b) Atomic dehydrogenation; gas-phase recombination of hydrogen
    (1) $B + L \rightleftharpoons BL$ (b1)
    (2) $BL + L \rightleftharpoons ML + HL$ (b2)
    (3) $ML + L \rightleftharpoons ML + HL$ (b3)
    (4) $DL \rightleftharpoons D + L$ (b4)
    (5) $2HL \rightleftharpoons H_2 + 2L$

IV. Some Experimental Results—Isothermal Reactors

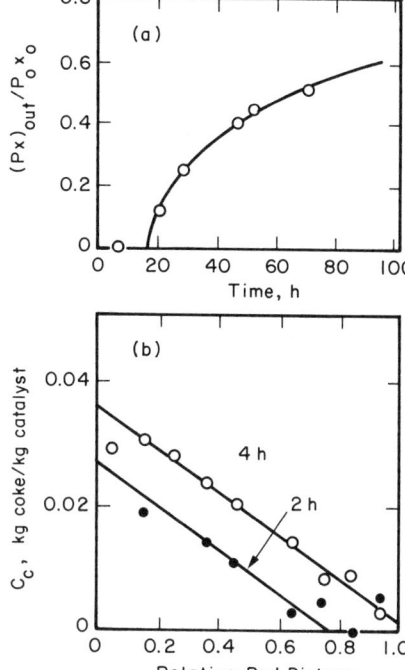

Fig. 10-11. (a) Partial pressure of unconverted n-hexenes at reactor outlet as a function of run time. $P = 46$ atm abs., $H_2/C_6^= = 3$, space velocity $= 3.1$ hr^{-1}. (b) Relation between coke content of catalyst and bed position. 1-hexene, 1 atm, space velocity $= 0.11$ hr^{-1}, $H_2/C_6^= = 4.7$. Parameter $a = 0.045$, $b = 0.24$ (atm-hr)$^{-1}$. [From Van Zoonen (36). Reproduced by permission of North-Holland Physics Publishing.]

(c) Molecular dehydrogenation
    (1) $B + L \rightleftharpoons BL$ (c1)
    (2) $BL + L \rightleftharpoons DL + H_2L$ (c2)
    (3) $DL \rightleftharpoons D + L$ (c3)
    (4) $H_2L \rightleftharpoons H_2 + L$

(d) Atomic dehydrogenation; intermediate complex with short lifetime; surface recombination of hydrogen
    (1) $B + L \rightleftharpoons BL$ (d1)
    (2) $BL + 2L \rightleftharpoons DL + 2HL$ (d2)
    (3) $DL \rightleftharpoons D + L$
    (4) $2HL \rightleftharpoons H_2L + L$
    (5) $H_2L \rightleftharpoons H_2 + L$

(e) Atomic dehydrogenation; intermediate complex with short lifetime; gas-phase hydrogen recombination
    (1) $B + L \rightleftharpoons BL$ (e1)
    (2) $BL + 2L \rightleftharpoons DL + 2HL$ (e2)
    (3) $DL \rightleftharpoons D + L$
    (4) $2HL \rightleftharpoons H_2 + 2L$

TABLE 10-3

Rate Equations for Butene Dehydrogenation[a]

$$r_H^0 = \frac{k_1 C_{tL}(p_B - p_H p_D/K)}{\left(1 + \dfrac{p_H p_D}{K_2 K_3 K_4 K_5 K_6} + \dfrac{p_D\sqrt{p_H}}{K_3 K_4 \sqrt{K_5 K_6}} + \dfrac{p_D}{K_4} + \dfrac{\sqrt{p_H}}{\sqrt{K_5 K_6}} + \dfrac{p_H}{K_6}\right)} \quad (a1)$$

$$r_H^0 = \frac{k_1 K_1 C_{tL}^2(p_B - p_H p_D/K)}{\left(1 + K_1 p_B + \dfrac{p_D\sqrt{p_H}}{K_3 K_4 \sqrt{K_5 K_6}} + \dfrac{p_D}{K_4} + \dfrac{\sqrt{p_H}}{\sqrt{K_5 K_6}} + \dfrac{p_H}{K_6}\right)^2} \quad (a2)$$

$$r_H^0 = \frac{k_1 K_1 K_2 \sqrt{K_5 K_6}\, C_{tL}^2(p_B - p_H p_D/K)}{\sqrt{p_H}\left(1 + K_1 p_B + K_1 K_2 \sqrt{K_5 K_6}\,\dfrac{p_B}{\sqrt{p_H}} + \dfrac{p_D}{K_4} + \dfrac{\sqrt{p_H}}{\sqrt{K_5 K_6}} + \dfrac{p_H}{K_6}\right)^2} \quad (a3)$$

$$r_H^0 = \frac{k_1 K_1 K_2 K_3 K_5 K_6 C_{tL}(p_B - p_A p_D/K)}{\left(p_A + K_1 p_B p_H + K_1 K_2 \sqrt{K_5 K_6}\, p_B\sqrt{p_H} + K_1 K_2 K_3 K_5 K_6 p_B + \dfrac{p_H^{3/2}}{\sqrt{K_5 K_6}} + \dfrac{p_H^2}{K_6}\right)} \quad (a4)$$

$$r_H^0 = \frac{k_1 C_{tL}(p_B - p_H p_D/K)}{\left(1 + \dfrac{p_H p_D}{K_2 K_3 K_4 K_5} + \dfrac{p_D\sqrt{p_H}}{K_3 K_4 \sqrt{K_5}} + \dfrac{p_D}{K_4} + \dfrac{\sqrt{p_H}}{\sqrt{K_5}}\right)} \quad (b1)$$

$$r_H^0 = \frac{k_1 K_1 C_{tL}^2(p_B - p_H p_D/K)}{\left(1 + K_B p_B + \dfrac{p_D\sqrt{p_H}}{K_3 K_4 \sqrt{K_5}} + \dfrac{p_D}{K_4} + \dfrac{\sqrt{p_H}}{\sqrt{K_5}}\right)^2} \quad (b2)$$

$$r_H^0 = \frac{k_1 K_1 K_2 \sqrt{K_5}\, C_{tL}^2(p_B - p_H p_D/K)}{\sqrt{p_H}\left(1 + K_B p_B + \dfrac{p_D}{K_4} + \dfrac{\sqrt{p_H}}{\sqrt{K_5}} + K_1 K_2 \sqrt{K_5}\,\dfrac{p_B}{\sqrt{p_H}}\right)^2} \quad (b3)$$

$$r_H^0 = \frac{k_1 K_1 K_2 K_3 K_5 C_{tL}(p_B - p_H p_D/K)}{\left(p_H + K_1 p_B p_H + K_1 K_2 K_3 K_5 p_B + K_1 K_2 \sqrt{K_5}\, p_B\sqrt{p_H} + \dfrac{p_H^{3/2}}{\sqrt{K_5}}\right)} \quad (b4)$$

$$r_H^0 = \frac{k_1 C_{tL}(p_B - p_H p_D/K)}{\left(1 + \dfrac{p_D p_H}{K_2 K_3 K_4} + \dfrac{p_D}{K_3} + \dfrac{p_H}{K_4}\right)} \quad (c1)$$

$$r_H^0 = \frac{k_1 K_1 C_{tL}^2(p_B - p_H p_D/K)}{\left(1 + K_1 p_B + \dfrac{p_D}{K_3} + \dfrac{p_H}{K_4}\right)^2} \quad (c2)$$

$$r_H^0 = \frac{k_1 K_1 K_2 K_4 C_{tL}(p_B - p_H p_D/K)}{\left(p_H + K_1 p_B p_H + K_1 K_2 K_4 p_B + \dfrac{p_H^2}{K_4}\right)} \quad (c3)$$

$$r_H^0 = \frac{k_1 C_{tL}(p_B - p_H p_D/K)}{\left(1 + \dfrac{p_B p_D}{K_2 K_3 K_4 K_5} + \dfrac{p_D}{K_3} + \dfrac{\sqrt{p_H}}{\sqrt{K_4 K_5}} + \dfrac{p_H}{K_5}\right)} \quad (d1)$$

IV. Some Experimental Results—Isothermal Reactors

$$r_H^0 = \frac{k_1 K_1 C_{tL}^2 (p_B - p_H p_D / K)}{\left(1 + K_1 p_B + \dfrac{p_D}{K_3} + \dfrac{\sqrt{p_H}}{\sqrt{K_4 K_5}} + \dfrac{p_H}{K_5}\right)^2} \qquad (d2)$$

$$r_H^0 = \frac{k_1 C_{tL}(p_B - p_H p_D / K)}{\left(1 + \dfrac{p_H p_D}{K_2 K_3 K_4} + \dfrac{p_D}{K_3} + \dfrac{\sqrt{p_H}}{\sqrt{K_4}}\right)} \qquad (e1)$$

$$r_H^0 = \frac{k_1 K_1 C_{tL}^2 (p_B - p_H p_D / K)}{\left(1 + K_1 p_B + \dfrac{p_D}{K_3} + \dfrac{\sqrt{p_H}}{\sqrt{K_4}}\right)^2} \qquad (e2)$$

[a] From Dumez and Froment (16). Reprinted by permission of the American Chemical Society.

Of course, there will be a corresponding number of descriptive rate equations based on LHHW analysis, for various postulates of reaction schemes and rate-determining steps. Table 10-3 shows this menu for such combinations; for example, the entry listed as (a2) is the rate equation corresponding to scheme a with step a2 rate-limiting. Hence, our initial task is to both discriminate among models and determine appropriate parameters.

The kinetic experiments were carried out in a differential reactor, with on-stream times up to 30 min. Since coking was occurring during the sequence of kinetic experiments, initial rates were determined by back-extrapolation from initial data points obtained after about 2 min. Dumez and Froment described and utilized a sequential experimental design procedure which provides a maximum divergence between rates measured in one experiment and the next, based on the performance index

$$D = \sum_{i=1}^{m} \sum_{\substack{j=1 \\ j \neq i}}^{m} |r_{Hi}^0 - r_{Hj}^0| \qquad (10\text{-}109)$$

where r_H^0 is the initial rate (extrapolated), i the model prediction, and j the experimental result[4]. Fourteen experiments were required to discriminate among models, with the winner (c2) of Table 10-3:

$$r_H^0 = \frac{k_H^0 K_B (P_B - P_H P_D / K)}{(1 + K_B P_B + K_H P_H + K_D P_D)^2} \qquad (10\text{-}110)$$

notation as detailed above, with $K_B = K_1$, $K_H = K_3^{-1}$, and $K_D = K_4^{-1}$. A good

---

[4] Details of the statistical procedure are beyond the scope of our present interest. However, the interested reader should consult Froment and Mezaki (21).

Arrhenius fit was obtained for $k_H^0 = A_H^0 \exp(-E_H/RT)$, and the parametric fit values obtained were $A_H^0 = 0.27$ kmol/kg cat-hr-atm, $E_H = 29{,}236$ kcal/kmol, $K_B = 1.73$ atm^{-1}, $K_H = 3.60$ atm^{-1}, and $K_D = 38.02$ atm^{-1}. Values for the adsorption coefficients are reported for experiments at 542°C and their temperature variation is not reported.

Turning now to the kinetics of coke formation, we recall that it is necessary to relate both coke to time on stream *and* activity to amount of coke. Some typical data on coke deposition kinetics are shown in Fig. 10-12. Here we see the influence of deactivation on the rate of coke formation directly, so one could use these data straightforwardly to form correlations of the sort:

$$\frac{dC_c}{dt} = r_c = r_c^0 \phi_c \tag{10-111}$$

where C_c is coke (weight/weight), r_c^0 the initial rate of coke deposition, and ϕ_c an activity variable corresponding to the s we described before. Again, we have the choice of a number of empirical forms, which must be integrated with respect to time and fit to the data of Fig. 10-12:

$$\phi_c = \exp(-\alpha C_c) \to \int (\ ) \, dt \to C_c = \frac{1}{\alpha} \ln(1 + \alpha r_c^0 t) \tag{10-112a}$$

$$\phi_c = 1 - \alpha C_c \to \int (\ ) \, dt \to C_c = \frac{1}{\alpha}[1 - \exp(-\alpha r_c^0 t)] \tag{10-112b}$$

$$\phi_c = (1 - \alpha C_c)^{-1} \to \int (\ ) \, dt \to C_c = \frac{1}{\alpha}[(2\alpha r_c^0 t + 1)^{1/c} - 1] \tag{10-112c}$$

and so on. Best fit of these possible forms to r_c versus (t) was obtained via the exponential model, Eq. (10-112a). Further variation of experimental parameters to determine a model for r_c^0 gave

$$r_c^0 = \frac{k_{cB}^0 P_B^{n_{cB}} + k_{cD}^0 P_D^{n_{cD}}}{(1 + K_{cH}\sqrt{P_H})^2} \tag{10-113a}$$

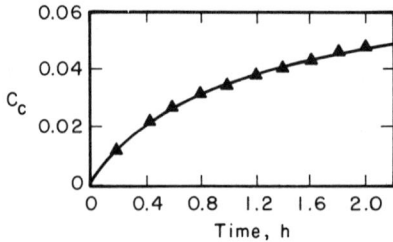

Fig. 10-12. C_c versus time as measured on the thermobalance. [From Dumez and Froment (16). Reproduced by permission of the American Chemical Society.]

IV. Some Experimental Results—Isothermal Reactors

so that overall

$$r_c^0 = \frac{dC_c}{dt} = \frac{k_{cB}^0 P_B^{n_{cB}} k_{cD}^0 P_D^{n_{cD}}}{(1 + K_{cH}\sqrt{P_H})^2} \exp(-\alpha C_c) \quad (10\text{-}113b)$$

Integrating with respect to time, we obtain

$$C_c = \frac{1}{\alpha} \ln\left\{1 + \alpha \left[\frac{k_{cB}^0 P_B^{n_{cB}} + k_{cD}^0 P_D^{n_{cD}}}{(1 + K_{CH}\sqrt{P_H})^2}\right] t\right\} \quad (10\text{-}114)$$

where

$$k_{cB} = A_{cB}^0 e^{-E_{cB}/RT}, \qquad k_{cD} = A_{cD}^0 e^{-E_{cD}/RT}$$

Values of the kinetic parameters for this coking model are given in Table 10-4.

To this point, then, we have developed a reaction model and a deactivation model and determined the associated parameters. Before going on to combining coking and reaction models, however, let us discuss another aspect of the analysis of this system—namely, let us attempt a coking correlation based on LHHW analysis rather than the empirical time-on-stream aproach of Eq. (10-112a). Assuming that both butene and butadiene are coke precursors, we may reasonably take the rate-determining steps in coke formation to be surface addition reactions of the form

$$I_{1B}L + (n_B D)L \rightarrow I_{nB}L$$

$$I_{1D}L + (n_D D)L \rightarrow I_{nD}L$$

TABLE 10-4

Kinetic Parameters for Coking

Parameter[a]	Value	Approximate standard deviation
α	45.53	1.08
$A_{cB}^0 (T_m = 822.9\ K)$	0.2917	0.0209
E_{cB}	32,800	758
n_{cB}	0.743	0.029
A_{cD}	1.3168	0.1158
E_D	21,042	501
n_{cD}	0.853	0.023
K_{cD}	1.695	0.076

[a] α = kg cat/kg coke; A_{cB}^0 etc. = kg/kg cat-hr; E_{cB} etc. = kcal/kmol.

where $I_{1B}L$ and $I_{1D}L$ are adsorbed intermediates of lower molecular weight in equilibrium with D, B, and higher intermediates $I_{nB}L$ and $I_{nD}L$; n_B and n_D are some appropriate stoichiometric coefficients. The corresponding surface rate equation is

$$r_c = k'_{cB} C_{I_{1BL}} C_{(n_BB)L} + k'_{cD} C_{I_{1BL}} C_{(n_DD)L} \tag{10-115}$$

Standard elimination of the surface intermediate concentrations in terms of free active sites and gas-phase partial pressures gives

$$r_c = (C_{TL} - C_{SL})^2 \frac{(k'_{cB} K_{I_{1BL}} P_B^{n_B+1} + k'_{cD} K_{I_{1DL}} P_D^{n_D+1})}{(1 + K_B P_B + K_H P_H + K_D P_D)^2} \tag{10-116}$$

where C_{TL} is the total number of active sites and C_{SL} the number of active sites covered by coke. A rough comparison of Eq. (10-116) with (10-113b) indicates that the deactivation function from the LHHW approach would be given by

$$\phi_c = \left(\frac{C_{TL} - C_{SL}}{C_{TL}}\right)^2 \tag{10-117}$$

which is, unfortunately, not compatible with the experimental best-fit exponential result. Now, we might go back and try variants on the assumptions of what constitute rate-determining steps in coke formation, but since the most reasonable one has failed it would not seem a particularly fruitful exercise at this point. Thus we will stay with the empirical result for the time being. The reader should compare this development with the discussion on activity functions given in Chapter 3.

Now we must turn to the coupling of coke formation with the kinetics of the main dehydrogenation reaction. Direct experimentation on this was carried out in a gravimetric balance, so that it was possible to measure the overall rate of reaction, and its change, in terms of C_c. Typical results are similar to those of Fig. 10-12, with ϕ_H, the relative activity measure for the main dehydrogenation reaction, best correlated as an exponential function of C_c:

$$\phi_H = \exp(-\alpha C_c) \tag{10-118}$$

The value of α determined in the dehydrogenation experiments, 42.1, is in good agreement with that for coking rates, 45.5; hence one is tempted to conclude that the coke formation reaction and the dehydrogenation reaction occur on the same active sites. Finally, then, the reaction–deactivation model is given by

$$(-r_H) = \frac{k_H^0 K_B (P_B - P_H P_D/K) \exp(-\alpha C_c)}{(1 + K_B P_B + K_H P_H + K_D P_D)^2} \tag{10-119}$$

IV. Some Experimental Results—Isothermal Reactors

and

$$(r_c) = \frac{(k_{cB}^0 P_B^{n_{cB}} + k_{cD}^0 P_D^{n_{cD}}) \exp(-\alpha C_c)}{(1 + K_{cH}\sqrt{P_H})^2} \quad (10\text{-}120)$$

with the same value of α in the two equations.

Under conditions of low conversion, thermal effects will be negligible, and if intrinsic kinetics are applicable we may formulate the fixed-bed deactivation model directly from Eqs. (10-5), (10-6), (10-119), and (10-120). Thus:

$$\frac{\partial x}{\partial z} = \frac{\Omega \rho_B d_p}{F} \left[\frac{\beta k_H^0 K_B (P_B) \exp(-\alpha C_c)}{(1 + K_B P_B + K_H P_H)^2} \right] \quad (10\text{-}121)$$

and

$$\frac{\partial C_c}{\partial \tau} = \left[\frac{k_{cB}^0 P_B^{n_{cB}} \exp(-\alpha C_c)}{(1 + K_{cH}\sqrt{P_H})^2} \right] \frac{\varepsilon \rho_A \Omega d_p}{F} \quad (10\text{-}122)$$

with β a conversion factor from molal to mass units, consistent with the notation of Eq. (10-5).

Dumez and Froment also reported the results of applying the overall model to experiments carried out in an integral reactor. Strictly speaking, these were not isothermal experiments since the temperature varied due to catalyst deactivation. However, this variation was apparently not large (experimentally measured profiles were used to adjust the constants), and a confrontation of the numerical integration using the conservation-rate model is shown in Fig. 10-13 for a typical run at $P_B = 0.22$ atm and $T = 595°C$. The agreement is truly remarkable; to emphasize this point we remind the

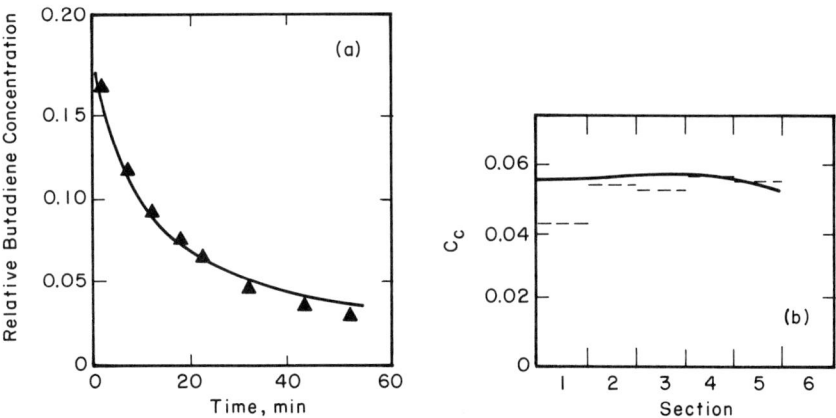

Fig. 10-13. (a) Butadiene yield versus time in the integral reactor. (b) Coke profile in the integral reactor. (---) Experimental, 52 min operation. [From Dumez and Froment (16). Reproduced by permission of the American Chemical Society.]

reader that these are not adjustable parameter fits to observation in the sense of the results obtained by Van Zoonen. Rather, this represents an impressive attempt at completely *a priori* simulation in which all reactor and kinetic parameters are independently determined apart from the simulation. It is also interesting to note that the end-of-run coke profile is nearly flat and that even this is predicted by the simulation model. The relative invariance of coke with position in the reactor is in accord with intuition gleaned from the combination of profile behavior when both reactant and products are coke precursors.

V. SCALE-UP TO A NONISOTHERMAL INDUSTRIAL REACTOR

The scale-up of the results just discussed for n-butene dehydrogenation is an extremely illuminating exercise. Several factors are involved. First, the reaction and deactivation kinetics were measured experimentally in the absence of any intraparticle or interphase gradients; to do so required the use of catalyst particle diameters in the range 0.4–0.7 mm. However, industrial reactors employing this process for manufacture of butadiene run with particles of ∼4 mm to limit bed pressure drop; consequently one must see how to map these kinetic correlations in the light of possible diffusional effects. A second factor is that industrial reactors are operated adiabatically, with characteristic operating cycles of about 15 min. This necessitates the inclusion of an energy balance and, to account for possible intraparticle diffusion, a two-phase model for the reactor. We conclude that scale-up is not just a bit more complicated than laboratory simulation, but is easily an order of magnitude more complicated. Some typical parameters associated with the scale-up are given in Table 10-5.

To investigate whether intraparticle diffusion limitations would be important in scale-up, the full-scale intraparticle diffusion–reaction problem, assuming quasi-steady-state conditions, no interphase gradients, and isothermal particles, was solved. Since these procedures have been treated in earlier chapters we will not redescribe the details here. The isothermal effectiveness factors, computed with effective diffusivities based on a Knudsen radius of 5000 Å and tortuosity factor of 5, were compared with experimental values obtained with larger particles. The computed values agreed reasonably well with the experimental ones and varied from about 0.15 to 0.8 in the particle size range of 2.3 to 0.35 mm, $0.20 < P_B < 0.25$ atm, and $500 < T < 600°C$. The good agreement demonstrated lends substance to the hope that one will be able to simulate diffusional effects in the industrial reactor, and the range of values obtained, all less than unity, shows the necessity to do so.

V. Scale-up to a Nonisothermal Industrial Reactor

TABLE 10-5

Characteristics of an Industrial Reactor for Butene Dehydrogenation

Length	0.8 m
Cross section	1 m^2
Catalyst and inert diameter	0.0046 m
Catalyst bulk density	400 kg cat/m^3 diluted reactor
Inert bulk density	900 kg cat/m^3 diluted reactor
Catalyst geometric surface area	274 m^2/m^3 diluted reactor
Inert surface area	411 m^2/m^3 diluted reactor
Inlet total pressure	0.25 atm
Inlet butene pressure	0.25 atm
Molar flow rate	15 kmol/m^2 cross section h
Feed temperature	600°C
Initial bed temperature	600°C

The fluid-phase continuity equations for the reactor, including the energy balance, are

$$\frac{\partial(u_s C_B)}{\partial z} + \varepsilon \frac{\partial C_B}{\partial t} = -a_k D_B \left(\frac{\partial C_{B,k}}{\partial r}\right)_{R_p} \tag{10-123}$$

$$\frac{\partial(u_s C_H)}{\partial z} + \varepsilon \frac{\partial C_H}{\partial t} = -a_k D_H \left(\frac{\partial C_{H,k}}{\partial r}\right)_{R_p} \tag{10-124}$$

$$\frac{\partial(u_s C_D)}{\partial z} + \varepsilon \frac{\partial C_D}{\partial t} = -a_k D_D \left(\frac{\partial C_{D,k}}{\partial r}\right)_{R_p} \tag{10-125}$$

$$\frac{\partial(u_s C_t C_{pf} T)}{\partial z} + \varepsilon \frac{\partial(C_t C_{pf} T)}{\partial t} = -[a_k h_k (T - T_k) + a_i h_i (T - T_i)] \tag{10-126}$$

where $C_t = C_B + C_H + C_D$. In the solutions to be given here, the unsteady-state terms are retained; however, Dumez and Froment pointed out that since the interstitial flow is 4 m/sec and the bed length only 0.8 m, they are, in fact, negligible. Also, in practice, the bed is diluted with inert particles to reduce the extent of the endotherm within the reactor; hence, Eq. (10-126) contains two terms on the right-hand side (RHS) for heat exchange between fluid and catalyst (k) and inert (i), respectively.

For the solid phase, we have the following:

$$\rho_{B,k} C_k \frac{\partial T_k}{\partial t} = a_k h_k (T - T_k) + a_{ik} h_{ik} (T_i - T_k)$$

$$+ a_k (-\Delta H) D_H \left(\frac{\partial C_{H,k}}{\partial r}\right)_{R_p} \tag{10-127}$$

$$\rho_{B,i} C_i \frac{\partial T_i}{\partial t} = a_i h_i (T - T_i) - a_{ik} h_{ik} (T_i - T_k) \tag{10-128}$$

where the second term on the RHS of both equations allows for the possibility of heat transfer between catalyst and inert via conduction and radiation. The intraparticle balances are

$$\frac{\partial^2 C_{B,k}}{\partial r^2} + \frac{2}{r} \frac{\partial C_{B,k}}{\partial r} - \frac{\varepsilon_p}{D_B} \frac{\partial C_{B,k}}{\partial t} = \frac{\rho_k}{D_B} \left(r_H + \frac{r_{cB}}{\psi_{cB} M_B} \right) \tag{10-129}$$

$$\frac{\partial^2 C_{H,k}}{\partial r^2} + \frac{2}{r} \frac{\partial C_{H,k}}{\partial r} - \frac{\varepsilon_p}{D_H} \frac{\partial C_{H,k}}{\partial t} = -\frac{\rho_k}{D_H} r_H \tag{10-130}$$

$$\frac{\partial^2 C_{D,k}}{\partial r^2} + \frac{2}{r} \frac{\partial C_{D,k}}{\partial r} - \frac{\varepsilon_p}{D_D} \frac{\partial C_{D,k}}{\partial t} = -\frac{\rho_k}{D_D} \left(r_H - \frac{r_{cD}}{\psi_{cD} M_D} \right) \tag{10-131}$$

$$\frac{\partial C_c}{\partial t} = r_{cB} + r_{cD} \tag{10-132}$$

where ψ_{cB} and ψ_{cD} are stoichiometric coefficients for coke formation from B and D.

The boundary and initial conditions for the entire set of Eqs. (10-123)–(10-132) are

$z = 0$, all t: $C_B = C_B^0$, $C_H = C_H^0$, $C_D = C_D^0$, $u_s = u_s^0$

$t = 0$, all z: $T_k = T_k^0$, $T_i = T_i^0$

$r = R_p$, all z and t: $(C_{B,k})_{R_p} = C_B$, $(C_{H,k})_{R_p} = C_H$, $(C_{D,k})_{R_p} = C_D$

$t = 0$, all z and r: $C_c = 0$

$r = 0$, all z and t: $\left(\frac{dC_{B,k}}{dr} \right)_{r=0} = 0$, $\left(\frac{dC_{H,k}}{dr} \right)_{r=0} = 0$, $\left(\frac{dC_{D,k}}{dr} \right)_{r=0} = 0$

A combination of mathematical methods was employed to solve this heroic set of equations: collocation for the particle equations, a Runge–Kutta method for the continuity equations in the fluid phase, and a semianalytical method for the energy equations. Full details of the method have been reported by Dumez (15). Heat transfer parameters were determined from the correlation of Handley and Heggs (22). Typical results for temperature profiles, coke profiles, and exit conversion behavior are shown in Fig. 10-14. Note in particular the very complex behavior of the intrareactor temperature profiles, with a temperature wave (endothermic) traveling very rapidly through the reactor to the exit under initial operation. Subsequently, a constant pattern for the profile is developed and the temperature gradually rises as the coke deposit builds up. The coke profiles shown in Fig. 10-14c

V. Scale-up to a Nonisothermal Industrial Reactor

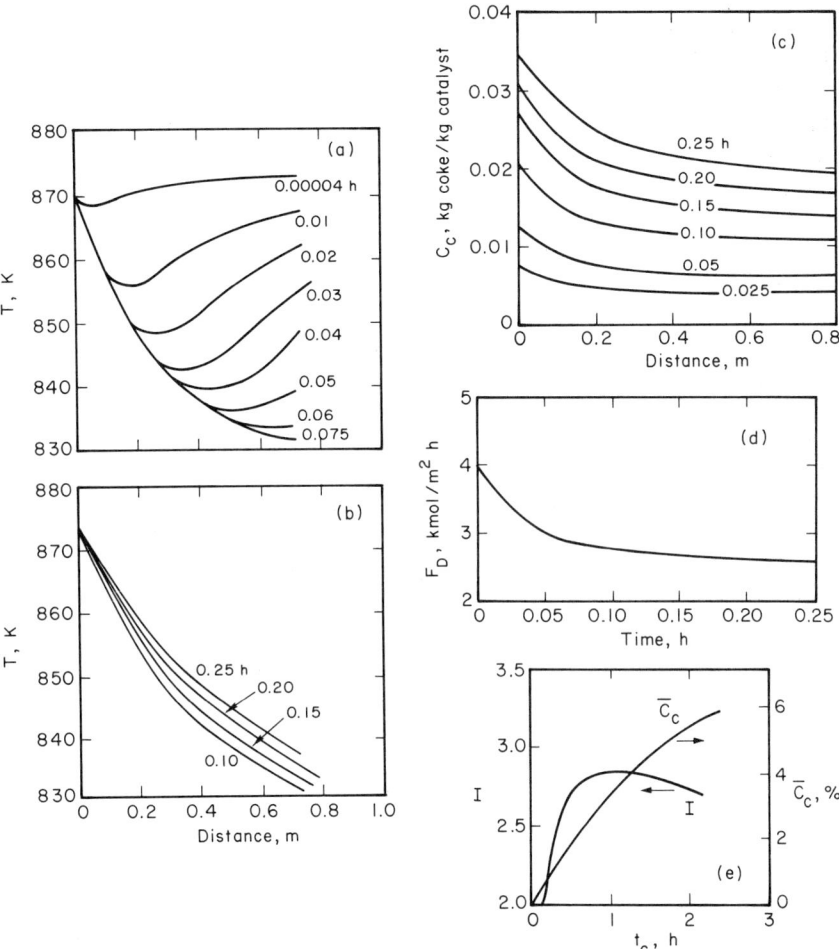

Fig. 10-14. (a and b) Intrareactor temperature profiles for butene dehydrogenation. (c) Intrareactor coke profiles for butene dehydrogenation. (d) Exit butadiene concentration versus time on stream. (e) Butadiene yield and average coke concentration versus total cycle time t_c. Parameters: bed depth = 0.80 m, bed area = 1.0 m², catalyst = 800 kg (no diluent), $T_0 = 600°C$, inlet $P_B = 0.30$ atm, $F_B^0 = 20.4$ kmol/m²-hr, A [Eq. (10-134)] = 0.05 hr/% coke, $t_p = t_e = 2$ min. [From Dumez and Froment (16). Reproduced by permission of the American Chemical Society.]

might be interpreted, in light of the previous results of Froment and Bischoff, as indicating the olefin to be the primary coke precursor. This would be incorrect, since the high temperature at bed inlet accelerates coke deposition there, and higher hydrogen partial pressures inhibit coke formation at positions farther down the bed [cf. Eq. (10-113b)]. Finally, there is an interesting type of compensation between the extent of coke deposition and the conversion of butene time-on-stream behavior. Figure 10-14d shows an initial rapid drop in conversion associated with the passage of the endotherm through the bed. Subsequently, even though coke deposition is occurring, the temperature is *rising*; the two factors tend to compensate each other and the net rate of deactivation is slowed down considerably, as evidenced by the nearly constant butene conversion after about 0.05 hr. It will be seen later that such compensation is not encountered in deactivating fixed beds with exothermic reactions.

A. A Little Optimization Problem

While the conversion of butene shown in Fig. 10-14d is relatively constant, there is no doubt where the trend illustrated will put it after sufficient time. We can formulate a nice optimization problem here which is concerned with the determination of optimal cycle time—reaction versus regeneration—for yield of butadiene. Longer operating times will produce larger amounts of product, but also larger amounts of coke which require longer to burn off the catalyst in a regeneration step. Hence the possibility exists that over the entire cycle of reaction plus regeneration there will exist a maximum in absolute yield of product, I. The expression for this is

$$I = \frac{\int_0^{t_f} F_D \, dt}{t_f + t_r} \qquad (10\text{-}133)$$

where t_f is the reaction time, t_r the regeneration time, and F_D the rate of production of butadiene. Now in general t_r will be a complex function of t_f, related through some function of coke content on catalyst. We illustrate here for the following relationship:

$$t_r = t_p + t_e + A\bar{C}_c \qquad (10\text{-}134)$$

where t_p is a reactor purge time, t_e an evacuation time, and $\bar{C}_c$ the average coke concentration corresponding to operation for t_f. The results of such a calculation with values of the associated parameters are given in Fig. 10-14. It is seen that there is a well-defined maximum in I corresponding to a total cycle time of about 1.1 hr. The corresponding value of $\bar{C}_c$ is 3.7%, hence the time for reduction from Eq. (10-134) is 15.1 min and the operating time 50.9 min.

It is important to remember that in actual operation there is another constraint which does not show up directly in the equations above. This constraint is that the amount of coke to be burned off at the end of the reaction cycle should be sufficient to reheat the catalyst bed back to the original initial temperature of operation. Conversely, coke in excess of this amount may require measures for removal of excess heat; in practice, as a result the optimization problem just discussed becomes muddied by the overall heat balance requirements and normal operating cycles are dictated by the time required to produce just that amount of coke required for reheating the deactivated bed. There is no further analysis of the delicate art of coke burning in fixed beds here, since there is sufficient justification to treat it as a separate topic in Chapter 11. Also, the entire question of optimization of reactor operation in the face of catalyst deactivation is a subject in itself. We do not attempt to treat it here, but some further comments are given at the end of the chapter.

VI. CATALYST POISONING AND NONISOTHERMAL FIXED-BED REACTOR DYNAMICS

We turn now to the discussion of some experiments and simulations similar to the work of Dumez and Froment but differing in two vital aspects:

(1) The mechanism of deactivation is by irreversible poisoning (basically a type I reaction scheme), not coking.
(2) The main reaction is exothermic, not endothermic.

The bulk of the work to be treated here consists of a series of studies of the deactivation of fixed beds of Ni/kieselguhr catalyst, used for the hydrogenation of benzene and poisoned by the rapid and irreversible chemisorption of thiophene. As in the work of Dumez and Froment, here we will attempt to determine all parameters—kinetic, poisoning, and reactor—either via separate experimentation or from reliable correlation, and then assemble all components into an *a priori* simulation of experimental results. The primary experimental data available are thermal profiles such as illustrated in Fig. 10-9 together with the exit conversion as a function of time of operation. In the experiments to be described it will be seen that the bed is "overdesigned" in the sense that at any given time the zone of total reaction is confined to only a portion of the total bed, and conversion is maintained at 100% until the thermal wave (corresponding to the zone of reaction) moves out the bed exit. Hence one may think of the situation as roughly analogous to the Wheeler-Robell model, although vastly more complicated owing to the nonisothermal environment.

For the kinetics of benzene hydrogenation under the conditions used in the reactor experiments, the following kinetic model is adequate to within

TABLE 10-6

Parameters of Eqs. (10-135) and (10-136)

Parameters	Conditions
(a) For benzene hydrogenation kinetics $E = 13{,}770$ kcal/mol $k_1^0 = 4.22$ kmol/kg-sec-torr $Q = -16{,}470$ kcal/kmol $K^0 = 4.22 \times 10^{-11}$ torr^{-1}	(1) Benzene from 0 to 0.20 mole fraction (2) Thiophene from 0 to 5,000 ppm (3) $65 < T < 200°C$ (4) Total pressure = 1 atm
(b) For thiophene poisoning kinetics $E_d = 1{,}080$ kcal/kmol $k_d^0 = 2.40 \times 10^{-2}$ (torr-sec)$^{-1}$	

±10% for temperatures below 220°C (26):

$$-r_\text{B} = \frac{k_1^0 K^0 \exp[(-Q-E)]RT/P^2 x_\text{B} x_\text{H}}{1 + K^0 \exp(-Q/RT) P x_\text{B}} \quad (10\text{-}135)$$

Parameters of Eq. (10-135) obtained by nonlinear least-squares fit to the data of Kehoe and Butt are given in Table 10-6a. The analysis of deactivation rates was based on a separable formulation of the rate equation for deactivation kinetics, linear in concentration of poison and availability of active sites. Thus:

$$-r_\text{d} = k_\text{d}^0 e^{-E_\text{d}/RT} P x_\text{T} \theta_\text{A}, \qquad k^0 = k_1^0 \theta_\text{A} \quad (10\text{-}136)$$

where θ_A is the scaled activity variable (equivalent to s used previously) and k^0 the rate constant to be used in place of k_i^0 in Eq. (10-135) as the catalyst is poisoned. Parameters of Eq. (10-136) obtained in independent deactivation experiments at differential conversion levels are given in Table 10-6.

In the following we shall look at fixed-bed dynamics, induced by thiophene poisoning, in both nonisothermal–nonadiabatic and adiabatic operation. The kinetic and deactivation models and the parameters above are, of course, the same for either type of operation.

A. Nonisothermal–Nonadiabatic Operation

The work of Weng et al. (38) involved the use of a reactor consisting of fore and aft packed inert sections with a central catalyst section with the same mesh size particles as the inerts. Experimentally, an initial steady state was established at a given inlet temperature and benzene and hydrogen concentrations (*no* thiophene present initially). In nonisothermal-nonadiabatic conditions, the reactor wall was exposed to ambient conditions and a suitable average tube wall temperature determined for subsequent use in the simulation. General experimental conditions are detailed in Table 10-7. Once the steady state had been established with the uncontaminated benzene feed, the reactor was switched to a second feed stream, identical in all respects to the first but now containing a fixed amount of thiophene, and the evolution of intrareactor temperature profiles was measured periodically together with exit conversion.

If we use run G3 as a typical base case, some qualitative observations may be made with respect to gross changes in experimental conditions. Comparison in Fig. 10-15 shows that, with respect to G1 (a run similar in all respects with the exception of about two-thirds the space velocity), the higher space velocity tends to locate the initial (steady state, $t = 0$) reaction zone farther into the active catalyst section and cause it to move through the reactor much more rapidly. Comparison of G4 with G1 (Figs. 10-15 and 10-16), where the only difference is a doubling of the inlet concentration of benzene in the former case, reveals higher hot spot temperatures and more rapid motion of the zone of reaction through the bed. Finally, if we compare G3 and G7, differing by one-half the inlet thiophene in the latter case, we find again a much slower passage of the thermal wave through the bed (i.e., longer effective life). Some other qualitative observations concern-

TABLE 10-7

Experimental Conditions for Fixed-Bed Poisoning Runs[a,b]

Operating condition	Run					
	G1	G2	G3	G4	G5	G7
Total pressure (inlet), torr	750.1	741.7	749.0	741.7	742.7	749.0
Total flow rate, ml/min (sc)	1034.2	1049.6	1551.5	1049.6	1050.4	1554.0
Space velocity, hr^{-1}	3190	3240	4790	3240	3240	4790
Ambient, °C	19.0	23.0	22.0	23.0	23.0	22.0
Inlet, °C	45.5	55.0	54.4	45.2	45.5	55.0
Mole fraction C_6H_6	0.028	0.043	0.033	0.043	0.043	0.035
Mole fraction H_2	0.972	0.957	0.967	0.957	0.957	0.965
Thiophene/$C_6H_2 \times 10^2$	1.136	1.136	1.136	1.136	0.565	0.565

[a] From Weng *et al.* (38). Reprinted by permission of Pergamon Press, Ltd.

[b] All experimental runs with reactor radius 0.822 cm, reactor length 50.00 cm, entrance section length 14.00 cm, catalyst section length 9.50 cm, catalyst wt. 6.876 g, catalyst bulk density 0.354 g/cm^3; 2:1 (volume) dilution with 12/20 mesh glass beads: 12/20 mesh Ni-0104T.

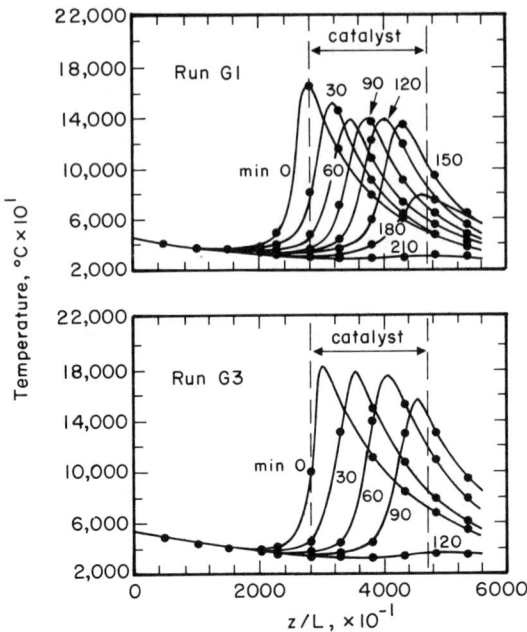

Fig. 10-15. Some typical thermal waves in the fixed-bed poisoning of an exothermic hydrogenation reaction: thiophene–benzene/Ni–kieselguhr. Run conditions as in Table 10-7. [From Weng *et al.* (38). Reproduced by permission of Pergamon Press, Ltd.]

ing the nature of these profiles we have remarked on before, in particular the relatively constant shape which is assumed after an initial transient (between 0 and 30 min here) and the constant velocity of propagation through the bed. It is also interesting that the magnitude of the hot spot temperature does not vary to a great degree as poisoning of the bed proceeds. To a large extent, it will be seen that this is a result of the fact that the poisoning process is almost thermally neutral (i.e., $E_d = 1080$ kcal/kmol) and this behavior would be expected on the basis of the results presented by Blaum. The low value for E_d essentially says that the uptake of thiophene is rapid and irreversible under all conditions of temperature employed in the experimentation.

Corresponding conversion behavior is shown in Fig. 10-17. As mentioned before, the bed is overdesigned in order to produce the narrow, moving reaction zone, hence the exit conversion remains at 100% until the hot spot emerges from the reactor. At this point, a sudden catastrophic decrease in conversion to an ultimate value of zero occurs. The relative time scales are shown very clearly in Fig. 10-17; this deactivating fixed bed is a rather

VI. Catalyst Poisoning and Nonisothermal Fixed-Bed Reactor Dynamics 399

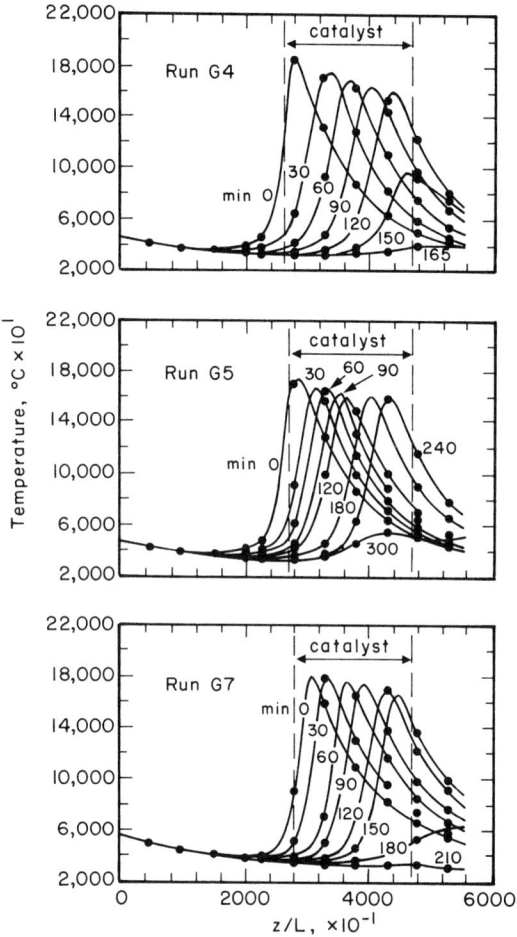

Fig. 10-16. More thermal waves in benzene hydrogenation. [From Weng et al. (38). Reproduced by permission of Pergamon Press, Ltd.]

delicate creature, since the relatively modest variations in experimental conditions detailed in Table 10-7 are sufficient to change bed life by a factor of three. This is one (among many) manifestations of the parametric sensitivity of the system.

The simulation model for these results is based on a one-dimensional axial dispersion formulation. Included are material balance equations for benzene and thiophene, an energy balance, and the rate equation for catalyst deactivation. Since hydrogen was present in great excess the hydrogen concentration was essentially constant and volume contraction is not

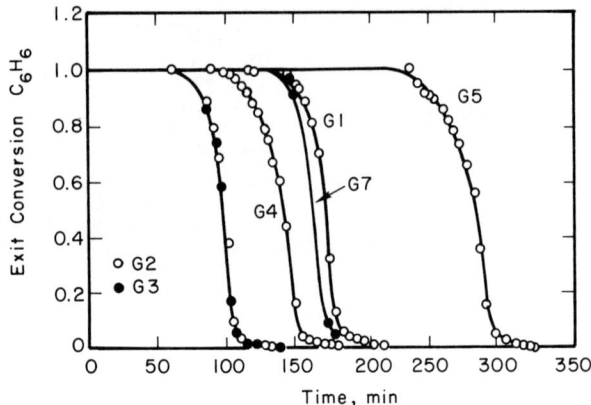

Fig. 10-17. Exit conversion versus time for poisoning of benzene hydrogenation in a fixed bed. [From Weng *et al.* (38). Reproduced by permission of Pergamon Press, Ltd.]

important. Thus we have:

(1) Mass balance, benzene

$$\varepsilon \frac{\partial C_B}{\partial t} = \varepsilon D_B \frac{\partial^2 C_B}{\partial z^2} - \varepsilon \frac{\partial v C_B}{\partial z} + \theta_A \rho_c r_B(C_B, T) \qquad (10\text{-}137)$$

(2) Mass balance, thiophene

$$\varepsilon \frac{\partial C_T}{\partial t} = D_T \frac{\partial^2 C_T}{\partial z^2} - \varepsilon \frac{\partial v C_T}{\partial z} + \rho_c r_T(C_T, T, \theta_A) \qquad (10\text{-}138)$$

(3) Energy balance

$$\frac{\partial T}{\partial t} = \frac{\lambda_{\text{eff}}}{\rho C_p} \frac{\partial^2 T}{\partial z^2} - \varepsilon v \frac{\rho_g C_{pg}}{\rho C_p} \frac{\partial T}{\partial z} + \frac{2\alpha}{R_r \rho C_p}(T_w - T) + \frac{\theta_A(-\Delta H_R)}{\rho C_p} \rho_c r_B(C_B, T)$$

$$(10\text{-}139)$$

(4) Activity decay

$$d\theta_A/dt = r_d(C_T, T, \theta_A) \qquad (10\text{-}140)$$

or

$$r_T = M_T r_d \qquad (10\text{-}141)$$

The boundary conditions corresponding are, at entrance:

$$\left.\frac{dY_i}{dz}\right|_{z=0} = \frac{v}{D_i}(Y_{i(z=0)} - Y_{i_0}) \qquad (10\text{-}142)$$

VI. Catalyst Poisoning and Nonisothermal Fixed-Bed Reactor Dynamics

where Y_{i_0} represents feed values of C_B, C_T, and T, and at exit:

$$\left.\frac{dY_i}{dz}\right|_{z=1} = 0 \qquad (10\text{-}143)$$

As can be seen on comparison, these equations are of the same general form as Eqs. (10-88)–(10-92). Intraparticle-intraphase resistances are excluded from this formulation, as experimental conditions in the laboratory can be controlled to eliminate such complications. However, it would be understood that if we were to attempt a scale-up of these results in the same way as illustrated by Dumez and Froment, such resistances would have to be considered and the pseudohomogeneous reactor model would probably be an inadequate one.

Now even in the case of very economical models, one is struck by the number of parameters that are involved in reaction–deactivation simulations via Eqs. (10-137)–(10-141) or (10-121)–(10-122). Aside from kinetic and deactivation parameters, reactor parameters such as porosity, wall heat transfer coefficient, and effective bed thermal conductivity must be determined. In fact, their accurate determination is vital, since we already know that the system is parametrically sensitive (cf. Fig. 10-17) and the propagation of small uncertainties in the combination of parameters may present great difficulties to successful simulation. These were determined by a combination of separate experimentation or from correlation and are detailed in Table 10-8. Even at this, we are left with one "float" in the simulation, namely T_w. In the results to be described, an average of the T_w values measured for the fore and aft inert sections was used for the wall temperature along the whole length of the reactor.

The model equations with parameter values as detailed here were solved using the Crank–Nicolson method with nonequidistant space steps and parabolic approximation of spatial difference quotients, as described previously. Initial results for the base case run G3 indicated that the match to the initial steady-state profile was good, but the calculated poisoning wave and corresponding hot spot location move through the bed much more slowly than was experimentally observed. In this calculation the zone of main reaction measured experimentally was at the end of the bed after 1.5 hr, while the corresponding calculation indicated 4 hr for this to occur. Such was the unfortunate result, in fact, for simulation of all the runs in the first attempt; calculated profiles, while agreeing in magnitude with experimentally observed values, passed through the bed much more slowly than the experimental observation[5], yet the fit to the initial steady state was

---

[5] Note that this is not only wrong, but unkindly so, since the simulation is being very optimistic about overall bed life.

TABLE 10-8

Parametric Quantities for Modeling Fixed-Bed Deactivation
of Ni Hydrogenation Catalyst by Thiophene[a]

Quantity	Evaluated from
Molecular weight	$MW = 2.106x_H + 78.12x_B$
ρ_g, kg/m^3	$\rho_g = (MW)(273.16)P/(22.41)(760)T_0$
$C_{pg}(350\ K)$, kcal/kmol-K[b]	$C_{pg} = 6.935x_H + 23.15x_B$
λ_{eff}, kcal/m-sec-K[c]	$\lambda_{eff} = 7\lambda_g + 0.8\rho_g C_{pg} v \in D_p \quad (\lambda_g \approx \lambda_{H_2})$
D_T, D_B, m^2/sec[d]	$D_T = D_B = D_{C_6H_6/H_2} = 0.4 \times 10^{-4}(45°C)$
α, kcal/m^2-K[e]	$\alpha = 2.6 \times 10^{-3}$
ε[e]	$\varepsilon = 0.6$
$\overline{\rho C_p}$, kcal/m^3-K	$\overline{\rho C_p} \approx \bar{\rho} \cdot \overline{C_p} = 175$
M_T, kmol/kg[e]	$M_T = 1.03 \times 10^{-3}$

[a] From Weng et al. (38). Reprinted by permission of Pergamon Press, Ltd.
[b] From API tables.
[c] From S. Yagi and D. Kuni, AIChE J. **3**, 373 (1957).
[d] From O. Krischer and K. Kroll, "Die wissenschaftlichen Grundlagen der Truckungstechnik," p. 176. Springer-Verlag, Berlin and New York, 1963.
[e] Determined in separate heat transfer and poisoning experiments.

always good. Weng et al. (38) originally speculated that the discrepancy could possibly arise from oversimplification of the deactivation model via Eq. (10-136) and proposed a two-site model for the deactivation of nickel by thiophene. This was based primarily on the observations of Lyubarskii et al. (31) that thiophene is strongly selective in deactivation of the hydrogenation activity for benzene of nickel catalysts—a little goes a long way—and one could envision uptake of thiophene on both hydrogenation-active sites and nonactive sites as discussed in Chapter 4. Hence, an overall measurement of the thiophene capacity would be in error (too high), and the discrepancy between experimentally observed and simulated bed life could be explained qualitatively on this basis. Analytical formulation of such a model, of course, puts another parameter, the ratio of hydrogenation-active sites to total sites available for thiophene chemisorption, into the simulation; this assumes the form of an adjustable parameter, so the essence of an *a priori* simulation of the fixed-bed dynamics is lost.

Subsequently, it was found that simply reducing the measured value of M_r by 60% resulted in excellent agreement between simulation and experiment using the single-site model. These results are illustrated in Fig. 10-18 for some selected runs. The details of the eventual resolution of this problem are a good lesson in the subtleties sometimes involved for those who would understand deactivation. It turns out that, while the kinetics of thiophene uptake on the Ni/kieselguhr are not very temperature-sensitive under the

VI. Catalyst Poisoning and Nonisothermal Fixed-Bed Reactor Dynamics

conditions employed, the total uptake is. At lower temperatures the thiophene molecule can be bonded through nonlocalized interactions with π bonds of the ring as well as with sulfur, while at higher temperatures the bonding is much more specific for Ni–S. The result of this is that at lower temperatures it takes fewer total thiophene molecules to block access to the

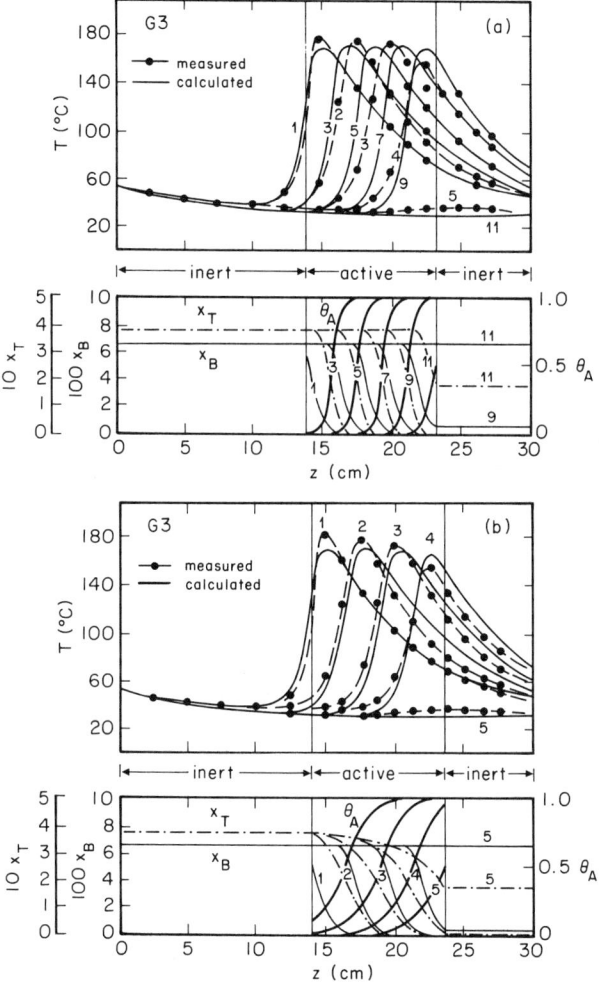

Fig. 10-18. Simulation results for the dynamics of fixed-bed hydrogenation of benzene with thiophene poisoning of Ni/kieselguhr. Comparison of profiles is shown for 30-min intervals starting with the initial steady state. (a) Single-site deactivation model; (b–d) two-site deactivation model. [From Weng *et al.* (38). Reproduced by permission of Pergamon Press, Ltd.]

(*Figure continues*)

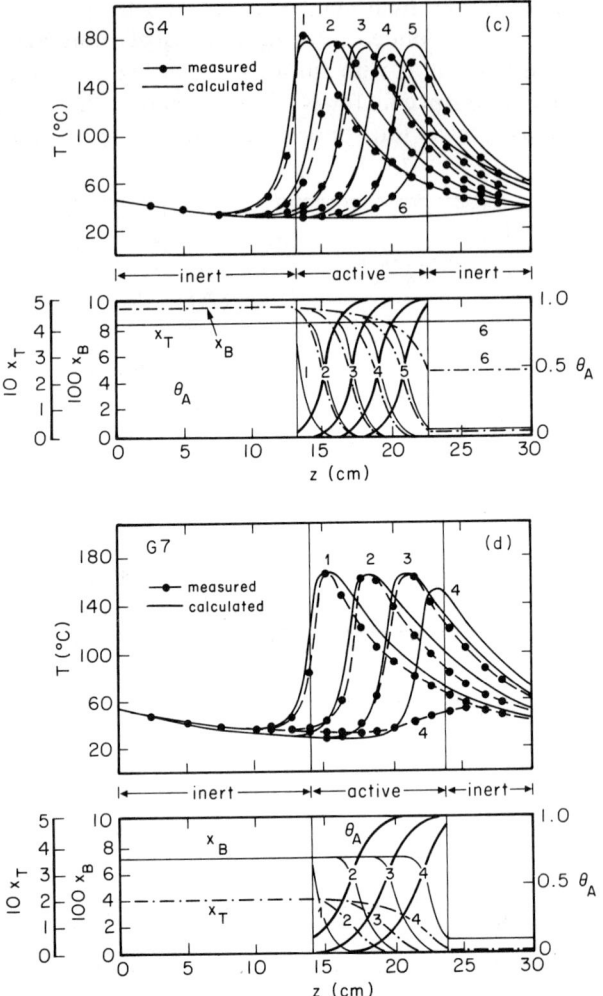

Fig. 10-18 (*continued*)

surface than at high temperatures; thus the apparent "poison adsorption capacity," M_T, required to completely deactivate the catalyst is temperature-dependent and will be incorrect unless one is at pains to duplicate the temperature conditions that will be involved in simulation in the poison capacity measurement. In fact, this was attempted by Weng *et al.*, since such capacity measures were undertaken under reaction conditions; one can only conclude, however, that the low-temperature adsorption (i.e., on the leading edge of the hot spot) is controlling in the overall poisoning mechanism. The situation is not aided by the fact that the simulation is parametrically very sensitive to the value of M_T, which in large measure dictates how fast the hot spot and associated zone of activity will move through the reactor.

Accepting the bittersweet result above concerning the parameter M_T, we may nonetheless conclude that the simulation is quite successful for a variety of experimental conditions and time scales of deactivation, as shown in Fig. 10-18. Somewhat of a surprise is provided by the computed activity profiles shown in the figure. We have made the point several times that the experiment was designed to produce a narrow well-defined zone of activity moving through the bed, the locus of which could be related to the locus of the hot spot. In fact, however, the zone of partial activity extends throughout most of the bed over a large portion of the time of operation. Corresponding concentration profiles for benzene and thiophene, while somewhat sharper, still do not correspond to what we would consider a narrow reaction zone.

A final note has to do with the axial dispersion terms included in the model equation. It is apparent from the temperature profiles that back-conduction of heat is a very important process and a corresponding dispersion term is included in the energy balance. Good agreement between experimentally measured values of λ_{eff} and those determined from the Yagi–Kunii correlation was obtained. This is a somewhat unexpected result since literature correlations very rarely work for small laboratory reactors; the fact that hydrogen was always present in large excess may be the reason for this. On the other hand, axial mass dispersion had only very small effects on the computed profiles and mass conservation equations could be fashioned according to the plug flow model with no detriment to the simulation results.

B. Adiabatic Operation

As stated before, strictly from a simulation point of view adiabatic operation is in some respects an easier problem than nonisothermal–nonadiabatic operation, since uncertainties associated with wall heat transfer coefficients and the like disappear. This, of course, pertains to reactor simulations whether or not catalyst deactivation is occurring. On the other hand, thermal responses and reactor dynamics in adiabatic reactors are likely to be even more rapid and parametrically sensitive than those in corresponding nonadiabatic systems, and the specific effects of deactivation are more easily identifiable.

Continuing with our example of benzene hydrogenation (34), in the adiabatic case the model equations consist of Eqs. (10-135)–(10-143), but with the heat transfer term $(2\alpha/R_r\rho\bar{C}_p)(T_w - T)$ of Eq. (10-134) missing. The reactor again consists of an active central section with inert packing fore and aft. Adiabatic operation in small-scale laboratory reactors is notoriously difficult to achieve; in this case the reactor tube was enclosed in an evacuated jacket and the entire assembly contained in a form-fit asbestos container. Even at this, operation could be considered only to be "nearly adiabatic"; some typical experimental temperature profiles are shown in Fig. 10-19a, where the tailing near the end of the aft section results from a finite amount of heat conduction along the reactor wall. An immediate contrast to nonadiabatic operation is the temperature jump which occurs at the adiabatic front once poisoning has begun. We will discuss this in some detail subsequently. Figure 10-19b shows a comparison of the steady-state simulation with the steady-state adiabatic profile; parameters employed in the simulation are the same as those used for the nonadiabatic simulation.

Now, in simulation of the dynamic response to poisoning, we may make some use of results obtained in prior work (14) for nondeactivating adiabatic

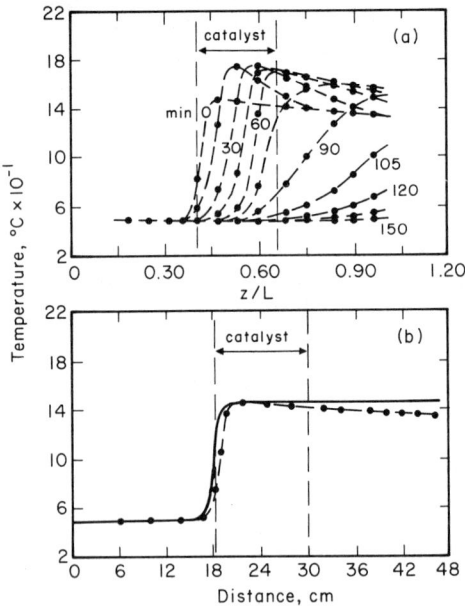

Fig. 10-19. (a) Typical measured temperature profiles for adiabatic operation. (b) Simulation of the steady-state adiabatic profile; (——) computed. Conditions: space velocity = 3475 hr^{-1}, $x_B = 0.014$, $x_T = 6.36 \times 10^{-4}$, $T_0 = 49.0°C$, total pressure = 1 atm. [From Price and Butt (34). Reproduced by permission of Pergamon Press, Ltd.]

VI. Catalyst Poisoning and Nonisothermal Fixed-Bed Reactor Dynamics

beds: (i) the thermal properties of the bed dominate the dynamic behavior, and (ii) the transient involves two distinct periods: a fast concentration response (FCR), where the temperature changes slowly but the concentration rapidly achieves a quasi-steady state, and a slow temperature response (STR), where both temperature and concentration slowly evolve toward a final steady state. For the simulation of transients due to poisoning we find that the dynamis are composed of three, rather than two, distinct time domains, illustrated in Fig. 10-20 for a typical calculation. The first domain is an FCR of thiophene, which very rapidly establishes a quasi-steady profile from its zero initial condition. This is not shown in the figure, since the profile is established within two residence times; during this period temperature, reactant, and activity profiles do not change significantly from their initial (steady-state) values. The other two domains can be classified as different types of STR. During the first of these (about 45 min in the example) the activity, reactant, and poison concentration profiles slowly

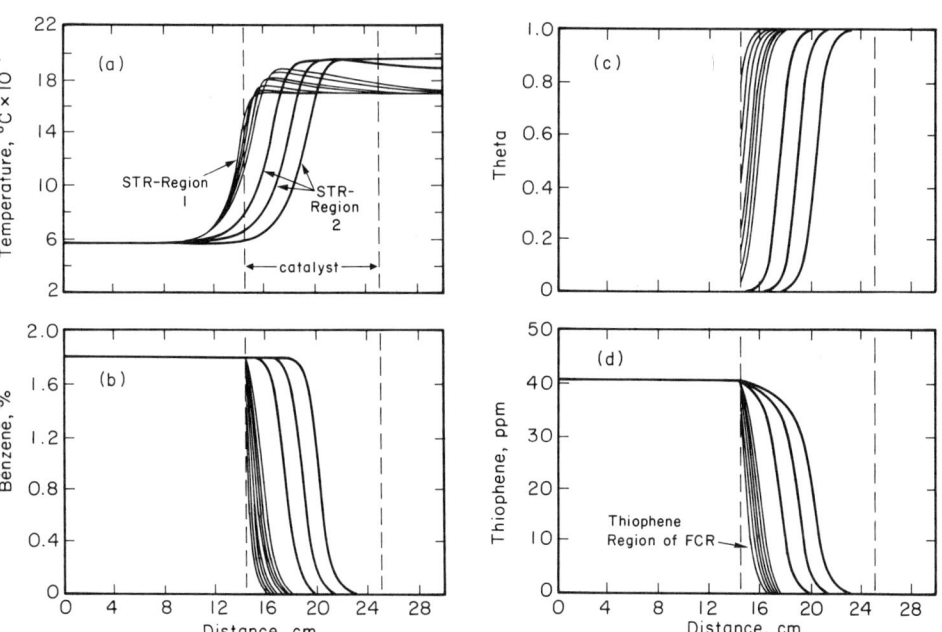

Fig. 10-20. General characteristics of reactor dynamics induced by catalyst poisoning in adiabatic operation. (a) Temperature response; (b) benzene concentration response; (c) catalyst activity response; (d) poison (thiophene) concentration response. Notes: Profiles plotted for STR region 1 are 3 min apart, for STR region 2 are 10 min apart. Conditions: space velocity = 2495 hr^{-1}, $x_B = 0.018$, $x_T = 4.04 \times 10^{-4}$, $T_0 = 56.0°C$, total pressure = 1 atm. [From Price and Butt (34). Reproduced by permission of Pergamon Press, Ltd.]

develop a characteristic shape, which, during the second period, remains unchanged as the profile moves down the bed with constant velocity.

A temperature maximum, in excess of the adiabatic temperature rise, is established in the first STR. This additional temperature rise results from the convection of heat accumulated in the front of the bed while the initial steady state was achieved. A similar phenomenon has ben noted in non-poisoned systems (17) when the reaction zone is shifted downstream by a sudden perturbation in operating conditions.

The second STR is the transient resulting from the poisoning of the catalyst. Temperature, activity, and concentration profiles move through the bed without changing shape; the maximum temperature remains near the value established during the first STR. These general characteristics are observed in the experimental temperature profiles and in the simulation. An example of comparison with experiment, using the previously described

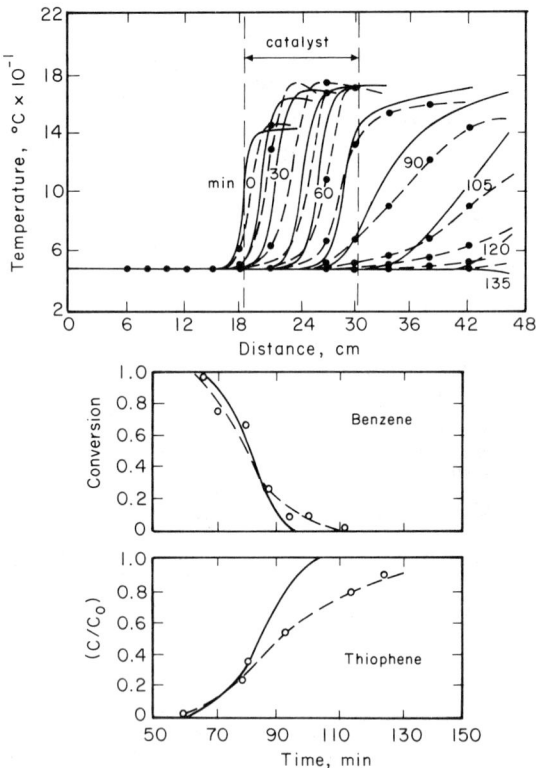

Fig. 10-21. Computed and experimental temperature and exit breakthrough profiles for adiabatic operation. Solid lines are computed. [From Price and Butt (34). Reproduced by permission of Pergamon Press, Ltd.]

VI. Catalyst Poisoning and Nonisothermal Fixed-Bed Reactor Dynamics

parameter values, is given in Fig. 10-21. In general, the computed profiles are slightly steeper than experimental ones and lag behind somewhat. This behavior is probably the result of a combination of three factors: (i) the temperature drops rapidly as the profiles exit the active section, which provides a very severe test of the precision of the hydrogenation and poisoning kinetics employed in the model; (ii) as in the nonadiabatic simulation the rate of passage of the profiles through the bed is very sensitive to the value of M_T, so that even relatively small experimental errors become important, and (iii) shell-progressive deactivation is occurring, so the bed appears to deactivate more rapidly than predicted by the uniform-poisoning model. Computed exit breakthrough curves for benzene and thiopene are also shown in comparison with experiment in Fig. 10-21. The agreement is about as good as one could wish for at this level of detail in modeling.

We pointed out earlier that during the first STR the reactor temperature maximum attains a value in excess of the adiabatic temperature rise established at steady state. Further, comparison with nonadiabatic operation under nearly the same conditions (cf. Fig. 10-15) indicates that no such behavior is observed in that case. This *dynamic temperature rise* associated with adiabatic operation thus must be the result of heat transfer from catalyst to gas phase upstream as the active zone moves down the bed. In Fig. 10-22a the shaded area represents the effect of heat transferred from the solid between time t_0 and t_1 with the zone moving down the bed; this heat can only supplement that released in the reaction and therefore must account for the increased temperature above ΔT_{Ad}. Over the lifetime of the bed, then, the following simple balance should apply:

(amount of heat accumulated at SS) = (heat released during deactivation)

or

$$\int_{Z_1}^{Z_2} [(\overline{\rho C_p})(T - T_0) \, dz]_{SS} = \int_0^{t_s} [(\overline{\rho C_p})_g U(T - T_A)] \, dt_p \quad (10\text{-}144)$$

where $(\overline{\rho C_p})$ is the volumetric heat capacity of the solid, $(\overline{\rho C_p})_g$ is that of the gas, U is the interstitial velocity, $T_a = T_0 + (\Delta T)_{Ad}$, and t_s is the time required to fully deactivate the bed. The value of t_s can be computed from experimental parameters as

$$t_s = \rho_B M_T (Z_2 - Z_1) / U C_T^0 \quad (10\text{-}145)$$

where ρ_B is the bulk catalyst density, M_T the poison adsorption capacity (weight/weight), C_T^0 the inlet thiophene concentration, and U again the interstitial velocity. For adiabatic operation a reasonable approximation is to assume that steady-state and transient temperature profiles are square waves, as shown in Fig. 10-22b. In this case Eq. (10-145) is easily integrated

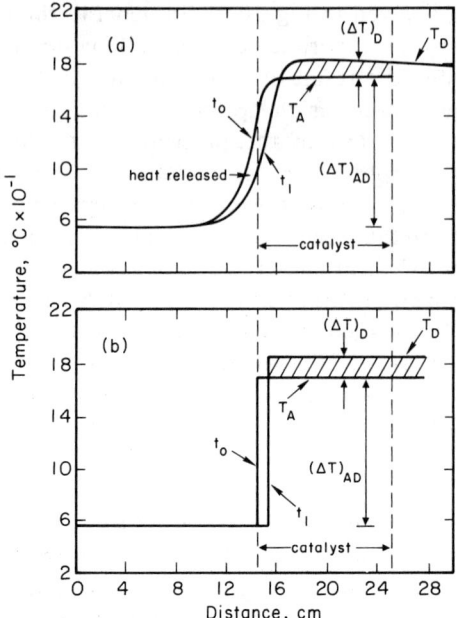

Fig. 10-22. (a) Dynamic temperature rise in adiabatic operation. (b) Square wave approximation to dynamic temperature rise. [From Price and Butt (34). Reproduced by permission of Pergamon Press, Ltd.]

to give, after substitution for t_s,

$$(\overline{\Delta T})_D = \frac{C_T^0}{\rho_B M_T} \frac{(\overline{\rho C_p})}{(\overline{\rho C_p})_g} (\Delta t)_{Ad} \quad (10\text{-}146)$$

where $(\overline{\Delta T})_D$ is the dynamic temperature rise above $(\Delta T)_{Ad}$. The average velocity of the activity front is

$$U_p = C_T^0 U / \rho_B M_T \quad (10\text{-}147)$$

which gives finally for $(\overline{\Delta T})_D$:

$$(\overline{\Delta T})_D = \left(\frac{U_p}{U}\right) \frac{(\overline{\rho C_p})}{(\overline{\rho C_p})_g} (\Delta T)_{Ad} \quad (10\text{-}148)$$

Several conclusions can be drawn, within the limits of the square wave approximation, regarding this temperature rise. It is

(1) Linear in all the poisoning variables,
(2) Dependent on inlet reactant concentration through (ΔT_{Ad}),
(3) Dependent on poison adsorption capacity, and

VI. Catalyst Poisoning and Nonisothermal Fixed-Bed Reactor Dynamics

(4) Independent of gas velocity through the bed (since U_p is proportional to U).

Note, however, that this analysis produces only a time-averaged value; from Fig. 10-20a it is apparent that the temperature behavior is somewhat different for each STR domain. A comparison of $(\overline{\Delta T})_D$ determined via the full-scale simulation, the approximation of Eq. (10-148), and experimental observation for our example run gives the following:

$$(\overline{\Delta T})_D \text{ (experimental)} = 24.2°C$$

$$(\overline{\Delta T})_D \text{ simulation} = 26.3°C$$

$$(\overline{\Delta T})_D \text{ [Eq. (10-148)]} = 20.8°C$$

Confirmation of the predicted trends for runs under other conditions was also obtained, with a similar range of agreement in predicted and experimental values of $(\overline{\Delta T})_D$.

Recently, Billimoria (2) extended the dynamic simulation model for adiabatic bed deactivation to include the problem of start-up transients in either fresh beds or partially deactivated beds—again for the benzene/thiophene/Ni/kieselguhr system. Typical results for the start-up of a partially deactivated bed[6] are shown in Fig. 10-23; again the same parameters employed for the nonadiabatic and adiabatic simulations are used, the only difference being that a curve fit value of α, the wall heat transfer coefficient, has been used to match the fall-off of the adiabatic profile in the aft section. The agreement is truly remarkable, particularly in view of the fact that the kinetic and poisoning models, based on steady-state information, are here being asked to model a most extreme case of reactor dynamics.

Finally, lest one become overly devoted to such vigorous analytical exercises as have been presented above, we can also give the following. It is based on the question of the extreme to which one wants to go in the simulation of such systems. So far, we have presented one extreme in which it is desired to model both the shapes of thermal and activity profiles and the rate at which they progress through the reactor. Suppose we are interested only in the very practical engineering question, how long will such a reactor endure before breakthrough of the temperature–conversion profiles? There is a simple way to go through such an analysis based on the idea that there must be some "work requirement" for the reactor, in the face of deactivation,

---

[6] The bed was deactivated under reaction conditions until the adiabatic temperature front was located near the center of the catalyst zone. At this point benzene and thiophene were eliminated from the feed stream and the reactor cooled in flowing hydrogen until the temperature was uniform at T_0 (~78°C in the illustration). Then benzene was reintroduced into the feed and the ignition of the reaction zone measured as a function of time.

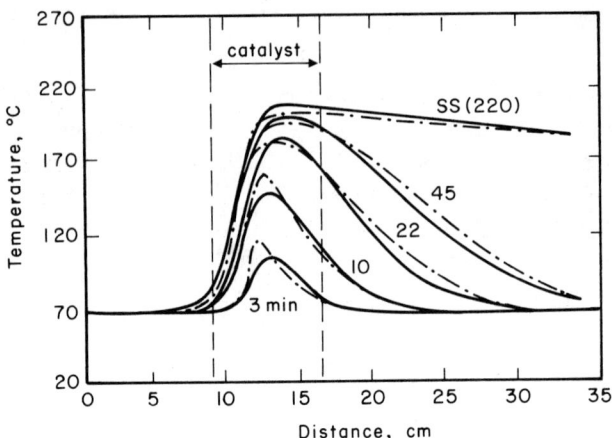

Fig. 10-23. Start-up profiles for benzene and hydrogen in a partially deactivated bed. Solid lines are experimental, dashed lines are computer simulation. [From Billimoria and Butt (2). Reproduced by permission of Elsevier Sequoia, SA.]

that would be related to the entrance concentration of poison, in this case given by x_T, and the difficulty of the reaction (given kinetics and conversion requirement) that is represented by the space velocity (SV). Thus the space velocity (SV) times the entering poison concentration (x_T) could be some measure of this work requirement as measured versus the time required for the poison wave to break out of the bed (initial time, not t for zero activity). There is, in fact, a good correlation of the breakthrough data of Weng *et al.* provided by this simple-minded idea, as shown in Fig. 10-24. One could

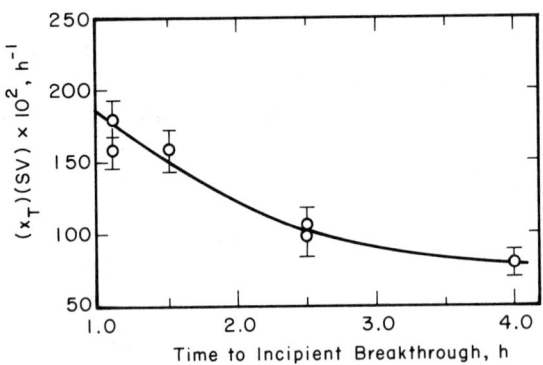

Fig. 10-24. A "work requirement" correlation for thiophene poisoning of benzene hydrogenation on an Ni/kiselguhr catalyst. [Data of Weng *et al.* (38). Reproduced by permission of Pergamon Press, Ltd.]

probably even evision working this into a kind of linear Voorhies-type correlation, but since the theoretical foundations of the work requirement concept are shaky at best, we shall avoid that temptation here.

VII. CONSTANT-CONVERSION OPERATION

In the foregoing we have focused on the concept of moving reaction zones and, in nonisothermal cases, the associated temperature waves which are produced in fixed beds by varying types of deactivation mechanisms. However, often large-scale reactors (e.g., hydrotreating processes of various sorts) are subject to relatively slow deactivation effects and constant-conversion operation is required in order not to upset subsequent processing units. Here, the temperature level of the reactor is used to compensate for the loss of intrinsic catalytic activity and the thermal parameters of the main and deactivation reactions, particularly activation energy, have a great influence on the operation. Further, it has been common practice in the petroleum industry for many years to evaluate catalyst activity and activity maintenance for these processes in laboratory experiments which are also conducted under constant-conversion conditions. In this procedure, catalyst deactivation effects are measured in terms of the rate of temperature increase needed to maintain constant conversion, that is, a *temperature increase required* (TIR).

In this section we shall consider primarily the analysis of constant-conversion operation in general terms and not try to detail explicitly what would be corresponding problems in interpretation of laboratory data. However, it should become apparent fairly soon that TIR operation is a positive feedback procedure and is probably the most difficult of all possible ways to obtain interpretable information on deactivation kinetics. Of course, sometimes one has no choice in the matter.

In simulation of constant-conversion operation, we will find it convenient to depart from the description of the reactor via (partial) differential equations, that is, continuum theory, and go to a mixing cell model. From the theory of chemical reactor analysis we know that by variation of the number of mixing cells in series it is possible to simulate various degrees of dispersion effects; however, the observations made with both the butene dehydrogenation and benzene hydrogenation examples indicate that plug flow mass balance approximations are generally adequate, so we will select a mixing cell sequence with a sufficient number of cells (say 40) to be representative of plug flow. A further simplification of the analysis will be provided for some examples in that we will take the bed at any given time to be isothermal so that back-conduction of heat will not be important (in any event, the simple mixing-cells-in-sequence model is incapable of representing this).

Let us start with a simple case of first-order kinetics for the main reaction and first-order kinetics for deactivation in which coke deposition via a reactant precursor mechanism (type IIa) is involved. For the main reaction we may write

$$\frac{C_0}{C_n} = \prod^n (1 + k_i s_i \theta) \qquad (10\text{-}149)$$

where k_i is the rate constant for the main reaction, s_i the scaled activity variable, and θ the holding time, equal in each of the n cells. Here we neglect the consumption of reactant in the coke formation reaction. Now:

$$\ln\left(\frac{C_0}{C_n}\right) = \sum^n \ln(1 + k_i s_i \theta) \qquad (10\text{-}150)$$

and we require conversion to be constant, hence

$$\frac{d[\ln(C_0/C_n)]}{dt} = 0 = \sum^n \frac{d[\ln(1 + k_i s_i \theta)]}{dt} \qquad (10\text{-}151)$$

Now:

$$\frac{d}{dt}[\ln(1 + k_i s_i \theta)] = \frac{\theta}{1 + k_i s_i \theta}\left(s_i \frac{dk_i}{dt} + k_i \frac{ds_i}{dt}\right) \qquad (10\text{-}152)$$

so for the sequence we have that

$$\sum^n \frac{s_i(dk_i/dt) + k_i(ds_i/dt)}{1 + k_i s_i \theta} = 0 \qquad (10\text{-}153)$$

where $k_i = k_i^0 e^{-E/RT_i}$. We may write

$$\frac{dk_i}{dt} = \frac{dk_i}{dT_i}\frac{dT_i}{dt} = \frac{k_i}{RT_i^2}\frac{dT_i}{dt} \qquad (10\text{-}154)$$

so

$$\sum^n \frac{s_i(k_i/RT_i^2)(dT_i/dt) + k_i(ds_i/dt)}{1 + k_i s_i \theta} = 0 \qquad (10\text{-}155)$$

Now for the deactivation model we have

$$ds_i/dt = k_{di} s_i C_i \qquad (10\text{-}156)$$

where k_{di} is the deactivation rate constant. Substituting and rearranging Eq. (10-155) gives us

$$\sum^n \frac{s_i(k_i/RT_i^2)\, dT_i/dt}{1 + k_i s_i \theta} = \sum^n \frac{k_i k_{di} s_i C_i}{1 + k_i s_i \theta} \qquad (10\text{-}157)$$

VII. Constant-Conversion Operation

For isothermal conditions, $dT_i/dt = dT/dt$ and we may solve for the TIR directly:

$$dT/dt = \sum^n kk_d s_i C_i / \sum_n s_i(k/RT^2) \qquad (10\text{-}158)$$

$$C_i/C_0 = (1 + ks_i\theta)^{-i} \qquad (10\text{-}159)$$

The simplifications realized in Eq. (10-158) arise from the isothermality, hence $k_i = k$, $k_{di} = k_d$, and $T_i = T$. If we wish to retain generality in the model insofar as the mechanism of deactivation is concerned, then

$$dT/dt = \sum^n k(ds_i/dt) / \sum_n s_i(k/RT^2) \qquad (10\text{-}160)$$

In simulation applications we have the initial conditions that

$$s_i = 1, \quad t = 0 \qquad (10\text{-}161)$$

and the temperature must satisfy the requirement:

$$\ln\left(\frac{C_0}{C_n}\right) = \text{set constant} = \sum^n \ln(1 + k_i^0 \theta\, e^{-E/RT}) \qquad (10\text{-}162)$$

Let us now turn to the construction of a similar simulation for constant-conversion operation in an adiabatic reactor, recognizing there may be some practical limitations in application because of the absence of a means for modeling the bed conductivity effect included in the continuum model. We will again take the case of first-order kinetics and first-order reactant deactivation. The conservation equations about cell i are

$$FC_{i-1} - FC_i = \bar{V} k_i s_i C_i \qquad (10\text{-}163)$$

and

$$FC_p T_i - FC_p T_{i-1} = (-\Delta H)\bar{V} k_i s_i C_i \qquad (10\text{-}164)$$

where F is the volumetric flow rate, $\bar{V}$ the reactor volume, and C_p a volumetric heat capacity. The direction of $T_i > T_{i-1}$ is implied in the energy balance, hence we are considering an exothermic reaction. Then

$$C_{i-1} - C_i = k_i s_i \theta C_i \qquad (10\text{-}165)$$

$$T_i - T_{i-1} = k_i s_i \beta C_i \qquad (10\text{-}166)$$

where $\theta = \bar{V}/F$ and $\beta = \bar{V}(-\Delta H)/FC_p$. As before,

$$C_n/C_0 = 1 / \prod^n (1 + k_i s_i \theta) \qquad (10\text{-}167)$$

and since temperature varies throughout the sequence, $k_i = k^0 e^{-E/RT_i}$.

Substituting for C_i in the energy balance of Eq. (10-166):

$$T_i - T_{i-1} = k^0 s_i \beta\, e^{-E/RT_i} / \prod^{i} (1 + k^0 s_i \theta\, e^{-E/RT_i}) \qquad (10\text{-}168)$$

Now the difficulty here is with the awkward form of Eq. (10-168) with respect to T_i. Ordinarily one computes in sequence through the series of cells, hence from the equation we face the implicit form of T_i as a function of T_{i-1}. For the case here, though, where we employ a relatively large value of the index n in approximation of a plug flow reactor, it can be reasonable to use the substitution $T_i \approx T_{i-1}$ in the exponentials and[7]

$$T_i = T_{i-1} + k^0 \beta\, e^{-E/RT_{i-1}} / \prod^{i} (1 + s_i k^0 \theta\, e^{-E/RT_{i-1}}) \qquad (10\text{-}169)$$

with the requirement that

$$C_n / C_0 = \text{constant} = 1 / \prod^{n} (1 + k^0 \theta\, e^{-E/RT_i}) \qquad (10\text{-}170)$$

to establish the initial adiabatic profile [$s_i = 1$ in Eq. (10-169)] for initial conditions. After computation of the initial set of T_i, we increment the activity:

$$ds_i / dt = -k_d^0\, e^{-E_d/RT_i} s_i C_i \approx \Delta s_i / \Delta t \qquad (10\text{-}171)$$

If j is a time subscript, with $j = 0$ initially, then

$$(s_i)_{j+1} = (s_i)_j - [k_d^0 (s_i)_j (C_i)_j\, e^{-E_d/R(T_i)_j}](t_{j+1} - t_j) \qquad (10\text{-}172)$$

Here $(s_i)_j$, $(C_i)_j$, and $(T_i)_j$ are all known from the previous iteration so we may solve directly for $(s_i)_{j+1}$. We recompute the temperature profile for the new t_{j+1} from

$$(T_i)_{j+1} = (T_{i-1})_{j+1} + \frac{k^0 \beta\, e^{-E/R(T_{i-1})_{j+1}}(s_i)_{j+1}}{\prod^{i}[1 + k^0 (s_i)_{j+1}\, e^{-E/R(T_{i-1})_{j-1}}]} \qquad (10\text{-}173)$$

with the requirement

$$C_n / C_0 = \text{constant} = 1 / \prod^{n}[1 + k^0 (s_i)_{j+1} \theta\, e^{-E/R(T_i)_{j+1}}] \qquad (10\text{-}174)$$

A computational example following the isothermal procedure outlined via Eq. (10-160) has been reported (9) for the type II reactant scheme. Concentration profiles for reactant A are

$$[a]_n = \prod^{n}[1 + (k_A^0 \Phi_{A,n} + k_{A,d}^0 \Phi_{d,n}) S_0 \theta (s)_n]^{-1} \qquad (10\text{-}175)$$

---

[7] If steep temperature gradients are encountered such that this is not a good assumption, then one would use this only as the first step in a successive substitution routine.

VII. Constant-Conversion Operation

where $\Phi_{i,n} = \exp[\gamma_i(1+1/\phi_n)]$, n is the cell number, θ is the holding time per cell, the parameter γ_i is the temperature sensitivity of reaction i, defined as E_i/RT_0, and $k_A^0 S_0$ and $k_{A,d}^0 S_0$ refer to the specific rate constants for unpoisoned catalyst; T_0 is a reference temperature, which can be taken as the initial operating temperature. In each cell deactivation occurs at a rate which is small compared with the residence time, so

$$r = d(s)_n/d\beta = -\tau(A_0/s_0)k_{A,d}^0 S_0 \Phi_{d,n}[a]_n[s]_n \qquad (10\text{-}176)$$

where τ is the total residence time in the reactor and β the total time of operation in terms of number of residence times. The kinetics of the main reaction incorporated in Eq. (10-175) include loss of reactant due to coke formation. The quantity S_0, which has not appeared in our previous discussion, is a scaling constant which essentially refers to the total number of active sites per unit area on the fresh catalyst surface.

Some characteristics of reactor behavior are shown in Fig. 10-25, where reactor temperature versus time of operation is illustrated for operation with constant conversion of 60% for the type II reactant fouling case. The parameter X of the figure is the residence time–concentration grouping appearing in the deactivation rate equation $X \equiv \tau(A_0/S_0)k_{A,d}^0 S_0$. The exponential nature of the increase is evident, representing the "activated" reaction process, though one obtains a near-linear correlation over a certain range of process time, depending on the parameters of the reaction system— principal of which are the intrinsic selectivity ψ between main and deactivation reactions, given by $(k_A^0/k_{A,d}^0)$, and the relative temperature sensitivity of the two reactions, given by $(\gamma_A/\gamma_{A,d})$. The initial temperature of operation indicated in Fig. 10-25 is determined wholly by the intrinsic activity of the catalyst for the main reaction; deactivation has no influence on initial operation. In these examples the reduced temperature ϕ is defined with respect to the reference temperature T_0 involved in the definition of γ;

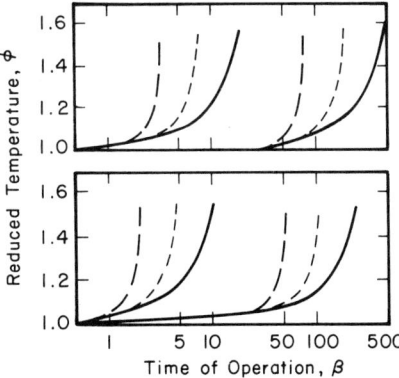

Fig. 10-25. Effect of catalyst deactivation on reactor operation: temperature increase at constant conversion for type II reactant deactivation. Top: $\psi = 20$, $\gamma_A = 10$; bottom: $\psi = 10$, $\gamma_A = 10$. Left set of curves: $X = 0.5$; right set of curves: $X = 0.02$. (———) $\gamma_{A,d} = 3\gamma_A$; (- - -) $\gamma_{A,d} = \gamma_A$; (———) $\gamma_{A,d} = 0.2\gamma_A$. [From Butt and Rohan (9). Reproduced by permission of Pergamon Press, Ltd.]

$\phi = T/T_0$. Operation at $\phi = 1$ then implies catalyst activity at a level just sufficient to give the specified conversion at an actual value of temperature, $T = T_0$.

A major qualitative result to remember from the examples of Fig. 10-25 is the very strong dependence of catalyst life here on the activation energy (i.e., temperature sensitivity) of the deactivation reaction. It is not surprising, at least in hindsight, to see that when $(\gamma_A/\gamma_{A,d})$ becomes on the order of or smaller than unity the deactivation rate becomes very large; in the limit this would represent an unstable process, since the control variable (temperature) which is being used to contain the effects of deactivation actually preferentially promotes that process, leading to a kind of kinetic positive feedback which can be modified only by allowing a certain drift to lower conversion levels with time or by restricting cycle times. The deactivation characteristics illustrated for type II reactant deactivation are also qualitatively the same for other mechanisms of deactivation including type I (feed impurity) and selective sintering. In observed behavior, then, the constant-conversion operation is similar to other modes of operation; the reactor acts as an efficient information filter, and no real detail concerning mechanisms or kinetic schemes of deactivation is available from such observation.

A type of linear activity correlation can be developed for constant-conversion operation which is roughly comparable to the Voorhies approach. It can be seen from Fig. 10-25 that the shapes of temperature–time of operation curves are essentially independent of the value of the parameter X, depending only on the value of $(\gamma_A/\gamma_{A,d})$. Hence a characteristic activity plot involving $\log \beta$ versus $\log X$, for constant parametric values of the reactor temperature, turns out to be linear. Such performance charts, then, offer the possibility of estimation of catalyst life (time of operation to a specified temperature level) in terms of the deactivation parameters.

It is also important to examine constant-conversion operation in terms of selectivity since it is apparent that if different reaction pathways involve different activation energies, changing temperatures will drive the overall selectivity in a direction dictated by differences in activation energy. Let us use as an example a bifunctional reaction employing the two functions U and V, with the following overall reaction scheme:

$$
\begin{array}{ll}
\text{U:} & A + S_1 \rightarrow B + S_1(k_A) \\
\text{V:} & B + S_2 \rightarrow C + S_2(k_B) \\
\text{U:} & L + S_1 \rightarrow L \cdot S_1(k_{A,d}) \\
\text{V:} & M + S_2 \rightarrow M \cdot S_2(k_{B,d})
\end{array}
\begin{array}{l}
\Big\} \text{ (main reaction)} \\
\Big\} \text{ (deactivation reaction)}
\end{array}
\qquad \text{III}
$$

Here we will have, in addition to the multiplication of parameters inherent with the additional complexity of the reaction–deactivation scheme, the important new variable of catalyst composition. An extensive series of

VII. Constant-Conversion Operation

parametric calculations has been reported for such systems (7), but we can summarize most of the important effects with the results shown in Fig. 10-26. As one can see, there are obviously many things going on at the same time here. First, let us deal with the catalyst composition, expressed as the fraction of U function, ε_U, and its relationship to reaction temperature. For the fresh catalyst this will be determined by the conversion required, the relative values of k_A and k_B, and the relative values of γ_A and γ_B. In the illustration $k_A = k_B$, but $\gamma_A = 4\gamma_B$; hence for a given temperature the rate on the U function will be more rapid than that on the V function (one might think of this in terms of the V function as rate-determining). There is, then, an optimal catalyst composition which results in a minimum start-of-run temperature for the required conversion level. In this case, the optimal composition is $\varepsilon_U = 0.4$[8]. The left-hand side of Fig. 10-26, then, shows the temperature history $(\phi - \beta)$ for the optimal initial catalyst composition. The selectivity behavior is shown in terms of the relative amounts of A and B in the exit versus time on stream for the specified conversion of 60% C. The temperature increase required to cope with deactivation drives the reaction $A \to B$ at ever increasing rates compared to $B \to C$ (since we have equal thermal responses of the deactivation reactions on the U and V functions), with the ultimate result that A eventually disappears from the product. Obviously, the reverse behavior would occur if $\gamma_B > \gamma_A$. The point here is not so much to describe in detail all possible trends via the various possible parametric combinations, but to point out that selectivities are being driven to a limiting case by the constant-conversion operation. The reason for this, of course, is the increasing temperature—but the reason for the increasing temperature is catalyst deactivation. The distortion of selectivity by the combined effects of temperature and deactivation, in even this simple example, is sufficiently complex that one could never reasonably expect to successfully interpret laboratory data based on this mode of operation.

One other aspect of the problem has to deal further with our new variable of catalyst composition. It is reasonable to ask whether the optimal initial catalyst composition, determined on the basis of minimum start-of-run temperature, will remain so in the sense of providing maximum time on stream to a specified temperature level. The answer in general is no, as shown in the right-hand side of Fig. 10-26. In the example, as temperature is increased, less and less of the U function is required to perform its required duty in getting A to B; hence, we may take some of the "pressure"

---

[8] The values of $\gamma_i = E_i/RT_0$ are quoted with respect to a reference temperature. In problems where catalyst composition is not variable, T_0 is conveniently selected as the start-of-run temperature and ϕ (initial) = 1. Here, since there are many start-of-run temperatures, depending on ε_U, T_0 is fixed at an arbitrary value and ϕ (initial) is generally not unity. Here, $T_0 = T_{\min}$ for $\varepsilon_U = 0.4$.

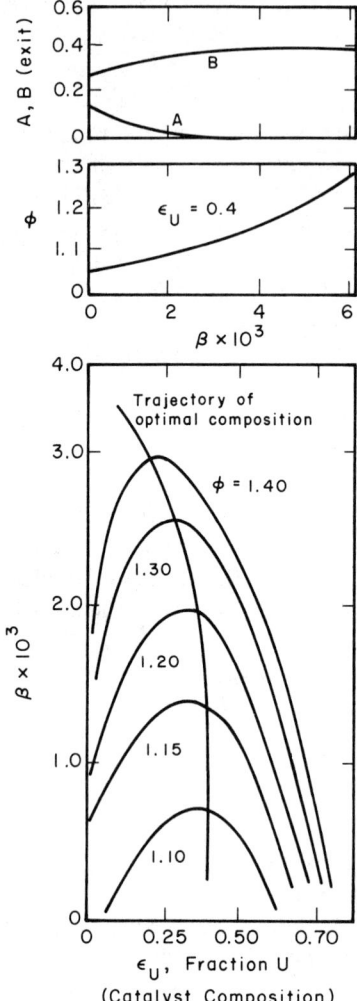

Fig. 10-26. Some results for constant-conversion operation in the deactivation of a bifunctional catalyst. Parameters: $X_U = X_V = 0.003$, $\gamma_A = 20$, $\gamma_B = 5$, $\gamma_L = \gamma_M = 5$, $\psi_U = \psi_V = 10$. Conversion to $C = 60\%$.

off the V function by increasing its relative amount. This not only facilitates B → C via an increase in the amount of V, it also decreases the relative impact of the chemisorption of a given amount of poison on the V surface, with the net effect that the TIR is reduced. The overall result is, as seen, a trajectory of optimal compositions curving gracefully off toward the northwest to lower values of ε_U as temperature and time on stream increase.

A simple continuum theory for certain classes of deactivation has been proposed by Krishnaswamy and Kittrell (28) that has promise for the interpretation of data from deactivation experiments utilizing the TIR

VII. Constant-Conversion Operation

approach. In the case that deactivation rates are functions of the activity level alone:

$$(-r_d) = k_d s^n = -ds/dt \qquad (10\text{-}177)$$

where s is the usual activity variable and n is any arbitrary power of deactivation kinetics, then the relationship between temperature and time on stream is obtained in the following way, assuming that conversion is time-independent. We have that

$$k_d = A_d \exp(-E_d/RT) \qquad (10\text{-}178)$$

and the normal constant-conversion requirement that

$$k \cdot s = k_0 \qquad (10\text{-}179)$$

where k_0 is the rate constant at start-of-run temperature T_0, and

$$k = A_A \exp(-E_A/RT) \qquad (10\text{-}180)$$

Combining the last two expressions gives

$$1/T = (R/E_A)\ln(s) + (1/T_0) \qquad (10\text{-}181)$$

as required for constant conversion. If we then substitute Eqs. (10-181) and (10-178) into the activity expression and integrate with respect to time, a time–activity relationship is obtained directly:

$$t = \frac{\exp(E_d/RT_0)}{A_d(E_d/E_A - n + 1)} (1 - s^{(E_d/E_A) - n + 1}) \qquad (10\text{-}182)$$

substitution of Eq. (10-181) into (10-182) gives the desired temperature–time relationship:

$$t = \frac{\exp(E_d/RT_0)}{A_d(E_d/E_A - n + 1)} \left\{1 - \exp\left[\frac{E_d - nE_A + E_A}{R}\left(\frac{1}{T} - \frac{1}{T_0}\right)\right]\right\} \qquad (10\text{-}183)$$

In the case that $n = 2$ and $E_d = E_A$, this expression must be rewritten to

$$t = \frac{\exp(E_d/RT_0)}{A_d} \frac{E_A}{R}\left(\frac{1}{T_0} - \frac{1}{T}\right) \qquad (10\text{-}184)$$

The central point of the analysis is to recognize that Eq. (10-183) is of the form

$$t = C(1 - e^{AY}) \qquad (10\text{-}185)$$

where

$$C = \frac{\exp(E_d/RT_0)}{A_d(E_d/E_A - n + 1)}, \quad A = \frac{(E_d - nE_A + E_A)}{R}, \quad Y = \left(\frac{1}{T} - \frac{1}{T_0}\right)$$

Thus, if one plots data of t versus $1/T$ and subdivides the reciprocal temperature axis into four equal segments, that is,

$$\left(\frac{1}{T_0}-\frac{1}{T_1}\right), \quad \left(\frac{1}{T_1}-\frac{1}{T_2}\right), \quad \left(\frac{1}{T_2}-\frac{1}{T_3}\right), \quad \left(\frac{1}{T_3}-\frac{1}{T_4}\right)$$

then

$$C = \frac{(t_1 t_4 - t_2 t_3)}{(t_1 + t_4) - (t_2 + t_3)} \tag{10-186}$$

where t_i are the times corresponding to $1/T_i$. Now, having obtained an estimate of C, one can rearrange Eq. (10-185) to the form

$$\ln(1 - t/C) = AY \tag{10-187}$$

It is then possible to plot the left-hand side of Eq. (10-187) versus Y and, of course, if the result is linear the model is a valid representation of the TIR data.

The application of this analysis to experimental data on both hydrocracking and reforming was demonstrated by Krishnaswamy and Kittrell, with excellent results. In the case of hydrocracking, for example, a simple deactivation model with $n = 1$ provided an excellent correlation of the data, although the analysis is not limited to linear deactivation kinetics. Once values of the constants A and C are known, and recognizing from the definition of A that

$$E_d = AR + E_A(n-1)$$

one can extract individual values for the activation energies. A further discussion of the analysis of constant conversion policies in catalyst deactivation has been given by Ho (24).

VIII. SUMMARY AND EVALUATION

It may seem strange that, after the extensive presentation of this chapter, the summary discussion is rather brief, but in fact it is not unreasonable. The story of deactivation in fixed beds is one of wave propagation. These waves can be very distinct and sharp, as seen in the results of Weng *et al.* (38) and Dumez and Froment (16), or more gradual, as distinguished by Froment and Bischoff (20). It really does not make too much difference how sharp these waves are, at least in mathematical description, since the governing equations are the same and the parametric values involved are most important. One way or another, in various disguises, the parameter

VIII. Summary and Evaluation

identified as M_T (which is a poison or coke adsorption capacity) in the analysis of Weng et al. (38) finds its way into almost all reaction–deactivation models (see also the results of Chapter 8). It is particularly important in the analysis of deactivation of fixed beds, since it is the "time driver" of the migration of profiles through the bed. It is also particularly important because of the fact that most simulations are very sensitive to the value of this parameter. Thus in reconciling simulation results with experiment, it is often an easier task to match the shape of various profiles than to predict their velocity or ultimately to develop accurate correlations for bed life. This can normally be traced back to uncertainties in the determination of a poison or coke capacity parameter such as M_T.

In certain aspects, the remarks above lead to a philosophical question concerning the modeling or simulation of such fixed-bed deactivation processes namely, to what extent do we wish to proceed? Certainly this is not a new question, and it is also not confined to the analysis of deactivation processes, but is related to the "principle of optimum sloppiness" proposed by Weisz and Prater and perhaps most clearly elucidated by Carberry (10). Simply stated, what is the number of fundamental physical laws compared to the number of empirical parameters that will be required to obtain some desired result of a simulation with a minimum of effort? This, of course, requires resolution of the most difficult task at the beginning, that is, "what is the desired result?" In one sense, the correlation of Fig. 10-24 is perfectly adequate if all one desires is some indication of bed life under very specific conditions of temperature, concentrations, and the like, but very little physical or chemical science is incorporated. Yet, need we really struggle for the detail of simulation sought in results such as those of Fig. 10-18? As pointed out in this chapter, even for very economical models the number of reactor/reaction/deactivation parameters can become quite large, and the parametric sensitivity among them is very unevenly distributed. This appears not only in the more fundamental approaches of Froment and co-workers or Weng et al., but also in the older superposition models such as those of Anderson and Whitehouse or Wheeler and Robell. A poison or coke capacity factor is always going to appear in one disguise or another, and the resulting time scale of the simulation is going to be sensitive to its absolute value. It is thus probably more important in such fixed-bed deactivation problems than in many other areas of chemical reaction engineering to have a firm idea beforehand of what the objectives of a simulation are. There is a vast difference between the prediction of bed life or average bed activity versus time and the detailing of the shapes and motions of temperature, concentration, or activity profiles.

In the past, it would probably have been appropriate to include here a discussion of the impact of computational requirements on the level of simulation. It is difficult to estimate that necessity now. Whether one considers bigger and bigger or smaller and smaller computers, it seems clear that the cost of modeling or simulation on the basis of some unit of information derived has been decreasing and will continue to do so. Thus the economics of computation are a diminishing contributor to the principle of optimum sloppiness and we can indeed afford to become more philosophical in our approach to such simulation. Precise determination of physical or chemical parameters appears not to be a glamorous or well-supported area of research, yet this appears (particularly in matters related to deactivation processes) to be a rate-limiting factor in what we are now able to do, especially in view of the parametric sensitivity of these simulations.

Constant-conversion operation with increasing temperature is important in industrial procedure, both in commerical operation and in the laboratory. The time–temperature history is very much a function of the relative activation energies of the main and deactivation reaction(s), as might be expected intuitively and certainly on the basis of the results given by Blaum. In reaction systems where a selectivity factor between two reactions is involved, the selectivity will also be driven in a certain direction unless the activation energies of the reaction steps involved are the same. All of this is to say that such operational policies make it difficult to obtain an unequivocal interpretation of experimental deactivation studies in the laboratory, although this can be done in certain instances as shown by Krishnaswamy and Kittrell (28).

Constant-conversion operation is also rather closely related to a number of reactor optimization problems balancing factors such as yield or selectivity against catalyst activity. An early study was presented by Szépe and Levenspiel (35) who considered the temperature policies required for optimization of the final conversion for a specified reaction time and final catalyst activity where the deactivation rate was independent of the concentration of any component of the reaction mixture. The results indicated that a constant-conversion policy was optimal. A considerable amount of work has since followed concerning various aspects of the constant-conversion policy for optimization of reactors with catalyst decay. Detailed discussion of this is beyond our present interests, but a review is available (27).

REFERENCES

1. R. B. Anderson and A. M. Whitehouse, *Ind. Eng. Chem.* **53**, 1101 (1961).
2. R. M. Billimoria and J. B. Butt, *Chem. Eng. J.* **22**, 71 (1981).
3. K. B. Bischoff, *Ind. Eng. Chem. Fundam.* **8**, 665 (1969).

References

4. E. Blaum, *Chem. Eng. Sci.* **29**, 2263 (1974).
5. G. Bohart and E. Adams, *J. Am. Chem. Soc.* **42**, 523 (1920).
6. J. B. Butt, *Adv. Chem. Ser.* No. 109, 259 (1972).
7. J. B. Butt, *Chem. React. Eng., Proc. Eur. Symp.*, 4th, 1968, p. 255 (1971); *Chem. Eng. J.* **2**, 90 (1971).
8. J. B. Butt and R. M. Billimoria, *ACS Symp. Ser.* No. 72, 288 (1978).
9. J. B. Butt and D. M. Rohan, *Chem. Eng. Sci.* **23**, 489 (1968).
10. J. J. Carberry, "Chemical and Catalytic Reaction Engineering." McGraw-Hill, New york, 1979.
11. G. F. Carey and B. A. Finlayson, *Chem. Eng. Sci.* **30**, 587 (1975).
12. A. Chow, W. H. Ray, and R. Aris, *Trans. Inst. Chem. Eng.* **45**, T153 (1964).
13. R. P. DePauw and G. F. Froment, *Chem. Eng. Sci.* **30**, 789 (1975).
14. H. V. Doesburg and W. A. DeJong, *Adv. Chem. Ser.* No. 133, 32 (1974).
15. F. J. Dumez, Ph.D. Thesis, Rijksuniversiteit, Gent, 1975.
16. F. J. Dumez and G. F. Froment, *Ind. Eng. Chem. Process Des. Dev.* **15**, 291 (1976).
17. G. Eigenberger, *Adv. Chem. Ser.* No. 133, 36 (1974).
18. G. Eigenberger and J. B. Butt, *Chem. Eng. Sci.* **31**, 681 (1976).
19. B. A. Finlayson, "The Method of Weighed Residuals and the Variational Principle." Academic Press, New York, 1972.
20. G. F. Froment and K. B. Bischoff, *Chem. Eng. Sci.* **10**, 189 (1961); **17**, 105 (1962).
21. G. F. Froment and R. Mezaki, *Chem. Eng. Sci.* **25**, 293 (1970).
22. D. Handley and P. Heggs, *Trans. Inst. Chem. Eng.* **46**, 251 (1968).
23. H. W. Haynes, Jr., *Chem. Eng. Sci.* **25**, 1615 (1970).
24. T. C. Ho, *J. Catal.* **86**, 48 (1984).
25. E. K. T. Kam and R. Hughes, *Chem. Eng. J.* **18**, 93 (1979).
26. J. P. Kehoe and J. B. Butt, *J. Appl. Chem. Biotechnol.* **23**, 22 (1972).
27. F. S. Kovarik and J. B. Butt, *Catal. Rev.-Sci. Eng.* **24**, 441 (1982).
28. S. Krishnaswamy and J. R. Kittrell, *Ind. Eng. Chem. Process Des. Dev.* **18**, 399 (1979).
29. L. Lapidus, "Digital Computation for Chemical Engineers." McGraw-Hill, New York, 1962.
30. E. S. Lee, "Quasilinearization and Invariant Embedding." Academic Press, New York, 1968.
31. G. D. Lyubarskii, L. B. Andeeva, and N. V. Kulkova, *Kinet. Katal.* **3**, (1962).
32. J. H. Olson, *Ind. Eng. Chem. Fundam.* **7**, 185 (1968).
33. Y. Ozawa, *Chem. Eng. Sci.* **25**, 529 (1970).
34. T. H. Price and J. B. Butt, *Chem. Eng. Sci.* **32**, 393 (1977).
35. S. Szépe and O. Levenspiel, *Chem. Eng. Sci.* **23**, 81 (1968).
36. D. D. Van Zoonen, *Proc. Int. Congr. Catal.*, 3rd, 1964, p. 1319 (1965).
37. V. W. Weekman, Jr., *AIChE J.* **20**, 833 (1974).
38. H. S. Weng, G. Eigenberger, and J. B. Butt, *Chem. Eng. Sci.* **30**, 1341 (1975).
39. A. Wheeler and A. J. Robell, *J. Catal.* **13**, 299 (1969).

CHAPTER 11

Regeneration of Fixed Beds

> Once was blind, but now I see...
> *Amazin' Grace, Spiritual*

In Chapter 10 we treated in some detail the deactivation behavior of fixed-bed reactors in terms of the motion of activity and thermal waves through the bed. Such waves could be characterized, at least in one respect, by the fact that the catalyst activity downstream of the wave was greater than that upstream of (behind) the wave. We may approach the topic of this chapter, regeneration, in the same way except that the activity patterns are reversed and now the inactive zone is downstream, the active zone upstream.

Regeneration is also a perhaps too-vague term to use for the process being considered here, since regeneration of fixed-bed reactors has been considered almost exclusively in terms of coke removal via oxidation. Analogous processes concerned with detoxification of poisoned catalysts, when conducted at all, are normally conducted external to the location of deactivation.

I. REGENERATION AS A DEACTIVATION PROBLEM IN REVERSE

The general objectives in analysis of fixed-bed regeneration are quite similar to our concerns with fixed-bed deactivation. Hence we are concerned with the extent of recovery of activity, the thermal history of the process, and the time required for a certain stage of regeneration. Thermal behavior is most important here, since coke burning is a very exothermic process and knowledge of the behavior of the time-dependent thermal waves developed and their dependence on operating parameters will allow one to avoid excessive exotherms with possible catalyst or reactor damage. We can, thus, think of regeneration as deactivation in reverse; the concerns of our analysis are similar to those for deactivation but the product is an active rather than an inactive catalyst.

Some further comparison with our previous discussion of fixed-bed deactivation is interesting as well. It will be recalled that there we treated a number of different deactivation mechanisms, as represented by specific examples of coking and poisoning, differing kinetics, and different modes of reactor operation, in terms of the moving-zone concept. While there was generally a common approach to the modeling involved, differences among the examples with respect to reaction kinetics and deactivation mechanisms may have left one with the vaguely uncomfortable feeling that an elephant (or several different ones) was (were) being examined in its (their) various aspects by differing observers who "knew in part and prophesied in part" and who communicated among themselves only imperfectly. Now, in the analysis of fixed-bed regeneration we have an opportunity to reexamine the elephant from a somewhat different point of view. For one thing, we now have the same basic elephant to examine—coke burning—and a uniform set of intrinsic kinetics. This allows us to focus in some depth on the influence, for example, of intraparticle transport limitation and compare examples with corresponding ones in which such transport effects are absent. Similarly, the comparison between adiabatic and nonadiabatic operation may be made more clearly. Generally, the basic reactor model considered in most reports concerned with regeneration is one-dimensional, plug flow, with no density gradients. The influence of interphase heat transfer is one point of departure among various models, since some include energy conservation for both fluid and solid phases and some (the majority) do not. However, the major differences in the various approaches are in how the kinetics of coke burning are treated, which is in turn a reflection of the operating conditions—particularly temperature level—under analysis. Some of the earlier studies (6) considered kinetics to be zero order and temperature-independent, that is, a constant burning rate model, which might be considered a rough approximation to high-temperature operation in excess oxygen with large amounts of coke on the catalyst. However, it seems reasonably well established that the best representation of intrinsic coke burning kinetics is given by (1, 7):

$$(-r)_c = k_c P y_{O_2} C_c \qquad (11\text{-}1)$$

as, indeed, we discussed in some detail earlier. A form of equation related to Eq. (11-1):

$$(-r)_c = k P y_{O_2} (C_c / C_c^0) \qquad (11\text{-}2)$$

normalized to the initial coke concentration, C_c^0, has been used by Johnson et al. (2). They claimed that this is an approximate representation of the rate when oxygen diffusion is the controlling mechanism. Equation (11-2) seems to differ only in some detail via k and k_c from the intrinsic kinetics of Eq. (11-1); however, it can be used as an approximate model for the diffusion-limited case in the event that k_d is taken to be independent of temperature. In the limit of complete diffusion-limited burning, though, one would observe the shell-progressive nature of the process; the operating conditions corresponding to these kinetics would generally include higher temperatures, and an appropriate kinetic model for shell-progressive burning is

$$(-r)_c = \frac{3\alpha D_e C_{O_2}}{R^2 \rho_g}\left[\left(\frac{C_c^0}{C_c}\right)^{1/3} - 1\right]^{-1} \tag{11-3}$$

where D_e is the effective diffusivity of oxygen in the catalyst particle, ρ_g the gas density, and α a stoichiometric coefficient relating grams of coke burned per gram mole of oxygen reacted (3).

II. SOME QUANTITATIVE STUDIES

A. Constant Burning Rate

As mentioned in the previous section, the original studies of van Deempter envisioned $(-r)_c$ independent of concentration and temperature, hence

$$(-r)_c = r_{O_2} = U \text{ per mole } O_2 \tag{11-4}$$

We shall not attempt here to set forth a complete exposition of the results of van Deempter, since the constant-rate approximation results in linear gradients, which we know from prior experience will be nonlinear moving boundaries in reality. However, some discussion of this simplified model is convenient to set the stage for other things to come. For the oxygen balance in the reactor we have then:

$$\varepsilon(\partial C_{O_2}/\partial \theta) + v(\partial C_{O_2}/\partial z) = -r_{O_2} = -U \tag{11-5}$$

where C_{O_2} is moles/volume, v superficial velocity, z bed length, and ε porosity. The heat balance is

$$[(1-\varepsilon)\rho_s c_s + \varepsilon \rho_g c_g](\partial T/\partial \theta) + \rho_g c_g v(\partial T/\partial z) = U(-\Delta H) \tag{11-6}$$

where $(-\Delta H)$ is the heat of reaction per mole of oxygen and ρ_g, ρ_s, c_g, and c_s are the appropriate densities and heat capacities for solid and gas phases.

Now the boundary conditions for solution of these equations in the initial stage of operation are

$$z = 0: \quad C_{O_2} = C_{O_2}^0, \quad T = T_0$$
$$\theta = 0: \quad (z > 0) C_{O_2} = 0, \quad T = 0$$

The solution to the heat balance is obtained, via the method of characteristics, as

$$T = T_0 + U \Delta H z / \rho_g c_g v \qquad (11\text{-}7)$$

where

$$\alpha = \rho_g c_g / [(1-\varepsilon)\rho_s c_s + \varepsilon \rho_g c_g] \qquad (11\text{-}8)$$

This depicts the oxidation under these initial conditions as being confined to an initial region of the bed, $z < \alpha v \theta$, and the heat produced in this region is transported through the bed with velocity αv. For $z > v\theta$, the temperature must correspond to the boundary condition $T = 0$. Also for the initial operation, the solution to the oxygen balance, Eq. (11-5), is

$$C_{O_2} = C_{O_2}^0 - Uz/v \qquad (11\text{-}9)$$

for $0 \leq z \leq \alpha v \theta$. The coke distribution follows from the kinetics of Eq. (11-4):

$$C_c = C_c^0 - U(\theta - z/\alpha v) \qquad (11\text{-}10)$$

Thus, all balance equations result in simple linear profiles. A zone $\alpha v \theta$ in length is the portion of the reactor in which regeneration is occurring. The behavior of the regenerating bed, quantitatively, is very much a function of the relative values of initial oxygen and coke concentrations (i.e., whether oxygen is present in stoichiometric excess or not). From the coke distribution equation, we may solve for the time for complete coke removal, θ_1, as

$$\theta_1 = C_c^0 / U \qquad (11\text{-}11)$$

Similarly, from Eq. (11-9) we can say that oxygen is completely reacted from the gas phase at a distance into the bed z_0, defined as

$$z_0 = C_{O_2}^0 v / U \qquad (11\text{-}12)$$

or at the corresponding time θ_0,

$$\theta_0 = z_0 / \alpha v = C_{O_2}^0 / \alpha U \qquad (11\text{-}13)$$

Two different types of behavior can be envisioned, dependent on the relative values of the "coke time," θ_1, and the "oxygen time," θ_0. When $\theta_0 > \theta_1$, oxygen is in excess and the coke at bed entrance will react completely with a finite concentration of oxygen remaining beyond the zone of reaction.

II. Some Quantitative Studies

This is depicted in Fig. 11-1a. Conversely, $\theta_1 > \theta_0$ defines oxygen as the limiting reactant and there will be an initial "induction period" in which the zone of combustion is stabilized within some entrance region until the coke disappears. In the intermediate case, there is a limiting lower bound to $C_{O_2}^0$ that corresponds to exact stoichiometry, which we may define as $\theta = \theta_0$. This intermediate case is shown in Fig. 11-1b and that for $\theta_1 > \theta_0$ in Fig. 11-1c.

The simplest case to look at in analysis is that in which oxygen is in stoichiometric excess, $\theta_0 > \theta_1$. Here both the front and back of the oxidation

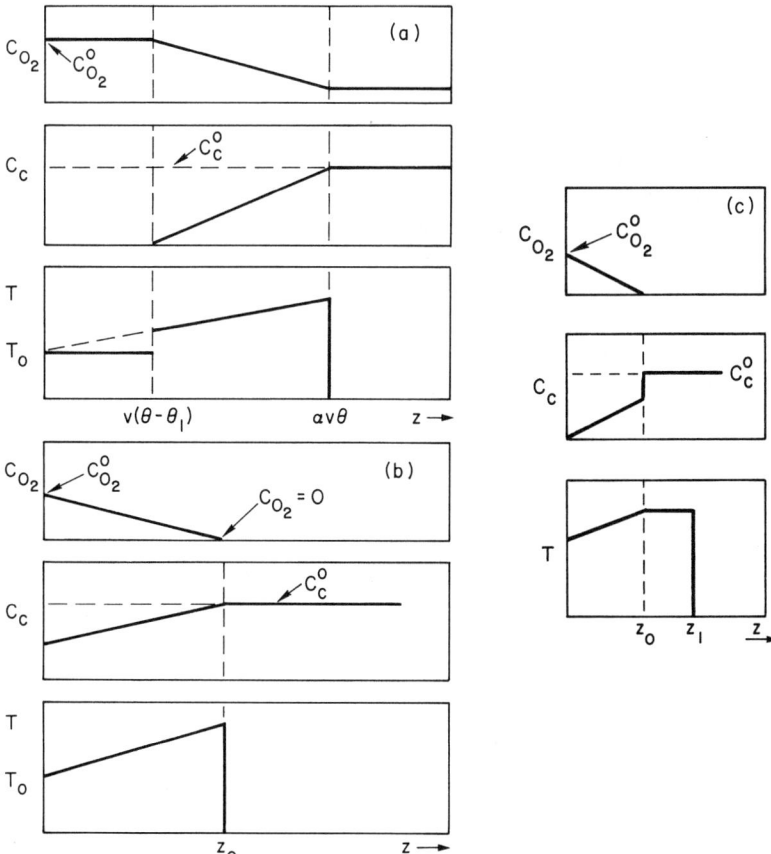

Fig. 11-1. Initial regeneration regimes for high and intermediate oxygen concentrations. (a) High oxygen concentration, $\theta_0 > \theta_1$. (b) Intermediate oxygen concentration, $\theta = \theta_0$. (c) Initial regeneration regime for low oxygen concentration, $\theta = \theta_1$. [From van Deempter (6). Reproduced by permission of the American Chemical Society.]

zone move with a velocity, αv, equal to that of heat transport. Thus the oxidation zone is constant in size as shown, and the temperature profile is

$$0 \leq z \leq \alpha v(\theta - \theta_1), \qquad T = T_0 \qquad (11\text{-}14)$$

$$\alpha v(\theta - \theta_1) \leq z \leq \alpha v\theta, \qquad T = T_0 + U\,\Delta Hz/\rho_g c_g v \qquad (11\text{-}15)$$

The depth of the oxidizing zone is

$$\alpha v \theta_1 = \alpha C_c^0 v / U \qquad (11\text{-}16)$$

and the time required for regeneration corresponds to that for the trailing edge of the oxidizing zone to pass through the bed:

$$z/\alpha v + \theta_1 = z/\alpha v + C_{O_2}^0/U \qquad (11\text{-}17)$$

Now, when $\theta_1 > \theta_0$, the situation becomes much more complicated and the location and nature of the reaction zone depend on the magnitude of regeneration time θ with respect to both θ_0 and θ_1. This is illustrated in Figs. 11-1b and 11-1c. At $\theta = \theta_0$ the reaction zone z_0 remains stationary in the bed until the coke at the bed entrance is depleted. At $\theta = \theta_1$ the coke at $z = 0$ has burned, and discontinuity in coke concentration results at z_0 (Fig. 11-1c). We recall that U, the burning rate, is nonzero only in the burning zone; the heat generated in this zone gives rise to the temperature profile of Eq. (11-7), and although during the interval from θ_0 to θ_1 the burning zone is stationary, the heat generated is removed from the zone at a velocity αv and a temperature level

$$T = T_0 + U\,\Delta H z_0/\rho_g c_g v \qquad (11\text{-}18)$$

as shown in Fig. 11-1c.

For $\theta > \theta_1$ the burning zone moves into the bed until the trailing edge coincides with the discontinuity of the coke concentration at z_0. In this interval the depth of the burning zone (determined only by the oxygen concentration) remains constant at z_0 and the thermal front passes through the bed such that

$$0 \leq z \leq \alpha v(\theta - \theta_1), \qquad U = 0, \qquad T = T_0$$

$$\alpha v(\theta - \theta_1) < z < z_0 + \alpha v(\theta - \theta_1), \qquad U \neq 0, \qquad T = T_0 + U\,\Delta Hz/\rho_g c_g v$$

$$z_0 + \alpha v(\theta - \theta_1) \leq z \leq \alpha v\theta, \qquad U = 0, \qquad T = T_0 + U\,\Delta H z_0/\rho_g c_g v$$

$$z > \alpha v\theta, \qquad U = 0, \qquad T = 0 \qquad (11\text{-}19)$$

The temperature always increases linearly in the burning zone while behind and ahead of it the temperature is constant.

When the trailing edge of the zone reaches the coke discontinuity at z_0, it again becomes stationary until at $\theta = 2\theta_1$ all the coke has been removed.

II. Some Quantitative Studies

Again heat is removed from the stationary zone, although this time at a temperature level $T = T_0 + 2U\,\Delta H z_0/\rho_g c_g v$, and entering gas at T_0 decreases the level of temperature in the region $z_0 \le z \le \alpha v \theta_1$ of the zone.

Obviously, the stop and go behavior of the constant-rate model is not going to be encountered in practice; what the model is trying to tell us within the girdle of its assumptions is that a thermal wave and a moving reaction zone are developing in either of the cases, the detailed nature of which depends on coke and oxygen concentration levels. Nonetheless, some approximate operating limits can be obtained from this crude model. The mean regeneration time is

$$z\theta_1/\theta_0 + \tfrac{1}{2}(\theta_0 + \theta_1) = C_c^0 z / C_{O_2}^0 v + \tfrac{1}{2}(C_c^0/U + C_{O_2}^0/\alpha U) \quad (11\text{-}20)$$

and the temperature profile is bounded by

$$T \le T_0 + U\,\Delta H z/\rho_g c_g v \quad (11\text{-}21)$$

so the simple theory does yield results good for rough estimates of regenerator design parameters. A number of other workers have confirmed the form of Eq. (11-21) for determining temperature rise inside the bed during regeneration. For the important case of low inlet oxygen concentration where the thermal transport velocity exceeds the combustion zone velocity, a boundary temperature may be established quantitatively:

$$\Delta T = T_{\max} - T_0 = \frac{C_{O_2}^0 \Delta H}{\rho_g c_g (1 - C_{O_2}^0/\alpha C_c^0)} \quad (11\text{-}22)$$

Such estimates, in terms of practical utility, will depend on the conservatism of the estimation of U and the validity of the approximation of constant U. Obviously, in order to build any more generality into the model we must be more specific concerning the regeneration kinetics.

B. A Kinetic Model with Temperature-Independent Rate Constant

A closer approach to reality, particularly that of diffusionally limited burning, is to employ the rate form of Eq. (11-2) where the rate constant k is independent of temperature. Hence, the kinetic model should display generally correct trends with respect to reaction orders in oxygen and coke and also reflect the small temperature dependence of diffusionally limited processes. The oxygen balance of Eq. (11-5) is thus modified to

$$\varepsilon(\partial C_{O_2}/\partial\theta) + \varepsilon v'(\partial C_{O_2}/\partial z) = -a(\rho_s/M_c)kPy_{O_2}(C_c/C_c^0) \quad (11\text{-}23)$$

in which
$$C_{O_2} = (\rho_g/M_g)y_{O_2} \qquad (11\text{-}24)$$
$$v = G/\varepsilon\rho_g = v'/\varepsilon \qquad (11\text{-}25)$$

where ρ_g and ρ_s are density of the gas and the bed (bulk), respectively, in mass units, M_g and M_c molecular weights of gas and carbon, respectively, P is total pressure, a is a stoichiometric factor having a value of unity for combustion to CO_2, y_{O_2} is the mole fraction of oxygen, and G is the gas mass velocity. The carbon balance is

$$(-r)_c = (\partial C_c/\partial\theta) = kPy_{O_2}(C_c/C_c^0) \qquad (11\text{-}26)$$

and the heat balance is

$$(\rho_s c_s + \varepsilon\rho_g c_g)\frac{\partial T}{\partial\theta} + c_g G\frac{\partial T}{\partial z} = \rho_s(-\Delta H)kPy_{O_2}\left(\frac{C_c}{C_c^0}\right) \qquad (11\text{-}27)^1$$

and boundary conditions are

$$z=0, \qquad y_{O_2} = y_{O_2}^0 \qquad (11\text{-}28)$$
$$C_c = C_c^0, \qquad \theta - z(\varepsilon\rho_g/G) \le 0 \qquad (11\text{-}29)$$

In nondimensional form, this set of equations can be written as

$$(\partial y_{O_2}/\partial w) = -ay_{O_2}(C_c/C_c^0) \qquad (11\text{-}30)$$
$$(y_{O_2}^0/C_c^0)(\partial C_c/\partial\tau) = -y_{O_2}(C_c/C_c^0) \qquad (11\text{-}31)$$
$$M(\partial t/\partial\tau) + H(\partial t/\partial w) = y_{O_2}C_c/y_{O_2}^0 C_c^0 \qquad (11\text{-}32)$$

where

$$\left.\begin{array}{ll} y_{O_2} = y_{O_2}^0, & w=0, \quad \tau \ge 0 \\ C_c = C_c^0, & \tau = 0, \quad w \ge 0 \\ t = 0, & w = 0 \end{array}\right\} \qquad (11\text{-}33)$$

$$H = \frac{M_g c_g C_c^0}{M_c c_s y_{O_2}^0}, \qquad M = \left(1 - \frac{c_g}{c_s}\frac{\varepsilon\rho_g}{\rho_s}\right)$$

$$\left.\begin{array}{l} t = T\dfrac{c_s}{(-\Delta H)C_c^0}, \qquad w = z\left(\dfrac{\rho_s}{G}\dfrac{M_g}{M_c}kP\right) \\ \\ \tau = \left(\dfrac{y_{O_2}^0}{C_c^0}kP\right)\left(\theta - z\dfrac{\varepsilon\rho_g}{G}\right) \end{array}\right\} \qquad (11\text{-}34)$$

and

---

[1] The formulation here is that of Johnson *et al.* (2). The value of $(-\Delta H)$ is in mass units, since C_c in the kinetic model is expressed by weight/weight catalyst; hence one must be careful in numerical comparison with the results of van Deempter, which employ molal concentrations. Of course, any consistent set of dimensions can be employed for any of the formulations discussed here.

II. Some Quantitative Studies

This set yields a straightforward solution. Since the reaction rate is independent of temperature, the mass balance equations can be solved separately as

$$C_c/C_c^0 = 1/[1 + e^{-aw}(e^\tau - 1)] \tag{11-35}$$

$$y_{O_2}/y_{O_2}^0 = 1/[1 + e^{-\tau}(e^{-aw} - 1)] \tag{11-36}$$

which result in the profiles (solid lines) shown for a typical calculation in Figs. 11-2b and 11-2c. Furthermore, we see from Eq. (11-32) that the product of Eqs. (11-35) and (11-36) is just the heat generation function; this result is plotted in Fig. 11-2a. Now the initial development of these profiles is not unlike the initial transients we described previously in the poisoning of a fixed-bed adiabatic reactor—corresponding in general to the fast concentration response (FCR) and slow temperature response (STR) periods in which the profiles evolve to a quasi-steady shape which is maintained thereafter

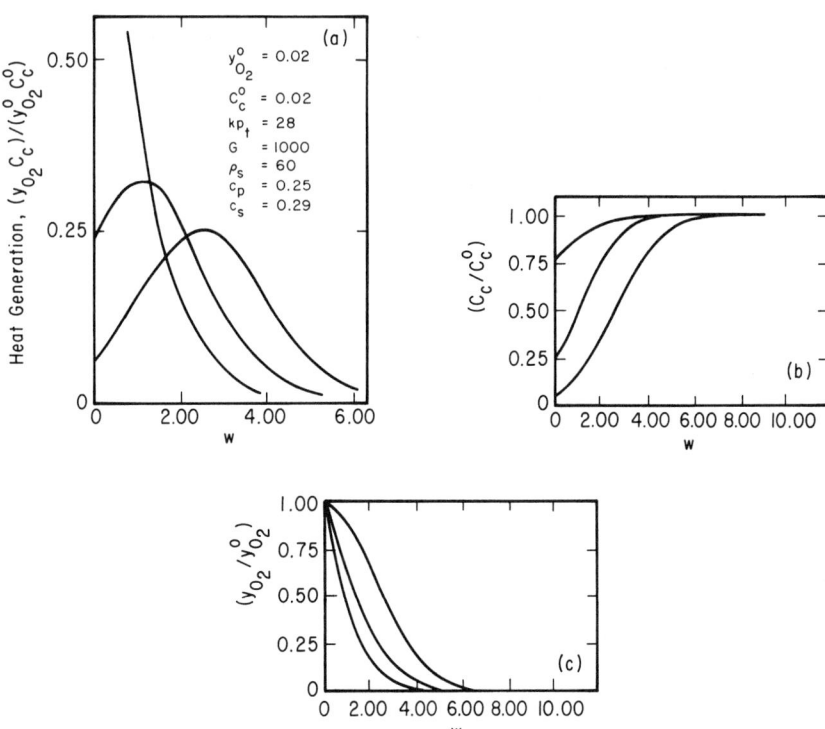

Fig. 11-2. Development of constant pattern profiles, carbon burning with temperature-independent kinetics. (a) Heat generation; (b) coke profiles; (c) oxygen profiles. [From Johnson et al. (2). Reproduced by permission of Pergamon Press, Ltd.]

as they pass through the bed. If we neglect the FCR–STR response regions and say the profiles are fixed at $\tau = 0$, then approximate solutions are given as

$$C_c/C_c^0 = 1/(1+e^{-aw+\tau}) \tag{11-37}$$

$$y_{O_2}/y_{O_2}^0 = 1/(1+e^{aw-\tau}) \tag{11-38}$$

The discrepancies at small τ are particularly large for the heat generation function, but they quickly die out. Hence, the quasi-steady solutions are entirely adequate for prediction of long-term bed behavior and the estimation of quantities such as regeneration cycle length. Using the approximations of Eqs. (11-37) and (11-38), we may solve the heat balance analytically to obtain

$$t = \frac{1}{2(aH-M)}\left[\tanh\frac{aw-\tau}{2} - \tanh\frac{w-(H/M)\tau}{2H/M}\right] \tag{11-39}$$

The first term measures the rate of translation of the burning zone, while the second term measures the convective transport rate of a temperature wave in the absence of heat generation. Two representations of temperature waves—temperature versus position with time as a parameter and temperature versus time with position as a parameter—are shown in Fig. 11-3. It is apparent from their shapes that temperature maxima are also an important consideration in fixed-bed regeneration. According to the two means of representing the results, two maxima are identifiable: $t_{max,w}$, which is the maximum temperature at a point w for any time τ, and $t_{max,\tau}$, which is the instantaneous maximum temperature along the length of the reactor. By approximate differentiation of Eq. (11-39) we have

$$t_{max,w} = \frac{1}{aH-M}\left[\tanh\frac{wM}{4H}(aH-1)\right] \tag{11-40}$$

$$t_{max,\tau} = \frac{1}{2(aH-M)}[\tanh(\Omega - \tfrac{1}{2}\ln X_1) - \tanh(-\Omega - \tfrac{1}{2}\ln X_2)] \tag{11-41}$$

in which

$$\left.\begin{aligned}\Omega &= \frac{w}{4}\left(\frac{aH-M}{H}\right) \\ X_1 &= \frac{\exp[(aH-M)/2Hw] - (aH-M)^{1/2}}{(aH/M)^{1/2}\exp[(aH-M)/2H] - 1} \\ X_2 &= \frac{\exp[(aH-M)/2Hw] - (aH-M)^{1/2}}{(aH/M)^{1/2}\exp[(aH-M)/2H] - 1}\end{aligned}\right\} \tag{11-42}$$

II. Some Quantitative Studies

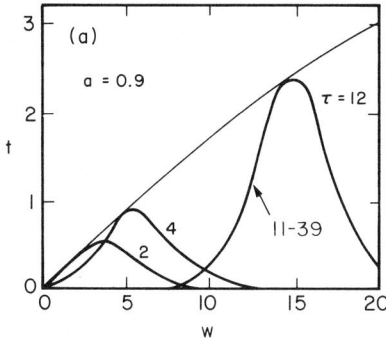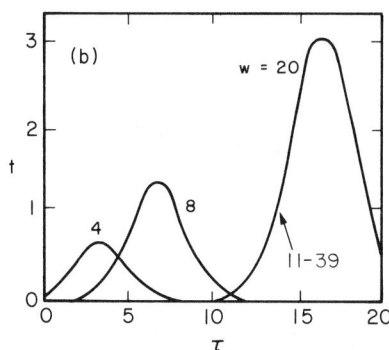

Fig. 11-3. Thermal waves computed from quasi-steady approximation, adiabatic regeneration. (a) Temperature versus position, time as parameter; (b) temperature versus time, position as parameter. [From Johnson *et al.* (2). Reproduced by permission of Pergamon Press, Ltd.]

The regeneration time can be computed from the time required for the midpoint of the burning zone (i.e., where $C_c/C_c^0 = 0.5$) to reach the end of the bed, plus the time required to reduce C_c/C_c^0 to some desired arbitrarily low level. The midpoint time is aw, and the additional time required to reduce coke to a final value of C_c/C_c^0 from 0.5 is proportional to the log ratio $\ln[1-(C_c/C_c^0)_F]/(C_c/C_c^0)_F$; thus:

$$\tau_{\text{regen}} = aw + \ln \frac{1-(C_c/C_c^0)_F}{(C_c/C_c^0)_F} \qquad (11\text{-}43)$$

Inlet oxygen concentration has a signal influence on the behavior of the bed. If $M \approx 1$, which is a reasonable approximation, it can be shown that for

$$aH > 1, \quad \text{that is;} \quad \left(\frac{y_{O_2}^0}{C_c^0}\right) < \frac{M_g}{M_c} \frac{c_g}{c_s} a \qquad (11\text{-}44)$$

then the temperature wave $t_{\max,w}$ travels down the bed faster than the burning zone; for $aH = 1$ the two velocities are the same, and for $aH < 1$ the burning zone precedes the temperature maximum. The physical implications of this are most significant, since when the temperature wave and the burning zone have the same velocity ($aH = 1$) a reinforcement between the two occurs, resulting in very large temperature rises in the bed. Now aH is determined by the ratio of inlet oxygen and carbon concentrations, so we can determine a critical inlet oxygen–carbon ratio from Eq. (11-44) as

$$\left(\frac{y_{O_2}^0}{C_c^0}\right)_{\text{crit}} = a\frac{c_g}{c_s}\left(\frac{M_g}{M_c}\right) \qquad (11\text{-}45)$$

Operation at $(y_{O_2}^0/C_c^0)$ *either* greater or lower than that specified by Eq.

(11-45) will avoid the problem of reinforcement of the hot spot. This, of course, is a much more refined result corresponding to trends only indicated in the van Deempter analysis.

We have already noted some similarities between the FCR–STR characteristics of the adiabatic fixed-bed regeneration problem and those of the poisoning of an adiabatic reactor. However, the temperature profiles shown in Fig. 11-3a are very different from those discussed previously for the poisoning problem. Part of this difference is the result of assuming the burning rate constant independent of temperature; hence, while there can be a reinforcement between reaction and thermal zones, this is not further intensified by the exponential amplification of an Arrhenius rate dependence. Another part of the difference has to do with the magnitude of the reaction zone and the steepness of the oxygen and carbon profiles within the bed during regeneration conditions. The profiles illustrated here (namely oxygen in Fig. 11-2c) are rather sharp, and once the limiting reactant (oxygen or carbon) is depleted, the reaction is quenched and the only processes occurring downstream aside from convection are those associated with the heat capacity of the solid phase. For less severe gradients the zone of combustion is much longer and the corresponding temperature profiles present something more representative of what one would consider an adiabatic front. Shulman (5), for example, presented the results of a numerical simulation of adiabatic fixed-bed regeneration in which characteristic thermal waves are very different from those of Fig. 11-3b; these are shown

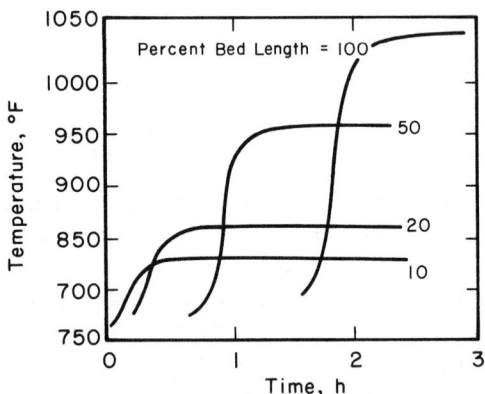

Fig. 11-4. Other types of thermal waves in adiabatic regeneration. Inlet temperature = 780°F; (x_0) inlet O_2 mole fraction = 0.01; (y_0) inlet weight fraction C on catalyst = 0.066; bulk density of bed = 55.2 lb/ft^3; specific heat of catalyst = 0.25 Btu/lb-°F; specific heat of gas = 0.25; bed length = 7.5 ft; superficial gas velocity = 2700 ft/hr; gas density = 0.212 lb/ft^3; (P) total pressure = 75 psia; gas MW = 30; (k_0) rate constant at inlet = 0.103; $dy/d\theta = -kPxy$. [From Shulman (5). Reproduced by permission of the American Chemical Society.]

in Fig. 11-4. The model employed is quite similar to that of Johnson et al., although parametric values and operating conditions are much different[2]. The profiles shown in Fig. 11-4 are qualitatively understood in comparing the velocity of the temperature wave versus the time required for complete carbon burning. For the parameters of this calculation, temperature breakthrough at bed exit is occurring at about 1.8 hr; however, Shulman reported that the time required for complete regeneration in this example is 30 hr. Hence, in the language of Johnson et al., the thermal wave velocity is much greater than the burning zone, that is, $aH \ll 1$.

C. Shell-Progressive Burning

While the temperature-independent rate coefficient is an approximation to diffusion-limited burning kinetics, the theory exists to formulate a completely detailed model based on shell-progressive burning of coke within the individual catalyst particles. However, since the shell-progressive theory is an intraparticle formulation, its inclusion in an overall model will require separate balances for gas and solid phases. This formulation has been reported by Olson et al. (3), using the Weisz–Goodwin model for shell-progressive burning kinetics, assuming an initial parabolic coke profile within the individual particles.

For the intraparticle situation, the motion of the dimensionless coked zone radius is ($s = r_c/R$)

$$(W\rho_s)(\partial s/\partial \theta) = \alpha D_e C_g / R^2 [(1 - 1/\text{Nu})s^2 - s] \quad (11\text{-}46)$$

where α is the stoichiometric coefficient (grams/gram O_2), W the coke concentration, and

$$\text{Nu} = KR_g T_s R / PD_e \quad (11\text{-}47)$$

with K an oxygen mass transfer coefficient, R_g the gas constant, and T_s the (uniform) temperature of the catalyst particle. The heat balance for the solid phase is

$$\rho_s c_s \frac{\partial T_s}{\partial \theta} = 3s^2 \frac{\partial s}{\partial \theta} \rho_s W(-\Delta H) - \frac{3}{R} h(T_s - T_g) \quad (11\text{-}48)$$

with h an interphase heat transfer coefficient. For the gas phase, we have

---

[2] In addition, the temperature dependence of the burning rate constant was included in the computation. It is not clearly stated what value was used for the activation energy, although one might infer ~ 5 kcal/mol.

for a heat balance:

$$\varepsilon \rho_g c_g \frac{\partial T_g}{\partial \theta} + G c_g \frac{\partial T_g}{\partial z} = \frac{3}{R}(1-\varepsilon)h(T_s - T_g) \qquad (11\text{-}49)$$

and mass balance (oxygen):

$$\varepsilon \frac{P}{R_g} \frac{\partial}{\partial \theta}\left(\frac{x_g}{T_g}\right) + \frac{G}{M} \frac{\partial x_g}{\partial z} = \frac{3}{R}(1-\varepsilon)\rho_s W s^2 \frac{\partial s}{\partial \theta} \qquad (11\text{-}50)$$

with x_g the mole fraction of oxygen, G the mass flow rate, and M the average gas molecular weight. Normal boundary conditions are

$$\left.\begin{array}{ll} x_g = x_g^0, & z = 0 \\ T_g = T_g^0, & z = 0 \\ s = 1, & \theta = 0 \\ T_s = T_g^0, & \theta = 0 \end{array}\right\} \qquad (11\text{-}51)$$

The parabolic coke concentration within the particle initially is given by

$$W = W^*[b + (r/R)^2] \qquad (11\text{-}52)$$

where W^* and b are scaling parameters. Olson *et al.* report a general numerical solution of this set under the assumption that the time derivatives in Eqs. (11-48)–(11-50) are negligible. The general results are in accord with what we saw previously, but with one major surprise. Very large temperature transients were found to occur at the initial stages of bed regeneration. These transients are the result of the initially high rate of reaction and heat generation occurring when the regeneration gas, relatively rich in oxygen at the bed inlet, contacts unregenerated catalyst in which the diffusion limitation characteristic of the shell-progressive mechanism has not yet had the opportunity to develop. Some details of this are shown in Fig. 11-5. It turns out that the development of the initial combustion zone and the associated temperature transients are well correlated as a function of the group $(PD_e/\sqrt{GR})$, where P is the total pressure, D_e the effective diffusivity of oxygen within the pellet, R the pellet radius, and G the gas mass flow rate based on void volume. This group essentially expresses an internal-to-external mass transfer ratio since the external mass transfer coefficient is proportional to $\sqrt{G}$ and can be thought of as a mass Biot number. Note that as the value of this group decreases, the transient temperature can exceed by a considerable amount the asymptotic maximum estimated from the van Deempter analysis, Eq. (11-22).

Aside from these initial temperature transients, the regeneration process is characterized, as in the other cases we have discussed, by the development

II. Some Quantitative Studies

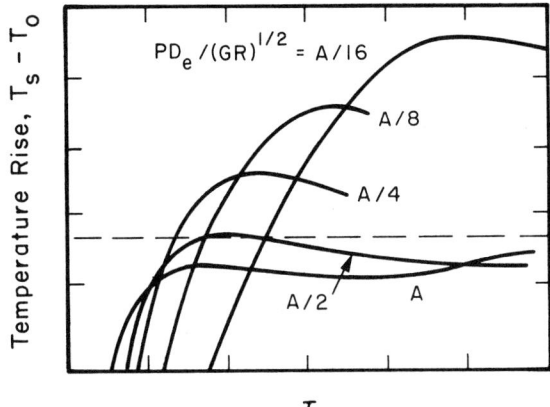

Fig. 11-5. Initial temperature transients; fixed bed regeneration with shell-progressive burning. Dashed line, Eq. (11-22). [From Olson *et al.* (3). Reproduced by permission of the American Chemical Society.]

of quasi-stationary profiles that pass through the bed with generally constant shape and velocity.

It is of interest here to make a direct comparison with the results of Johnson *et al.* to explore directly the adequacy of their approximation to diffusion-limited burning. The two sets of kinetics are

$$(-r)_c = (dC_c/d\theta) = kPy_{O_2}(C_c/C_c^0) \tag{11-53}$$

and

$$(-r)_c = (dC_c/d\theta) = (3\alpha D_e C_{O_2}/R^2 \rho_g)[(C_c^0/C_c)^{1/3} - 1]^{-1} \tag{11-54}$$

where Eq. (11-54) is an equivalent representation of the coke zone model written in terms of coke concentration, and where the coke concentrations on the RHS of Eqs. (11-53) and (11-54) refer strictly to average values.

A comparison of temperature profiles computed from the quasi-stationary model of Olson *et al.*, with kinetics modeled via Eq. (11-54), was made with the results of Johnson *et al.* for a specific numerical example. The k of Eq. (11-53) was determined by requiring the axial distance between $(y_{O_2}/y_{O_2}^0)$ values of 0.25 and 0.75 to be the same for both calculations. On this basis the comparison of results shows the two computations to be essentially indistinguishable. The major difference seems to be a slightly more diffuse profile of both oxygen concentration and temperature near the leading edge of the burning zone in the Johnson *et al.* calculation—hardly of major importance. It should be noted that the Johnson *et al.* model does not predict the initial temperature transients noted by Olson *et al.*; however, inclusion of separate heat balances for solid and gas phases makes little difference in the overall computed results in the quasi-stationary region.

D. Low-Temperature Regeneration

We have discussed in the foregoing examples various approaches to the fixed-bed regeneration problem when the burning rate is controlled by the mass transfer of oxygen, in the limit the shell-progressive situation just discussed. At lower temperatures ($\leq 450°C$ from the data of Weisz and Goodwin) the coke burning reaction will be controlled by the intrinsic kinetics where the burning rate is uniform throughout the individual catalyst pellets. In both cases, the burning rate can be expressed in terms of the fraction of carbon remaining and the oxygen partial pressure; however, the effective activation energy is in the range 2–8 kcal/mol in the mass transfer-controlled region, but is ~37.5 kcal/mol for intrinsic kinetics (8).

In the case of regeneration of some zeolite or other temperature-sensitive catalysts, the regeneration process may be carried out at lower temperatures where burning rates are controlled by intrinsic kinetics, and the process is much more temperature-sensitive owing to the approximately fourfold increase in activation energy. This low-temperature regeneration process was studied by Ozawa (4), via reformulation of the familiar mass and energy balances including intrinsic coke burning kinetics. In nondimensional form, we obtain a set of equations somewhat reminiscent of Eqs. (11-30)–(11-32).

Oxygen balance:

$$(\partial X/\partial t) + (\partial X/\partial w) = -A \exp(-1/T) XY \tag{11-55}$$

with

$X = 0$ at $t = 0$ for $1 \geq w \geq 0$

$X = 1.0$ at $w = 0$ for $t > 0$

Coke kinetics:

$$(\partial Y/\partial t) = -B \exp(-1/T) XY \tag{11-56}$$

with

$Y = 1.0$ at $t = 0$ for $1 \geq w \geq 0$

Heat balance:

$$D(\partial T/\partial t) + H(\partial T/\partial w) = \exp(-1/T) XY \tag{11-57}$$

with

$T = T_0$ at $t = 0$ for $1 \geq w \geq 0$

$T = T_0$ at $w = 0$ for $t > 0$

II. Some Quantitative Studies

where

$$\left.\begin{aligned} X &= x/x_0 \\ Y &= y/y_0 \\ T &= R\theta/\Delta E \\ w &= \xi/L \\ t &= \tau G/L\rho_g\varepsilon \end{aligned}\right\} \quad (11\text{-}58)$$

$$\left.\begin{aligned} A &= (\alpha k_0 t_c) \frac{M_g}{M_c} \frac{\rho_s}{\rho_g} \frac{1-\varepsilon}{\varepsilon} \\ B &= k_0 t_c \frac{X_0}{Y_0} \\ D &= \frac{1}{k_0 x_0 R t \rho_s} \left[\frac{\varepsilon}{1-\varepsilon} \rho_g c_g + \rho_s c_s\right]\left(\frac{\Delta E}{\Delta H}\right) \\ H &= \frac{\rho_g c_g \varepsilon}{k_0 x_0 R t_c \rho_s (1-\varepsilon)} \frac{\Delta E}{\Delta H} \\ t_c &= L\rho_g \varepsilon / G \end{aligned}\right\} \quad (11\text{-}59)$$

This model, in fact, differs only in minor detail from those discussed previously, except that the rate constant embedded in the parameters D and H is now a very sensitive function of temperature.

A numerical scheme based on the characteristics which can be used to convert the partial differential equations into ordinary differential equations is conveniently applied to problems such as this if full transient solutions are desired. The gas bulk velocity and the heat propagation velocity along the characteristics are as follows.

Gas bulk velocity:

$$U_{O_2} = FL/V\varepsilon \quad (11\text{-}60)$$

Oxygen transit time (required for oxygen front to reach end of reactor):

$$t_{O_2} = L/U_{O_2} = V\varepsilon/F = t_c \quad (11\text{-}61)$$

Heat propagation velocity:

$$U_T = LF\rho_g c_g / [\varepsilon \rho_g c_g + (1-\varepsilon)\rho_s c_s] V \quad (11\text{-}62)$$

Heat transit time (required for heat front to reach end of reactor):

$$t_T = L/U_T = \left[1 + \frac{1-\varepsilon}{\varepsilon} \frac{\rho_s c_s}{\rho_g c_g}\right] \frac{V\varepsilon}{F} \quad (11\text{-}63)$$

TABLE 11-1

Parameters for Numerical Solution of the Low-Temperature Regeneration Problem

$x_0 = 0.0175$	$c_s = 0.292$ cal/g-°C
$y_0 = 0.06$	$c_g = 0.294$ cal/g-°C
$\theta_0 = 800°F$	$\Delta E = 37{,}500$ cal/gmol
$\rho_s = 1.0$ g/cm^3	$\Delta H = 83{,}100$ cal/gmol
$\rho_g = 4.52 \times 10^{-4}$ g/cm^3	$k_0 = 1.80 \times 10^{11}$ hr^{-1}
$\varepsilon = 0.35$	$V = 300$ cm^3
$M_c = 12$ g/gmol	$F = 24.0$ SCF/hr
$M_g = 28$ g/gmol	

Typical parameters for a numerical solution to these equations are given in Table 11-1. Under these conditions, the oxygen and heat transit times are $t_{O_2} = 0.20$ sec and $t_T = 13.7$ min.

The heat transit time is much longer than the oxygen transit time because of the greater density of a catalyst particle compared to a gas. This results in two widely divergent characteristic directions, which require a small mesh size to attain numerical convergence. The unsteady-state profiles so computed are shown in Fig. 11-6. The most interesting feature here, and one which is unique to the low-temperature problem, is the minimum in the coke profile caused by a maximum in the rate of coke burning. This

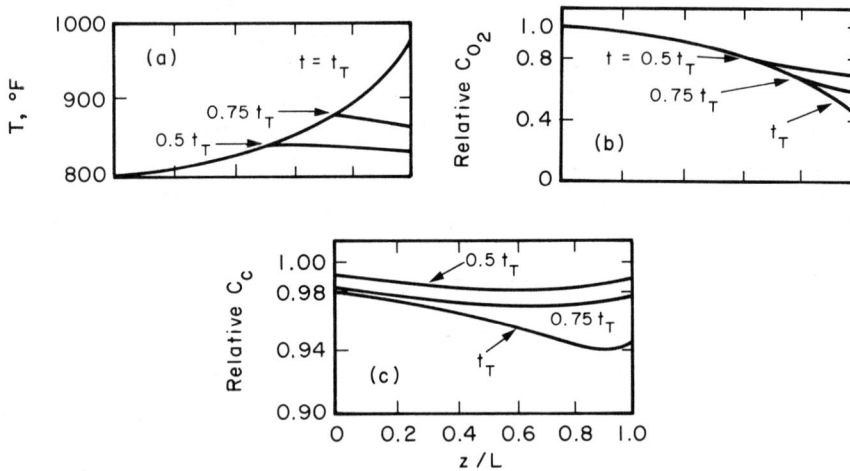

Fig. 11-6. Transient profiles in low-temperature regeneration (experimental points omitted in reproduction). (a) Temperature; (b) oxygen; (c) coke. [From Ozawa (4). Reproduced by permission of the American Chemical Society.]

II. Some Quantitative Studies

maximum occurs in the axial direction, because the effect of temperature on the burning rate increases while the oxygen contribution decreases, leading to a maximum in the burning rate.

For the quasi-steady-state model a transient temperature peak travels through the bed after one heat transit time and the temperature and concentration profiles approach with time the characteristic quasi-steady state asymptotically. In the numerical solutions it was found that the contribution of the first term (time derivative term) in Eqs. (11-55) and (11-57) becomes smaller as time increases after a certain period.

It is important to know the temperature and concentration profiles for longer times in order to evaluate the time required to reduce the amount of coke to the desired level.

Now the steady-state forms of Eq. (11-55) and (11-57), together with the kinetics of Eq. (11-56), give a set of the form

$$
\begin{aligned}
(\partial X/\partial z) &= aXY \\
(\partial Y/\partial t) &= bXY \\
(\partial T/\partial z) &= cXY
\end{aligned}
\tag{11-64}
$$

where a, b, c are functions of temperature. This is of the form that we showed in Chapter 10 to be susceptible to solution via a Legendre transform method. Details as applied to the present problem are discussed by Ozawa. Unfortunately, the results do not come out explicitly in terms of X and Y, but the other way around. Hence:

$$
\begin{aligned}
t = 1/B\{&[e^{-\xi}Ei(W'+\xi) - e^{-\zeta}Ei(w'+\zeta)] \\
&- [e^{-\xi}Ei(w+\xi) - e^{-\zeta}Ei(w+\zeta)]\}
\end{aligned}
\tag{11-65}
$$

where

$$
\left.\begin{aligned}
w' &= 1/\{T_0 + (C/A) - C[(u-1/A)/v + 1/A]\} \\
w &= 1/(T_0 + C/A - Cu) \\
\xi &= -1/T_0 \\
\zeta &= -A/(C + AT_0) \\
Ei(w) &= \int_{-\infty}^{w} \frac{e^u}{u}\, du \\
u &= X/A, \quad v = Y/B
\end{aligned}\right\}
\tag{11-66}
$$

and

$$z = \frac{1}{BC}\left\{-\frac{1}{w'}[e^{-\xi}Ei(w'+\xi) - e^{-\zeta}Ei(w+\zeta)]\right.$$
$$+ \frac{1}{\xi}[Ei(w') - e^{-\xi}Ei(w'+\xi)] - \frac{1}{\zeta}[Ei(w')$$
$$\left. - e^{-\zeta}Ei(w'+\zeta)]\right\} - \frac{u - 1/A}{Bv}[e^{-\xi}Ei(w'+\xi)$$
$$- e^{-\zeta}Ei(w'+\zeta)] + \frac{1}{AB}[e^{-\zeta}Ei(\zeta-\xi) - Ei(-\xi)] \qquad (11\text{-}67)$$

To obtain corresponding values of X and Y (u and v), a set of t-w is selected and the corresponding w' evaluated from Eq. (11-65). For this w-w' combination, the z corresponding is evaluated from Eq. (11-67). As seen from the parametric definitions in Eq. (11-66), determination of w and w' sets a corresponding pair of u and v. A comparison of this quasi-steady solution with the full transient solution is shown in Fig. 11-7. In general, there is good agreement between the two for operating times $\geq 5t_T$. Note that the coke minimum has disappeared (it has passed out of the end of the bed). Also, as a reflection of the low-temperature operations, the profiles are now quite diffuse, the reaction zone is large, and coke burning is occurring throughout the length of the bed. In the terminology we used earlier, then, low-temperature regeneration is characterized by a thermal wave which travels much more rapidly than does the combustion zone. In fact, it is difficult to identify a specific "combustion zone" here, since it extends throughout the reactor.

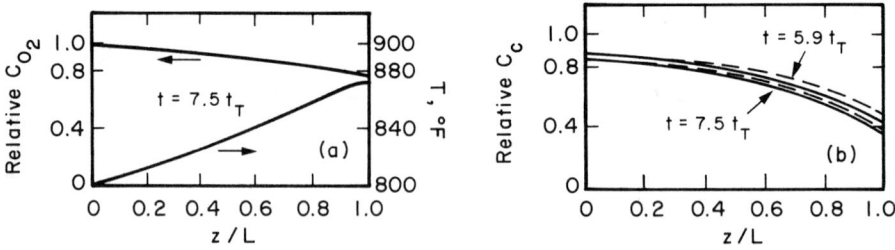

Fig. 11-7. Comparison of transient and quasi-steady solutions for low-temperature regeneration. (a) Solutions for temperature and oxygen concentration profiles. The two methods give essentially the same results. (b) Solutions for coke profiles. Dashed lines are unsteady-state solutions, solid lines quasi-steady-state solutions. [From Ozawa (4). Reproduced by permission of the American Chemical Society.]

III. SUMMARY—DEACTIVATION IN REVERSE

We promised at the beginning of this chapter an opportunity to examine essentially the same elephant in which only one parameter, the kinetics of coke burning, was changed among the various studies. Now that this has been done, does any more consistent picture emerge, and how does it fit in with the concept of catalyst deactivation in reverse?

Indeed, it does seem that some internal consistency exists among these studies. In particular, there is the quasi-stationary behavior demonstrated in the various investigations for diffusion-limited burning. This behavior is consistent with the STR behavior observed for adiabatic (and, in fact, nonisothermal) reactor poisoning and provides some rather simple ways in which overall behavior such as cycle time, net regeneration, and the like can be estimated. In fact, one is tempted to say that the model of Johnson, Froment, and Watson has withstood the test of time relatively well; most of the essential features of the fixed-bed regeneration process are reproduced by that model if the kinetic constant k is visualized as a specific parameter of the particular coke-catalyst system under consideration. Little really seems to be gained here by going to two-phase heat balance models; space velocities involved in actual operation—if the examples described here are at all realistic in the conditions illustrated—are sufficiently high that interphase temperature gradients are insignificant in comparison to the overall exotherm associated with coke burning.

Low-temperature burning provides an interesting contrast. We again can see the development of quasi-stationary behavior, but it is difficult to view this in terms of wave or thermal front motion as in the high-temperature case, and the mathematics are much less convenient. Perhaps the profiles of Fig. 11-7 can best be thought of as "constant pattern," developing regularly with time of operation, but certainly here the zone of reaction is the entire bed.

In comparison with the fixed-bed deactivation experiments discussed in Chapter 10, there also seems to be a general parallel. High-temperature adiabatic regeneration is very similar to rapid poisoning of exothermic reactions in adiabatic beds. Low-temperature regeneration is very similar in many respects to the examples involving coke deposition—although perhaps even more leisurely if direct comparison is made.

REFERENCES

1. A. Bondi, R. S. Miller, and W. G. Schlaffer, *Ind. Eng. Chem. Process Des. Dev.* **1**, 196 (1962).
2. B. M. Johnson, G. F. Froment, and C. C. Watson, *Chem. Eng. Sci.* **17**, 835 (1962).

3. K. E. Olson, D. Luss, and N. R. Amundson, *Ind. Eng. Chem. Process Des. Dev.* **7**, 96 (1968).
4. Y. Ozawa, *Ind. Eng. Chem. Process Des. Dev.* **8**, 378 (1969).
5. B. L. Shulman, *Ind. Eng. Chem.* **55**, 44 (1963).
6. J. J. van Deempter, *Ind. Eng. Chem.* **45**, 1227 (1953); **46**, 2300 (1954).
7. P. B. Weisz and R. B. Goodwin, *J. Catal.* **2**, 397 (1963).
8. P. B. Weisz and R. B. Goodwin, *J. Catal.* **6**, 227 (1966).

CHAPTER 12

A Case History: Kinetic Lumping, Deactivation, and Reactor Models for Catalytic Cracking

> Ignorance is only degrading when found in company with riches.
>
> *Schopenhauer*

Over the years there has appeared in the literature a series of reports on the development, application, and refinement of a model describing the interactions of reaction kinetics and deactivation in reaction/reactor analysis as applied to catalytic cracking. The work involves an integration of many of the concepts we have been concerned with throughout this book, hence it seems particularly appropriate to present this summary as a capping example of the things we have discussed as reduced to practice. The material has been abstracted from the following papers, which we shall not cite individually unless particular identification is required:

1) V. W. Weekman, Jr., *Ind. Eng. Chem. Process Des. Dev.* **7**, 90 (1968).
2) V. W. Weekman, Jr., *Ind. Eng. Chem. Process Des. Dev.* **8**, 385 (1969).
3) V. W. Weekman, Jr. and D. M. Nace, *AIChE J.* **16**, 397 (1970).
4) D. M. Nace, S. E. Voltz, and V. W. Weekman, Jr., *Ind. Eng. Chem. Process Des. Dev.* **10**, 530 (1971).
5) S. E. Voltz, D. M. Nace, and V. W. Weekman, Jr., *Ind. Eng. Chem. Process Des. Dev.* **10**, 538 (1971).
6) S. E. Voltz, D. M. Nace, S. M. Jacob, and V. W. Weekman, Jr., *Ind. Eng. Chem. Process Des. Dev.* **11**, 261 (1972).
7) B. Gross, D. M. Nace, and S. E. Voltz, *Ind. Eng. Chem. Process Des. Dev.* **13**, 199 (1974).
8) S. M. Jacob, B. Gross, S. E. Voltz, and V. W. Weekman, Jr., *AIChE J.* **22**, 701 (1976).
9) V. W. Weekman, Jr., *AIChE Monogr. Ser.* **75**, 11 (1979).

In development of the overall model, we must first identify the kinetics of the (complex) reactions involved and integrate these with the deactivation mechanism. Since the reaction network in catalytic cracking involves many hundreds of individual reactions with associated rate constants, it is obviously impractical to attempt kinetic analysis on an individual basis. Similarly, there would be associated a vast number of individual reactions which are involved in the process of coke deposition and subsequent deactivation, and again one would not seek to treat these on an individual basis. Hence, the initial task here is to develop a suitable lumping model, in which a generic class of compounds may be treated as a kinetic entity with respect to both the cracking reactions and the deactivation behavior.

I. KINETIC LUMPING AND DEACTIVATION

It is well known that when an assembly of first-order reactions proceeds with a very wide variety of individual rates of reaction, the ensemble appears second order with respect to the total concentration of reactants. This is partially because the more refractory materials in essence skew the distribution of velocity constants in a direction toward net second-order behavior. In catalytic cracking, this behavior is also accentuated by the fact that there is a volume change on reaction which must be accounted for. With this in mind, we can start by writing a very simple model for the kinetic lumps in the cracking process; the gas oil charge is cracked to a gasoline fraction together with low molecular weight products and coke directly, and the gasoline fraction itself can be cracked to low molecular weight products and coke. Schematically:

$$C_1 \xrightarrow{k_0} a_1 C_2 + a_2 C_3$$
$$C_2 \xrightarrow{k_1} C_3$$
                    I

where C_1 is the gas oil charge, $C_2 = C_5 - 410°F$ gasoline, and $C_3 = C_4$'s, dry gas, and coke. Now, in this reaction sequence, the C_1 fraction represents the major lump with respect to a wide range of molecular weights, hence one would logically expect the reaction of the C_1 fraction to be second order. However, before worrying further at this point about details of the reaction kinetics, let us begin to be concerned about how deactivation will be included in the model.

Figure 12-1 displays some typical results on the deactivation of catalytic cracking catalysts, in this instance an X-sieve, with a representative feedstock and conditions of operation. Disregarding for the purposes of this analysis

I. Kinetic Lumping and Deactivation

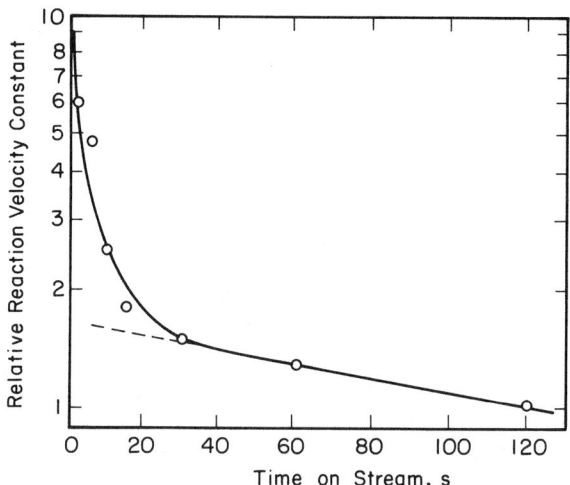

Fig. 12-1. Rate of catalyst decay, LETGO feed, REHX catalyst, 325 LHSV, 900°F. The linear portion of the plot is correlated by $k = k_0 \exp(-18 t_c)$. [From V. W. Weekman, Jr., *Ind. Eng. Chem. Process Des. Dev.* **7**, 90 (1968). Reproduced by permission of the American Chemical Society.]

the initial very rapid activity decay, we see that the net activity is describable in terms of an exponential function of the time on stream, t_c hr. We have earlier discussed the role of time-on-stream correlations in deactivation by coke deposition, primarily in terms of Voorhies-type relationships between coke on catalyst and time:

$$C_c = a t_c^b \qquad (12\text{-}1)$$

and separate relationships between activity and coke on catalyst. In the present case, if we write this as

$$s = 1/C_c^m \qquad (12\text{-}2)$$

then the corresponding activity–time relationship is

$$s \approx 1/t_c^n \qquad (12\text{-}3)$$

where

$$n = mb \qquad (12\text{-}4)$$

In fact, the power law relationship of Eq. (12-3) is well represented by

$$s = e^{-\alpha t_c} \qquad (12\text{-}5)$$

where α is some characteristic decay constant. Hence, the exponential model here is consistent with Voorhies-type behavior with activity inversely dependent on coke content.

Now assuming that we may write the deactivation in terms of separable kinetics, and further that the overall rate of gas oil cracking is second order (to be verified later), we can write the kinetics for **I** as

$$(-r_1) = k_0 s_1 y_1^2 \tag{12-6}$$

$$(-r_2) = k_1 s_1 y_1^2 - k_2 s_2 y_2 \tag{12-7}$$

where we allow for the possibility that the activity functions for overall cracking (gasoline formation), s_1, and for gasoline cracking, s_2, may be different.

Now, in order to verify the kinetic scheme of Eqs. (12-6) and (12-7), we must develop the appropriate reactor models. In the following we shall be concerned not only with the fixed-bed model, the topography of which we have already mapped in detail, but also with moving-bed and fluidized-bed reactors. In the latter two cases, catalyst is added and removed continuously and a steady-state activity level prevails within the reactor at any given time. Comparison among the three types in this service (rapid deactivation due to coking) is informative, since it will be seen that disguises of activity and selectivity can arise in fixed-bed applications because of the transient nature of their operation.

II. REACTOR MODELS—FIXED, MOVING, AND FLUIDIZED BEDS

A. Fixed Beds

The major difference between our fixed-bed modeling here and prior applications we have discussed is the necessity to account for the volume change on reaction. Also, slightly different units are employed in deference to the original derivations; in particular, mass units will be employed for reactants and products, and residence times will be reported in terms of liquid hourly space velocity (LHSV), volume/volume-hour. Then, for isothermal reaction conditions with no diffusional limitations, the fixed-bed plug flow reactor (PFR) mass balance is

$$\frac{1}{\rho}\frac{\partial(\rho y)}{\partial t} + u\frac{\partial y}{\partial x} = (-r_1) \tag{12-8}$$

where y is the fraction of gas oil unconverted, ρ the vapor density in mass volume, u the linear velocity, x the axial dimension, and $(-r_1)$ the kinetic

II. Reactor Models—Fixed, Moving, and Fluidized Beds

form for gas oil cracking. Density variation with conversion is given by

$$\rho = \rho_0/[y + a(1-y)] \tag{12-9}$$

where a is the ratio of reactant molecular weight to product molecular weight. If we define the LHSV as

$$S = F_0/V_r\bar{\rho} \tag{12-10}$$

where F_0 is the mass of reactant fed per hour, V_r the reactor volume (void fraction e), and $\bar{\rho}$ the liquid reactant density (STP), then Eqn. (12-8) can be written in the following nondimensional form:

$$B\frac{\partial y}{\partial \theta} + \frac{\partial y}{\partial x} = \frac{e\rho}{\bar{\rho}S}(-r_1) \tag{12-11}$$

where

$$X = x/L, \qquad B = e\rho_0 a/\bar{\rho} St_c[y + a(1-y)]^2 \tag{12-12}$$

The quantity B represents physically the ratio of oil transit time to a characteristic time of catalyst decay, t_c (used in definition of the dimensionless time variable θ). As we discussed previously, the importance of the time derivative term in modeling is determined by the magnitude of these relative time scales. Also, as we have done in large measure before, we will make the quasi-steady-state approximation in which $B \approx 0$ and write the PFR model as

$$\frac{dy}{dX} = \frac{e\rho}{\bar{\rho}S}(-r_1) \tag{12-13}$$

Now, for the reaction kinetics we have a combination of Eqs. (12-5) and (12-6):

$$(-r_1) = k_0 e^{-\alpha t_c} y^2 \tag{12-14}$$

or in terms of the characteristic decay time:

$$(-r_1) = k_0 y^2 \exp(-\lambda\theta) \tag{12-15}$$

with $\lambda = \alpha t_c$. We may consider the parameter λ to define a characteristic length of decay. Thus Eq. (12-13) becomes

$$\frac{dy}{dX} = -\frac{e\rho k_0}{\bar{\rho}S} y^2 \exp(-\lambda\theta) = -Ay^2 \exp(-\lambda\theta) \tag{12-16}$$

with

$$A = e\rho k_0/\bar{\rho}S = K_0/S \tag{12-17}$$

representing a severity or extent of reaction factor. Straightforward integration of Eqn. (12-17) yields

$$y = 1/[1 + A\exp(-\lambda\theta)] \tag{12-18}$$

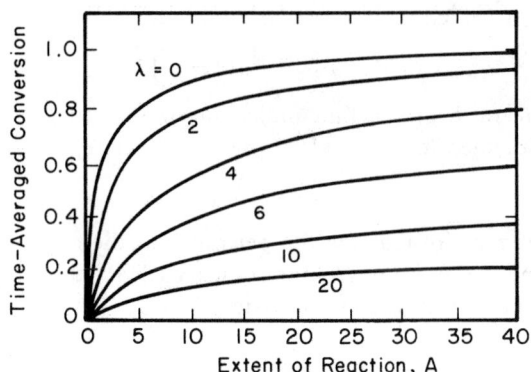

Fig. 12-2. Time-averaged conversion in fixed-bed catalytic cracking. [From V. W. Weekman, Jr., *Ind. Eng. Chem. Process Des. Dev.* 7, 90 (1968). Reproduced by permission of the American Chemical Society.]

or, for conversion,

$$\varepsilon = 1 - y = 1 - [1 + A \exp(-\lambda\theta)]^{-1} \quad (12\text{-}19)$$

However, as pointed out above, the fixed-bed operation is long-term unsteady state while the conversion above is an instantaneous value. Given chemical analytical capabilities which are less than instantaneous, an observer would measure a time-averaged value of conversion, that is,

$$\bar{\varepsilon} = \int_0^1 (1-y)\, d\theta \quad (12\text{-}20)$$

or

$$\bar{\varepsilon} = \frac{1}{\lambda} \ln\left[\frac{1+A}{1+A\exp(-\lambda)}\right] \quad (12\text{-}21)$$

The result here, then, is essentially a two-parameter model for the PFR operation in terms of extent of reaction A and extent of decay λ. A performance diagram according to this model is given in Fig. 12-2.

B. Moving Beds

In the moving bed, we have continuous introduction of fresh catalyst at the bed entrance and continuous withdrawal of spent catalyst at the exit. Hence, while a nonuniform catalyst activity profile exists within the bed at all times, it is invariant and the time-averaging effect seen in fixed-bed operation does not exist here. If we assume plug flow for both solid and gas phases, basically the same formulation can be employed as for the PFR

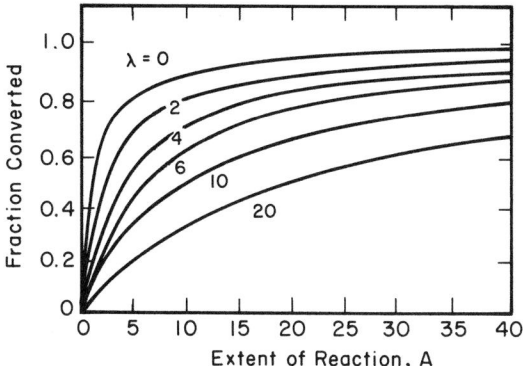

Fig. 12-3. Coversion in moving-bed catalytic cracking. [From V. W. Weekman, Jr., *Ind. Eng. Chem. Process Des. Dev.* **7**, 90 (1968). Reproduced by permission of the American Chemical Society.]

fixed bed with a slightly different interpretation of the parameters. The most important change is that the position in the bed now determines time on stream, so X will replace θ in the argument of the exponential in the decay function. This, in turn, means that the characteristic decay time λ is just the total residence time of the catalyst in the moving bed, t_c. With this simple change, the reactor mass balance is

$$(dy/dX) = -Ay^2 \exp(-\lambda X) \tag{12-22}$$

The corresponding conversion at bed exit is

$$\varepsilon = \frac{A[1-\exp(-\lambda)]}{\lambda + A[1-\exp(-\lambda)]} \tag{12-23}$$

This result is shown in Fig. 12-3.

C. Fluidized Beds

For fluid-bed operations, we must establish a model with specified residence time distributions of both gas and solid phases (in fact, we just did this with the moving bed, where both phases were taken to be in plug flow). The analysis here will be of the simplest possible situation, where the gas is in plug flow and the solid is perfectly mixed. The perfect mixing in the catalyst phase imposes an average activity on the reactor which, since catalyst is continuously introduced and removed, is time-invariant. Recalling that the internal age distribution $I(\theta)$ for a perfect mixer is identical to the exit age distribution $E(\theta)$, we may determine the average activity level from

$$\bar{k} = \int_0^\infty k_0 \exp(-\lambda\theta) E(\theta)\, d\theta \tag{12-24}$$

and for perfect mixing

$$E(\theta) = \exp(-\theta) \tag{12-25}$$

Hence:

$$\bar{k} = k_0 \int_0^\infty \exp(-\theta) \exp(-\lambda\theta)\, d\theta \tag{12-26}$$

and

$$\bar{k} = k_0/(1+\lambda) \tag{12-27}$$

The reactor mass balance equation is now

$$(dy/dX) = Ay^2/(1+\lambda) \tag{12-28}$$

and

$$\varepsilon = A/(1+\lambda+A) \tag{12-29}$$

This result is shown in Fig. 12-4.

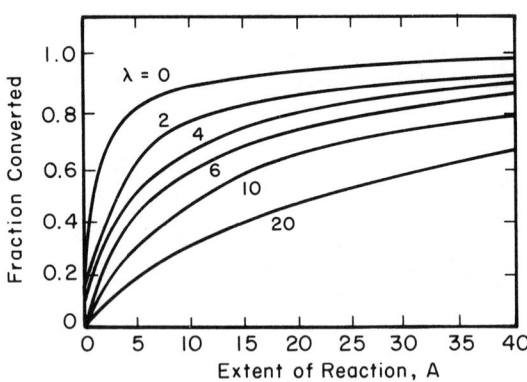

Fig. 12-4. Conversion in fluid-bed catalytic cracking. [From V. W. Weekman, Jr., *Ind. Eng. Chem. Process Des. Dev.* 7, 90 (1968). Reproduced by permission of the American Chemical Society.]

III. REACTION KINETICS AND MODEL EVALUATION

The previous sections have seen the assembly of different pieces of the deactivation, kinetic lumping, and reactor models, and we are now in a position to evaluate the validity of the assumptions concerning kinetics that were written down without justification in Eqs. (12-6) and (12-7). Note first, however, that the performance of all three reactor types is described in terms of the two parameters A and λ or, more precisely, in terms of the overall cracking rate constant k_0 (contained within A) and the decay constant α (contained within λ). These, of course, are not quantities that we know on an *a priori* basis, so any use we make of the models in interpretation of either kinetics or laboratory reactor performance will be in the sense of two-parameter curve fits. In the end, then, we can justify the validity of the kinetic–deactivation model only via internal consistency and demonstrated application over very wide ranges of parameter space.

With these thoughts in mind, let us now look at an example of the evaluation of the kinetic model. From the solution to the moving-bed problem, we have for the fraction unreacted at bed exit:

$$y = \frac{\lambda}{\lambda + A[1 - \exp(-\lambda)]} \tag{12-30}$$

Inverting this and writing λ and A in terms of the kinetic and deactivation parameters gives

$$\frac{1}{y} = 1 + \left(\frac{K}{\alpha t_c}\right)\frac{1}{S}[1 - \exp(-\alpha t_c)] \tag{12-31}$$

hence a plot of $(1/y)$ versus $(1/S)$ from moving-bed experimental results should be linear, and the catalyst residence time t_c can be employed as a further parameter. Linearity of the $(1/y)$-$(1/S)$ plot for a fixed t_c implies obedience to the kinetic model, while conservation of this linear relationship over a range of values of t_c is verification of the decay model. Such an interpretation of a typical set of experimental results is given in Fig. 12-5 for roughly a 6-fold range of $(1/S)$ and a 10-fold range of t_c. A number of additional examples of agreement of experiment with model are provided in the original references; we shall not pursue the point further here other than to say that excellent results pertain to operation over a wide variety of conditions in all reactor configurations, and the requirement of internal consistency over large ranges of the A-λ parameter space is amply demonstrated.

Now, having established the validity of the kinetic–decay–reactor model, we are in a position to do quite a number of things. Certainly, one of the

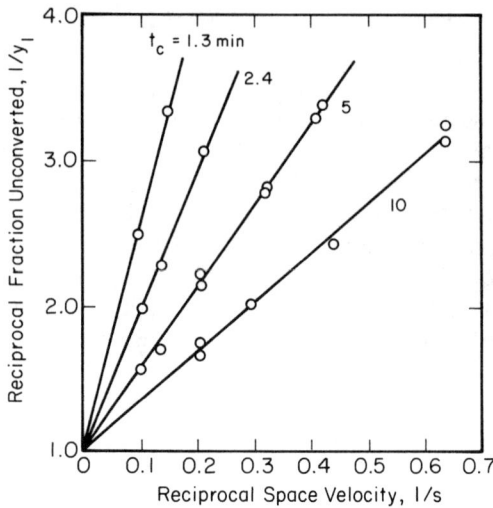

Fig. 12-5. Second-order test of moving-bed data; MCGO feed, Durabead 5 catalyst, 900°F. Parameter is catalyst residence time. [From V. W. Weekman, Jr., *Ind. Eng. Chem. Process Des. Dev.* **7**, 90 (1968). Reproduced by permission of the American Chemical Society.]

most important of these tasks is to use the model fit to experimental data to extract the values of the kinetic and deactivation parameters, k_0 and α. This application is most importantly involved in catalyst testing and evaluation, where the influence of the type of reactor employed in experimentation has been accounted for in the model equations and the intrinsic properties of the catalyst have been evaluated directly. Figures 12-6a and 12-6b give examples of such application; note in particular the variety of circumstances encompassed by these two figures. In Fig. 12-6a we have time-averaged conversion in a fixed bed over a sevenfold range of space velocities and almost 100-fold variation of time on stream. The parameter fit values of k_0 and α are representative of a very active catalyst which deactivates at a relatively moderate rate. In Fig. 12-6b we have conversion data from a moving bed over roughly comparable ranges of S and t_c, but here for a catalyst which is rather inactive and which deactivates rapidly. In sum, evaluation of the parameters k_0 and α permits side-by-side comparison of various catalyst formulations as to *both* intrinsic activity and deactivation behavior.

In addition to comparison of catalysts in given operation, the model also permits discrimination of performance with various feedstocks. Consider the comparison of the X- and Y-zeolite formulations (REX and REY, respectively) given in Table 12-1. The performance of the X-zeolite in terms of intrinsic activity for Mid-Continent gas oil (MCGO) cracking is much

III. Reaction Kinetics and Model Evaluation

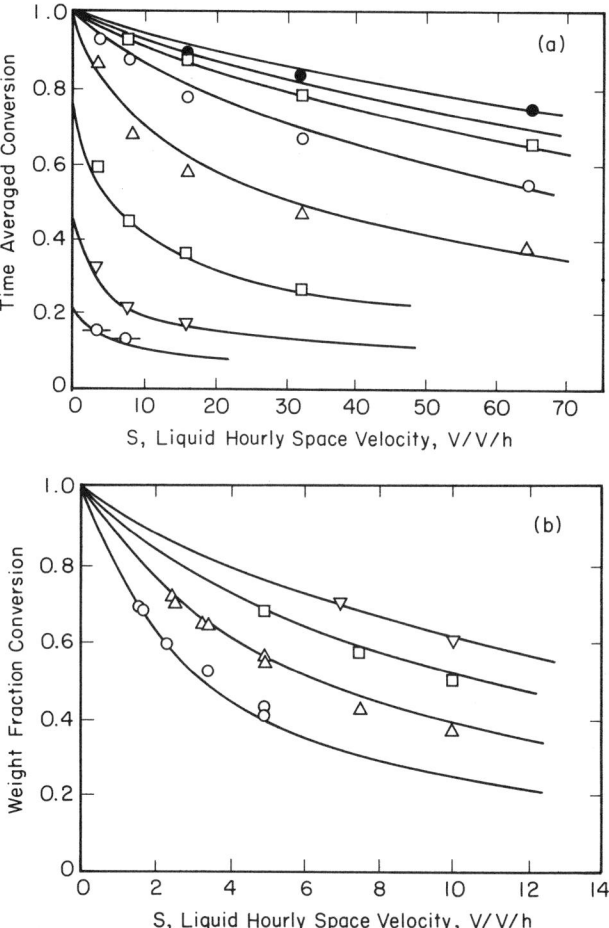

Fig. 12-6. Comparison of fixed-bed model to fixed-bed data for MCGO, aluminosilicate catalyst, 900°F. Lines show model prediction. (—, top solid line) 0 min, (●) 1.25 min, (□) 2.5 min, (○) 5 min, (△) 10 min, (□) 20 min, (▽) 40 min, (—○—) 80 min; $K_0 = 194.0 \text{ hr}^{-1}$, $\alpha = 21.8 \text{ hr}^{-1}$. (b) Comparison of model to moving-bed conversion data; t_c of (▽) 1.3 min, (□) 2.4 min, (△) 5.0 min, (○) 10.0 min; $K_0 = 22.9 \text{ hr}^{-1}$, $\alpha = 42.7 \text{ hr}^{-1}$. [From V. W. Weekman, Jr., *Ind. Eng. Chem. Process Des. Dev.* **7**, 90 (1968). Reproduced by permission of the American Chemical Society.]

superior to that of the Y, yet for the cycle oil stock the Y sieve is more active. In terms of deactivation behavior, if we use the ratio (k_0/α) as a rough figure of merit, the two are about equivalent for MCGO, while the Y sieve is not only more active but a little more resistant to deactivation for the cycle oil stock. On balance, then, one would select REX as the

TABLE 12-1

Constants for Various Charge Stocks and Catalysts[a]

Catalyst	Feedstocks	Temperature (°F)	α (hr^{-1})	k_0 (hr^{-1})
REX	MCGO	900	18.8	143.0
REY	MCGO	900	7.5	62.0
REX-SiO$_2$/Al$_2$O$_3$[b]	MCGO	900	10.7	23.4
REX	TCC cycle	900	28.8	32.1
REY	TCC cycle	900	25.3	39.2

[a] From V. W. Weekman, Jr., *Ind. Eng. Chem. Process Des. Dev.* 7, 90 (1968). Reprinted by permission of the American Chemical Society.

[b] 25% REX

TABLE 12-2

Comparison of Two Catalysts in Fixed-, Moving-, and Fluid-Bed Reactors[a]

Catalyst[b]	S	t_c	A	λ	ε Fixed	ε Moving	ε Fluid
1	1	0.5	30	10	0.34	0.75	0.73
2	1	0.5	2.5	1	0.59	0.61	0.55

[a] From V. W. Weekman, Jr., *Ind. Eng. Chem. Process Des. Dev.* 7, 90 (1968). Reprinted by permission of the American Chemical Society.

[b] Catalyst 1: $k_0 = 30$ hr^{-1}, $\alpha = 20$ hr^{-1}; catalyst 2: $k_0 = 2.5$ hr^{-1}, $\alpha = 2$ hr^{-1}.

catalyst of choice for MCGO, and REY for TCC cycle. Note also what happens to catalytic properties when the pure zeolite is embedded in a silica/alumina matrix[1]. Unfortunately, the overall cracking activity is much more severely affected than the deactivation behavior in this instance, and rational selection of a catalyst for commercial application to MCGO cracking would await comparison with similar REY-SiO$_2$/Al$_2$O$_3$ formulations.

We have alluded several times to the fact that fixed-bed results are time-averaged and the possibility consequently exists that blind comparison among results obtained using different reactors for experimentation may lead to erroneous conclusions regarding intrinsic activity and activity maintenance. This is shown in Table 12-2. We assume there are the two catalysts, 1 and 2, with set values of the intrinsic parameters k_0 and α as

---

[1] Commercial formulations require this, since the pure zeolites are not sufficiently attrition-resistant to survive the rigors of life in normal fluid-bed operation.

shown. Thus, catalyst 1 has a high intrinsic activity but also deactivates rapidly. Catalyst 2 is of much lower activity but deactivates much more slowly. Now, for set values of S and t_c as shown, let us compare the conversions evaluated for fixed, moving, and fluid beds via Eqs. (12-21), (12-23), and (12-29), remembering that $\bar{\varepsilon}$ from Eqn. (12-21) is time-averaged and the others are not. It is clear from the moving-bed and fixed-bed results that catalyst 1 is superior, yet if a selection were made on the basis of fixed-bed observations alone we would pick the wrong catalyst. Bad as this is, we shall see presently that time-averaging disguises on selectivity can be even worse.

IV. SOME COMPARISONS OF REACTOR PERFORMANCE

While our attention to this point has been devoted to the overall model development, its evaluation, and its application to catalyst testing, the degree of success achieved would suggest that more general applications can be envisioned. An important one is in the comparison of reactor performance among the three types considered for the catalytic cracking model. Since all three models are expressed in terms of the extent-of-reaction and extent-of-decay parameters, direct comparison is easily accomplished. One such comparison of the various types is shown in Fig. 12-7, where the ratios of conversion are set forth as a function of extent of reaction and extent of decay. The results in general suggest the expected, in that the moving bed, with continuous introduction of fresh catalyst, is generally superior to the fixed bed with continuously decaying catalyst. Also, the moving bed is superior to the fluid bed, which is the consequence of the difference between the PFR solid-phase behavior in the one case and the CSTR solid-phase behavior in the other. An interesting result is the comparison between fixed and fluid beds in Fig. 12-7c. For low values of A and λ the fixed bed gives slightly more conversion but falls rapidly from this position as λ increases, particularly for larger values of A. The superiority of the fluidized bed in coping with deactivation is particularly evident here for high values of the extent-of-decay parameter.

The generalized plots for the three reactor types given in Figs. 12-2 to 12-4 can also be used for the evaluation of processing conditions required for attaining equivalent conversion levels. Since the plots are functions only of the reaction and deactivation parameters, such evaluation reduces to appropriate comparisons of these groups. Four examples set forth in the

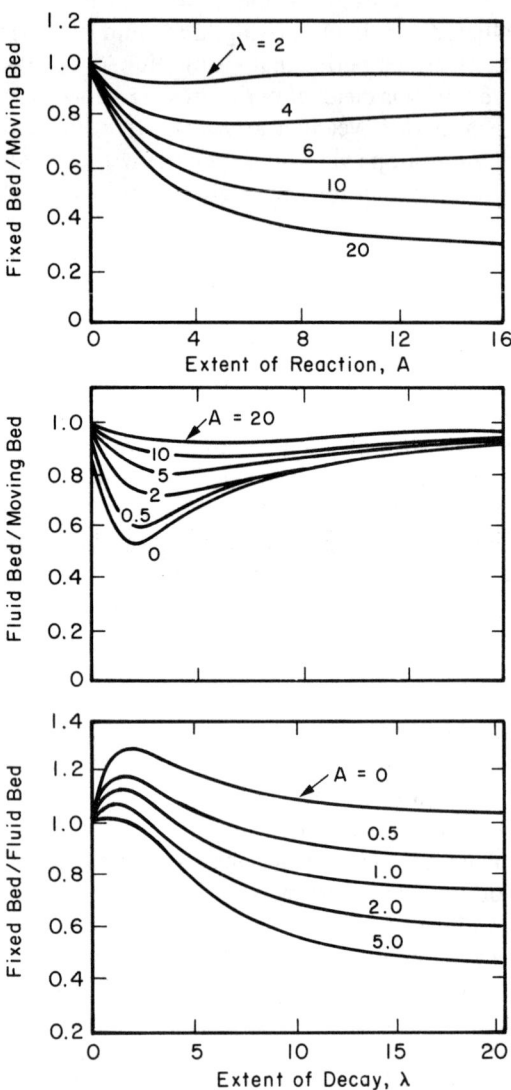

Fig. 12-7. Comparisons among fixed-, moving-, and fluid-bed performance in catalytic cracking in terms of conversion ratios. [From V. W. Weekman, Jr., *Ind. Eng. Chem. Process Des. Dev.* **7**, 90 (1968). Reproduced by permission of the American Chemical Society.]

original papers are:

(1) Compare reactor volumes required for the same conversion, catalyst residence time (constant λ), and oil throughput:

$$V_1/V_2 = A_1/A_2 \qquad (12\text{-}32)$$

(2) Compare catalyst/oil ratios required for the same conversion and space velocity (constant A):

$$(\text{cat/oil})_1/(\text{cat/oil})_2 = \lambda_2/\lambda_1 \qquad (12\text{-}33)$$

(3) Compare space velocities required for the same conversion and catalyst residence time (constant λ):

$$(\text{space velocity})_1/(\text{space velocity})_2 = A_2/A_1 \qquad (12\text{-}34)$$

(4) Compare catalyst residence times required for the same conversion and space velocity (constant A):

$$(\text{catalyst residence time})_1/(\text{catalyst residence time})_2 = \lambda_1/\lambda_2 \qquad (12\text{-}35)$$

Direct measures of catalyst activity loss are also provided by these generalized plots. Two experiments are required, one with the fresh catalyst and the second with the aged sample under inspection. Measurement of the fractional activity loss would be given by the ratio of the activity parameters A from the appropriate plot.

As a final point to emphasize the interaction of the decay with reactor performance, evaluate the conversion predicted by Eqs. (12-21), (12-23), and (12-29) when the catalyst is not deactivating, that is, as $\lambda \to 0$. In each case, the result is

$$\varepsilon = A/(1+A) \qquad (12\text{-}36)$$

The fact that conversion is identical for the three types is due to the assumption of plug flow in the vapor phase; hence conversion is dependent only on the magnitude of the reaction rate constant and vapor residence time, that is, A.

V. ANOTHER OPTIMIZATION PROBLEM

In the use of fixed-bed reactions for applications such as the above, a complete process design would specify an operation–regeneration cycle. In processes in which the catalyst decay time is long (i.e., hydrotreating), the regeneration cycle occupies only a small portion of total operating time and close control of the cycle length may not be critically important to the overall operating strategy. However, the rapid decay encountered in catalytic cracking implies that operating and regeneration cycles may be of the same order of magnitude in length and a well-specified optimization problem appears. The situation is similar to that which we discussed for butene dehydrogenation in Chapter 10.

As usual, the first requirement is to define a performance criterion by which the overall operation may be evaluated. A logical one is the ratio of total product obtained in actual operation to that obtained at 100% conversion with no catalyst deactivation. The corresponding reactor efficiency is

$$E_r = \frac{NF_0 t_0 \bar{\varepsilon}}{F_0 t_t} = \frac{\text{total actual product}}{\text{total ideal product}} \qquad (12\text{-}37)$$

where N is the number of operation cycles of duration t_0 each, and t_t the total operating cycle length. Note also that $\bar{\varepsilon}$ is the time-averaged conversion corresponding to the end-of-cycle time, t_0. Total cycle length is

$$t_t = N(t_0 + t_R) \qquad (12\text{-}38)$$

where t_R is regeneration. Combining Eqs. (12-37) and (12-38) and the expression for $\bar{\varepsilon}$, Eq. (12-21) gives

$$E_r = \frac{1}{\alpha(t_0 + t_R)} \ln\left[\frac{1 + A}{1 + A \exp(-\alpha t_0)}\right] \qquad (12\text{-}39)$$

where $\lambda = -\alpha t_0$.

For the case in which regeneration time is independent of the operating time per cycle (for example, catalyst discharge for external regeneration), it is apparent from inspection of Eq. (12-39) that an optimum value of t_0 will exist. Figure 12-8a presents the results of parametric calculations[2] with Eq. (12-39) for representative values of k_0 and α and a space velocity of 8 vol/vol-hr, while Fig. 12-8b presents similar results for fixed regeneration time and variable space velocity. While the particular values of the optima shown are dependent on the parameters used in the calculation, the trends depicted are not. For fixed space velocity, as one increases the regeneration time, optimum cycle length increases. For constant regeneration time, as space velocity is decreased the value of E_r increases and occurs at larger t_0. However, it should be remembered that for a fixed feed rate as space velocity is decreased the size of the reactor increases; hence the trend shown in Fig. 12-8b may be a bit deceptive, since it does not reflect the increased

---

[2] Analytical differentiation with respect to t_0 to determine the extremum results in an awkward transcendental form, so it is easiest in this case to proceed via direct numerical evaluation.

V. Another Optimization Problem

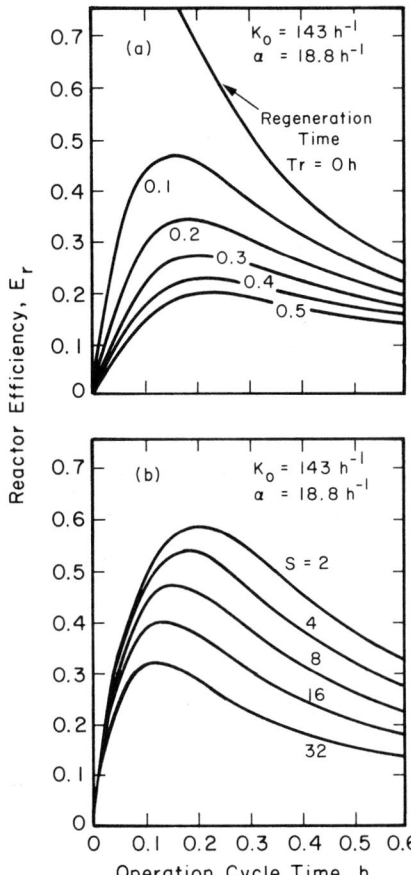

Fig. 12-8. (a) Reactor efficiencies for regeneration time independent of operation time. Space velocity = 8.0 vol/vol-hr. (b) Reactor efficiencies at various space velocities for fixed regeneration time. Regeneration time = 0.1 hr. [From V. W. Weekman, Jr., *Ind. Eng. Chem. Process Des. Dev.* **7**, 252 (1968). Reproduced by permission of the American Chemical Society.]

capital cost of equipment required to attain the improvement in E_r. It is interesting to note that in both Figs. 12-8a and 12-8b, the locus of the maximum in E_r does not vary much with respect to the abscissa even for relatively large ranges of the parameter (save the trivial case of $t_r = 0$ Fig. 12-8a).

Variation of the optimum cycle length with respect to the catalyst activity and decay parameters is shown in Figs. 12-9a and 12-9b. For a fixed decay velocity (12-9a) the operation time decreases as the catalyst is less active in initial (intrinsic) operation, and, of course, for a catalyst of initial infinite activity no regeneration is required. For a fixed intrinsic activity (12-9b) cycle length decreases as the decay constant increases. Unlike the other examples here, in this case there is a pronounced change in cycle length

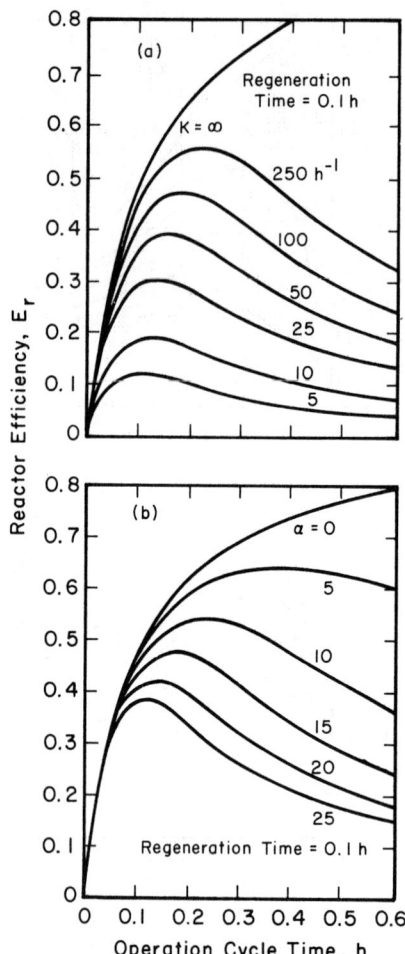

Fig. 12-9. (a) Effect of intrinsic activity on reactor efficiency. Reaction velocity $K_0 = 100 \text{ hr}^{-1}$; space velocity 8.0 vol/vol-hr. (b) Effect of decay velocity on reactor efficiency. Reaction velocity $K_0 = 100 \text{ hr}^{-1}$; space velocity 8.0 vol/vol-hr. [From V. W. Weekman, Jr., *Ind. Eng. Chem. Process Des. Dev.* **7**, 252 (1968). Reproduced by permission of the American Chemical Society.]

with the variation of α, and as one might expect this appears to be the single most sensitive parameter indicating relative operative and regeneration times.

When regeneration time and operating cycle time are not independent, a somewhat different analysis may be used. Regeneration and operating cycles become interdependent in a process such as catalytic cracking, where the amount of coke deposited on the catalyst is dependent on the length of the operating cycle and the time required for regeneration is in turn dependent on the amount of coke. We can start with the Voorhies relationship for coke content:

$$C_c = at_0^b \qquad (12\text{-}40)$$

V. Another Optimization Problem

which is obtained from integration of

$$dC_c/dt = k_c C_c^{-n} \tag{12-41}$$

with initial conditions $C_c = 0$ when $t_0 = 0$. In the integrated form of Eq. (12-40) the constants are

$$a = [(1+n)k_c]^b \tag{12-42}$$

and

$$b = 1/(1+n) \tag{12-43}$$

Further, for the kinetics of coke burning we can use the model

$$dC_c/dt_R = k_b C_{O_2} C_c^m \tag{12-44}$$

where normally $m = 1$ but can be less than this for high concentrations of coke (multilayer deposits). Now for constant oxygen concentration we pose the question of what values of m and n will define an optimum operation–regeneration cycle. Solving Eq. (12-44) for C_{O_2} = constant and $C_c = at_0^b$ at $t_R = 0$ gives

$$t_R = \frac{b^{1-m}(t_0)^{(1-m)/(n+1)}}{k_b C_{O_2}(1-m)}, \quad m \neq 1 \tag{12-45}$$

Now from the reactor efficiency Eq. (12-39) it can be shown that if $E_r \to 0$ as $t_0 \to 0$ an optimum value for E_r will exist. Hence, substituting t_R from Eq. (12-45) into Eq. (12-39) and evaluating the limit as $t_0 \to 0$ we obtain

$$\lim_{t_0 \to 0}(E_r) = \lim_{t_0 \to 0}(A)\left[\frac{1 + b^{1-m}(t_0)^{-(m+n)/(n+1)}}{(1+n)k_b C_{O_2}}\right]^{-1}(1+A)^{-1} \tag{12-46}$$

For the efficiency to approach zero as $t_0 \to 0$, the exponent on t_0 must be negative; therefore if $n > 0$ and $m \geq 0$, an optimum in E_r with t_0 will exist. Thus, for zero- or positive-order coke burning kinetics an optimum operation–regeneration cycle combination will be found provided the Voorhies parameter b is between zero and unity. Figure 12-10 shows just such a relationship, illustrated for $m = 1$, $b = 0.5$, and a final coke content at the end of regeneration of 0.03 wt. % coke on catalyst. The locus of the maximum in E_r is space velocity-dependent and is moderately sensitive to that variable. Corresponding regeneration times are not shown on the figure but can be calculated directly from Eq. (12-45) using the optimum value for t_0.

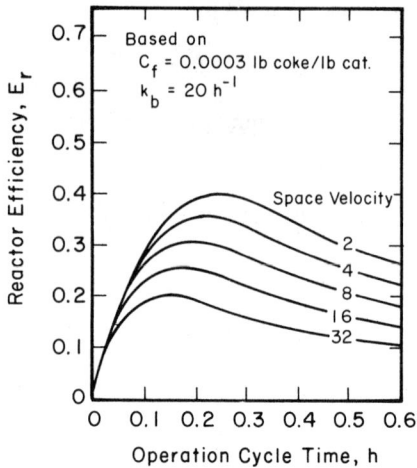

Fig. 12-10. Reactor efficiency with regeneration time dependent on operation time. Voohries coefficient $a = 0.5$; first-order coke burning, $m = 1$. [From V. W. Weekman, Jr., *Ind. Eng. Chem. Process Des. Dev.* **7**, 252 (1968). Reproduced by permission of the American Chemical Society.]

VI. EFFECTS OF DEACTIVATION ON SELECTIVITY

To this point we have been concerned with the coupling of catalyst deactivation and reactor type and performance, using the overall conversion of feedstock as a norm. Obviously the kinetics of the overall reaction scheme I invite scrutiny as to selectivity behavior in the face of deactivation, since the intermediate product C_2 is desired and we would like to maximize its production in the face of both reactor and deactivation factors. In the kinetic formulation of Eqs. (12-6) and (12-7), separate activity functions s_1 and s_2 were carried for the feedstock cracking (s_1 for both gasoline and direct yield of coke) and gasoline cracking (s_2 to coke). Since both of the reaction steps are acid-catalyzed and differ only in the extent of cracking, it seems reasonable that $s_1 = s_2$, at least as a first approximation. In this case, dividing Eq. (12-7) by (12-6) we obtain

$$\frac{dy_2}{dy_1} = \left(\frac{k_2}{k_0}\right)\frac{y_2}{(y_1)^2} - \left(\frac{k_1}{k_0}\right) \tag{12-47}$$

This may be solved analytically for the conversion to y_2 in terms of y_1. The results are

$$y_2 = (k_1 k_2 / k_0^2) \exp(-k_2 / k_0 y_1)[(k_0 / k_2) \exp(k_2 / k_0)$$
$$- (k_0 y_1 / k_2) \exp(k_2 / k_0 y_1) - E_i(k_2 / k_0) + E_i(k_2 / k_0 y_1)] \tag{12-48}$$

VI. Effects of Deactivation on Selectivity

where

$$E_i(x) = \int_{-\infty}^{x} \frac{e^x}{x} dx \qquad (12\text{-}49)$$

and is a tabulated function (1). Now Eq. (12-48) is essentially a four-parameter model, involving the three rate constants k_0, k_1, and k_2 and the decay velocity constant α. In testing the assumption concerning the equality of s_1 and s_2, however, we may use experimental data on conversion and the appropriate reactor model to determine k_0 and α; hence the selectivity determination is again a two-parameter fit. The validity of the assumption will rest on confirmation of the internal consistency of the model in comparison with experimental data over a wide range of conditions.

Typical selectivity data are given in Figs. 12-11 and 12-12. Characteristic of the plot of weight fraction gasoline versus conversion is the curve shown in Fig. 12-11, where the diagonal represents 100% selectivity for gasoline formation. Deviations below this, then, represent the extent of reaction to undesirable products, and it is seen that there is a well-defined maximum in gasoline selectivity. In Fig. 12-12 is shown the variation in gasoline yield with space velocity, and we again note the existence of a maximum. These two figures are representative of a large amount of data, and we will take them as providing support for the internal consistency of the model with

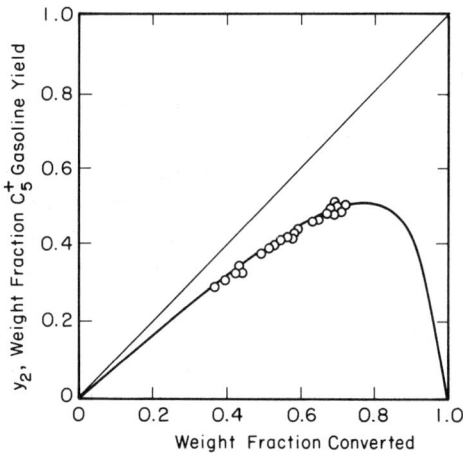

Fig. 12-11. Comparison of model to moving-bed data; MCGO with Durabead 5 catalyst, 900°F. $k_0 = 22.9$, $k_1 = 18.1$, $k_2 = 1.7$, $\alpha = 42.7$. Space velocity = 1.5–10, catalyst/oil = 1.2–8.6. Diagonal line represents 100% gasoline efficiency. Experimental data represent catalyst residence times from 1.2 to 10.0 min. [From V. W. Weekman, Jr., *Ind. Eng. Chem. Process Des. Dev.* **8**, 385 (1969). Reproduced by permission of the American Chemical Society.]

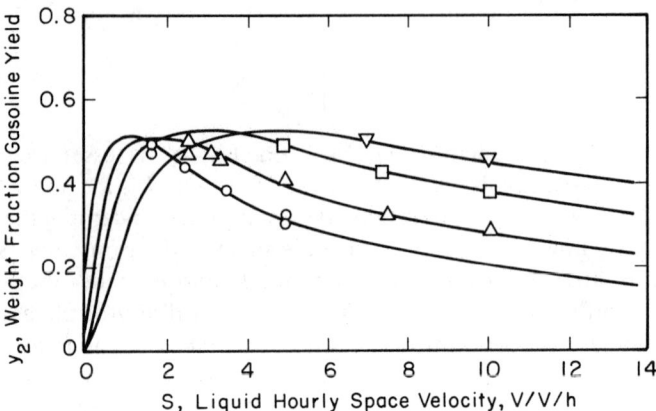

Fig. 12-12. Comparison of t_c^n model gasoline yield to experimental moving-bed data. MCGO, DB-5, 900°F. $k_0 = 0.266$, $k_1 = 0.214$, $k_2 = 0.0188$, $n = 0.719$. The t_c values are ($\triangledown$) 1.25 min, ($\square$) 2.50 min, ($\triangle$) 5.00 min, ($\bigcirc$) 10.0 min. [From V. W. Weekman, Jr., *Ind. Eng. Chem. Process Des. Dev.* **8**, 385 (1969). Reproduced by permission of the American Chemical Society.]

$s_1 = s_2$ both with respect to intrinsic selectivity (Fig. 12-11) and on incorporation into the reactor model (Fig. 12-12)[3]. Figure 12-12 is yet another, somewhat indirect, manifestation of activity waves in a reactor experiencing deactivation. In application to the present example, the situation is depicted more clearly in Figs. 12-13a and 12-13b for a fixed bed and a typical set of parameters. The first illustrates the value of y_2 as a function of position in the bed at increasing times of operation. We notice at time zero a well-defined maximum in yield at about 40% bed length; for the initial undeactivated catalyst gasoline is formed rapidly, but it remains in the bed long enough to crack to the undesirable C_3 fraction further. As time progresses, the catalyst becomes less active and we see the movement of maximum y_2 through the bed and its eventual disappearance. This selectivity behavior is quite similar to that discussed earlier with respect to activity in the findings of Froment and Bischoff, but, in the context of this practical example, a further important consequence is illustrated in Fig. 12-13b. Here the instantaneous gasoline yield curve versus θ reflects that which would be observed at the bed exit were one to have available an instantaneous measure of y_2. One sees, as expected, the emergence of the maximum y_2 after a period of operation and then subsequent decrease in conversion as the bed deactivates: in practice, however, one is not likely to have an instantaneous measure of y_2, but rather that reflective of bed operation over some period of time.

---

[3] Selectivity variation with temperature, not shown here, is well represented.

VI. Effects of Deactivation on Selectivity

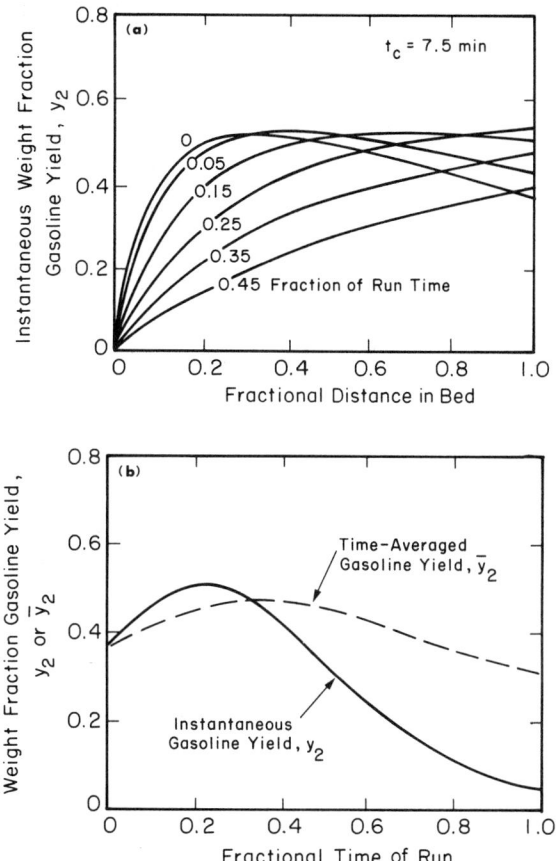

Fig. 12-13. (a) Instantaneous gasoline profiles in a decaying fixed-bed reactor at various times. MCGO, DB-5, 900°F, $SV = 2.0$ vol/vol-hr. $k_0 = 22.9$, $k_1 = 18.1$, $k_2 = 1.7$, $\alpha = 42.7$. (b) Comparison of instantaneous and time-averaged gasoline yields in fixed-bed cracking. Parameters as for part (a). [From V. W. Weekman, Jr., *Ind. Eng. Chem. Process Des. Dev.* **8**, 385 (1969). Reproduced by permission of the American Chemical Society.]

Indeed, if we are talking about laboratory operation in which the activity/selectivity characteristics of a particular catalyst are being examined, then product may well be collected continuously and analyzed only from time to time. In that event, the composition is time-averaged by the collection procedure, and one would then observe the time-averaged yield shown in Fig. 12-13b. Note that both the magnitude of maximum yield and its position with respect to θ are altered with respect to intrinsic values; here, then, the existence of deactivation now produces an indirect disguise of catalytic selectivity via this time-averaging procedure.

The results of this selectivity disguise are direct and potentially disastrous in catalyst screening. Figure 12-14a shows typical data with associated parameters from a run in which average yield is measured versus average conversion for various catalyst residence times. It is seen that as the residence time (run length) is increased, the time-averaging disguise becomes more and more important. While the model is able to fish out this behavior and track it well with the four parameters shown, consider the quandary of the observer not quite so well equipped, as shown in Fig. 12-14b. Here we have generated via the model some "experimental" data for two hypothetical

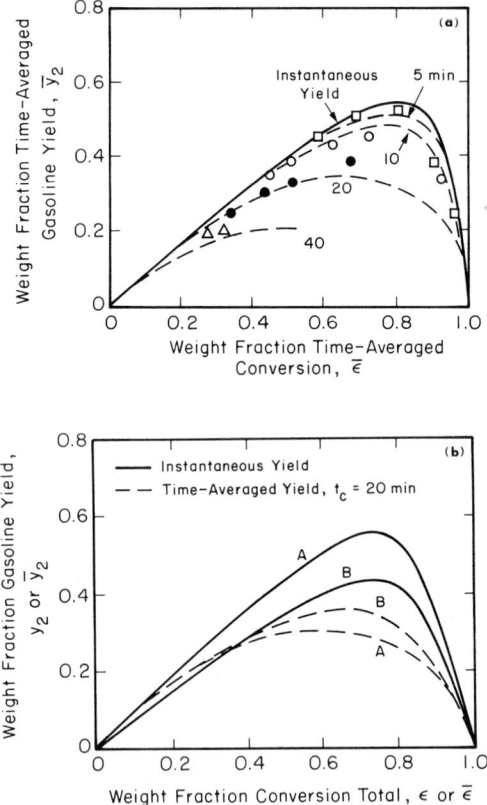

Fig. 12-14. (a) Comparison of model to experimental data—constant catalyst residence time. MCGO, 30-60 mesh pure zeolite, 900°F. $k_0 = 210.7$, $k_1 = 164.8$, $k_2 = 8.9$, $\alpha = 18.2$. Experimental catalyst residence time: (□) 5 min, (○) 10 min, (●) 20 min, (△) 40 min. (b) Effect of time averaging on catalyst selectivity evaluation. Catalyst A: $k_1/k_0 = 0.9$, $k_2/k_0 = 0.1$, $\alpha = 20$; catalyst B: $k_1/k_0 = 0.7$, $k_2/k_0 = 0.1$, $\alpha = 10$. [From V. W. Weekman, Jr., *Ind. Eng. Chem. Process Des. Dev.* **8**, 385 (1969). Reproduced by permission of the American Chemical Society.]

catalysts, A and B. Catalyst A has a higher intrinsic selectivity for gasoline than B, but it also decays faster. Now for the magnitude of the parameters shown, the instantaneous yield of A is clearly better than that of B and hence it is intrinsically the superior performer in terms of selectivity behavior, yet the faster deactivation of A means that time-averaging disguises will be more severe. If we look to the calculation ("experiment") at t_c of 20 min, it is clear that the unwitting observer would pick B as the superior catalyst—with very sad results if the experimental operation were then translated into other regions of t_c for design purposes.

This particular form of time-averaging disguise, as pointed out before, pertains to situations in which short time, repetitive analysis of product streams is impractical—at least in the context of experimental evaluation of catalytic properties. As a result, there may be many cases, in which product analysis can be conducted conveniently and rapidly, where this may not be a problem in the laboratory. If we turn to the process design aspect, however, it is quite possible that one might wish to base such a design on maximum time-averaged values even though intrinsic data are available experimentally. In our discussion of Fig. 12-14b, for example, although catalyst A demonstrated superior intrinsic selectivity, one would obviously choose B if the process design called for operation with t_c of 20 min. The important point is that the kinetic, lumping, deactivation, reactor model allows one to go either way—interpretation of experiment or extension to process analysis and design.

VII. SOME FEEDSTOCK CORRELATIONS

While the procedures we have described so far have considerable utility in catalyst evaluation and design extrapolations, there is yet another aspect of the problem which is important to explore. It has to do with the fact that, in all the illustrations we have used, the values of the pertinent reaction and decay parameters have been determined on a one-case-at-a-time basis. That is, there is nothing predictive about the model, and it is reasonable to ask if we can use some of the insights it provides to develop more general correlations with respect to the behavior of various catalysts for a specified feedstock or various feedstocks for a given catalyst. Since there is nothing predictive about the model as far as its parameters are concerned, we must rely on the results of a large amount of experimentation to develop correlations more or less by hindsight. The first type of correlation mentioned above amounts to the prediction of catalytic properties and is something which remains probably beyond the state of the art for now and for the foreseeable future; however, the second type is much more tractable and

TABLE 12-3
Composition and Parameters of Feedstock Investigation[a,b]

Feed	P (wt. %)	N (wt. %)	A (wt. %)	α^c	k_0	k_1	k_2	n^c	a^c	b^c
P1	51.9	33.7	14.4	24.8	31.8	26.3	1.83	0.58	0.22	0.30
P2	40.9	36.5	22.6	29.9	32.7	26.2	1.09	0.71	0.46	0.12
P3	46.2	35.1	18.6	30.5	34.0	28.0	1.86	0.72	0.41	0.16
N1	11.3	68.8	19.9	18.5	39.2	33.5	1.54	0.42	0.23	0.27
N2	8.6	59.4	32.4	28.7	34.2	29.4	2.35	0.68	0.46	0.18
N3	9.8	64.0	26.3	25.5	36.2	31.0	2.02	0.60	0.36	0.23
PN33	27.8	49.9	22.5	27.4	36.4	30.9	1.94	0.64	0.32	0.24
PA31	33.8	26.1	40.1	31.4	24.7	21.0	2.87	0.72	0.42	0.21
PA32	32.1	31.9	36.0	31.6	22.9	18.6	1.95	0.71	0.47	0.21
PA33	31.3	30.4	38.3	33.9	22.1	17.6	1.48	0.77	0.54	0.24
PA331	17.7	26.2	56.1	36.4	15.5	12.6	2.66	0.78	0.51	0.24
PA34	34.9	28.6	36.5	34.4	22.1	17.8	1.78	0.78	0.64	0.21
PA37	30.2	23.7	46.1	37.7	10.3	7.71	2.18	0.77	0.69	0.17
PA38	32.5	26.5	41.0	34.7	21.1	16.6	1.66	0.78	0.75	0.20
AA45	11.0	14.2	74.8	40.1	12.3	9.30	2.28	0.83	0.78	0.21
PC32	37.3	29.5	30.0	31.6	19.3	15.0	1.15	0.71	0.52	0.22

[a] From D. M. Nace, S. E. Voltz, and V. W. Weekman, Jr., *Ind. Eng. Chem. Process Des. Dev.* **10**, 530 (1971). Reprinted by permission of the American Chemical Society.

[b] Catalyst: "commercial FCC zeolite"; Temperature: 900°F. General properties of the feedstocks: average molecular weight, 206–402; sulfur, 0.01–3.77 wt. %; nitrogen, 0.001–0.14 wt. %; hydrogen, 10.87–13.85 wt. %; Conradson carbon, 0.01–0.73 wt. %; bromine number, 0.0–5.3.

[c] n from $s = t_c^{-n}$, α from $s = e^{-\alpha t_c}$, a, b from $C_c = at_c^b$.

VII. Some Feedstock Correlations

is very useful in practical application. Thus we now follow the development of a feedstock correlation based on the model.

The basis of the investigation was a series of 16 feed materials, blended from various paraffinic, naphthenic, and aromatic fractions, ranging widely in composition and properties as shown in Table 12-3. Individual conversion and selectivity experiments were carried out for each of these feedstocks and the corresponding model parameters determined in the manner discussed previously. Typical fits for a series of feeds with increasing aromatic content are shown in Fig. 12-15. The values of α, k_0, k_1, and k_2 obtained from fits such as these are also tabulated in Table 12-3. There are some general trends discernible with composition; for example, α is generally higher for feeds of high aromatics composition, reflecting their propensity for making coke, and k_0 is smallest for these feeds, reflecting their generally refractory nature. The relatively high values of k_0 and low values of α for the paraffinic feed indicate easier cracking and less coke formation, and so on.

Also given in Table 12-3 are some parameters of related correlations: n for activity versus time on stream and a and b from the Voorhies correlation of coke content versus time on stream. Trends in n and α are generally correlated; feeds with high α also have high n. Interestingly, the constant a of the Voorhies correlation seems to be a rather good indicator of the coking tendency of a particular feed. The order for a is P2 > P3 > P1 and

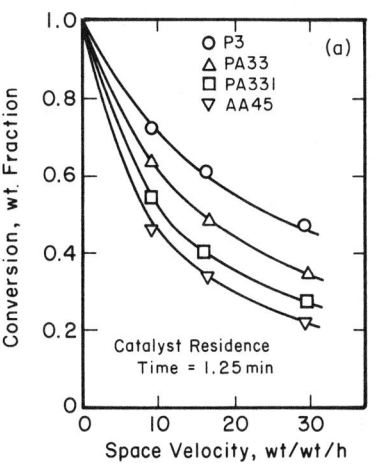
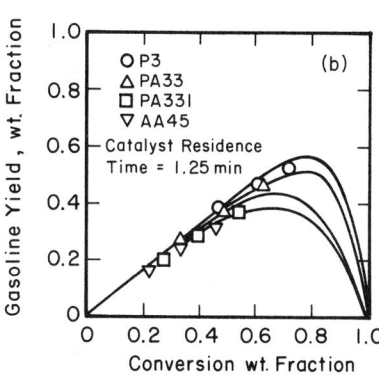

Fig. 12-15. Conversion and selectivity fits for a series of charge stocks of increasing aromatics composition, 900°F. (a) Comparison of conversions for different charge stocks. (b) Comparison of gasoline selectivities for different charge stocks. [From D. M. Nace, S. E. Voltz, and V. W. Weekman, Jr., *Ind. Eng. Chem. Process Des. Dev.* **10**, 530 (1971). Reproduced by permission of the American Chemical Society.]

N2 > N3 > N1, which is the order of increased concentrations of high molecular weight aromatics. Values of α also order in the same fashion as a for these feeds, and note that b increases in the reverse order of a. If we go to a higher content of aromatics, the trend persists: a increases sharply for P3, PA33, PA331, and AA45; however, b is relatively insensitive to aromatics content.

All of the feeds listed in Table 12-3 were virgin stocks, not subject to any prior catalytic reaction, with the exception of PA37 and PC32. PA37 was a blended recycle stock which contained a higher proportion of refractory aromatic compounds, and PC32 was prepared by blending P3 with a heavy coker stock. The fact that these materials have undergone prior chemical reaction will have an effect on subsequent correlation, as we shall see. With respect to the data of Table 12-3, both are characterized by relatively high values of α and a and low values of k_0.

The general trend of increasing α and decreasing k_0 and k_1 with increasing aromatics content makes this an attractive first possibility at correlation of feedstock reactivity; however, aromatics content characterizes only the most refractory compounds and by default ignores the influence of the more reactive naphthenes and paraffins. Now the naphthenic charge stocks listed in Table 12-3 have in general the lowest values of α and highest k_0 and k_1, so the ratio of aromatics to naphthenes, A/N, was selected as a possible correlating variable for feedstock properties. The striking result of such a correlation is shown in Fig. 12-16 for both rate constants and the decay parameter.[4] Note that while the plots of log k_0 and log k_1 versus log(A/N) are reasonably linear for all the virgin stocks, the two feeds previously treated chemically do not fall on the correlation. From a practical point of view this has important consequences, since in plant operation one does not expect to run exclusively virgin feedstocks. We will return to this point later. No correlation is given for the constant k_2 representative of the cracking of gasoline to dry gas and coke; the representation in terms of (A/N) is not as good as for the other constants; however, inspection of Table 12-3 reveals that variation of k_2 with feedstock is relatively small, so the gasoline fractions produced from the various charges are, to a first approximation, chemically the same.

The relationships between catalyst coke and constants α, k_0, and k_1 are shown in Fig. 12-17. These are essentially linear (no log-log needed here) and follow expectations with α increasing and k_0 and k_1 decreasing with amount of coke. An interesting result is that the slopes of k_0 and k_1 versus

---

[4] One more example of the old adage in chemical reaction engineering that "if we can find the log of the right thing to plot versus the log of the right other thing, we are bound to get a straight line."

VII. Some Feedstock Correlations

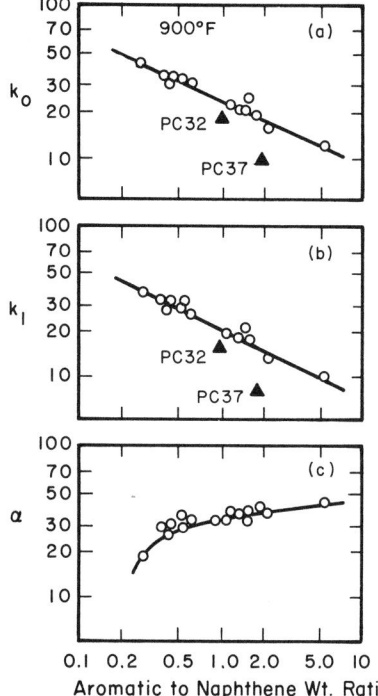

Fig. 12-16. Correlations of the catalytic cracking model parameters in terms of the aromatic-to-naphthene ratio. (a) Gas oil cracking rate constant; (b) gasoline formation rate constant; (c) catalyst decay constant. [From S. E. Voltz, D. M. Nace, and V. W. Weekman, Jr., *Ind. Eng. Chem. Process Des. Dev.* **10**, 538 (1971). Reproduced by permission of the American Chemical Society.]

C_c are approximately the same, indicating that the selectivity factor (k_1/k_0) is independent of coke or catalyst. It is also interesting that in this system the relationship between activity, as expressed by the relative k values, and amount of coke is *linear*, not exponential or hyperbolic as used in some previous examples.

In the above we have discussed at some length the successes of the correlation for feedstock effects on activity and deactivation; now let us dwell on its failures. We noted that the linear log–log relationship of Fig. 12-16 pertained only to virgin stocks; those chemically treated (PC32 and PA37) did not obey the correlation. To explore this further a number of additional experiments were run with a MCGO containing approximately equal amounts of paraffins, naphthenes, and aromatics—which should be of the same class as the virgin stocks of Table 12-3—and the MCGO with various amounts of quinoline added. Also, a number of fluid catalytic cracker (FCC) feedstocks consisting of FCC fresh feed and fresh feed blended in different amounts with recycle stocks were investigated. In the case of the MCGO–quinoline feeds, the basic quinoline would be expected to poison the acidic cracking sites of the catalyst, while with the FCC feeds

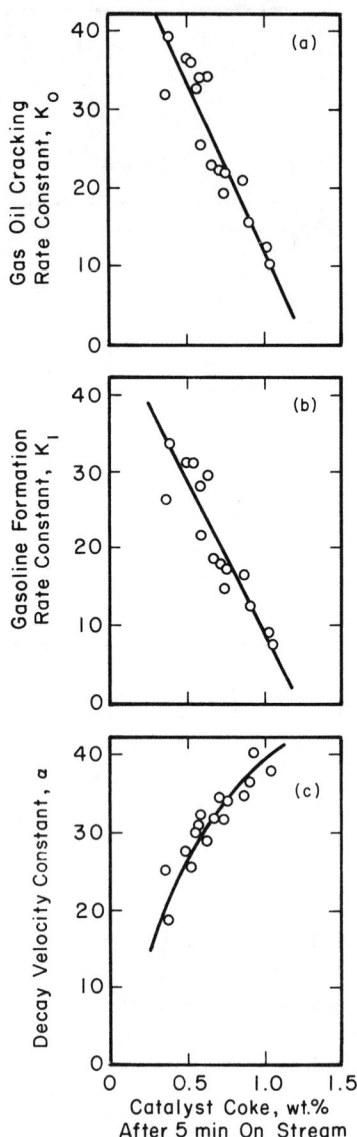

Fig. 12-17. Relationships between model constants and coke on catalyst at 900°F. (a) Gas oil cracking rate constant; (b) gasoline formation rate constant; (c) decay velocity constant. [From S. E. Voltz, D. M. Nace, and V. W. Weekman, Jr., *Ind. Eng. Chem. Process Des. Dev.* **10**, 538 (1971). Reproduced by permission of the American Chemical Society.]

the increased aromatic content relative to MCGO or other virgin feeds would lead to lower cracking rates and higher coke makes. In either case, MCGO-quinoline or FCC-recycle, the composition of the feed has been tampered with sufficiently that one would expect quite different catalytic chemistry and a corresponding severe test of the correlation. We may still

VII. Some Feedstock Correlations

evaluate the quantities α, k_0, k_1, and k_2 by fit of the kinetic model to experimental data for these feeds; the results of doing so, cast in the form of the correlations of Fig. 12-16, are shown in Fig. 12-18. From the point of view of predictive capability, these results must be considered at least a minor disaster, if not even the Gettysburg of our correlation attempt. MCGO fits well, as expected, but for any feeds in which the chemistry is altered the experimental points begin to stray in a sharp southeasterly direction.

What is to be done with this? At this point it might be well to review what has been assembled in the entire analysis and what the weak points might be. Briefly, there are three major components here: (i) the kinetic lumping model, reaction scheme **I**, (ii) the deactivation model, Eq. (12-5),

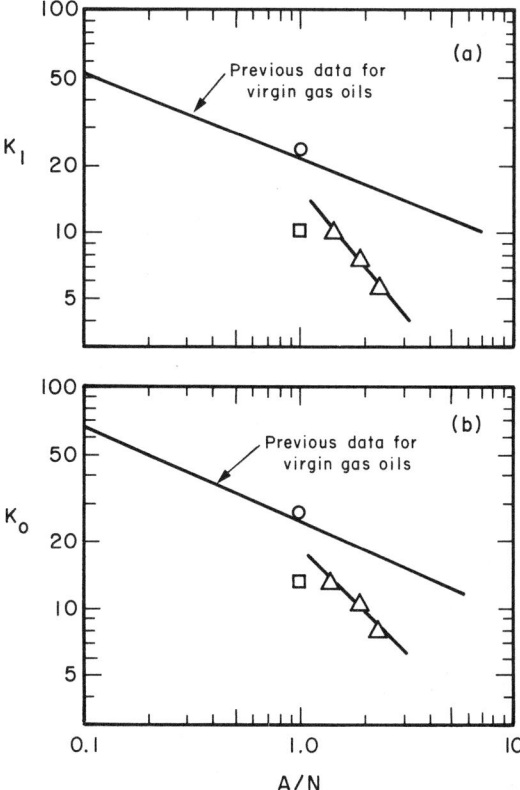

Fig. 12-18. Correlation attempts for nitrogen-containing and recycle feedstocks. (a) Gasoline formation rate constant; (b) gas oil cracking rate constant. (○) MCGO, (□) MCGO + 0.1 wt. % quinoline; (△) FCC fresh feed and combined feeds. [From S. E. Voltz, D. M. Nace, S. M. Jacob, and V. W. Weekman, Jr., *Ind. Eng. Chem. Process Des. Dev.* **11**, 261 (1972). Reproduced by permission of the American Chemical Society.]

and (iii) the various reactor models for fixed, moving, and fluidized beds. Now the form of the deactivation model, while admittedly empirical, has been shown to represent decay for a wide range of feedstocks and catalysts over considerable variations in experimental conditions, and in view of this it can probably be considered to be above suspicion. Similar thoughts pertain to the reactor models, since there is clearly no reason why a reactor model would be adequate to the task for one feedstock but not for another. Thus circumstantial evidence points to the kinetic lumping model, and how decay is built into it, as the likely source of trouble.

Over the years a considerable literature has been devoted to the process and associated theories of lumping kinetics in complex reaction systems. In the present case we have a reaction system in which the feed material consists of literally thousands of individual components, all reacting in parallel. It is likely that the cracking kinetics of any individual compound will be first order, but there will of course be a very wide distribution in the values of the associated rate constants owing to the wide differences in reactivity of individual compounds, and overall this was taken to be second order. This is reasonable, since it has been shown that many parallel reactions with a wide range of individual reactivities, each of first order, lead to apparent reaction orders greater than unity (2). Some attention has been given to the development of analytical criteria for lumping kinetics (3, 4), but much of this art has been developed on the basis of chemical intuition, which is pretty much the procedure we will employ here.

In the work on feedstock characterization we have discussed, it was seen that the major components aligned themselves in order of increasing difficulty for cracking as paraffins, naphthenes, and aromatics. Within each of these groups there will also be a wide range of compounds of differing reactivities, and a reasonable first approach might be to subdivide the three major lumps further into the cracking of "easy" and "hard" fractions, which could be roughly identified with low- and high-boiling-range materials. Of course, caution must be exercised in this approach, particularly in identification of chemical lumps via physical properties such as boiling range; as it turns out, we will see that such an approach works reasonably well for paraffins and naphthenes but is not a sufficient description of the behavior of aromatics.

For the paraffin and naphthene fractions, assuming that we will be able to identify easy and hard, we can try a variation on the theme of reaction scheme I. Within each fraction, the following sequence occurs:

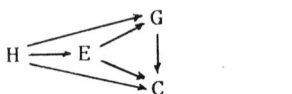

II

VII. Some Feedstock Correlations

where H = hard, E = easy, G = gasoline, and C = coke and low molecular weight gases. The aromatic fraction, as we can probably guess from the feedstock studies, may not be viewed quite as simply, even for a first approximation. The nonvirgin stocks investigated previously were much more difficult to crack, and one chemical fingerprint identified was that they generally possess higher aromatic contents—however, *not in every case*. Working from this clue, then, we can say that not only is the aromatics composition important (again lumped into easy and hard), but also some types of aromatic compounds exhibit a chemistry which is not represented by **II**. To anticipate the final result, this indeed is the case; hard aromatics themselves display two types of chemistry. Those with substituent groups are able to crack to final products in a scheme much akin to **II**, but those with none (i.e., bare rings) cannot go to G, only to C.

If it is assumed that there are no interactions between the paraffin, naphthene, and aromatic fractions, then the 10-lump scheme shown in Fig. 12-19 is obtained. It is seen that hard and easy fractions are defined as $650°F^+$ and 430–$650°F$, respectively; the gasoline lump, G, is C_5^+ to $430°F$ and coke, C, includes C_1 to C_4 as well as carbon or catalyst. The structures on the left side of the figure, corresponding to P_h and N_h, are of the form of reaction scheme **II**. Modification for the aromatic fraction is seen in the reactions starting with C_{Ah} (high-boiling bare-ring aromatics), which can

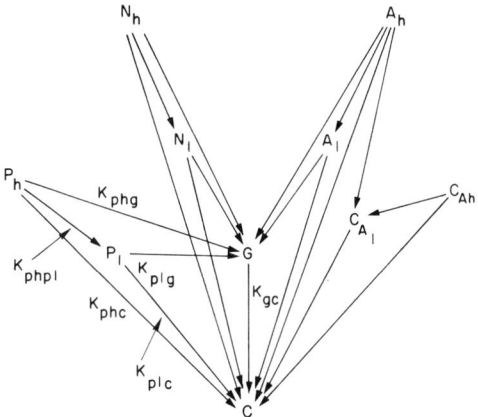

Fig. 12-19. A 10-lump scheme for catalytic cracking. P_l = wt. % paraffinic molecules [mass spectral (MS) analysis], 430–650°F; N_l = wt. % naphthenic molecules (MS analysis), 430–650°F; C_{Al} = wt. % carbon atoms among aromatic rings, 430–650°F; A_l = wt. % aromatic substituent groups (430–650°F); P_h = wt. % paraffinic molecules (MS analysis), 650°F$^+$; N_h = wt. % naphthenic molecules (MS analysis), 650°F$^+$; C_{Ah} = wt. % carbon atoms among aromatic rings, 650°F$^+$; A_h = wt. % aromatic substituent groups (650°F$^+$); G = G lump (C_5–430°F); C = C lump (C_1 to C_4+coke); $C_{Al} + P_l + N_l + A_l$ = LFO (430–650°F); $C_{Ah} + P_h + N_h + A_h$ = HFO (650°F$^+$). [From S. M. Jacob, B. Gross, S. E. Voltz, and V. W. Weekman, Jr., *AIChE J.* **22**, 701 (1976). Reproduced by permission of the American Institute of Chemical Engineers.]

TABLE 12-4
Matrix of Rate Constants K

	P_h	N_h	A_h	C_{Ah}	P_l	N_l	A_l	C_{Al}	G	C
P_h	$-(K_{phpl}+K_{phg}+K_{phc})$	0	0	0	0	0	0	0	0	0
N_h	0	$-(K_{nhnl}+K_{nhg}+K_{nhc})$	0	0	0	0	0	0	0	0
A_h	0	0	$-(K_{ahal}+K_{ahg}+K_{ahc}+K_{ahcal})$	0	0	0	0	0	0	0
C_{Ah}	0	0	0	$-(K_{cahcal}+K_{cahc})$	0	0	0	0	0	0
P_l	$\nu_{hl}K_{phpl}$	0	0	0	$-(K_{plg}+K_{plc})$	0	0	0	0	0
N_l	0	$\nu_{hl}K_{nhnl}$	0	0	0	$-(K_{nlg}+K_{nlc})$	0	0	0	0
A_l	0	0	$\nu_{hl}K_{ahal}$	0	0	0	$-(K_{alg}+K_{alc})$	0	0	0
C_{Al}	0	0	$\nu_{hl}K_{ahcal}$	$\nu_{hl}K_{cahcal}$	0	0	0	$-K_{calc}$	0	0
G	$\nu_{hg}K_{phg}$	$\nu_{hg}K_{nhg}$	$\nu_{hg}K_{ahg}$	0	$\nu_{lg}K_{plg}$	$\nu_{lg}K_{nlg}$	$\nu_{lg}K_{alg}$	0	$-K_{gc}$	0
C	$\nu_{hc}K_{phc}$	$\nu_{hc}K_{nhc}$	$\nu_{hc}K_{ahc}$	$\nu_{hc}K_{cahc}$	$\nu_{lc}K_{plc}$	$\nu_{lc}K_{nlc}$	$\nu_{lc}K_{alc}$	$\nu_{lc}K_{calc}$	$\nu_{gc}K_{gc}$	0

where ν_{hl} = stoichiometric coefficient (MW of heavy fuel oil/MW of light fuel oil)
ν_{hg} = stoichiometric coefficient (MW of heavy fuel oil/MW of gasoline)
ν_{hc} = stoichiometric coefficient (MW of heavy fuel oil/MW of C lump)
ν_{lg} = stoichiometric coefficient (MW of light fuel oil/MW of gasoline)
ν_c = stoichiometric coefficient (MW of light fuel oil/MW of C lump)
ν_{gc} = stoichiometric coefficient (MW of gasoline/MW of C lump)
MW = molecular weight

[a] From S. M. Jacob, B. Gross, S. E. Voltz, and V. W. Weekman, Jr., *AIChE J* **22**, 701 (1976). Reprinted by permission of the American Institute of Chemical Engineers.

VII. Some Feedstock Correlations

go either directly to C or to C via the intermediate C_{Al}, lower-boiling bare-ring compounds. Each of the reaction steps in the reaction scheme is now first order and irreversible; mathematical representation of the overall scheme gives the matrix of rate constants shown in Table 12-4. In order to complete the model, we still have to incorporate the deactivation effects on these individual rate constants; this will include not only coke but nitrogen compounds as well, since the results shown in Fig. 12-18 demonstrate a strong poisoning effect of quinoline in the feed. Further, the increased sophistication of the reaction scheme allows corresponding increases in the sophistication of reaction kinetics modeling. Thus, the rate of disappearance of species j in a given reaction step is written as

$$(-r_j) = k'_j C_j \left(\frac{\rho c}{\varepsilon}\right) \frac{1}{1 + K_h C_{Ah}} \tag{12-50}$$

where C_j is the molar concentration, ρc is grams catalyst per cubic centimeter of bed, ε is void fraction, and $K_h C_{Ah}$ is an adsorption inhibition term (*not* a deactivation function) related to the strong chemisorption of heavy refractory aromatic rings on the catalyst. For $(-r_j)$ in units of moles per cubic centimeter per hour, the rate constant k'_j has units of grams catalyst per cubic centimeter per hour. If we incorporate these kinetics into a fixed-bed reactor model (in fact, a fixed fluidized dense bed was used in experimentation; however, this is equivalent to a fixed bed if the fluid is in plug flow) and assume that the time scale of decay is small in relation to residence time, then

$$G_v \left(\frac{da_j}{dx}\right) = -k'_j \rho a_j \left(\frac{\rho c}{\varepsilon}\right) \frac{1}{1 + K_h C_{Ah}} \tag{12-51}$$

where G_v is the superficial mass flow, grams per square centimeter per hour, and $\rho a_j = C_j$, where a_j is moles per gram of fluid and ρ is grams of fluid per cubic centimeter. Rewriting Eq. (12-51) in terms of normalized reactor length, $X = x/L$, and weight hourly space velocity, S_{WH} grams total feed per gram catalyst per hour, gives

$$\frac{da_j}{dX} = \frac{1}{1 + K_h C_{Ah}} \left(\frac{k'_j \rho a_j}{S_{WH}}\right) \tag{12-52}$$

where

$$G_v = S_{WH} \rho c L / \varepsilon$$

The decay model will follow prior practice, in which separable kinetics are assumed and the activity variable s is the same for all reactions, hence:

$$k'_j = k_j s \tag{12-53}$$

where k_j is the intrinsic rate constant for the undeactivated catalyst. Now the fluid density ρ will change with conversion, since the average molecular weight is a function of conversion, hence:

$$\rho = P\overline{MW}/RT \qquad (12\text{-}54)$$

and the reactor model equation becomes

$$\frac{da_j}{dX} = -\frac{1}{1+C_{Ah}K_h}\left(\frac{P\overline{MW}}{RT}\right)\left(\frac{k_j s a_j}{S_{WH}}\right) \qquad (12\text{-}55)$$

Recalling that we are dealing with the 10-lump kinetic model with $j = 1, 2, \ldots, 10$, then Eq. (12-55) can be written in the general matrix form

$$\frac{da}{dX} = -\frac{1}{1+K_h C_{Ah}}\left(\frac{P\overline{MW}}{RT}\right)\left(\frac{s}{S_{WH}}\right)Ka \qquad (12\text{-}56)$$

with

$$\overline{MW} = \sum a_j M_j / \sum a_j$$

The deactivation function s employed in the 10-lump model is also more sophisticated than that of the 3-lump model, primarily in the fact that an effect of oil partial pressure is included in addition to time-on-stream parameters. Higher pressures produce higher rates of deactivation by coke formation. The function for coke formation is

$$s_c = \alpha/(P)^m(1+\beta t_c \gamma) \qquad (12\text{-}57)$$

The deactivation function s_c, however, does not include effects due to poisoning of the acidic surface by basic nitrogen compounds. This is carried along in the model by an inhibition function of the form

$$s_N = 1/(1+K_n N) \qquad (12\text{-}58)$$

where N is a cumulative amount of nitrogen to which the catalyst has been exposed up to catalyst residence time t_c, in grams per gram of catalyst. Overall activity s in Eq. (12-56) is thus a scalar multiplier on the rate constant matrix with

$$s = s_c s_N \qquad (12\text{-}59)$$

We will recall from the three-lump model that for the fixed bed, after the conversion expression was derived from the differential equation, it was necessary to compute a time-averaged value representative of the mixed average reactor effluent. It is convenient to define a new variable ω as

$$d\omega = (sP/S_{WH}RT)dX \qquad (12\text{-}60)$$

VII. Some Feedstock Correlations

such that

$$\omega = (sP/S_{WH}RT)X \qquad (12\text{-}61)$$

In terms of ω the model of Eq. (12-56) becomes

$$\frac{da}{d\omega} = \frac{1}{1 + K_n C_{an}} Ka \qquad (12\text{-}62)$$

In the interpretation of experimental data, it is necessary to determine from this model mixed average concentrations. From the composition of a given feedstock Eq. (12-62) can be integrated to give a as a function of ω. Then, for the reactor exit, $X = 1$ and

$$\omega = sP/S_{WH}RT \qquad (12\text{-}63)$$

where all is known except s. In computation a number of run times from 0 to t_c were chosen and $a(\omega)$ was evaluated and then time-averaged.

Evaluation of both kinetic and decay parameters for this model is much more complex than for the previous one simply because there are so many

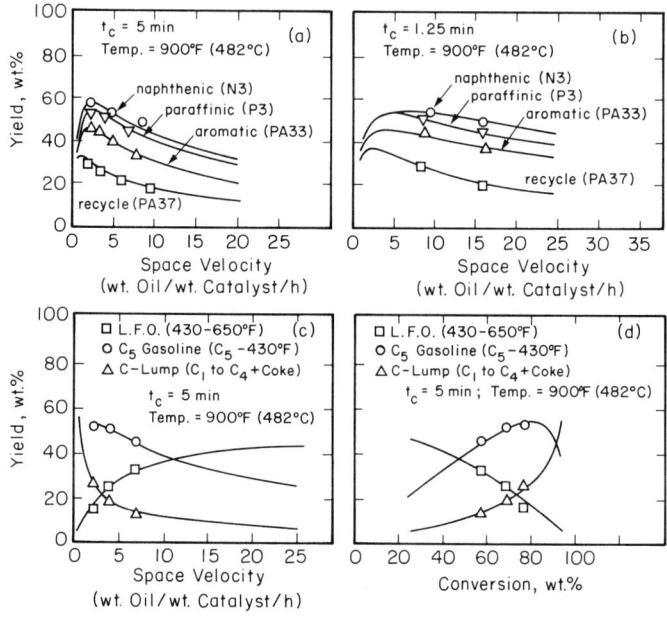

Fig. 12-20. Typical fits to yield data by the 10-lump model for a number of feedstocks. (a and b) Observed versus computed time-averaged gasoline yields as a function of charge stock. (c and d) Cracking yields for a paraffinic feed (P3) as a function of space velocity. [From S. M. Jacob, B. Gross, S. E. Voltz, and V. W. Weekman, Jr., *AIChE J.* **22**, 701 (1976). Reproduced by permission of the American Institute of Chemical Engineers.]

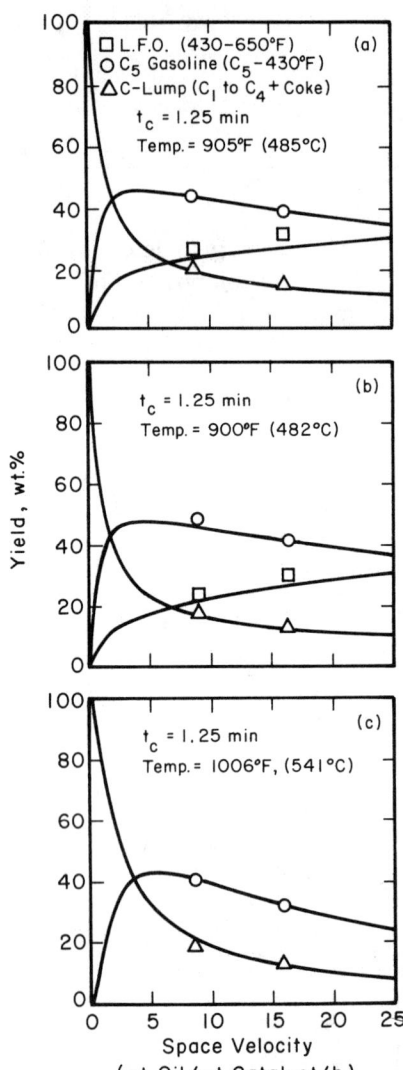

Fig. 12-21. Some *a priori* model predictions. (a) Cracking yields for a feedstock with high-boiling aromatics (PA38). (b) Cracking yields for WCMCGO. (c) Cracking yields for WCMCGO with 0.2 wt. % addition of basic N as quinoline. [From S. M. Jacob, B. Gross, S. E. Voltz, and V. W. Weekman, Jr., *AIChE J.* **22**, 701 (1976). Reproduced by permission of the American Institute of Chemical Engineers.]

more of them. However, note that the rate constants derived from a fit to any given feedstock experiment should be able to predict the behavior of any other feedstock. Many comparisons are provided in the original reference; we abstract only a few here. Figure 12-20 provides an indication of the fit to yield data for a number of the feedstocks of Table 12-3. Such a fit is, in the words of one of the authors (V. W. Weekman, Jr.), "like a wet sheet." Figure 12-21a indicates the predictive capabilities of the model in terms of yield structure for two feedstocks which were *not* used in the evaluation of the rate constant matrix; in each case there is a significant extrapolation from the original data base and the agreement is excellent. Figure 12-21b extends the extrapolation to include the effects of basic nitrogen on cracking yield structure—again excellent agreement is obtained. Finally, activation energies were obtained for the various reactions in several groups, as follows:

(1) Gasoline (G) formation from paraffins and naphthenes.
(2) Coke (C) formation from paraffins and naphthenes.
(3) Gasoline (G) formation from A_h and A_1.
(4) Coke (C) formation from A_h, A_1, C_{ah}, and C_{al}.
(5) Coke (C) formation from gasoline G.

The activation energy for coke from gasoline (5) is 20 kcal/mol. Activation energies for (3) and (4) range from 14 to 18 kcal/mol, while those of (1) and (2) are lower at 5 to 9 kcal/mol. Again, reasonable prediction of the temperature dependence of yield and selectivity patterns was obtained from the model using these values of activation energy.

VIII. SUMMARY REMARKS

The net impact of the predictive capabilities of such a model on a process as important and as widely applied as is catalytic cracking is difficult to overemphasize. The scheme here permits direct and unequivocal evaluation of laboratory and pilot plant data and, as incorporated into the appropriate reactor model, also permits detailed prediction of commercial operation with various feedstocks and process conditions. This, in turn, forms the nucleus of operating strategies that can be used for large-scale economic optimization. In spite of our preoccupation in the last section of the chapter with feedstock correlations, it must be kept in mind that the single most crucial factor in any specific application remains the deactivation problem. It must also be remembered that the appreciation of the chemistry involved in the primary cracking reactions here is at least an order of magnitude more advanced than the deactivation model, which remains on an essentially empirical basis. Largely because of this, the catalyst remains the independent variable. The large investment of time and experiment that is encompassed in this example gives a unique set of parameters for a *single* catalyst. Change the catalyst formulation and a large portion of the exercise must be repeated. A completely *a priori* correlation would avoid this, of course, but this is unfortunately clearly beyond present capabilities.

REFERENCES

1. M. Abramowitz and H. Stegun, "Handbook of Mathematical Functions." Dover, New York, 1965.
2. P. Hutchinson and D. Luss, *Chem. Eng. J.* **1**, 129 (1970); **2**, 172 (1971).
3. Y. Ozawa, *Ind. Eng. Chem. Fundam.* **12**, 191 (1973).
4. J. Wei and J. C. W. Kuo, *Ind. Eng. Chem. Fundam.* **8**, 114 (1969).

INDEX

A

Acidity, 6, 130, 132
  correlations 68
Active centers, 4, 29, 32, 35, 75
Active fraction, 30
Activity, 28, 38, 42, 105, 386
  approximate, 73
  characterization, 2
  correlation, 418
  global, 249
  profiles, 348
  ratio, 28
  relations, 55
Activity-time ratio, 53
Adsorption
  inhibition, 125
  poison, 51
Aluminas, 132
Anchor position, 139
Antiselective, 148
Atomic migration, 172
Atomic mobility, 171
Auguer Electron Spectroscopy (AES), 139
Automobile, emissions catalysts, 21, 271
Averaging
  coke, 366
  rate constant, 367
Axial dispersion, 368

B

Basicity correlation, 87
Benzene hydrogenation, 39
Berzelius, 19
β-hydride, palladium, 168
Bifunctional catalysts, 43, 418
Bimetallic catalysts, 113
Blaum model, 372
Bohart–Adams model, 355
Brønsted acid, 4, 5, 6
Brownian motion, 173

C

Calcination, 133
Capacity parameter, 370, 409
Carbide, 15
Carbon profiles, 72, 99
Carbonium ion, 95, 130, 146
Centered structure (2 x 2), 140
Characteristic times, 341
Chemically induced sintering, 144
Chemisorption
  bonding, 125, 145
  stoichiometry, 175
Chloride
  modification, 152
  sintering, 226
Coagulation model, 173
Coal liquid, 105
Coke
  activity relationship, 81, 388
  burning, 430, 434, 439
  concentration profiles, 98, 352, 439
  correlations, 64
  definition, 14, 63

deposition in HDS, 282
inhibition, 321
precursors, 47
refractory, 326
yield, 67
Coking–fouling, 22
Collocation, 378
Combined action, reforming, 116
Condensation, 95
polymerization, 14
Configurational diffusion, 284
Conradson carbon, 286
Constant conversion operation, 413
Constant rate period, 335
Contrast, 174
Coordination, 124
Core poisoning, 246, 251
Covalent bond, 126

D

d-band, 142, 143
d-character, 142
d-orbitals, 124
Dangling orbitals, 124
Deactivation rate equations, 48, 108
Dealkylation, 44
Deasphalted oil, 285
Defect structure, 8
Demetallation, 279
Detoxification, 18
Diffusivity, 105, 108
configurational, 284
Dispersion
coke deposits, 329
metals, 174, 181
Disproportionation, 95
Distributions
bimodal, 193
Gaussian, 39
inner, middle, outer, 274, 275
log-normal, 209
tailoring of, 158
Dual sites, 29, 43, 51, 137, 360
Dynamic temperature rise, 410

E

Effective diffusivity, 81, 101
Effectiveness
bed, 365

factor, 106, 242
Electron
configuration, 125
spin resonance (ESR), 325
Electronegativity, 126
Electronic
factors, 142
interactions, 139
properties, 4
structure, 143
Elementary steps, 31, 41
Elovich equation, 54
Ensembles, 37, 114, 140
Exponential
correlation, 33, 50, 53
decay, 451
Extent
deactivation, 453
reaction, 453
Extrapolation, 59, 60

F

Falling rate period, 337
Fast concentration response (FCR), 407
fcc crystal, 137
Feed rate, coke dependence, 69, 71
Feedstock, 66, 473
Finite elements, 378
Fischer–Tropsch synthesis, 15
Fixed-bed deactivation
activity functions
exponential, 353
hyperbolic, 352
arbitrary profiles, 348
experimental results, 378
quasi-steady-state assumption, 382
reactor dynamics, 395
scaling isothermal data, 390
temperature increase (constant conversion), 413
thermal waves
adiabatic, 405
nonisothermal–nonadiabatic, 396
transport resistances, inter- and intraphase, 362
various reaction systems, 380
Fixed-bed modeling
moving zones
adiabatic, 371
Bischoff–Ozawa analysis, 358

Index 491

hot spot motion, 371
nonisothermal, 369
Wheeler–Robell analysis, 357
poison wave analysis, 354
Fixed-bed regeneration
constant burning rate, 429
deactivation in reverse, 427
low-temperature, 442
shell-progressive, 439
temperature-independent, 433
Flash desorption, 141
Fluidized-bed modeling, 165
catalytic cracking, 455
Foulant profiles
parallel, 256
series, 260
Fouling, 10, 14
definition, 22, 63
mechanism of, 108
reforming catalysts, 112
Four-site concept, 161
Fractional coverage, 41
Freundlich, 39, 46

G

Gasoline yield, 166
Geometric effects, 4, 142
configuration, 137
crystallite size, 9
Global activity
foulant distribution, 259
with Thiele parameter
impurity poisoning, 263
parallel fouling, 255
series fouling, 263
time-dependent deactivation, 267
Growth models, 205

H

Heat
chemisorption, 126, 148
sublimation, 197
Heterogeneous
sites, 37, 52
surfaces, 146, 160
Homogeneous, surfaces, 38
Hot spot, 312
motion, 371

Hydrocracking, 44
Hydrodesulfurization, 63, 144, 279
Hydrogen
chemisorption, 178
effect of, 73
strongly bound, 178, 186
Hydroxylation, 5, 134
Hyperbolic function, 34, 50, 52

I

Ideal
reactor system, 56
surface, 36
Inaccessible platinum, 197
Incorporation, 17
Infrared spectroscopy (IR), 87, 126
Inhibitor, 34, 74, 77
Initial activity, determination of, 58
Inseparability, 35
Insulators, 6
Integral reactor, 55
Internal diffusion, 97
Intraparticle
coke, 98, 352, 439
distribution, metals in HDS, 280
transport, model for, 207
Intra- and interphase gradients, 311
Isomerization, 130, 133

K

Kerosine gas-oil, 97
Kinetic
ensembles, 37
models, 44
properties, 38
reaction paths, 46
Kinetics
coke burning, 328, 333, 428
interpretation of, 458

L

L_{III} absorption edge, 143
Langmuir–Hinshelwood–Hougen–Watson (LHHW) kinetics, 12, 29, 41, 46, 52
Lattice properties, 4
Lead poisoning, 148

Legendre transformation, 359, 445
Lewis acid, 5
Light East Texas Gas Oil (LETGO), 74, 77
Light-off, 277
Limiting coke formation, 97
Linear activity function, 50
Low-energy electron diffraction (LEED), 126, 139
Low-index planes, 142

M

Macropores, 237
Mathematical forms, activity, 49
Mercury poisoning, 163
Metals deposition, 165
Metal poisoning
  d-orbitals, 124
  strength of bond, 125
Methanation catalysts, poisoned by hydrogen sulfide, 146
Methylcyclohexane, dehydrogenation, 109
Methylcyclopropane, hydrogenolysis, 156
Microcombustion, 98
Micropores, 237
Model
  components, 47
  discrimination, 383
  evaluation, 457
Molecular surface area, 130
Molybdena–alumina catalysts, coke formation, 72
Monofunctional catalysts, 126
Monolayer coverage, 48
Monte Carlo simulation, 134
Mortality, 19
Moving beds, cracking, 454
Multifunctional catalysts, 126, 160, 268
Multifunctional oxides, 161
Multiple reactions, 41
Multiple sites, 36

N

Nickel catalyst, thiophene poisoning, 138, 396
Nitrogen-containing organics, 129
Nonideal surfaces, 39
Nonmetallic poisons, 125

Nonselective poisoning, 122, 161
Nonuniform
  coking, 102
  distributions of catalyst, 270
  sites, 139

O

Optimization, fixed-bed reactors, 394, 463
Organic bases, 64, 129
Ostwald ripening, 173
Overlapping sites, 137
Oxidation, coke, 328
Oxides
  poisoning, 128
  surfaces, 5
Oxygen
  enhancement, 156
  modification, 154
  transport, 99

P

Parabolic approximation, 376
Parallel
  deactivation, 31, 58
  fouling, 252, 296, 305
  reactions, 110
Parametric sensitivity, 315
Particle
  growth model, 211
  size, 174
  splitting, 220
Pellet deactivation, 249, 255
  hydrodesulfurization, 279
  nonuniform distributions
    automobile catalysts, 276
    general strategies, 275
    inpurity mechanism, 273
    series mechanism, 271
  nonuniform fouling, experimental, 291
Percentage exposed, 174, 181
Performance index, 385
Physical adsorption, 5
Platinum catalyst deactivation, 109, 164
  arsine, 144, 148
  carbon monoxide, 147
  lead, 148
  oxygen, 143

Index

sulfur, 140
Platinum-oxygen system, 143, 221
Platinum-rhenium catalyst
 ensemble mechanism, 115
 hydrogen spillover, 115
 mechanisms of fouling, 114
 physical mixtures, 114
 reforming, 114
 titrations, 115
Plug flow model, 370
Poison
 definition, 11
 multisite adsorption, 10, 136
Poisoning
 metals, 122
 nonuniform surfaces, 130
 oxide catalysts, by metals, 165
 pyridine, 11
 single crystals, 140
 structure-sensitive, 155
 thiophene, 39, 138, 396
 uniform surfaces, 139
 versus inhibition, 121
 waves, 355
Poisons, nonselective versus selective, 12
Polycyclic aromatics, 83
Polymerization, 14, 34, 84
Population balance, 33, 44
Pore blockage, 13, 103, 106
Pore mouth poisoning, 246, 250
Power law model, 110
Precursor, 108, 387
Predepositied poisons, 153
Primitive structure (2×2), 140
Purgatory, 20

Q

Quasi-steady-state, 240, 439, 441

R

Rate surface, 354
Rate-limiting step, 31
Reactivity
 coke, 323
 lumping, 323, 450
Reactors
 fixed-bed deactivation
  activity profiles, 347
  adiabatic, 371, 405
  continuity equations, 391, 400, 453
  isothermal, 378
  nonisothermal-nonadiabatic, 369, 390
 fluid-bed, 455
 ideal, 56
 integral, 55
 moving bed, 454
 performance comparison, 461
 plug flow, 350
 recycle, 56
 variable flow recycle, 57
Reciprocal forms, 53
Redispersion, 17, 214
 effect of chloride, 226
 experimental observations, 215
 metal-support interactions, 225
 platinum in oxygen, 216
 theoretical proposals, 220
 thin films, 217
Reforming, 44, 112
 catalysts, 160, 162
  fouling, 112
  poisoning, 162
  sintering, 198
 reaction, 46
 sulfur, 116
Regeneration, 13, 99, 427
 coked catalysts, 321
 low-temperature, 442
Rejuvenation, 18
Relaxation time, 34
Resid, 166
Runge-Kutta method, 392

S

Scale-up, 390
Scherrer equation, 179
Second-order lumping, 286
Selective
 chemisorption, 175
 poisoning, 145, 160, 163
Selectivity, 42, 146
 cracking, 468
 reaction, 161
Self-inhibition, 12
Self-poisoning, 12, 14, 16
Self-preservation, size distribution, 213

Index

Separability, 24, 30, 35, 38
Series fouling, 252, 259, 297, 306
Shell model, 99, 101
Shell-progressive
  burning, 338, 429
  deactivation, 362
Shielded structure, 125
Single-pellet reactor
  isothermal, 292
  nonisothermal, 301
Sintering, 16, 171
  activation energy, 189
  atomic metal migration, 172
  chemically assisted, 196
  constant temperature–variable time, 184
  correlation with activity, 188, 198, 200, 203
  crystallite migration, 173
  effect of atmosphere, 182
  effect of melting point, 197
  formation of metal oxides, 195
  intraparticle transport, 207
  kinetic order, 189
  metal–support interactions, 225
  model systems, 202
  Ostwald ripening, 173
  oxide stabilization, 229
  particle growth model, 205
  stability of noble metals, 195
  studies, experimental characterizations, 174, 175, 177, 179
  variable temperature
    constant time, 184
    variable time, 152
Site
  density, 41
  distribution functions, 39
  population, 37
  strength, 126
Slow temperature response (STR), 407
Solid-state reactions, 17
Soluble platinum, 215
Strong adsorption, 125
  hydrogen ("strongly bound"), 178, 186
  oxygen, 156
Strong metal–support interactions, 160, 179, 186
Structure-insensitive reactions, 149
Structure-sensitive
  reactions, 149, 151
    primary, 151
    secondary, 153
  poisoning, 155
Surface
  area correlation, 211
  coverage, 140, 148
  diffusion, 207
  elements, 37
  intermediates, 4, 7
  reconstruction, 144
    effect of oxygen, 157
  stabilization, 229
  stoichiometry, 9

T

Tagged reactants, 89
Temperature increase requirement (TIR), 284
  data interpretation, 414, 421
Temperature
  coke formation, 66
  desorption and reduction via programming, 114, 126
  profiles, 306, 398
  sensitivity, 418
Tempkin isotherm, 39
Thermal waves, 373
Thiele parameter, 241, 246, 263, 292
Thiele–Zeldovich problem, 236, 292
Thermal waves, fixed-bed modeling, 373
Time scale of deactivation, 240
Time-on-stream correlations, 15, 386
Titration, 5, 9
  oxygen–hydrogen, 177
Transients
  intraparticle, 310
  fixed-bed reactor, 391, 400
  measurement, 106
Transmission electron microscopy (TEM), 171, 174
Triangular fouling, 298
Turnover frequency, 38, 149, 150
Two-lump model, 450
Type I, 23
Type II, 24

U

Ultraviolet photemission spectroscopy (UPS), 142
Uniform
  coking, 102

Index

poisoning, 246
sites, 126
Unshielded structure, 125

V

Vapor-phase diffusion, 208
Volatilazition, 18
Volcano correlation, 37, 145, 148
Voorhies correlation, 64, 73, 77, 451

W

Warren-Averbach method, 180
Water vapor poisoning, 163
Wheeler model, 246
Wheeler-Robell model, 357
Work requirement, 411

X

X-ray
 absorption, 143
 diffraction (XRD), 174
 scattering, 182

Y

Yagi-Kunii correlation, 405

Z

Zeolites
 mordenite, 79, 93
 Y, 93, 460

DATE DUE